This book is to be returned on or before
the last date stamped below.

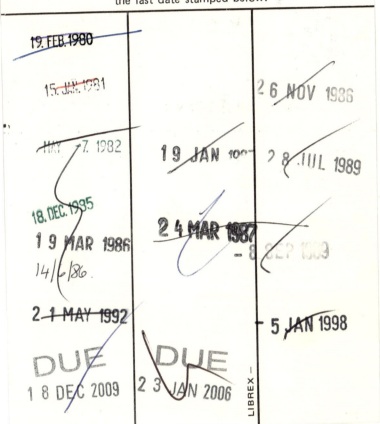

19. FEB. 1980		
15. JAN. 1981		2 6 NOV 1986
MAY 7. 1982	1 9 JAN 1987	2 8 JUL 1989
18. DEC. 1985	2 4 MAR 1987	
1 9 MAR 1986		- 8 SEP 1989
14/6/86.		
2 1 MAY 1992		- 5 JAN 1998
DUE	DUE	
1 8 DEC 2009	2 3 JAN 2006	LIBREX —

HANDBOOK
OF AEROSOL
TECHNOLOGY

HANDBOOK OF AEROSOL TECHNOLOGY

Second Edition

PAUL A. SANDERS, Ph.D.

Formerly Research Associate
Freon® Products Laboratory
E. I. du Pont de Nemours & Company
Wilmington, Delaware

 VAN NOSTRAND REINHOLD COMPANY
New York Cincinnati Atlanta Dallas San Francisco
London Toronto Melbourne

Van Nostrand Reinhold Company Regional Offices:
New York Cincinnati Chicago Millbrae Dallas

Van Nostrand Reinhold Company International Offices:
London Toronto Melbourne

Manufactured in the United States of America

Published by Van Nostrand Reinhold Company
135 West 50th Street, New York, N.Y. 10020

Published simultaneously in Canada by Van Nostrand Reinhold Ltd.

15 14 13 12 11 10 9 8 7 6 5 4 3 2 1

Library of Congress Cataloging in Publication Data

Sanders, Paul Amsdon, 1913-
    Handbook of aerosol technology.

    First ed. published in 1970 under title: Principles
of aerosol technology.
    Includes index.
    1. Aerosols. I. Title.
TP244.A3S35 1978   660.2'9'4515   78-18428
ISBN 0-442-27348-7

To Alice, Alan and Rick

PREFACE TO THE SECOND EDITION

From its inception in the 1940's, the aerosol industry has always been extraordinarily dynamic. It has been characterized by rapid expansion, the continuing introduction of new products into the market, intense competition, and the development of a completely new science—the science of aerosol packaging and technology. A magazine devoted entirely to aerosols (*Aerosol Age*) appeared as early as 1956. A number of technical books on aerosols have been published, and numerous articles about different aspects of the aerosol industry have appeared in a wide variety of magazines. The last several years, however, have been by far the most turbulent in the aerosol industry because of the fluorocarbon/ozone controversy. As a result of this issue, propellant manufacturers instituted contingency alternative fluorocarbon programs anticipating that fluorocarbons 11, 12, and 114 would be regulated or restricted. Toxicological studies on these alternatives have been accelerated. The investigation of nonfluorocarbon alternatives such as hydrocarbons and carbon dioxide has been intensified. Considerable federal and state legislative activity has occurred. Consequently, in order to cover all of these subjects, it was necessary to revise extensively the first edition of this book. Several chapters have been eliminated ("Food Aerosols," "Miscellaneous Aerosol Systems," and "Triangular Coordinate Charts") and two new chapters have been added ("Fluorocarbons in the Atmosphere" and "Technical Programs in the Atmospheric Sciences"). Particular emphasis has been placed upon alternative fluorocarbons and their properties, fluorocarbon toxicology, hydrocarbons and compressed gases, flammability, and new developments in aerosol valves, containers, and filling equipment. The discussions of vapor pressure, spray characteristics, and solvency have been expanded considerably.

The author has been particularly fortunate in the help provided by his associates in the Du Pont Company. Special thanks go to Dr. John J. Daly, Senior Technical Supervisor, Industrial Products, Freon® Products Division, for his chapter on fluorocarbon propellants and his suggestions on other chapters. The contribution of Dr. Frank A. Bower, Environmental Manager, Freon® Products Division, who provided both a chapter on programs in environmental science and help in editing, was invaluable. The author also would like to express his appreciation to the other Du Pont Company writers whose chapters are a necessary and important part of this book. They are Dr. Richard B. Ward, Public Affairs Department; Mr. Thomas D. Armstrong, Supervisor, Physical Properties Section, Freon® Products Division; Ruth R. Montgomery, Information Specialist (M.S.); and Dr. Charles F. Reinhardt (M.D.), Director, Haskell Laboratory of Toxicology and Industrial Medicine. The constructive suggestions of Dr. Raymond L. Mc-Carthy, Director, Freon® Products Laboratory, during the preparation of the book, were particularly useful.

One of the most valuable contributions that any technical book can make is to provide new experimental data. Much of the data in the present book were developed by the following Freon® Products Laboratory personnel: Dr. James A. Thompson, Senior Research Chemist; Mr. Camiel de Brabander, Senior Chemical Assistant; Margaret C. McConnell, Chemical Assistant; and Mr. Toivo Ratsep, Laboratory Technician.

Finally, very special thanks to my secretary, Elizabeth Lloyd, who volunteered to type the entire manuscript.

No book is complete without figures to illustrate data. Those in this edition were prepared by Mr. T. Howell Rickerson and Mr. Bruce P. Esdale.

PAUL A. SANDERS

Wilmington, Delaware

PREFACE TO THE
FIRST EDITION

The Freon®* Products Division of the Du Pont Company has provided a course in aerosol technology for over ten years. The course, designed to present a comprehensive picture of the technical areas in the aerosol field, consists of lectures combined with laboratory experiments. The lectures have been given to a cross section of aerosol technical personnel, ranging from those who needed basic instruction to experienced aerosol chemists who wished to discuss specific areas in detail. The course has been well received by the aerosol industry; therefore, it seemed desirable to make the lecture material more readily available to the academic and industrial scientific community in the form of a textbook. The present volume on aerosol technology is a modified and expanded version of the lectures.

Particular emphasis in the book has been placed upon the fundamental principles that govern the characteristics and behavior of aerosols. Only through an understanding of these principles will many of the problems in the aerosol field be solved. For this reason, the basic properties of aerosols, such as spray characteristics, vapor pressure, solubility, and flammability, have been treated as individual subjects. Aerosol emulsion and foam technology, particularly in the fields of cosmetic and pharmaceutical products, has become increasingly important in recent years. Many of the fundamental concepts of surface chemistry have been reviewed in some detail in the chapters of Part Two.

In any field, there is a considerable amount of information useful to laboratory chemists which is difficult to find in the literature. Much of this type of information has been included in the present volume. Examples in the aerosol field which fall into this category are methods for determining the compositions of propellant blends with a specified density or vapor pressure, and the pitfalls

---

*Freon® is Du Pont's registered trademark for its fluorocarbons.

that may be encountered in the measurement of properties such as solubility, flammability, and vapor pressure.

The class of fluorinated hydrocarbon propellants is generally referred to in this text by the Freon® nomenclature. This seemed appropriate because most of the data listed in the text for the fluorinated hydrocarbon propellants were obtained with the Freon® propellants. The Freon® compounds were the first fluorinated hydrocarbon propellants to appear on the market and were the only fluorinated hydrocarbon propellants available for a considerable number of years. At the present time, these same chemical compounds are also manufactured by a number of other companies in the world who use the same numerical nomenclature prefixed by their respective trade names.

The preparation of a book is a difficult undertaking under the best of conditions, but the task is made much easier by the assistance of others. The author has been fortunate in the help extended by his associates in the Du Pont Company. It is impossible to list all those who have contributed to the book. However, the author would like to express his particular appreciation to the following four associates: Mr. N. W. Kent, Manager, Special Services and Training, for his assistance in preparing the original manuscript; Dr. R. L. McCarthy, Director of the Freon® Products Laboratory, for his continued encouragement during the preparation of the manuscript; Dr. R. P. Ayer, Aerosol Project Leader; and the late Dr. F. T. Reed, Division Head, Aerosol Section, Freon® Products Laboratory, for their careful and patient reading of the manuscript. In addition, the comments and suggestions of the following Freon® Products Division personnel were helpful during the preparation of the manuscript: Mr. C. E. Kimble, Manager, Aerosol Customer Services; Dr. D. E. Kvalnes, Manager, Technical Section; Mr. A. H. Lawrence, Jr., Marketing Manager, Aerosol Propellants; and Mr. D. C. Miller, Assistant Director of Sales. Finally, the author would like to extend his thanks to the three contributors for their chapters: Mr. T. D. Armstrong, Jr., Project Leader, Freon® Products Laboratory; Dr. J. Wesley Clayton, Jr., Hazleton Laboratories, Inc., formerly Assistant Director of the Haskell Laboratory for Toxicology and Industrial Medicine; and Dr. J. H. Fassnacht, Technical Associate in the Technical Section.

PAUL A. SANDERS

Wilmington, Delaware
December 1969

CONTENTS

III MISCELLANEOUS

I

HOMOGENEOUS SYSTEMS AND THEIR PROPERTIES

1

HISTORICAL
BACKGROUND

The word "aerosol" was first employed in the field of colloid chemistry to describe a suspension of small particles in air or gas in which the radius of the particles was less than 50 μ.[1] (A micron μ is 0.001 mm or 0.000039 in.). The suspended particles could be either solid or liquid. Dust, smoke, and fog are examples of this class of aerosols. The particles in dusts (suspensions of solids in air) may have diameters as small as 0.1 μ or less. These suspensions produce haze. Smoke is an aerosol consisting of solid particles, usually carbon, in air. Carbon smoke is composed of small particles with a radius of about 0.01 μ. These will coagulate into long, irregular filaments that may reach several microns in length. The particle sizes in fogs (suspensions of water droplets in air) are larger and range from about 4 to 50 μ in diameter.

In the aerosol industry, the term "aerosol" initially had about the same meaning as in colloid chemistry. The first aerosol of any commercial significance, the aerosol insecticide, was defined in 1949[2] as a system of particles suspended in air where 80% of the particles were less than 30 μ in diameter and no particles were larger than 50 μ. This definition developed from observations that aerosol insecticides with this particle size were particularly effective because the particles remained suspended in air for approximately one hour.

In 1955, the Chemical Specialties Manufacturers Association (CSMA) published a tentative glossary of terms for the aerosol industry. An aerosol was still defined as a suspension of fine solid or liquid particles in air or gas, for example, smoke, fog, or mist. A distinction was made between the terms aerosol and aerosol product. The latter was defined as a "self-contained sprayable product in which the expelling force was supplied by a liquefied gas." This classification included space sprays, residual sprays, surface coatings, foams, and various other products. Products pressurized with compressed gases, such as whipped cream, were not included.

The word aerosol has now become a familiar term throughout most of the

3

world. According to a recent CSMA glossary of terms,[4] aerosol packaging is defined as "pressurizing sealed containers with liquefied or compressed gases so that the product is self-dispensing." The term aerosol, as used here, is not confined to the scientific defintion, that is, a suspension of fine solid or liquid particles in air or gas. The distinction between aerosol product, or pressurized product has disappeared as far as the general use of the term is concerned.

The present aerosol industry is generally considered to have received its stimulus from the development of the aerosol insecticides used in World War II. Although there were few aerosol products on the market before World War II, a considerable amount of information had already been published about aerosols. During the period from 1862 through 1942, patents granted to various workers revealed that many of the basic principles of aerosols were already recognized. These included the convenience provided by aerosol packaging, fundamentals of aerosol formulation, advantages of liquefied gas propellants over compressed gases, use of three-phase systems for discharging water, value of an expansion chamber in a valve for decreasing particle size in a spray, and the concept of a mechanical break-up valve.

One of the first patents in the aerosol field was granted to Lynde[5] in 1862 for a valve with a dip tube for closing bottles. The valve was constructed so that a bottle filled with mineral water could be pressurized with gas. The water could be discharged from the bottle merely by opening the valve.

After this initial work, there was little activity in the aerosol field until 1899. In that year, Helbing and Pertsch[6] described a method for producing a fine jet or spray from a solution of materials such as gums, resins, nitrocellulose, etc., in methyl or ethyl chloride. The heat of the hand caused enough pressure to develop in the container so that the solution would eject through an orifice in a fine jet or spray. The main object of the invention was to apply a thin, uniform coating of a product, such as a collodion, to the skin.

Gebauer discovered the value of an expansion chamber for producing a finer spray in 1901.[7] He was granted a patent in that year for an improved receptacle for containing and discharging volatile liquids. When the valve on the receptacle was opened, the pressure inside the container forced the liquid into a capillary tube. The capillary tube led to an expansion chamber which ended in a nozzle open to the atmosphere. The diameter of the capillary tube was smaller than that of the outer nozzle. Gebauer recognized that the partial vaporization of the liquid in the expansion chamber produced a finer spray. He also stated that the design of the expansion chamber affected spray characteristics. The section of the unit containing the capillary tube, expansion chamber, and outer nozzle could be removed for cleaning when the capillary tube became clogged.

In 1902, Gebauer improved the design of his receptacle.[8] The capillary tube leading into the expansion chamber was eliminated and the valve between the expansion chamber and the main body of the receptacle was used as the inlet orifice. When the valve was barely opened, the diameter of the valve orifice was smaller than that of the orifice of the outer nozzle. Under these conditions, the liquid discharged as a spray. At valve openings where the diameter of the valve orifice was larger than that of the outer nozzle, the product was emitted as a stream. The discharge characteristics of the product, therefore, were controlled by the ratio of the diameters of the inner and outer orifices. This, in turn, was controlled by the degree of valve opening. Gebauer's observations on both the effect of the relative ratios of the two orifice diameters and the important effect of the expansion chamber upon spray characteristics sound very familiar in view of present-day knowledge about aerosol valves.

Gebauer carried the aerosol idea even further. He packaged whipped colloidal graphite lubricant as an aerosol for application to locks. Gebauer also put soap into ethyl chloride for the medical profession and developed a mixture to remove adhesive tape from the skin.[9]

In 1903, Moore[10] obtained a patent on an atomizer for perfumes in which carbon dioxide was employed as the propellant. The atomizer was constructed so that carbon dioxide had two functions. The pressure of the carbon dioxide forced the solution of perfume up the dip tube and through a nozzle. Carbon dioxide was also discharged from the vapor phase as a gas through a separate nozzle. The latter was arranged so that carbon dioxide passed over the tip of the perfume nozzle, creating a partial vacuum. This produced a fine spray or stream, depending upon the diameter of the opening of the discharge valve for the perfume solution.

Carbon dioxide was utilized by Mobley in 1921 for dispensing liquid antiseptics.[11] One of the most interesting features of Mobley's apparatus was the discharging device, which contained the equivalent of four orifices. After the carbonated liquid passed through the first metering orifice, it entered a chamber equipped with what Mobley described as a block-partition containing two small orifices. The construction of the block-partition was such that the two orifices imparted a spirally rotating motion to the jet stream before it was discharged through the outer orifice. This may have been one of the first mechanical break-up valves. A patent on the use of carbon dioxide for atomizing perfumes was also granted to Lemoine in 1926.[12]

An event that played a major part in the future development of the aerosol industry occurred in the refrigeration field in 1928.[13] The refrigerants used up to that time were toxic gases such as sulfur dioxide, methyl chloride, and ammonia. Leakage of these gases from refrigeration systems sometimes caused fatalities; consequently, the need for a refrigerant of low toxicity was recognized.[14] To meet this demand, Midgley, Henne, and McNary synthesized dichlorodifluoromethane (CCl_2F_2), a fluorocarbon of very low toxicity. By 1929, experimental quantities were being produced in a small plant. Midgley presented the results of his work at a meeting of the American Chemical Society in 1930.[15] During his presentation, he demonstrated the nonflammability and low toxicity of dichlorodifluoromethane by inhaling sufficient quantities of the new refrigerant so that he could extinguish a burning candle when he exhaled. Dichlorodifluoromethane was destined to become one of the most commonly used aerosol propellants.

Du Pont and General Motors formed a new company, Kinetic Chemicals, Inc., in 1930 to manufacture the new refrigerant. In 1931, both dichlorodifluoromethane and another new refrigerant, trichloromonofluoromethane (CCl_3F), became commercially available. Dichlorotetrafluoroethane ($CClF_2CClF_2$) and trichlorotrifluoroethane (CCl_2FCClF_2) appeared on the market in 1934. Along with dichlorodifluoromethane, trichloromonofluoromethane and dichlorotetrafluoroethane would become widely used as aerosol propellants in the future.

Chlorodifluoromethane, $CHClF_2$ (Freon® 22), became commercially available in 1936. However, production did not become significant until World War II created a demand for Teflon® fluorocarbon resin. Teflon® was discovered by Plunkett in 1938. It was prepared from the monomer tetrafluoroethylene, which, in turn, was obtained by the thermal cracking of chlorodifluoromethane. After the war, production of chlorodifluoromethane increased rapidly as the refrigerant properties of Freon® 22 became known. Applications in home and commercial air conditioning and in low-temperature freezers and display cases expanded. The existing production capacity was strained by wartime demands for the refrigerant. In 1944, a new plant was built by the Government at East Chicago, Indiana and operated by Kinetic Chemicals for about one year. The plant was dismantled when the war ended in 1945. A new plant was erected by Kinetic Chemicals at the same site three years later.

Another significant advance in the field of aerosols was achieved by Rotheim[16] in 1931. Rotheim reported a method for spraying coating compositions which involved dissolving materials such as lacquers, soaps, resins, cosmetic products, etc., in liquefied dimethyl ether in a closed container and discharging the solution through a valve. Rotheim observed that the spray characteristics of the products could be varied from coarse to fine by changing the percentage of dimethyl ether in the formulation. He also noted that the volume of gas resulting when dimethyl ether vaporized was about 350 times the volume of the initial liquefied dimethyl ether.

In a subsequent patent,[17] Rotheim reported that a valve with an expansion chamber gave a considerably finer spray from a liquefied gas aerosol than one with a single orifice. He concluded that the orifice leading into the expansion chamber should have a smaller diameter than the orifice that opened to the atmosphere. These principles of valve construction were recognized much earlier by Gebauer. Rotheim also disclosed that other compounds, such as methyl chloride, isobutane, methyl nitrite, vinyl chloride, and ethylene dichloride, could be used as liquefied gas propellants.

Fluorocarbon refrigerants likewise were utilized as aerosol propellants in this period. In 1933, a patent was granted to Midgley, Henne, and McNary[18] on the use of these compounds in fire extinguishers. The function of the propellant was to create sufficient pressure to expel itself and a flame arrester from a container. Dichlorodifluoromethane was specifically mentioned.

The use of fluorocarbons as propellants for fire extinguishers was extended by Bichowsky in 1935.[19] Dichlorodifluoromethane, dichloromonofluoro-methane, and dichlorotetrafluoroethane were used in combination with fire-extinguishing agents such as trichloromonofluoromethane, carbon tetrachloride, sodium bicarbonate, and water. Bichowsky disclosed that combinations such as dichlorodifluoromethane and carbon tetrachloride maintained a relatively con-stant pressure during discharge. He may have been the first to recognize this im-portant property of liquefied gas aerosols.

Bichowsky also described systems containing mixtures of propellant and water in which the water was layered over the propellant. The dip tube extended only to the bottom of the aqueous phase. The function of the propellant was to pro-vide sufficient pressure to discharge the water from the container. This is a rather accurate description of the so-called "three-phase system" used about 20 years later for formulating aqueous solutions as aerosols.

Rotheim continued to work on aerosols and in 1938[20] was granted a patent covering insecticidal compositions using propellants boiling below $-20°F$. He claimed that such compounds provided more efficient atomization of the aerosols. Propellant mixtures disclosed included combinations of butane and ethane, propane and methyl chloride, ethane and dimethyl ether, propane and methane, etc. Iddings[21] mentioned ethane, propane, and ethylene as potential propellants for spraying liquid insecticides, mothproofers, and other products.

One of the most important developments in the aerosol field occurred during World War II. The need for a portable insecticide dispenser became imperative because of the disease caused among overseas troops by insects. The prior work of Goodhue, Sullivan, and Fales[22-25] on aerosol insecticides was extremely timely and became the basis for the aerosols used during the war. Westinghouse signed the first contract for the World War II insecticides in 1942. From July of that year, when the requirement was 10,000 units per day, until the end of the war, Westinghouse alone supplied over 30 million aerosols to the armed forces. Other companies that manufactured aerosol insecticides for the military included Airosol, Inc., Bridgeport Brass, Industrial Management, and Virginia Smelting.

The aerosol insecticides were packaged in 1-lb heavy steel containers pat-terned after the small Freon® 12 units produced by Westinghouse for charging refrigerators. The containers were made of two shells, drawn of 0.044-in. steel and welded together. The cylinders were fitted with an oil-burner-type valve with a swirl chamber.[26] The containers had a capacity of 16 fluid ounces (fl. oz) and were filled with a mixture of 90 wt. % Freon® 12 and 10 wt. % of a pyrethrin and sesame oil concentrate. A 4-in. metal dip tube, 0.017 in. in diameter, was attached to the valve, inside the container. The pressure in the aerosols was about 70 pounds per square inch gage (psig) at 70°F, and the blow-off release was set at 300 psig. The aerosol insecticides were named "bug bombs" by Westinghouse employees because of their resemblance to a small bomb.[23]

The World War II insecticides were heavy, unattractive, difficult to operate, and expensive. Because of this, there was considerable speculation during the war whether aerosol insecticides could be successfully marketed to the general public.[27-29] One factor in favor of the insecticides was that millions of Americans and others in foreign countries had used them and were impressed with their efficiency. The manufacturers of the aerosol insecticides received thousands of letters after the war from people who wanted to purchase "bug bombs" for their own use. This suggested that a sizable civilian market existed for the insecticides.

Westinghouse was still interested in this potential market after the war and carried out a consumer test in Jacksonville, Florida in September 1945.[30] The appearance of the "bug bomb" was improved, the release mechanism simplified for home use, and a set of instructions supplied with each unit. The insecticides were sold at 80 locations, including department stores, groceries, supermarkets, hardware stores, and filling stations. After 30 days, enough insecticides had been sold to supply 8 out of every 10 homes. More than 90% of the purchasers indicated that they intended to buy additional units. Regardless of the success of this test, Westinghouse decided in 1946 not to become involved in the civilian aerosol industry.

Although initial public acceptance of the aerosol insecticides after the war was encouraging, it was recognized as early as 1942 that ultimately the civilian consumer would demand aerosols that were cheaper, lighter, and easier to use.[31] Subsequent considerations indicated that if such a product were to become a reality, four needs had to be met. These were:

1. A low-cost, lightweight, disposable container.
2. Amendment of the Interstate Commerce Commission (ICC) regulations to permit pressures greater than 25 psig at 70°F in the lightweight containers.
3. A satisfactory low-pressure propellant.
4. A reliable valve.

The four requirements were all achieved during the latter part of the 1940's. This made possible the production of aerosol products at a price within reach of the general American public. The first commercial shipment of a low-pressure aerosol insecticide, trademarked "Jet," occurred on March 7, 1947. By 1949, deodorant aerosols, perfume sprays, a suntan stimulant, and paints and lacquers were on the market. Aerosol products under development included antiseptics, cosmetics, lotions, creams, disinfectants, fungicides, germicides, hair dressings, and shampoos. As the advantages of aerosol packaging became more and more apparent, the list of products likewise continued to expand. The phenomenal growth of the aerosol industry during the next 25 years has been well publicized. Estimated aerosol production in 1947 was 4.3 million units and that in 1973 was almost 3 billion units.[32]

The peak year was 1973. Production in 1974 dropped to 2.721 billion units and was estimated to be 2.354 billion units in 1975.[33] The market now appears to be leveling off with estimated fillings of 2.290 billion units in 1976.[34] The decrease since 1973 has been attributed to a combination of factors, including shortage of propellant, the economic climate, the fluorocarbon/ozone controversy, and adverse aerosol publicity.

REFERENCES

1. D. Sinclair, "Handbook on Aerosols," Washington, D.C., 1950, p. 64.
2. R. A. Fulton and S. A. Rohwer, *Soap Sanit. Chem.* **25**, 122 (February 1949).
3. "Glossary of Terms Used in the Aerosol Industry," Chemical Specialties Manufacturers Association (CSMA), August 25, 1955.
4. "Aerosol Guide," Aerosol Division, Chemical Specialties Manufacturers Association (CSMA), Inc., 6th ed., March 1971.
5. J. D. Lynde, U. S. Patent 34,894 (1862).
6. H. Helbing and G. Pertsch, U. S. Patent 628,463 (1899).
7. C. L. Gebauer, U. S. Patent 668,815 (1901).
8. C. L. Gebauer, U. S. Patent 711,045 (1902).
9. *Aerosol Age* **19**, 26 (May 1974).
10. R. W. Moore, U. S. Patent 746,866 (1903).
11. L. K. Mobley, U. S. Patent 1,378,481 (1921).
12. R. M. L. Lemoine, D. R. Patent 532,194 (1926).
13. R. C. Downing, "History of the Organic Fluorine Industry," in Kirk-Othmer, Encyclopedia of Chemical Technology, Vol. 9, 2nd ed., John Wiley & Sons, Inc., New York, 1966, p. 704.
14. T. Midgley, *Ind. Eng. Chem.* **29**, 239 (1937).
15. T. Midgley and A. L. Henne, *Ind. Eng. Chem.* **22**, 542 (1930).
16. E. Rotheim, U. S. Patent 1,800,156 (1931).
17. E. Rotheim, U. S. Patent 1,892,750 (1933).
18. T. Midgley, A. L. Henne, and R. R. McNary, U. S. Patent 1,926,396 (1933).
19. R. R. Bichowsky, U. S. Patent 2,021,981 (1935).
20. E. Rotheim, U. S. Patent 2,128,433 (1938).
21. C. Iddings, U. S. Patent 2,070,167 (1937).
22. W. N. Sullivan, L. D. Goodhue, and J. H. Fales, *J. Econ. Ent.* **35**, 48 (February 1942).
23. L. D. Goodhue, *Ind. Eng. Chem.* **34**, 1456 (December 1942).
24. L. D. Goodhue and W. N. Sullivan, U. S. Patent 2,321,023 (1943).
25. E. R. McGovern, J. H. Fales, and L. D. Goodhue, *Soap. Sanit. Chem.* **19**, 99 (September 1943).
26. W. C. Beard, "Valves," in H. R. Shepherd (Ed.), "Aerosols: Science and Technology," Interscience Publishers, Inc., New York, 1961.

27. L. D. Goodhue, *Aerosol Age* **10,** 30 (May 1965)
28. J. H. Fales and L. D. Goodhue, *Soap Sanit. Chem.* **20,** 107 (July 1944).
29. W. W. Rhodes, *Soap Sanit. Chem.* **20,** 108 (July 1944).
30. R. E. Ditsler, *Soap Sanit. Chem.* **22,** 131 (July 1946).
31. L. D. Goodhue, "Aerosol Technicomment," Vol. XII, No. 2.
32. Du Pont Company 1974 Aerosol Market Report.
33. "Pressurized Products Survey," Chemical Specialties Manufacturers Association (CSMA), May 1975.
34. "Pressurized Products Survey," Chemical Specialties Manufacturers Association (CSMA), May 1976.

2

THE HOMOGENEOUS AEROSOL SYSTEM

Most aerosol products contain three major components: propellants, solvents, and active ingredients. Depending upon the manner in which these components are combined in the product, aerosols can be divided into two broad classes: homogeneous or heterogeneous. Products are homogeneous or have been formulated as homogeneous systems when all the components are mutually soluble. Homogeneous products are two-phase systems, consisting of a single liquid phase in equilibrium with a vapor phase. Examples of common homogeneous aerosols are hair sprays, personal deodorants, and colognes. One of the advantages of products formulated as homogeneous systems is that they do not need to be shaken before use.

Products in which all the components are not mutually soluble are classified as heterogeneous. There are several types of heterogeneous aerosols. Powder sprays and dry-type antiperspirants are formulated as suspensions of solids in a liquefied gas propellant. Other heterogeneous systems contain several immiscible liquids combined as an emulsion (a dispersion of one liquid in another). Many space deodorants, insecticides, and foams are formulated as emulsion systems.

Heterogeneous systems contain at least three phases. Suspension systems consist of a solid phase and liquid phase in equilibrium with a vapor phase. Emulsion systems contain two liquid phases in equilibrium with a vapor phase. Heterogeneous systems are discussed in detail in Part II of this book.

PROPERTIES OF LIQUIDS AND GASES

The behavior of propellants in aerosol products is determined by the same fundamental laws that apply to other gases and liquids. A knowledge of these laws is helpful in explaining how propellants function in aerosol products. Phenomena such as the liquefaction of gases, formation of sprays from aerosol systems, effect of solvents upon the vapor pressure of propellants, etc., are easier to understand when the general properties of liquids and gases are known. Therefore, some of these are reviewed briefly before discussing actual aerosol systems.

The structures and properties of gases are explained by the kinetic theory. According to this theory, gases are considered to consist of a number of small particles called molecules. Under normal conditions of room temperature and atmospheric pressure, the average distance between the molecules is large compared to the size of the molecules themselves. The molecules are in constant motion. As a result of this motion, gas molecules collide with one another and also strike the walls of any vessel in which they are confined. It has been estimated that a gas molecule collides and is halted about 60,000 times in travelling a distance of 1 cm.[1] The average distance a molecule travels between collisions is referred to as the mean-free-path of the molecule.[2] As a result of these collisions, the speeds of the molecules range from zero to very high values. Molecules of hydrogen, for example, have been reported to move at an average speed of over 1 mile/sec at 0°C.[3] A consequence of the motion of the molecules is the property of gases called pressure. The pressure of a gas results from the impact of the gas molecules on the object against which the pressure is being exerted. This was first suggested by Bernoulli in 1740.

The properties of gases have been investigated for many years. In 1662, Boyle discovered that the volume of a gas is inversely proportional to the pressure if the temperature is constant. This is known as Boyle's Law and can be illustrated as follows, where V = volume, P = pressure, T = temperature, and $\propto$ is a proportionality symbol:

$$V \propto \frac{1}{P} \ (T \text{ is constant})$$

Later, Charles and Gay-Lussac found that the volume of a gas is directly proportional to its temperature if the pressure remains constant:

$$V \propto T \ (P \text{ is constant})$$

These discoveries lead to the derivation of an expression that relates the pressure, volume, and temperature. This is called the ideal gas law and is illustrated mathematically as follows:

$$PV = nRT \qquad (2\text{-}1)$$

In this equation, P is the pressure exerted by the gas, V is the volume occupied by the gas, n is the number of moles of gas (which indicates its concentration), T is the absolute temperature, and R is a number which is called a gas constant.

Initially, it was assumed that Equation 2-1 could be used to calculate the exact properties of any gas if a sufficient number of the other variables were already known. For example, if V, n, R, and T were known, then P could be cal-culated. However, careful experiments by a number of investigators showed that the properties of the gases that were calculated using Equation 2-1 did not agree exactly with the experimentally determined values. It was also noted that the differences between the calculated and experimental values were the least under conditions of high temperature and low pressure. The fact that gases do not fol-low the ideal gas law (Equation 2-1) is referred to as the deviation of the gas from the gas laws.

Budde[4] was the first to suggest that one reason for the deviation of actual gases from the ideal gas law was that the volume V in Equation 2-1 did not take into account the volume occupied by the gas molecules themselves. If a correc-tion for the volume of the gas molecules, designated by the letter b, is applied, then the preceding equation becomes

$$P(V - b) = nRT \qquad (2\text{-}2)$$

Equation 2-2 was found to be satisfactory for hydrogen, but not for any other gas. In 1879, van der Waals pointed out that the ideal gas law also did not take into account the intermolecular forces of attraction that exist between all molecules. This attractive force is electrical in nature and tends to bring the molecules closer together.[5]

The intermolecular forces of attraction are designated by the term a/V^2, where a is the constant of intermolecular attraction and V is the volume oc-cupied by the gas. When the term a/V^2 is added to Equation 2-2, the following equation, known as the van der Waals equation, is obtained:[4,5]

$$(P + a/V^2)(V - b) = nRT \qquad (2\text{-}3)$$

The van der Waals equation not only applies to gases, but also to many liquids.

The intermolecular forces of attraction between the molecules only operate over a very short distance of several molecular diameters.[2] When the gas is at a comparatively high temperature, the velocity of the gas molecules is sufficient to overcome the attractive forces between the molecules. Also, at low pressures, there are so few molecules present that the volume occupied by the gas mole-cules themselves is small compared to the total volume occupied by the gas. Therefore, under conditions of low pressure and high temperature, the correc-tions b and a/V^2 become negligible. The gas then approaches ideal behavior.

The liquefaction of gases can be explained from the preceding discussion. The attractive force between molecules, designated by a in the van der Waals equation, is responsible for the condensation of gases to liquids.[4,5] When pres-sure is applied to a gas, the molecules are brought closer together and the attrac-tive forces between the molecules become stronger. Likewise, if the temperature of the gas is lowered, the velocity of the gas molecules is decreased and this also increases the molecular attraction. If the pressure is high enough and the temperature low enough, the attractive forces between the molecules become so predominant that the molecules begin to cling to each other.[1] These clumps of molecules grow larger and ultimately settle to the bottom. The gas then changes to a liquid.

According to Richards, the molecules of a liquid are in actual contact.[1] This is why liquids are so difficult to compress compared to gases. However, the molecules in a liquid still possess a limited amount of motion and, therefore, a liquid is capable of flowing and taking the shape of the vessel the liquid is in. The strength of the attractive forces between molecules depends upon the structure of the molecules. The main difference between gases and liquids is the different order of magnitude of the intermolecular forces of attraction in the two cases.[2] In compounds such as water and ethyl alcohol which are liquids under normal conditions of room temperature and atmospheric pressure the attractive forces between molecules are very strong.

In all liquids, there is a continuous flight of molecules from the surface of the liquid into the vapor space above the liquid. In order for the molecules to escape from the surface of the liquid and enter the vapor phase, the molecules must overcome the intermolecular forces of attraction in the liquid. Only the molecules with the highest speeds, that is, the highest kinetic energy, are capable of doing this. If the liquid is open to the air, these molecules are lost to the atmosphere and the liquid evaporates. Since these molecules have the highest kinetic energy, their loss results in a decrease in the total energy in the liquid. When the kinetic energy decreases, the liquid cools. This is why a liquid cools during evaporation, unless heat is supplied by an external source.

If the liquid is confined in a closed vessel, there is also a continuous flight of molecules from the surface of the liquid into the vapor space. However, at the same time, a reverse process of condensation of vapor molecules at the surface of the liquid takes place. In time, a condition of equilibrium will be established in the container when the rates of vaporization and condensation are equal. The vapor is said to be saturated at this point. The pressure exerted by the molecules of the vapor in equilibrium with the liquid is known as the vapor pressure of the liquid. The concentration of molecules in the vapor phase at equilibrium is constant for any liquid at a given temperature and, therefore, the vapor pressure remains constant. When the temperature is increased, both the number of molecules in the vapor phase and their velocities are increased. Therefore, the pressure increases when temperature is increased.

Aerosol propellants have boiling points below room temperature and are gases under conditions of room temperature and atmospheric pressure. However, when the gaseous propellants are cooled or compressed sufficiently, they are converted into liquids because of the intermolecular forces of attraction, as previously discussed. If the propellants are cooled below their boiling points, they can be poured from one container to another, just as any normal liquid.

However, at room temperature they must be confined in a closed container in order to keep them from changing into vapor. When the liquefied gas propellants are discharged into the atmosphere, they again vaporize and become gases.

AEROSOL COMPONENTS

Propellants

Propellants provide the pressure that forces the product out of the container when the valve is opened. They also influence the form in which the product is discharged—foam, stream, or spray. Variations in the type and concentration of a propellant can change a coarse, wet spray to a fine, dry spray.

There are two general types of aerosol propellants: liquefied gases and compressed gases. In the aerosol industry, a liquefied gas is defined as a compound having a vapor pressure greater than atmospheric pressure (14.7 psia) at 105°F.[1] The purpose of the particular pressure and temperature limitations of the definition was to include liquids such as fluorocarbon 11 (trichloromonofluoromethane) and methylene choride. Neither of these compounds can function alone as a propellant because they have too low a vapor pressure at room temperature. However, they give satisfactory propellant blends when mixed with other liquefied gases having higher vapor pressures. Methylene chloride has a vapor pressure of 14.7 psia at 103.6°F and just falls within the definition. The two major classes of liquefied gas propellants are fluorinated hydrocarbons (fluorocarbons) and hydrocarbons.

The distinction between liquefied and compressed gases is somewhat artificial. Gases such as carbon dioxide and nitrous oxide can be liquefied at very low temperatures. They are excluded from the definition of liquefied gases, however, because they have pressures in excess of 700 psig at 70°F in the liquefied state. These pressures are far too high for any practical consideration of these compounds as liquefied gas propellants for aerosol products. Therefore, they are used only in the compressed state in aerosol systems, generally at pressures less than 100 psig.

General Properties

Some general properties of liquefied gas propellants are summarized below.

Boiling Point and Vapor Pressure. A liquefied gas propellant (or propellant blend) must boil enough below room temperature so that the pressure in the aerosol container is sufficient to expel the contents when the valve is opened. The pressure cannot exceed certain limits, however.

Cost. The manufacturing cost of a propellant has to be low so that the ultimate price of the aerosol product is within reach of the general public.

Flammability. Nonflammability is not a necessity. It is, however, an extremely desirable property for an aerosol propellant. This is particularly true for the propellant manufacturer and aerosol filler who must store and handle large quantities of the propellants. Flammable propellants require explosion-proof filling equipment. The flammability of an aerosol product is important to marketers because of regulations.

Odor. The odor of aerosol products such as hair sprays, deodorants, perfumes, colognes, antiperspirants, etc., is important to the consumer. Therefore, propellants with very low odor levels are desirable. Propellants for aerosol perfumes and colognes must have essentially no odor.

Purity. Propellants must be very pure and their quality must be consistent. Variable amounts of impurities could result in a change in perfume odor, inoperable valves, and container corrosion. These adverse effects could lead to withdrawal of the product from the market.

Stability. Propellants must be stable in the aerosol formulation in which they are used. Any reaction with other components of the formulation, or propellant decomposition alone, could lead to product deterioration, change in perfume odor, and container corrosion.

Toxicity. Low toxicity is essential. Inhalation of many aerosol products, such as hair sprays, antiperspirants, room deodorants, perfumes, etc., is difficult to avoid. Other products, both cosmetic and pharmaceutical, are applied to the skin. Some pharmaceutical aerosols are formulated for inhalation therapy.
  Since liquefied gas propellants are liquids, they are part of the solvent system in homogeneous products. Therefore, they influence the solvent properties of the liquid phase. The solvent characteristics of propellants in the liquid state must be taken into consideration in formulating products.

Solvents

Typical solvents, such as ethyl alcohol, isopropyl alcohol, methylene chloride, methyl chloroform, or odorless mineral spirits, are present in many aerosols. Like the propellants, solvents generally perform a number of functions. One of the major uses of solvents is to bring active ingredients into solution with the propellants. Most propellants have poor solvent characteristics; in many cases, active ingredients are not soluble in propellants. In order to obtain a homogeneous mixture, it is necessary to add a liquid with the necessary solvent properties. It is sometimes desirable to have another liquid present which is not miscible with the propellant, for example, water or propylene glycol. In these cases, a co-solvent such as ethyl alcohol is added to obtain a homogeneous system.

Another function of solvents is to help produce a spray with a particle size that is most effective for the particular application involved. Aerosol insecticides, for example, are most efficient when the particle size is neither too large nor too small. If no solvents were present, the propellants would evaporate completely in air shortly after discharge, leaving extremely small particles. High-boiling solvents evaporate only slowly in the air once the propellant has vaporized; therefore, droplets of a specified size can be obtained by proper formulation.

Still another use of solvents is to reduce the vapor pressure of the propellant. There are various pressure regulations governing the packaging of aerosol products. Some propellants, such as fluorocarbon 12, have too high a vapor pressure to be used alone. In such a case, a vapor pressure depressant is added to reduce the vapor pressure to an acceptable level. The vapor pressure depressant may be a solvent such as ethyl alcohol or odorless mineral spirits, or it may be a high-boiling propellant like fluorocarbon 11, which can be considered both a solvent and a propellant.

Active Ingredients

Active ingredients are the materials (perfumes, pharmaceuticals, insecticides, deodorants, hair spray resins, etc.) essential for the specific application for which the aerosol was designed. It is the job of the aerosol chemist to combine the active ingredients, solvents, and propellants so that an efficient, attractive, and acceptable product is obtained.

In addition to the three major components (propellants, solvents, and active ingredients), aerosols usually contain a number of other materials that have been added to improve the product. Some of these, such as perfumes, enhance the aesthetic appeal, while others increase the performance or efficiency of the product. Typical additives in this group include moisturizers, emollients, plasticizers, stabilizers, surfactants, and skin penetrants. Often the formulation of a successful aerosol product is quite difficult and may require the services and knowledge of many different suppliers and companies. The perfuming of aerosols, for example, is an art that requires extensive experience and skill. A reputable fragrance house should be consulted when questions regarding perfuming of aerosol products arise.

OPERATION OF AN AEROSOL

Basically, a homogeneous aerosol is an intimate mixture of liquefied gas propellant, solvents, and active ingredients packaged under pressure in a suitable con-

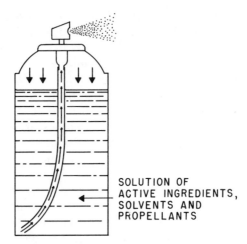

Figure 2-1.   Operation of a homogeneous aerosol.

tainer equipped with a valve. When the valve is opened, the internal pressure forces the aerosol up the dip tube and out the valve. As soon as the aerosol leaves the container and enters the atmosphere, part of the propellant vaporizes and the propellant molecules fly off into space. In doing so, they collide with solvent molecules and cause collisions among the solvent molecules themselves, disrupting the attractive forces between them. As the propellant vaporizes in air, the solvent molecules are left almost isolated in relatively small groups of particles. The liquefied gas propellant, therefore, not only provides the pressure to force the aerosol out of the container when the valve is opened, but also breaks up the concentrate into small droplets or particles once the aerosol has reached the atmosphere.

This operation is illustrated in Figure 2-1.

REFERENCES

1. M. C. Sneed and J. L. Maynard, "General Inorganic Chemistry," 2nd ed., Chap. 8, D. Van Nostrand and Company, Inc., Princeton, N.J., 1942.
2. F. A. Saunders, "A Survey of Physics for College Students," Chap. 10, Henry Holt & Company, New York, 1930.
3. F. H. MacDougall, "Physical Chemistry," Chap. 3, The Macmillan Company, New York, 1936.
4. F. H. Getman and F. Daniels, "Outlines of Theoretical Chemistry," 5th ed., Chap. 2, John Wiley & Sons, Inc., New York, 1931.
5. F. Daniels and R. A. Alberty, "Physical Chemistry," Chap. 2, John Wiley & Sons, Inc., New York, 1955.
6. "Aerosol Guide," Aerosol Division, Chemical Specialties Manufacturers Association (CSMA), Inc., 6th ed., March, 1971.

3

FLUOROCARBON PROPELLANTS – CURRENT AND ALTERNATIVE

John J. Daly, Jr.

Prior to World War II, fluorocarbons were used commercially only in the refrigeration and air conditioning industries. The development of aerosol insecticides during the war resulted in the first large-scale use of a fluorocarbon as an aerosol propellant. The fluorocarbon was Freon[®] 12, dichlorodifluoromethane, which has a pressure of 70.2 psig at 70°F. After the war, a major effort by such pioneers as Peterson[1] and Goodhue[2] was directed toward finding a suitable low-pressure propellant for the nonrefillable containers being developed by the Continental Can Company and Crown Cork & Seal Company. Various ratios of Freon[®] 12 and Freon[®] 11, trichloromonofluoromethane, were evaluated by Young,[3] who finally settled on a Freon[®] 12/Freon[®] 11 (54/46) blend, which had a vapor pressure of about 40 psig at 70°F. This was desirable because the ICC regulations had been amended in July 1947 to allow pressures up to 40 psig in the nonrefillable containers. Ultimately, the Freon[®] 12/Freon[®] 11 (50/50) blend became the most popular because it was slightly less expensive and gave excellent results with valves developed in later years.

The first aqueous foam product, an aerosol shaving lather, appeared on the market in 1950. Freon[®] 11 could not be used because of its instability in the presence of water and metal. Freon[®] 114, dichlorotetrafluoroethane, proved to be an excellent vapor pressure depressant for Freon[®] 12 in aqueous systems because of its stability. Blends of these two propellants have been used in aqueous-based products since the early 1950's.

Fluorocarbons 12, 11, and 114 are the most commonly used fluorocarbon aerosol propellants. In recent years, however, they have come under considerable investigation from an ecological viewpoint (ecological effects, for example, the fluorocarbon/ozone controversy, are discussed in detail in Chapter 22, "Fluorocarbons in the Atmosphere"). Fluorocarbon manufacturers have undertaken programs to develop alternative fluorocarbon propellants anticipating that legislative restrictions would be passed. Potential alternative propellants, their properties and uses, are covered in the second section of this chapter.

TERMINOLOGY

Fluorocarbons are divided into two groups, depending on their flammability. Nonflammable fluorocarbons marketed by the Du Pont Company are given the trademark Freon® to distinguish them from flammable fluorocarbons. The term "fluorocarbon" (FC) is broadly used for all types of organo-fluorine compounds employed as propellants regardless of their flammability or commercial status. A further refinement of the term fluorocarbon can be made to distinguish chlorofluorocarbons (which contain only carbon, chlorine, and fluorine) from hydrochlorofluorocarbons (which contain hydrogen in addition to carbon, chlorine, and fluorine) and hydrofluorocarbons (which contain carbon, hydrogen, and fluorine).

Fluorocarbons are manufactured worldwide. Producers and the trade names of their products are listed in Table 3-1.

THE NUMBERING SYSTEM

The various fluorocarbon propellants are usually distinguished only by a number after the trade name or term "fluorocarbon," for example, Freon® 12, FC-11, or fluorocarbon 114.

The numbering system used for the different fluorinated compounds was developed by the Du Pont Company in 1929 so that the chemical formula of a fluorocarbon could be determined from the number of the compound alone. Thus, it is possible to write both the empirical and structural formulas for fluorocarbons 11, 12, 114, etc., from the numbers of these products. The numbering system can also be applied to hydrocarbons and chlorinated hydrocarbons. The system was offered to the refrigeration industry in 1960 to establish uniformity in numbering refrigerants.[4] It is now used by all U.S. manufacturers of fluorocarbons and applies to all products—propellants, refrigerants, blowing agents, and solvents—regardless of end use.

The formulas of aerosol propellants can be written from the propellant numbers by following seven rules.

RULE 1. The first digit on the right signifies the number of fluorine (F) atoms in the compound.

RULE 2. The second digit from the right is numerically one more than the number of hydrogen (H) atoms in the compound.

TABLE 3-1. WORLDWIDE FLUOROCARBON PRODUCERS

Country	Company	Trade Name
U.S.A.	E. I. du Pont de Nemours & Company, Inc., Wilmington, Del.	Freon®
	Allied Chemical Corp., Morristown, N.J.	Genetron
	Kaiser Chemicals, Oakland, Calif.	Kaiser
	Pennwalt Corp., Philadelphia, Pa.	Isotron
	Racon, Inc., Wichita, Kans.	Racon
	Union Carbide Corp., New York, N.Y.	Ucon
Argentina	Ducilo S.A., Buenos Aires	Freon®
	Tool Research S.A., Buenos Aires	Algeon
	I. R. A., S.A., Buenos Aires	Frateon
Australia	Australian Fluorine Chemicals Pty. Ltd., Rozelle, N.S.W.	Isecon
	Pacific Chemical Industries Pty. Ltd., Camelia, N.S.W.	Forane
Brazil	Du Pont do Brasil S.A., Industrias Quimicas, Sao Paulo	Freon®
	Hoechst do Brasil, Sao Paulo	Frigen
Canada	Du Pont of Canada Ltd., Toronto, Ontario	Freon®
	Allied Chemical (Canada) Ltd., Amherst, Ontario	Genetron
Czechoslovakia	Slovek Pro Chemickov A Hutni, Bohemia	Ledon
England	Imperial Chemical Industries Ltd., London	Arcton
	I.S.C. Chemicals Ltd., London	Isceon
France	Rhone-Poulenc Chemiefine, Paris	Flugene
	Produits Chimiques Ugine Kuhlman, Paris	Forane
W. Germany	Farbwerke Hoechst AG, Frankfurt	Frigen
	Kali-Chemie Pharma GmbH, Hannover	Kaltron
E. Germany	VEB Chemiewerk Nunchritz, Kreisriesa	Fri-dohna
Greece	C.I.N.G. Chemical Industries of Northern Greece, Athens	Flugene
India	Everest Refrigerants Ltd., Bombay	Everkalt
	Navin Fluorine Industries, Nariman Point Bay	Mafron
Italy	Montedison, Milano	Edifrene
Japan	Mitsui Fluorochemicals Company, Ltd., Tokyo	Freon®
	Daikin Kogyo Company, Ltd., Osaka	Daiflon
	Asahi Glass Company, Ltd., Tokyo	Asahiflon
	Showa Denko K.K., Tokyo	Flon-Showa
	Nippon Halon Company, Ltd., Tokyo	Nippalon 1301
Mexico	Halocarburos S.A., Mexico City	Freon®
	Quimobasicos S.A., Monterrey	Genetron
Netherlands	Du Pont de Nemours (Nederland) N.V., Dordrecht	Freon®
	AKZO Chemie B.V., Amersfoort	FCC
Russia	Not known	Eskimon
Spain	Ugine Quimica de Halogenos S.A., Madrid	Forane
	Kali-Chemie Iberia S.A., Madrid	Kaltron
	Electro-Quimica de Flix S.A., Barcelona	Frigen
South Africa	Africa Explosives & Chemical Industries Ltd., Johannesburg	Arcton
Korea	Korea Fluorochemicals Industry Company, Ltd., Seoul	Korfron
Venezuela	Products Halogenados de Venezuela S.A., Caracas	Forane

RULE 3. The third digit from the right is numerically one less than the number of carbon (C) atoms in the compound. When this digit is zero, it is omitted from the number.

RULE 4. The number of chlorine (Cl) atoms in the compound is found by subtracting the sum of the fluorine and hydrogen atoms from the total number of atoms that can be connected to carbon. When only one carbon atom is involved, the total number of attached atoms is four. When two carbon atoms are present, the total number of attached atoms is six.

An application of the first four rules in determining the formula for fluorocarbon 12 is as follows:

Derivation of the Formula for Fluorocarbon 12

1. The first digit on the right in fluorocarbon 12 is 2. Therefore, the compound contains two fluorine atoms (Rule 1).
2. The second digit from the right is 1. Therefore, there are no hydrogen atoms in the compound (Rule 2).
3. The third digit from the right is zero and has been omitted from the number. The compound therefore contains one carbon atom (Rule 3).
4. Since fluorocarbon 12 contains one carbon atom, there are four other atoms attached to the carbon atom. The number of chlorine atoms is found by subtracting the sum of the fluorine atoms and the hydrogen atoms from four. There are two fluorine atoms and no hydrogen atoms. Therefore, there are two chlorine atoms in the compound (Rule 4).

Summarizing the above information, fluorocarbon 12 contains one carbon atom, two fluorine atoms, and two chlorine atoms. Therefore, its chemical formula is CCl_2F_2.

Additional rules are necessary to designate symmetry, cyclic character, and unsaturation.

RULE 5. In the case where isomers exist, each has the same number, but the most symmetrical is indicated by the number alone. As the isomers become more and more unsymmetrical, the letters a, b, c, etc., are appended. For ethanes, symmetry is determined by dividing the molecule in two and adding the atomic weights of the groups attached to each carbon atom. The closer the total weights are to each other, the more symmetrical the product. For propanes and higher carbon series, symmetry rules become quite complex. Since the latter compounds are not normally used in the aerosol industry, symmetry rules for them will not be discussed in this chapter. An illustration of symmetry determination for isomers of fluorocarbon 114 is as follows:

Derivation of the Formula for Fluorocarbons 114 and 114a

1. The first digit on the right is 4. Therefore, the compound contains four fluorine (F) atoms (Rule 1).
2. The second digit from the right is 1. Therefore, there are no hydrogen atoms in the compound (Rule 2).
3. The third digit from the right is 1. Therefore, the compound contains two carbon atoms (Rule 3).
4. Since the compound contains two carbon atoms, there are six other atoms attached to the carbon atoms. Since there are four fluorine atoms and no hydrogen atoms, the compound contains two chlorine atoms (Rule 4). The empirical formula for fluorocarbon 114 is, therefore, $C_2Cl_2F_4$. The most symmetrical, $CClF_2CClF_2$, is designated fluorocarbon 114 and has the following structure:

$$Cl - \overset{\overset{\displaystyle F}{|}}{\underset{\underset{\displaystyle F}{|}}{C}} - \overset{\overset{\displaystyle F}{|}}{\underset{\underset{\displaystyle F}{|}}{C}} - Cl$$

Fluorocarbon 114

The unsymmetrical form, CCl_2FCF_3, is designated fluorocarbon 114a and has the following structure:

$$Cl - \overset{\overset{\displaystyle Cl}{|}}{\underset{\underset{\displaystyle F}{|}}{C}} - \overset{\overset{\displaystyle F}{|}}{\underset{\underset{\displaystyle F}{|}}{C}} - F$$

Fluorocarbon 114a

RULE 6. For cyclic derivatives, the letter C, followed by a dash, is used before the identifying number. Thus, by applying the first six rules, the formula for fluorocarbon C-318 is found to be

$$\begin{array}{ccc} CF_2 & \!\!\!\!-\!\!\!\!\! & CF_2 \\ | & & | \\ CF_2 & \!\!\!\!-\!\!\!\!\! & CF_2 \end{array}$$

Fluorocarbon C-318

RULE 7. If the compound is unsaturated, the number one (1) is used as the fourth digit from the right to indicate an unsaturated double bond. Using all rules above, the number for vinyl chloride is 1140; the number for tetrafluoroethylene is 1114.

COMMERCIAL FLUOROCARBON PROPELLANTS

The propellants discussed in this section are fluorocarbons 11, 12, 113, 114, 115, and C-318. Other fluorocarbons which have been considered as alternative propellants to the above are covered in a later section of this chapter.

General Physical Properties

Formulas, molecular weights, boiling points, and vapor pressures of fluorocarbons currently used in the aerosol industry today are listed in Table 3-2.[5-13] All but fluorocarbon C-318 are chlorofluorocarbons. Fluorocarbon 113, although strictly not a propellant because of its high boiling point, was included because it is used in aerosol formulations. The compounds are listed in the table in order of increasing boiling point and, therefore, in order of decreasing vapor pressure.

Pressures in the aerosol industry are commonly expressed as pounds per square inch gage (psig). Although the air pressure at sea level is 14.7 psi, most gages are set to read zero. Thus, the gages contain air at a pressure of about 14.7 psi, although the pointer reads zero. In order to convert gage pressure into absolute pressure, it is necessary to add 14.7 to the gage reading:

$$\text{Gage pressure (psig)} + 14.7 = \text{absolute pressure (psia)}$$

At 70°F the propellants listed in Table 3-2 have vapor pressures from subatmospheric to 103.0 psig. Since they are completely miscible with each other, any desired pressure within this range may be obtained. Propellant mixtures are commonly used in the formulation of aerosols, principally because they allow flexibility in adjusting spray characteristics and meeting pressure regulations. Some fluorocarbons cannot be used as the sole propellant for specific aerosol products because of their high vapor pressures.

Vapor pressure curves for the most widely used combinations of fluorocarbons 11 and 12, and fluorocarbons 12 and 114, over a range of temperatures are given in Reference 13.

TABLE 3-2. PHYSICAL PROPERTIES OF THE FLUOROCARBONS

Fluorocarbon Number	Formula	Molecular Weight	Boiling Point (°F)	Vapor Pressure (psig)		Density (g/cc)	
				70°F	130°F	70°F	130°F
115	$CClF_2CF_3$	154.5	−37.7	103.0	254.1	1.31	1.11
12	CCl_2F_2	120.9	−21.6	70.2	181.0	1.32	1.19
C-318	C_4F_8 (cyclic)	200.0	21.1	25.4	92.0	1.51	1.37
114	$CClF_2CClF_2$	170.9	38.4	12.9	58.8	1.47	1.36
11	CCl_3F	137.4	74.8	13.4 (psia)	24.3	1.48	1.40
113	CCl_2FCClF_2	187.4	117.6	5.5 (psia)	3.7	1.57	1.49

Densities of the fluorocarbons are tabulated in Table 3-2.[5-13] All are non-flammable, that is, they do not form flammable mixtures at any concentration in air.

Additional properties and uses of specific fluorocarbons listed in Table 3-2 are summarized below.

Summary of Specific Properties

Fluorocarbon 115

Fluorocarbon 115 is an extremely stable compound having such a low order of toxicity that it is approved for use as a food propellant and aerating agent by the U.S. Food and Drug Administration (FDA). When its low toxicity is coupled with its nonflammability and extremely low water solubility, the result is a compound of considerable interest to the food industry. Fluorocarbon 115, in combination with nitrous oxide, is used in about 75% of the whipped toppings marketed in this country. Propellant blends containing fluorocarbon 115 have also been evaluated in a number of nonfood aerosol products, even though the propellant has a higher cost than fluorocarbons 11, 12, or 114. Fluorocarbon 115 is not used alone as a liquefied gas propellant because of its high vapor pressure.

Fluorocarbon 12

Fluorocarbon 12 was the first fluorocarbon propellant to be used in aerosol products and is the most widely used fluorocarbon propellant today. It has application in the majority of personal and household aerosol products, either alone or in combination with fluorocarbon 11 or fluorocarbon 114. Such products include hair sprays, personal deodorants, antiperspirants, colognes, pan sprays, insecticides, industrial cleaners, and lubricants. Fluorocarbon 12 is stable to hydrolysis under mild acidic and alkaline conditions and is therefore suitable for use with aqueous-based products. It is essentially odorless and is used in blends with fluorocarbon 114 for many fragrance aerosols. Fluorocarbon 12 has a very low order of toxicity. The FDA has approved its use as a direct-contact freezing agent for food products.

Fluorocarbon C-318

Fluorocarbon C-318 is one of the most chemically stable propellants and, like fluorocarbon 115, has been approved by the FDA as a propellant and aerating agent for foods. It was recommended by Du Pont for this use in blends with fluorocarbon 115.

Fluorocarbon 114

Fluorocarbon 114 is the third most commonly used fluorocarbon propellant in aerosol products. It is very stable to hydrolysis. Because of its low vapor pressure, fluorocarbon 114 is normally blended with fluorocarbon 12 for alcohol- and aqueous-based aerosols. Fluorocarbon 114 has little odor and is used extensively for aerosol colognes and perfumes. Strobach[14] has proposed a fragrance system in which the perfume oil is dissolved directly in fluorocarbon 114. The fluorocarbon serves both as a solvent and aerosol propellant; no alcohol is present. Fluorocarbon 114 has found use in other aerosol cosmetics and pharmaceuticals because of its high stability and very low toxicity.

Fluorocarbon 11

Fluorocarbon 11 is the least expensive fluorocarbon and, next to fluorocarbon 12, the most widely used. It has too low a vapor pressure to be used alone as a propellant and is generally combined with the higher vapor pressure fluorocarbon 12. Fluorocarbon 11 has excellent solvent properties and is useful in the formulation of many products, such as hair sprays, where the active ingredient must be dissolved in the propellant system. Fluorocarbon 11 decomposes in the presence of water and certain metals at room temperature over long periods of time. For this reason, its use in aqueous-based aerosols is limited.

Under certain conditions, fluorocarbon 11 can undergo a free radical reaction with ethyl alcohol to produce acetaldehyde, hydrogen chloride, and fluorocarbon 21. The reaction is easily inhibited with stabilizers such as oxygen or nitromethane, thus eliminating a potential problem in the use of the propellant in personal products.

Fluorocarbon 113

Fluorocarbon 113 is not strictly a propellant because of its high boiling point. However, it is a good solvent and is used in aerosol formulations designed for cleaning applications. Fluorocarbon 113 has been proposed for use in alcoholic lotions.[15] It is normally used in combination with fluorocarbon 12.

Specifications

The fluorocarbons are among the purest organic compounds manufactured commercially on a large scale. Specifications for Freon® propellants are shown in Table 3-3.

Manufacture

Early methods for preparing fluorocarbons involved the replacement of chlorine in halogenated compounds with fluorine using metallic fluorides as catalysts.[16] The reactions were slow and incomplete. Around 1890, Swartz found that the

TABLE 3-3. FREON® PROPELLANT SPECIFICATIONS

	Freon® Propellant		
	11	12	114
Maximum water content (ppm by wt.)	10	10	10
Maximum nonabsorbable gas (vol. % in the vapor)	–	1.5	1.5
Boiling point at 1 atm (°F)	74.8	−21.6	38.4
Maximum boiling range (°F)	0.5	0.5	0.5
Maximum high-boiling impurities (vol. %)	0.01	0.01	0.01
Chloride ion content	None	None	None
Minimum organic purity (wt. %)	99.8	99.8	99.8

addition of pentavalent antimony to SbF_3 produced a catalyst which gave rapid and complete replacement of other halogens by fluorine. Most commercial processes for preparing the fluorocarbons today are based upon the Swartz reaction.

In a typical commercial process, fluorocarbons 11 and 12 are prepared from carbon tetrachloride in the presence of hydrogen fluoride and an antimony catalyst:

$$2CCl_4 + 3HF \xrightarrow{\text{catalyst}} CCl_3F + CCl_2F_2 + 3HCl$$

A flow diagram for the manufacture of fluorocarbons 11 and 12 is shown in Figure 3-1. Carbon tetrachloride and hydrogen fluoride are fed continuously into the reactor. The latter contains the antimony catalyst maintained in the pentavalent state by addition of chlorine. The mixture of fluorocarbons is led

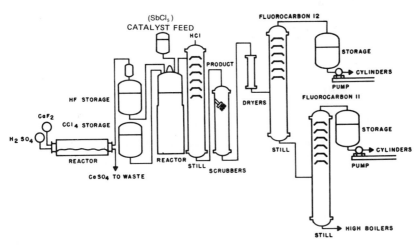

Figure 3-1. Typical flow diagram for the manufacture of fluorocarbons 11 and 12.

from the top of the reactor to a distillation column where hydrogen chloride is removed. The fluorocarbons are scrubbed with caustic to remove traces of acids, dried with silica or alumina, and fractionally distilled. Different ratios of fluorocarbons 11 and 12 can be obtained by varying the reaction conditions.

In a process similar to that for fluorocarbons 11 and 12, fluorocarbons 113 and 114 are prepared by reacting chlorine and hydrogen fluoride with perchloroethylene:

$$CCl_2{=}CCl_2 + Cl_2 + 3HF \xrightarrow{\text{catalyst}} CCl_2FCClF_2 + 3HCl$$

$$CCl_2{=}CCl_2 + Cl_2 + 4HF \xrightarrow{\text{catalyst}} CClF_2CClF_2 + 4HCl$$

The ratio of fluorocarbons 113 and 114 can be altered by changing the reaction conditions.

Fluorocarbon 115 is prepared by reacting fluorocarbon 114 with hydrogen fluoride:

$$CClF_2CClF_2 + HF \xrightarrow{\text{catalyst}} CClF_2CF_3 + HCl$$

ALTERNATIVE FLUOROCARBON PROPELLANTS[17-20]

During the past several years, fluorocarbons 11, 12, and 114, the most widely used aerosol propellants, have come under considerable scrutiny from an ecological viewpoint. During 1974, a theory was postulated that fluorocarbons in the stratosphere could lead to depletion of the ozone layer. The theory and its ramifications are covered in detail in Chapter 22, "Fluorocarbons in the Atmosphere."

Fluorocarbon suppliers have been exploring the development of alternative fluorocarbons as replacements for fluorocarbons now in commercial use, anticipating that the latter products would be restricted by regulation. The potential alternative propellants under consideration, their properties and uses, are the subject of this section of the chapter. Their toxicology is summarized in Chapter 25.

Criteria for Development

An important first step in the selection of alternative candidates is the establishment of criteria for their selection. Criteria establishment allows a narrowing of the field of candidates to be selected for further investigation. The following requirements can be logically imposed upon candidates for technical and toxicological development.

Environmental Acceptability

Present evidence indicates that hydrogen-containing fluorocarbons (hydrofluorocarbons or hydrochlorofluorocarbons) are less stable in the atmosphere than saturated perhalocarbons. Reaction with hydroxyl radicals results in the formation of water and the degradation of the fluorocarbon before it reaches the stratosphere. Lifetimes for hydrogen-containing fluorocarbons are calculated to be less than corresponding lifetimes for fluorocarbons 11, 12, and 114 by factors ranging from 5 to over 100 (Chapter 22).

Technical Performance

Alternative candidates must adequately replace present fluorocarbons in a variety of aerosol end-uses from the viewpoints of efficacy and stability. They must have the proper vapor pressures and be compatible with active ingredients, solvents, containers, and valves. Nonflammability is desirable, but not strictly required.

Low Toxicity

The inhalation toxicities of alternative propellants should be comparable to those of fluorocarbons 11, 12, and 114.

Manufacturing Capability

Processes must be invented or developed for the manufacture of alternative candidates. Following this, present plants must be modified or new plants designed and constructed for these products. An adequate supply of raw materials must be assured prior to commercial manufacture.

Cost

The cost of an alternative propellant cannot be known until basic process and plant data are established. In general, however, it should not greatly exceed present propellant costs in aerosol formulations.

General Physical Properties

With a view toward the above criteria, consideration of alternative fluorocarbon candidates for the aerosol industry has focused on fluoro- or chlorofluoromethanes and ethanes containing hydrogen. The physical properties of these compounds are listed in Table 3-4. Standard fluorocarbon nomenclature, as described in an earlier part of this chapter, is used to identify each compound listed. Propanes and butanes were excluded from consideration as alternatives because their physical properties, for example, boiling point, were not suitable or lack of known manufacturing processes prohibited their consideration for aerosol use. Specific isomers listed in the chart, for example, fluorocarbon 142b, represent the most attractive products among the various isomeric possibilities.

TABLE 3-4. PHYSICAL PROPERTIES OF THE ALTERNATIVE FLUOROCARBONS

Fluorocarbon Number	Formula	Molecular Weight	Boiling Point (°F)	Vapor Pressure (psig)		Density (g/cc)		Flammability
				70°F	130°F	70°F	130°F	
METHANES								
21	$CHCl_2F$	102.9	48.1	8.4	50.5	1.32	1.29	Nonflammable
22	$CHClF_2$	86.5	−41.4	121.4	296.8	1.21	1.06	Nonflammable
31	CH_2ClF	68.5	15.6	31.0	105.8	1.20	—	Flammable
32	CH_2F_2	52.0	−61.0	206.3	—	1.10	—	Flammable
ETHANES								
123	$CHCl_2CF_3$	152.9	82.3	11.4 (psia)	21.0	1.47	—	Nonflammable
124	$CHClFCF_3$	136.5	12.2	34.1	115.6	1.38	—	Nonflammable
125	CHF_2CF_3	120.0	−55.3	163.8	430.2	1.23	—	Nonflammable
132b	$CH_2ClCClF_2$	134.9	115.5	5 (psia)	4.3	1.41 (77°F)	—	Nonflammable
133a	CH_2ClCF_3	118.5	44.5	11.0	55.9	1.33	1.23	Nonflammable
134a	CH_2FCF_3	102.0	−15.7	70.9	199.2	1.22[a]	—	Nonflammable
141b	CH_3CCl_2F	117.0	89.6	10 (psia)	14.3[a]	1.26[a]	—	Flammable
142b	CH_3CClF_2	100.5	14.4	29.1	97.3	1.12	1.03	Flammable
143a	CH_3CF_3	84.0	−53.7	165[a]	410	0.94 (86°F)	—	Flammable
152a	CH_3CHF_2	66.1	−13.0	62.5	176.3	0.91	0.81	Flammable

[a] Estimated.

Present Status of Alternative Fluorocarbons

The status of fluorocarbons considered as alternatives to fluorocarbons 11, 12, and 114 is summarized in Table 3-5. The viability of the candidates as alternatives is based on technical properties, toxicity data, and process information developed over the last several years.

Summary of Physical Properties

Fluorocarbons 21 and 22[21-23]

Fluorocarbons 21 and 22 provide technically satisfactory replacements for fluorocarbons 11 and 12, principally because of the range of vapor pressures possible—from 8.4 to 121.4 psig at 70°F. Fluorocarbon 21 has the highest solvent power of the compounds listed in Table 3-5. It readily attacks elastomers used in valve gaskets, thus limiting its concentration in aerosol formulations. Butyl rubber is the preferred valve gasket elastomer for fluorocarbon 21; its use minimizes leakage problems.

Fluorocarbon 22 alone is a particularly good propellant for alcohol-based

TABLE 3-5. STATUS OF FLUOROCARBON ALTERNATIVES[a]

Fluorocarbon Number	Formula	Status
METHANES		
21	$CHCl_2F$	Low potential because of adverse toxicity.
22	$CHClF_2$	Indefinite potential; toxicity needs clarification.
31	CH_2ClF	Low potential because of adverse toxicity.
32	CH_2F_2	Low potential; vapor pressure too high.
ETHANES		
123	$CHCl_2CF_3$	Potentially useful, but toxicity information inadequate.
124	$CHClFCF_3$	Potentially useful, but toxicity information inadequate.
125	CHF_2CF_3	Low potential; same comments as for fluorocarbons 123 and 124; vapor pressure too high.
132b	$CH_2ClCClF_2$	Potentially useful, but toxicity information inadequate; vapor pressure too low.
133a	CH_2ClCF_3	Low potential because of adverse toxicity.
134a	CH_2FCF_3	Potentially useful, but toxicity information inadequate.
141b	CH_3CCl_2F	Same as for fluorocarbon 134a.
142b	CH_3CClF_2	Potentially useful; toxicity information needs to be clarified. Product presently manufactured.
143a	CH_3CF_3	Low potential; toxicity information inadequate; vapor pressure too high.
152a	CH_3CHF_2	Potentially useful; may require additional toxicity information. Currently used as a propellant.

[a]See also Chapter 25 for summary of toxicity of fluorocarbons.

products. Personal products can be formulated with spray characteristics essentially equivalent to those obtained with fluorocarbon 11/12 or 12/114 blends. In many instances, the high vapor pressure and low molecular weight of fluorocarbon 22 make it possible to use less of this propellant than fluorocarbon 12 alone or fluorocarbon 11/12 or 12/114 blends. With colognes, for example, the low concentration of fluorocarbon 22 (about 20-25%) provides a potential cost savings over products formulated with the conventional fluorocarbon 12/fluorocarbon 114 (10/90) blend.

Both fluorocarbon 22 and fluorocarbon 21 are unstable in aqueous alkaline systems. The resulting hydrolytic products cause container corrosion. Therefore, these fluorocarbons are not suitable for formulating foams with alkaline surfactants.

Fluorocarbons 31 and 32[24]

Fluorocarbon 31, with a vapor pressure of 31 psig at 70°F, is a possible propellant candidate for dry-type antiperspirants, even though it is flammable. Fluorocarbon 31 cannot be used in the formulation of alcohol-based products because it reacts with ethyl alcohol to form acidic by-products which cause corrosion in tinplate containers. The reaction cannot be easily inhibited:

$$CH_2ClF + 2C_2H_5OH \longrightarrow C_2H_5OCH_2OC_2H_5 + HF + HCl$$

Fluorocarbon 32, with a vapor pressure of 206 psig at 70°F, is essentially a compressed gas. It has the further undesirable feature of flammability.

Fluorocarbons 123, 124, and 125[25]

Technically, fluorocarbons 123 and 124 are potential alternative propellant candidates. Practically all aerosol products can be formulated using either fluorocarbon 124 alone or in combination with fluorocarbon 123. Both compounds are nonflammable. Fluorocarbon 124 is stable in alkaline systems and may be used to formulate aqueous aerosol foams. Fluorocarbon 123 hydrolyzes under aqueous alkaline conditions, causing can corrosion. Neither fluorocarbon reacts with ethyl alcohol and, therefore, both can be used to formulate alcohol-based products without the use of inhibitors.

Fluorocarbon 123 and 124 blends are particularly useful for dry-type antiperspirants. The nonflammable, relatively high-boiling fluorocarbon 123 can be used for preparing a slurry of active ingredients in the same manner that fluorocarbon 11 is now used.

Fluorocarbon 125 has too high a vapor pressure to be a general substitute for fluorocarbon 12 and, therefore, seems impractical as an aerosol propellant.

Fluorocarbons 132b, 133a, and 134a

Fluorocarbon 132b has not yet been seriously investigated in aerosol formulations. It has the highest boiling point of all alternatives considered, is nonflammable, and might be a suitable vehicle for preparing slurries of active ingredients for dry-type antiperspirants. Fluorocarbon 132b would have to be used in combination with a higher vapor pressure propellant because of its low vapor pressure.

Fluorocarbon 133a is an alternative to fluorocarbon 11. It has a vapor pressure of 11.0 psig at 70°F, is nonflammable, and has a density of 1.33 g/cc at 70°F. Fluorocarbon 133a must be blended with a higher vapor pressure propellant for aerosol use. Its low vapor pressure, in combination with its nonflammability, makes fluorocarbon 133a particularly useful for the preparation of slurries for powder aerosols.

Fluorocarbon 133a offers flame depressant properties when blended with hydrocarbons. In addition, fluorocarbon 133a forms azeotropes with selected hydrocarbons, thus offering the possibility of constant boiling fluorocarbon/hydrocarbon blends in aerosol propellant systems. Azeotropic compositions are particularly important when vapor tap valves are used.

Fluorocarbon 133a has a unique solubility for carbon dioxide, which allows its use at relatively low concentrations in alcohol-based products. Fluorocarbon 133a/carbon dioxide propelled alcohol-based products can be formulated with spray characteristics comparable to totally liquefied-gas-propelled systems.

Fluorocarbon 133a, like fluorocarbon 11, is not stable in aqueous systems in tinplate containers. Hydrolysis is sufficient to cause container corrosion. Fluorocarbon 133a should be used with valves containing either neoprene or butyl gaskets.

Fluorocarbon 134a, with a vapor pressure of 70.9 psig at 70°F, is a possible direct substitute for fluorocarbon 12. It contains no chlorine (and is therefore excluded from the ozone depletion hypothesis), is nonflammable, and probably is stable to hydrolysis. Little additional technical information is available at the present time.

Fluorocarbons 141b, 142b,[26] and 143a

Little work has been done on the evaluation of fluorocarbon 141b as an alternative aerosol propellant. Like fluorocarbons 123 and 133a, fluorocarbon 141b has physical properties similar to those of fluorocarbon 11. It is flammable, which may limit its use as a vehicle for the preparation of slurries for powder sprays and antiperspirants.

Fluorocarbon 142b was one of the most promising alternative propellants evaluated during the period from 1975 to 1976. It was of primary interest as a propellant for dry-type antiperspirants, although it could also be used to formulate other aerosol products.

Fluorocarbon 142b forms flammable mixtures with air at concentrations of 6.0–15 vol. %. However, it has no flash point and gives negative results in the closed and open drum tests. Fluorocarbon 142b does not extend a flame with an antiperspirant valve, and flashback is zero.

The vapor pressure of fluorocarbon 142b (29 psig at 70°F) is in the right range for antiperspirants, and its density of 1.1 g/cc results in a lower rate of settling of active ingredients than occurs with hydrocarbons. Because it can be used alone, it has the advantage (over blends) that the properties of formulations propelled with this fluorocarbon do not change during use with a vapor tap valve.

Fluorocarbon 143a is flammable and has a very high vapor pressure at room temperature, causing it to be of low interest as an alternative fluorocarbon.

Fluorocarbon 152a [27]

Fluorocarbon 152a is a stable, virtually odorless propellant with a density of .91 g/cc at 70°F. It has a vapor pressure of 62.5 psig at 70°F and, like fluorocarbon 12, must be used in combination with a lower vapor pressure propellant or a solvent such as ethyl alcohol. It is currently used in low volume in specialized aerosol personal products.

Fluorocarbon 152a forms flammable mixtures with air at concentrations of 3.9–16.9 vol. %. However, it has zero flame extension and flashback when used with an antiperspirant valve.

Fluorocarbon 152a does not react with alcohol via a free radical reaction as does fluorocarbon 11. It can be packaged with valves containing Buna N, neoprene, or butyl gaskets with acceptable leakage rates.

The utility of fluorocarbon 152a has been technically demonstrated in a wide variety of personal products, including colognes and hair sprays, and in insecticides, medical-pharmaceutical products, and industrial formulations. Commercial expansion of the use of this propellant in aerosols will depend, in part, on cost and flammability comparisons with other flammable propellants.

Fluorocarbon 152a contains no chlorine and, therefore, is not implicated in the ozone depletion hypothesis.

Manufacture

Although processes have not yet been completely developed for any alternative fluorocarbons and, in many cases, are still in the research laboratory phase, it is reasonable to predict that they will use present technology to the maximum extent possible. This technology involves the reaction of hydrogen fluoride with chlorocarbons—trichloroethylene, tetrachloroethylene, and 1,1,1-trichloroethane—in the presence of selected catalysts in the liquid or gas phase. The process selected would depend upon the product distribution desired.

A somewhat different route is currently used for the production of fluorocarbon 152a, viz., the addition of hydrogen fluoride to acetylene.

The process for fluorocarbon 22, a currently commercial product, involves the reaction of hydrogen fluoride with chloroform.

REFERENCES

1. H. E. Peterson, Personal Communication, Peterson/Puritan Filling Corp.
2. L. D. Goodhue, *Aerosol Age* **15,** 16 (August 1970).
3. E. G. Young, Personal Communication, E. I. du Pont de Nemours & Company.
4. "American Standard Number Designation of Refrigerants," ASHRAE Standard 34-57 and ASA Standard B 79.1, 1960.
5. Freon® Technical Bulletin B-2, "Properties and Applications of the Freon® Fluorinated Hydrocarbons."
6. Freon® Technical Bulletin B-18, "Freon® C-318."
7. Freon® Technical Bulletin B-18B, "Physical Properties of Freon® C-318 Perfluorocyclobutane."
8. Freon® Technical Bulletin D-6, "Comparative Stability of Freon® Compounds."
9. Freon® Technical Bulletin T-115, "Thermodynamic Properties of Freon® 115 Monochloropentafluoroethane."
10. Freon® Aerosol Report A-63, "Food Propellant Freon® 115."
11. F. T. Reed, "Fluorocarbon Propellants," in A. Herzka (Ed.), "International Encyclopaedia of Pressurized Packaging" (Aerosols), Pergamon Press, Inc., New York, 1966.
12. F. T. Reed, "Fluorocarbon Propellants," in H. R. Shepherd (Ed.), "Aerosols: Science and Technology," Interscience Publishers, Inc., New York, 1961.
13. P. A. Sanders, "Propellants and Solvents," in J. J. Sciarra and L. Stoller (Eds.), "The Science and Technology of Aerosol Packaging," John Wiley & Sons, Inc., New York, 1974.
14. D. R. Strobach, *Cosmet. Perfum.* **89,** 46 (June 1974).
15. K. Humpfner, *Cosmet. Perfum.* **90,** 24 (December 1975).
16. R. C. Downing, "Simple Organic Compounds Containing 'Fluorine'," in J. S. Sconce (Ed.), "Chlorine, Its Manufacture, Properties and Uses," American Chemical Society Monograph 154, Reinhold Publishing Corp., New York, 1962, p. 864.
17. J. J. Daly, Jr., "Alternative Fluorocarbon Propellants," Aerosol Institute of Mexico, February 1976. Published in *Perfumia Moderna* **86,** 23 (1976).
18. J. J. Daly, Jr., "Alternative Fluorocarbon Propellants," Chemical Specialties Manufacturers Association (CSMA) Spring Meeting, Chicago, May 1976.
19. J. J. Daly, Jr., "Alternative Fluorocarbon Propellants" (An Update), Third National Seminar of the Aerosol Industry, Mexico City, August 1976.
20. J. J. Daly, Jr., "Fluorocarbon/Ozone Status and Alternative Propellants," Annual Scientific Meeting of the Society of Cosmetic Chemists, New York City, December 1976.
21. Freon® Technical Bulletin DP-5, "Freon® 21 Fluorocarbon."
22. Freon® Technical Bulletin T-21, "Thermodynamic Properties of Freon® 21."
23. Freon® Technical Bulletin FA-7, "Vapor Pressures of Fluorocarbon 22."
24. Freon® Technical Bulletin D-11a, "Physical and Thermodynamic Properties of Fluorocarbon 31."
25. Freon® Technical Bulletin X-150, "General Properties of Freon® 123 Fluorocarbon."
26. Freon® Technical Bulletin D-60, "Properties of FC-142b, 1-Chloro-1,1-difluoroethane CH_3CClF_2."
27. Freon® Technical Bulletin D-62, "Properties of 1,1-Difluoroethane (DFE) CH_3CHF_2."

4

HYDROCARBON PROPELLANTS, COMPRESSED GASES, AND PROPELLANT BLENDS

HYDROCARBON PROPELLANTS

Although isobutane was suggested as an aerosol propellant by Rotheim as early as 1933,[1] large-scale commercial use of hydrocarbons as propellants did not occur until 1954. After World War II, there was an increasing interest in hydrocarbons. Goodhue, at Airosol, Inc., worked with hydrocarbon blends in 1947. The following year, Shepherd evaluated hydrocarbons as propellants for various aerosols. The hydrocarbons were supplied by the Phillips Petroleum Company. Hydrocarbons were investigated as propellants for shaving lathers by the Colgate-Palmolive Company in 1950.[2] Flammability and odor were the main objections to hydrocarbons.

The use of hydrocarbons accelerated when the Spitzer, Reich, and Fine shaving cream patent was issued in 1953.[3] Since the patent claimed fluorocarbons but not hydrocarbons, there was a general shift from fluorocarbons to hydrocarbons by shaving cream loaders and marketers. The Phillips Petroleum Company had succeeded in eliminating most of the odor from the hydrocarbons by this time. The first tank car shipment of hydrocarbons was made to the Lenk Manufacturing Company in 1954. The propellant was intended for an aerosol shaving lather marketed by the Colgate-Palmolive Company. Since then, hydrocarbons have been used to an increasing extent as aerosol propellants. Johnsen[4] reported that in 1971, 100,000,000 lb of hydrocarbons was used as aerosol propellants compared to 420,000,000 lb of fluorocarbons. Hydrocarbons, therefore, accounted for about 19% of the total aerosol market for liquefied gas propellants. In 1974, Brock[5] estimated that approximately 164,000,000 lb of hydrocarbons was used as aerosol propellants. The total fluorocarbon market for that year was 616,000,000 lb,[6] so hydrocarbon penetration increased to about 21%. Brock[7] has also reported that by the end of 1974, more than 55% of all aerosol packages contained hydrocarbon propellants. More recent estimates indicate that in 1978 at least 70–80% of the aerosol products will be formulated with hydrocarbons.

Terminology

The four major suppliers of hydrocarbon propellants are Aeropres Corp., Diversified Chemicals and Propellants Company, Industrial Hydrocarbons, Inc., and Phillips Petroleum Company. These companies provide various grades of aerosol hydrocarbon propellants for different uses. Aeropres markets a cosmetic and aerosol grade,[8] Diversified Chemicals an aerosol grade,[9] Industrial Hydrocarbons an aerosol and pure grade,[10] and Phillips Petroleum an aerosol and pure grade.[10a] The aerosol grade is used for general aerosol applications. Specifications for the hydrocarbon grades vary slightly depending upon the company and are available in the various company technical bulletins. The pure grade is used where a very high mole purity is desired.

Aerosol grade hydrocarbons are designated by the letter "A." This letter is followed by a number which indicates the pressure of the propellant (in psig) at 70°F. Thus, isobutane, which has an approximate vapor pressure of 31 psig at 70°F, is designated Hydrocarbon Aerosol Propellant A-31 or simply A-31. Propane, with a vapor pressure of about 108 psig at 70°F, is A-108. Mixtures of hydrocarbons with intermediate vapor pressures are treated similarly. Propellant A-46, with a vapor pressure of 46 psig at 70°F, is a blend of approximately 84 wt. % isobutane and 16 wt. % propane. The cosmetic grade of hydrocarbons supplied by Aeropres is indicated by the letter "C" followed by a number indicating the vapor pressure of the propellant, for example, C-46. Diversified Chemicals' propellants have the trade name "Aeron."

Properties

The major hydrocarbon propellants are propane and isobutane. n-Butane, isopentane, and n-pentane have been used very little. However, in the aerosol industry there is increasing interest in these compounds as low-vapor-pressure propellants. The hydrocarbons are stable, noncorrosive, essentially odorless liquefied gases. They have low toxicity and are environmentally acceptable. A primary advantage of hydrocarbons, in comparison with fluorocarbons, is their lower cost. The main problem is their flammability.

Physical properties of the pure grade of propane, isobutane, and n-butane[10] and the cosmetic grade of isopentane and n-pentane[8] are listed in Table 4-1. Solvent properties and flammability are discussed in Chapter 10, "Solvency," and Chapter 14, "Flammability." Vapor pressures range from about 8 psia to 110 psig at 70°F. Since the hydrocarbons are all miscible with each other, a propellant blend with any desired vapor pressure between the two limits can be obtained with the proper mixture. Vapor pressures of mixtures of propane,

TABLE 4-1   PHYSICAL PROPERTIES OF THE HYDROCARBON
              PROPELLANTS

| Property | Hydrocarbon | | | | |
	Propane	Isobutane	n-Butane	Isopentane	n-Pentane
Formula	C_3H_8	C_4H_{10} (iso)	C_4H_{10} (n)	C_5H_{12} (iso)	C_5H_{12} (n)
Molecular weight	44.1	58.1	58.1	72.1	72.1
Boiling point (°F)	−43.7	10.9	31.1	82	97
Vapor pressure (psig)					
70°F	108.1	31.1	16.9	11.5 (psia)	8.6 (psia)
130°F	262.3	96.8	65.5	19.0	11.6
Density (g/cc)					
70°F	0.50	0.56	0.58	0.62 (60°F)	0.63 (60°F)
130°F	0.44	0.51	0.54	0.58	0.59

isobutane, and n-butane are available in References 8-10a. Approximate vapor pressures of blends can be calculated using the procedure outlined in Chapter 8.

Hydrocarbon propellants have lower molecular weights than fluorocarbons. At the same weight percent concentration in an aerosol, more moles of hydrocarbon than moles of fluorocarbon are present. The volume of gas resulting when liquefied propellants vaporize is directly proportional to their molar concentration. Therefore, at the same weight percent concentration, hydrocarbons produce a greater volume of gas than do fluorocarbons. The particle size of an aerosol spray is a function of the volume of gas resulting when liquefied propellants vaporize. Consequently, at equal weight percent concentrations, hydrocarbons produce a finer spray than fluorocarbons. It also follows that less hydrocarbon than fluorocarbon is required to produce the same spray characteristics.

Hydrocarbons have lower densities than fluorocarbons. The lower density does not permit as much hydrocarbon as fluorocarbon to be packaged in a given volume. Therefore, at the same volume fill, products formulated with hydrocarbons do not weigh as much as fluorocarbon aerosols.

Hydrocarbon propellants are very flammable. The potential hazard resulting from the transportation, handling, and storage of the propellants must be recognized and adequate precautions observed. There have been several serious hydrocarbon explosions at aerosol fillers in recent years. Frauenheim[11] suggests that aerosol fillers store hydrocarbon propellants above ground in tanks which are installed according to national, state, and local restrictions. The guidebook for this type of installation is the National Fire Prevention Association pamphlet #58. Isolation requirements specified by the National Fire Prevention Association, Factory Mutual Insurance, and Factory Insurance Association are listed in Reference 11. Filling equipment must be explosion-proof. The hydrocarbons are heavier than air and areas at the floor level in the loading area should be well ventilated. Ford[12] has emphasized the need for proper design and continued attention to good safety practices in plants filling hydrocarbon propellants.

Safety aspects of hydrocarbons in aerosols have been discussed by Fiero and Johnsen.[13] Another potential hazard—leakage of flammable propellants from aerosols in closed confined storage areas—has been covered by Johnsen[14] and Parks.[15]

Uses

In the past, hydrocarbon propellants were used mostly in aqueous-based products such as shaving lathers, window cleaners, starch sprays, room deodorants, and space insecticides, or in blends with the fluorocarbons. In aqueous-based products, flammability was minimized by the water and relatively minor propellant concentrations. The concentration of hydrocarbons in fluorocarbon/hydrocarbon blends usually was sufficiently low so that the blend was nonflammable, that is, did not form a flammable mixture with air. Hydrocarbons in higher concentrations are now used alone as propellants in some formulations which already are quite flammable as a result of the solvents present. According to Brock,[7] 97% of the paint and coating aerosols in 1975 contained hydrocarbons as the only propellant. Frauenheim[11] lists a wide variety of typical hydrocarbon-propelled aerosols. Brock[7] has suggested formulations for hair sprays, personal deodorants, and even dry-type antiperspirants with hydrocarbons as the propellant. These products are reported to have flame extensions less than 18 in. and essentially no flashback with the recommended valves. The extent to which hydrocarbons will penetrate the propellant market is difficult to predict. It certainly will be affected by future costs of fluorocarbons and hydrocarbons and the ultimate outcome of the fluorocarbon/ozone controversy.

PROPELLANT BLENDS

Fluorocarbon/Hydrocarbon Blends

Fluorocarbon/hydrocarbon blends have been investigated for many years. During World War II, Jones and Scott, at the U.S. Bureau of Mines, evaluated mixtures of Freon® 12 with propane and butane in an attempt to extend the supply of Freon® 12 for aerosol insecticides. Subsequently, Goodhue and his co-workers formulated aerosol insecticides using blends of Freon® 12 with various hydrocarbons.[16]

Fluorocarbon/hydrocarbon blends are commonly used in the aerosol industry at the present time. One objective in blending propellants is to reduce the cost of fluorinated propellants by addition of lower cost flammable hydrocarbons. In the past, the concentration of hydrocarbons in the blends was generally maintained at a low enough level, for example, about 10 wt. %, so that the blend was nonflammable. Examples of such blends are Propellant A (fluorocarbon 12/ fluorocarbon 11/isobutane) (45/45/10) and fluorocarbon 12/propane (91/9). Today, flammable blends with much higher concentrations of hydrocarbons are under consideration. A fluorocarbon 22/isopentane (35/65) blend was of interest for a time as a propellant for dry-type antiperspirants.

Fluorocarbon/hydrocarbon blends are discussed below.

Propellant A

Propellant A was developed in 1956 by Reed.[17] It was one of the first commercial blends of nonflammable and flammable propellants and is still one of the most widely used. Reed used the term "Propellant A" initially to cover a range of nonflammable compositions of fluorocarbon 12, fluorocarbon 11, and isobutane with varying vapor pressures. However, the name now indicates only a blend with a composition of 45 wt. % fluorocarbon 12, 45 wt. % fluorocarbon 11, and 10 wt. % isobutane.

Propellant A has about the same vapor pressure as a fluorocarbon 12/fluorocarbon 11 (50/50) blend and is used in products such as hair sprays as a replacement for the 50/50 blend because it is less expensive. Propellant A has slightly poorer solvent properties than the fluorocarbon 12/fluorocarbon 11 (50/50) blend. Propellant A does not form flammable mixtures with air.

Fluorocarbon 12/Propane (91/9)

This nonflammable blend has been used primarily in coating compositions. It has a somewhat higher vapor pressure than fluorocarbon 12, but the solvents in coating compositions function as vapor pressure depressants.

Fluorocarbon 22/Isopentane (35/65)

This blend was promoted extensively during 1976 as a propellant for dry-type antiperspirants. The blend is significantly less expensive than fluorocarbon propellants alone because of the lower cost of isopentane. However, the blend has several disadvantages for use as a propellant for antiperspirants. It is flammable and has a relatively low density. The low density has two adverse effects: (1) The settling rate of active ingredients is rapid and (2) for the same volume in the container, the weight of product with the fluorocarbon 22/isopentane (35/65) blend is only about one-half that with the usual fluorocarbon 12/ fluorocarbon 11 (30/70) blend. Blends whose components have considerably different vapor pressures, for example, fluorocarbon 22 and isopentane, show a significant change in composition during discharge (see Chapter 9). Vapor pressure and spray rate decrease as a result of the composition change.

TABLE 4-2   BINARY AZEOTROPES OF FLUOROCARBONS 152a AND
133a WITH VARIOUS HYDROCARBONS[22]

Binary	Composition (wt. %)	Vapor Pressure (psig) at 70° F	Flammability
Fluorocarbon 152a/propane	45/55	130	Flammable
Fluorocarbon 152a/isobutane	75/25	72	Flammable[a]
Fluorocarbon 152a/n-butane	85/15	68	Flammable[a]
Fluorocarbon 133a/isobutane	33/67	32	Flammable[a]
Fluorocarbon 133a/n-butane	55/45	23	Flammable[a]
Fluorocarbon 133a/isopentane	94/6	12	Flammable[a]
Fluorocarbon 133a/n-pentane	96.5/3.5	12	Nonflammable[a]

[a] Flame extension is less than 18 in. with a dry-type antiperspirant valve.

Binary Azeotropes

Binary azeotropes of fluorocarbons 152a and 133a with various hydrocarbons are listed in Table 4-2. The composition of the vapor phase in these blends is the same as that of the liquid phase. Therefore, when the azeotropes are used with a vapor tap valve, the composition of the liquid phase remains constant during discharge. This is a considerable advantage for products such as dry-type antiperspirants, since the properties do not change with use.

The binary azeotropes, with the exception of the fluorocarbon 133a/n-pentane blend, are all flammable, that is, form flammable mixtures with air. Five of the azeotropes give a flame extension less than 18 in. when discharged with a dry-type antiperspirant valve.

Fluorocarbon Blends with Chlorinated Hydrocarbons and Dimethyl Ether

Dimethyl ether and ethyl chloride have been used in blends with fluorocarbons to only a very limited extent. Methylene chloride has been of interest as a component of aerosols for years because it is an excellent solvent and is less expensive than fluorocarbons. Methyl chloroform, although not defined as a propellant, has been included in this section because it is used in a number of aerosol products.

Dimethyl ether, ethyl chloride, methylene chloride, methyl chloroform, and their blends are discussed below. Some common properties of the compounds are listed in Table 4-3 and those of their blends in Table 4-4.

Dimethyl Ether (DME)

Dimethyl ether has been used only as a component of Propellant P, a blend developed specifically for paints by Union Carbide.[18] Propellant P has a com-

TABLE 4-3 PROPERTIES OF MISCELLANEOUS PROPELLANTS

Property	Dimethyl Ether	Ethyl Chloride	Methylene Chloride	Methyl Chloroform
Formula	$(CH_3)_2O$	CH_3CH_2Cl	CH_2Cl_2	CH_3CCl_3
Molecular weight	46.1	64.5	84.9	133.4
Boiling point (°F)	-12.7	54.5	103.6	165.2
Vapor pressure (psig)				
70°F	63.0	5.3	7.1 (psia)	2.2 (psia)
130°F	171.7	38.3	9.0	7.2 (psia)
Density (g/cc)				
70°F	0.66	0.91	1.32	1.34 (68°F)
130°F	–	0.84	1.24	–
Solubility in water at 1 atm (wt. %)	7% at 64.4°F 31.5% at 70°F at its own vapor pressure	0.57% (68°F)	1.32% (77°F)	–
Flammability[a]	Flammable	Flammable	Flammable	Flammable

[a]Forms flammable mixtures with air.

position of 75 wt. % fluorocarbon 12, 10 wt. % fluorocarbon 11, and 15 wt. % dimethyl ether. Dimethyl ether was included in the blend because it is a better solvent than either fluorocarbon 12 or fluorocarbon 11. Dimethyl ether is very flammable and must be handled accordingly. It is unique among the propellants in its high water solubility. Possibly, it may find use in the future as a propellant

TABLE 4-4 PROPERTIES OF PROPELLANT BLENDS[22]

Propellant Blend	Vapor Pressure (psig) at 70°F	Density (g/cc) at 70°F	Flammability
Propellant A (fluorocarbon 12/fluorocarbon 11/ isobutane) (45/45/10)	38	1.22	Nonflammable[a]
Fluorocarbon 12/ethyl chloride (65/35)	47	1.15	Flammable
Fluorocarbon 12/ethyl chloride (80/20)	56	1.22	Nonflammable[a]
Fluorocarbon 12/propane (91/9)	82	1.15	Nonflammable[a]
Propellant P (fluorocarbon 12/fluorocarbon 11/ dimethyl ether) (75/10/15)	62	1.21	Nonflammable[a]
Fluorocarbon 22/isopentane (35/65)	49.8	0.75	Flammable

[a]Does not form flammable mixtures with air.

for aqueous-based aerosols. Dimethyl ether has the disadvantage of having a rather penetrating, ethereal odor which some people find disagreeable. The physical properties and flammability of fluorocarbon blends containing dimethyl ether have been investigated in detail by Scott and his co-workers.[19-21]

Ethyl Chloride

Ethyl chloride has too low a vapor pressure to be used alone as a propellant. It has been used to a limited extent in combination with fluorocarbon 12 for coating compositions. Ethyl chloride is flammable. However, concentrations not exceeding 20 wt. % can be used with fluorocarbon 12 while still maintaining nonflammability of the blend. Ethyl chloride cannot be used for alcohol-based products because it reacts with alcohol to form diethyl ether and hydrogen chloride.[22] This reaction is not free radical in nature and cannot be inhibited. The hydrogen chloride can cause corrosion in metal containers. Aqueous-based products should also be avoided with ethyl chloride blends, since ethyl chloride hydrolyzes in the presence of water.

Methylene Chloride

Methylene chloride is present in many aerosol formulations. Its main use has been as a less expensive substitute for fluorocarbon 11. It has excellent solvent properties (see Chapter 10) and has been a component in such products as hair sprays, insecticides, room deodorants, mothproofers, artificial snow, paints, oven cleaners, and paint removers. High concentrations of methylene chloride require the use of valves with special gaskets.

At the present time, methylene chloride is actively being promoted for use in aerosols by the Dow Chemical Company under the trade name "Aerothene" MM.[23] The advantages claimed for "Aerothene" MM are as follows:

1. Environmentally acceptable.
2. Good toxicological properties.
3. Proven inhibitor system.
4. Reduces flammability.
5. Low cost.
6. Fast drying rate.

The toxicological properties of methylene chloride are under intensive investigation at the present time.

Methyl Chloroform

Methyl chloroform has also been used in aerosols, although to a lesser extent than methylene chloride. It is a good solvent and less expensive than fluorocarbon 11. Methyl chloroform has a distinctive odor that is somewhat difficult to mask. In some cases the cost of masking the odor is sufficiently high so that the cost advantage compared to fluorocarbon 11 is lost.

Methyl chloroform is also being promoted by the Dow Chemical Company as a component of aerosols under the trade name of "Aerothene" TT.[23]

COMPRESSED GAS PROPELLANTS

Compared to liquefied gases, the compressed gases—carbon dioxide, nitrous oxide, and nitrogen—are used to only a minor extent as aerosol propellants. They have inherent disadvantages as propellants that limit the products in which they are used. Until recently, compressed gases were employed in essentially two types of nonfood aerosols: those in which a coarse spray was acceptable and those designed for use at low temperatures. However, during the past several years, there has been an increased interest in compressed gases as propellants for personal products, particularly alcohol-based aerosols such as hair sprays and underarm deodorants. This is the consequence of increased cost of fluorocarbon propellants and questions concerning their effect upon the environment. New low-spray-rate valves designed specifically for compressed gas systems produce far better aerosols than was possible with the older standard valves. Commercial filling equipment with high production rates has been developed. All of these factors resulted in an increase in the number of compressed gas aerosol products.

It seems fairly certain that carbon dioxide will obtain a larger share of the propellant market than it held previously. However, the extent to which it will penetrate the market is difficult to predict at the present time. Several carbon dioxide propelled hair sprays have not performed well in the test market. The carbon dioxide products are definitely inferior to fluorocarbon hair sprays. The fundamental difference between a compressed gas and a liquefied gas propellant cannot be changed. The latter achieves break-up of the concentrate by changing from liquid to vapor, and a compressed gas causes spray formation by mechanical break-up action. A number of problems with carbon dioxide still remain regardless of the improvements in valves and filling equipment.

Historically, compressed gases have been considered for aerosol propellants far longer than liquefied gases. Carbon dioxide was used for charging beverages as far back as 1869.[24] Subsequently, a number of patents were granted to various workers on the use of carbon dioxide as a propellant.[25-27] However, no commercial compressed gas aerosols appeared on the market in any significant quantity until the 1950's. Nitrogen was one of the first gases used during this period. Initially, it was greeted with enthusiasm, but customer complaints about the "Nitrosols" soon forced a revision of the attitude toward nitrogen.[28-31] One of the main reasons for customer dissatisfaction was the effect of "misuse." Nitrogen was essentially insoluble in the concentrates, and the propellant was confined to the vapor phase. If the product were accidentally inverted and discharged vapor phase, most of the propellant was lost. The product remaining in the container could not be discharged because of lack of propellant. Today, nitrogen is used very little for compressed gas aerosols. Carbon dioxide and nitrous oxide have been accepted to a much greater extent than nitrogen because they are more soluble in the concentrates.

Properties and Uses

Physical Properties

Some common physical properties of the compressed gases are given in Table 4-5.[32, 33] All of the gases are nonflammable.

Solubility

The solubility of a compressed gas in the liquid concentrate is one of its most important properties. The higher the solubility, the better the initial spray and the less the change in spray properties and pressure during discharge. Gas solubility affects the extent to which misuse will be a factor in product acceptance by the consumer.

Nitrogen has a low solubility in water and most organic liquids. Carbon dioxide and nitrous oxide, on the other hand, are considerably more soluble. Webster, Hinn, and Lychalk[34] determined the solubility of the three gases in a variety of liquids. Some of their data, shown in Table 4-6, illustrate the difference in solubility properties between nitrogen and the other two compressed gases.

The solubilities of carbon dioxide and nitrous oxide in a variety of solvents at a pressure of 100 psig have been reported by Johnsen and Haase.[35] The solubilities of the two gases in fluorocarbon propellants 11, 12, and 114 have been determined by Dantzler, Holler, and Smith at several different temperatures and pressures.[36]

Methylene chloride and methyl chloroform are excellent solvents for nitrous oxide and carbon dioxide. Carbon dioxide is more soluble in methylene chloride

TABLE 4-5 PHYSICAL PROPERTIES OF THE COMPRESSED GASES

Property	Carbon Dioxide	Nitrous Oxide	Nitrogen
Formula	CO_2	N_2O	N_2
Molecular weight	44.0	44.0	28.0
Boiling point (°F)	-109^a	-127	-320
Vapor pressure (psig) at 70°F	837	720	477^b
Toxicity (Underwriters' Laboratories Classification)	5	–	6

[a]Sublimes.
[b]At critical temperature, $-233°$F.

TABLE 4-6 SOLUBILITIES OF THE COMPRESSED GASES IN LIQUIDS

	Solubility of Gas[a]		
Liquid	Nitrous Oxide	Nitrogen	Carbon Dioxide
Water	0.6	0.016	0.82
Acetone	5.3	0.15	6.3
Amyl acetate	5.9	0.15	4.1
Ethyl alcohol	2.8	0.14	2.6
Chloroform	5.2	0.13	3.4
Methyl alcohol	3.2	0.14	3.8

[a]Solubility is expressed as volume of gas per volume of liquid at $25°C$ and 1 atm.

than in ethyl alcohol.[23] Propellant systems based upon combinations of methylene chloride, methyl chloroform, and nitrous oxide (the "Aerothene" system) were suggested by Anthony in 1967 for use with such products as hair sprays.[37,38] More recently, the Dow Chemical Company has suggested hair sprays and underarm deodorant formulations with various combinations of methylene chloride ("Aerothene" MM), methyl chloroform ("Aerothene" TT), and carbon dioxide.[23] The use of methylene chloride in hair sprays increases the solubility of both the compressed gas and the hair spray resin.

The Ostwald solubility coefficient (λ) is a useful concept in dealing with the solubility of compressed gases in liquids. It is defined as the ratio between the volume of dissolved gas and the volume of the liquid:

$$\lambda = \frac{\text{Volume of dissolved gas}}{\text{Volume of liquid}} \tag{4-1}$$

The Ostwald solubility coefficients of some common solvents are shown in Table 4-7.[39]

TABLE 4-7 OSTWALD SOLUBILITY COEFFICIENTS OF SOME COMMON
SOLVENTS

	Ostwald Solubility Coefficient at $70°F$	
Solvent	Nitrous Oxide	Carbon Dioxide
Water	0.6	0.85
Acetone	5.92	6.95
Ethyl alcohol	2.96	2.84
n-Pentane	4.13	1.94
Methyl chloroform	4.96	3.74
Methylene chloride	5.53[a]	4.62[a]

[a]Data from Reference 22.

Hsu[39] has shown that the Ostwald solubility coefficient can be obtained by pressurizing a liquid in a container with a compressed gas and determining the weight of gas, temperature, pressure, liquid fill, and total container volume. The coefficient is calculated using the following equation:

$$\lambda = \frac{1}{X} \left(\frac{WRT}{V_c MP} + (X - 1) \right) \qquad (4\text{-}2)$$

where

λ = Ostwald solubility coefficient
X = Volume of liquid/volume of container
W = Weight of compressed gas (g)
R = Gas constant, 82.06 ml $\cdot$ atm/mole $\cdot$ $^{\circ}$K
T = Temperature ($^{\circ}$F)
V_c = Volume of container (ml)
M = Molecular weight of compressed gas
P = Total pressure (abs. atm)

The Ostwald solubility coefficients for a variety of liquefied gases (fluorocarbons and hydrocarbons) were determined at the Freon® Products Laboratory using the above procedure. The data are listed in Table 4-8. Fluorocarbons 21, 123, 132b, and 133a all have significantly higher solubility coefficients than methylene chloride.

The Ostwald solubility coefficients for carbon dioxide in ethyl alcohol and in mixtures of fluorocarbon 133a, fluorocarbon 12, methylene chloride, and isobutane with ethyl alcohol are listed in Table 4-9. The Ostwald solubility coefficient is highest with fluorocarbon 133a.

The Ostwald solubility coefficient for a concentrate with active ingredients will differ slightly from that for pure liquids. However, the latter can be used as an approximation. Gregg[40] has pointed out several slight corrections to be made in calculating the Ostwald solubility coefficient. These include the change in volume of the liquid under pressure and the increase in volume when carbon dioxide is dissolved. Empirically, Gregg found that 1 g of carbon dioxide increases the liquid volume by about 1 cc.

After the Ostwald solubility coefficient has been obtained for a given pressure and temperature, the weight of compressed gas required for any other pressure and container fill can be calculated. The following rearranged form of Equation 4-2 is used:

$$W = \frac{V_c MP}{RT} (X\lambda + 1 - X) \qquad (4\text{-}3)$$

Hsu also presents equations for calculating (1) the variation of pressure with temperature and (2) the variation of pressure with discharge. Bucaro[41] has pub-

TABLE 4-8 OSTWALD SOLUBILITY COEFFICIENTS OF THE LIQUEFIED GASES[22]

| Liquefied Gas | Formula | Ostwald Solubility Coefficient at 70°F | |
		CO_2	N_2O
Fluorocarbons			
11	CCl_3F	3.96	6.05
12	CCl_2F_2	0.60	0.55
21	$CHCl_2F$	5.65	6.94
22	$CHClF_2$	0.50	–
113	CCl_2FCClF_2	3.62	5.55
114	$CClF_2CClF_2$	2.94	3.96
123	$CHCl_2CF_3$	5.43	7.01
124	$CHClFCF_3$	2.67	2.81
132b	$CH_2ClCClF_2$	5.43	7.31
133a	CH_2ClCF_3	5.49	6.07
141b	CH_3CCl_2F	4.79	6.58
142b	CH_3CClF_2	3.58	4.02
152a	CH_3CHF_2	1.55	1.44
Hydrocarbons			
Isobutane	C_4H_{10}	1.86	2.71
Isopentane	C_5H_{12}	2.58	4.28
Chlorocarbons			
Methylene chloride	CH_2Cl_2	4.64	5.53

TABLE 4-9 OSTWALD SOLUBILITY COEFFICIENTS FOR ETHYL ALCOHOL SOLUTIONS[22]

Solution	Ostwald Solubility Coefficient for CO_2 at 70°F
Ethyl alcohol alone	2.86
Ethyl alcohol/fluorocarbon 133a (80/20)	3.14
Ethyl alcohol/methylene chloride (80/20)	2.98
Ethyl alcohol/fluorocarbon 12 (80/20)	2.13
Ethyl alcohol/isobutane (80/20)	2.07

lished several detailed articles covering calculations applicable to the use of compressed gases as propellants.

COMPRESSED GAS AEROSOLS

Types of Products

Until last year, the major use for carbon dioxide was in industrial and household products. These included furniture polishes, window cleaners, starch sprays, residual insecticides, car waxes, lubricants, degreasers, deicers, tire traction sprays, and rust removers. Hair sprays have now appeared on the market and other alcohol-based personal products are under study.

Properties of Compressed Gas Aerosols

Advantages

The compressed gases are essentially odorless, colorless, low in toxicity, non-flammable, inexpensive, and environmentally acceptable. They are not involved in the present fluorocarbon/ozone controversy. These are the advantages that have made compressed gases of increased interest to the aerosol industry in the last several years. Gregg[40] estimates that propellant costs for compressed gases range from about 0.1 to 0.6 cents per container. These costs are considerably less than those for fluorocarbon propellants. Compressed gases have always had a considerable price advantage over liquefied gases.

Another advantage of compressed gases is that fairly large variations in temperature cause only relatively minor changes in pressure. This property makes compressed gases ideal for products such as windshield deicers and tire traction sprays which must operate at low temperatures. These aerosols are normally kept in the automobile during the winter season. Windshield deicers were originally formulated with fluorocarbon 12. When left in a car at subfreezing temperatures, the pressure inside the container became too low to discharge the contents. Windshield deicers and tire traction sprays are interesting examples of products where compressed gases are superior to liquefied gases as propellants.

The change in pressure of a compressed gas with temperature compared to that for fluorocarbon 12, assuming that both have an initial pressure of 70.2 psig at $70°F$, is shown in Table 4-10.

Disadvantages

In considering a compressed gas as a propellant, advantages must be balanced against disadvantages. The disadvantages of a compressed gas are listed below.

TABLE 4-10   COMPARATIVE PRESSURES OF FLUOROCARBON 12 AND
A COMPRESSED GAS PROPELLANT AT DIFFERENT
TEMPERATURES

	Pressure (psig)	
Temperature ($^\circ F$)	Compressed Gas	Fluorocarbon 12
130	79.8	181.0
70	70.2	70.2
32	64.1	30.1
0	59.0	9.2

Initial Spray Characteristics. Compressed gases are soluble to only a limited extent in concentrates. Their main function as propellants is to provide the pressure that forces the concentrate out of the container and through the actuator. They do not break up the concentrate by changing from a liquid to a vapor as do liquefied gases. Therefore, mechanical break-up actuators are necessary. As a result, spray characteristics with compressed gases are inferior to those with liquefied gases. The sprays are wetter and less forceful. They become even wetter after most of the product has been discharged. Wetness of spray was one of the reasons for several of the carbon dioxide propelled hair sprays being unsuccessful in the test market.

The new low-spray-rate valves and actuators have stem and orifice diameters varying from about .008 to .013 in. (see Chapter 6, "Valves and Actuators," for a more detailed discussion). These valves have a marked effect upon spray characteristics. For example, alcohol-based products formulated with carbon dioxide at a pressure of 90 psig at 70°F have a soft, misty spray. The spray rate is about 0.5-0.7 g/sec. The spray with a so-called standard valve having a stem orifice of .018 in. and an actuator having a .016-in. orifice is forceful, harsh, and coarse. The delivery rate is about 1.5 g/sec. Products based upon higher boiling solvent systems such as industrial aerosols give coarser sprays than alcohol-based aerosols.

The lower spray rate has proven to be a disadvantage for professional hair sprays in some beauty salons. The low rate requires too long a time for application.

Change in Spray Characteristics during Discharge. The rate at which the concentrate is forced through the mechanical break-up actuator is an important factor in determining the type of spray. The rate is a function of the pressure in the container. Since the pressure decreases in compressed gas aerosols during use, the rate also drops and the spray becomes increasingly coarser. The change in spray characteristics during use resulting from the decrease in pressure is not nearly as noticeable with the new low-spray-rate valves. The pressure drops as rapidly, but the change in spray characteristics is less than with standard valves having the larger orifices.

The decrease in pressure is more noticeable with nitrogen, which is practically insoluble in the concentrate and, therefore, is confined to the vapor phase. When an aerosol product is used, the volume of the liquid phase decreases and that of the vapor phase correspondingly increases. As the volume of the vapor phase increases, the pressure in a product with nitrogen will decrease essentially according to the relationship

$$P_2 = P_1 \frac{V_1}{V_2}$$

since $P_1V_1 = P_2V_2$, where P_1 and P_2 are the initial and final pressures, and V_1 and V_2 are the initial and final volumes in the vapor phase. Therefore, if the vapor phase in an aerosol product pressurized initially with nitrogen to 100 psi doubles in volume as the product is discharged, the pressure will drop to about 50 psi.

The greater solubilities of carbon dioxide and nitrous oxide in concentrates provide somewhat of a reservoir for these gases. Therefore, the pressure would not be expected to drop as rapidly during use with carbon dioxide and nitrous oxide because some of the gas in the vapor phase is replaced by gas from the liquid phase. However, this is only partially true. The gas dissolved in the liquid phase is released at a relatively slow rate. Equilibrium between the gas in the vapor and liquid phases usually is not established while the product is being used.[42] The pressure decreases more rapidly than would be expected because the drop in pressure, while results from an increase in the volume of the vapor phase, is not completely compensated for by the release of gas from the liquid phase. The decrease in pressure also causes a decrease in spray rate, although not in direct proportion.

The comparative changes in pressure and spray rate with discharge for ethyl alcohol propelled with nitrogen and carbon dioxide are shown in Table 4-11. The alcohol was pressurized to approximately 95 psig with the compressed gases. The percent volume fill was 80%. The samples were equipped with an ARC (now the Dispenser Products Division of Ethyl Corp.) valve having a 0.013-in. stem orifice and an actuator having an external orifice of 0.013 in.[22] A fluorocarbon-propelled aerosol would show essentially no change in vapor pressure or spray rate with discharge under the same conditions.

Misuse. Misuse occurs when the consumer discharges the container with the end of the dip tube in the vapor phase. This could occur in a horizontal as well as a vertical position. Under these conditions, the gas in the vapor phase is lost immediately. If misuse occurs often, enough compressed gas may be lost so that the product will no longer discharge.

TABLE 4-11   CHANGE IN PRESSURE AND SPRAY RATE WITH
DISCHARGE FOR ETHYL ALCOHOL PRESSURIZED WITH
COMPRESSED GASES[22]

Property	Nitrogen	Carbon Dioxide
Pressure (psig) at 70°F		
Initial	95	94
90% Discharge	20	55
99% Discharge	16	50
Spray Rate (g/sec) at 70°F		
Initial	0.55	0.50
90% Discharge	0.26	0.43
99% Discharge	0.22	0.39

The extent to which misuse will affect consumer acceptance of products equipped with the low-spray-rate valves remains to be determined by market tests. The effect of misuse with the new valves will be less than with the standard valves because of the lower spray rate. The effect of misuse upon product performance has been investigated by Hinn[42] and Webster.[43]

Valve and Actuator Clogging. Clogging is an ever present possibility with the new valves and actuators. The smaller the orifices, the more prone the valves are to clogging. Meuresch has pointed out that an orifice with a 0.10 in. diameter has a circular area only about 25% that of an orifice with a 0.20 in. diameter.[44] Rus[45] does not recommend orifice diameters as low as .008 in. Clogging occurs for two reasons: suspended particles in the product and filming over the actuator orifice. The Seaquist Valve Company has developed several filter systems for the valve body which decrease clogging due to particulate matter. The second cause, filming over of the actuator orifice, is more difficult to prevent. This occurs with products such as hair sprays which contain film-forming resins. According to Murphy,[46] it occurs after spraying has terminated. The carbon dioxide left in the expansion chamber bubbles out but does not clear the actuator orifice. The residual resin then clogs the orifice after drying. With fluorocarbon-propelled hair sprays, sufficient liquefied gas remains in the valve expansion chamber after spraying so that subsequent vaporization clears the actuator orifice of any remaining product.

Corrosion. Carbon dioxide forms carbonic acid in the presence of moisture. This creates a potential for corrosion. According to Johnsen,[4] the pH of an aqueous solution of carbon dioxide at a pressure of 88.2 psig is about 3.2 at 70°F. Aluminum is not affected, but rusting and perforation may occur in steel containers. Lacquer-lined containers usually are satisfactory. The container companies should be consulted for advice on containers for any carbon dioxide propelled product.

Neutralization of Hair Spray Resins. Some hair spray resins are acidic and are neutralized with amines in aerosol formulations. The acidic nature of carbon dioxide must be taken into account when formulating such products. Additional base may be required to counteract the acidity of carbon dioxide.

Reactivity of Nitrous Oxide. Nitrous oxide is an oxidizing agent. This property must be considered when nitrous oxide is used as a propellant. Mixtures of nitrous oxide and alcohols can be particularly dangerous. Johnsen[4] cites an example of a disastrous fire that resulted from a reaction between nitrous oxide and methyl alcohol in a saturator. However, nitrous oxide has been safely used for years as a food propellant.

Filling Equipment. New filling equipment has been developed for high-speed production of carbon dioxide propelled products (see Chapter 7, "Filling Methods"). However, the problem of changing to carbon dioxide is complex, as Sievers has reported.[47] Reformulation of the product may be necessary and new equipment required.

Quality control problems may arise because the amount of propellant added is too small to be checked by weight. Pressure checkers are recommended.

Compressed Gas/Liquefied Gas Systems

There has been interest in these systems since 1963. Webster, Hinn, and Lychalk[48] disclosed an insecticide formulated with 20 wt. % concentrate, 55 wt. % methyl chloroform, and 25 wt. % fluorocarbon 12. It was pressurized to 80 psig with nitrous oxide. The product had to be classified as a pressurized spray rather than an aerosol insecticide because the particle size was not small enough to meet the requirements for a space insecticide.

Alcohol-based formulations with 10 wt. % fluorocarbon 12 or fluorocarbon 22 and pressurized to 90 psig with carbon dioxide have sprays superior to those with either propellant alone, that is, carbon dioxide at 90 psig or 10 wt. % fluorocarbon. Carbon dioxide is less soluble in both fluorocarbon 12 and fluorocarbon 22 than in ethyl alcohol. However, the contribution to spray characteristics of these two fluorocarbons, because they are liquefied gases, is greater than the loss in carbon dioxide solubility when they are mixed with ethyl alcohol.

Carbon dioxide is more soluble in methylene chloride than in ethyl alcohol. Hair sprays utilizing combinations of methylene chloride, ethyl alcohol, and carbon dioxide have been suggested.[23] Methylene chloride increases the solubility of carbon dioxide in the system and decreases flammability.

Fluorocarbons 123, 132b, and 141b should be useful in combination with ethyl alcohol and carbon dioxide, since they have carbon dioxide solubilities higher than that of methylene chloride. However, they have not been tested as yet. Combinations involving fluorocarbon 133a, ethyl alcohol, and carbon dioxide were evaluated and found to be particularly effective. Fluorocarbon 133a not only has better solubility for carbon dioxide than methylene chloride (see Tables 4-8 and 4-9), but it also functions as a liquefied gas. The combination of the increased solubility for carbon dioxide and liquefied gas property produces excellent spray characteristics.

Fluorocarbon 133a is superior in the following respects to methylene chloride for use in alcohol-based products pressurized with carbon dioxide:

1. Higher solubility for carbon dioxide. This not only results in a finer spray, but also decreases the effect of misuse to a greater extent.
2. Nonflammability (methylene chloride is flammable).
3. Fluorocarbon 133a is a liquefied gas and thus is a propellant by itself.
4. Fluorocarbon 133a has a lower leakage rate through valve gaskets than methylene chloride. This permits higher concentrations of fluorocarbon 133a to be used in the products.
5. The change in spray characteristics during use is significantly less with fluorocarbon 133a than with methylene chloride.

In comparison with hydrocarbons, fluorocarbon 133a has a much higher solubility for carbon dioxide and is nonflammable.

REFERENCES

1. E. Rotheim, U. S. Patent 1,893,750 (1933).
2. G. F. Ford, Personal Communication, Phillips Petroleum Company.
3. J. G. Spitzer, I. Reich, and N. Fine, U. S. Patent 2,655,480 (October 1953).
4. M. A. Johnsen, W. E. Dorland, and L. K. Dorland, "The Aerosol Handbook," Chap. 9, Wayne E. Dorland Company, Caldwell, N.J., 1974.
5. G. H. Brock, Personal Communication.
6. Report EPA - 560/1-76-002, Midwest Research Institute, February, 1976.
7. G. H. Brock, *Aerosol Age* **21,** 22 (July 1976).
8. Technical Data Sheets, "Physical Constants of Hydrocarbon Propellants and Pentanes," Aeropres Corp., 1967.
9. Technical Brochure, " 'Aeron' Hydrocarbon Aerosol Propellants," Diversified Chemicals and Propellants Company.
10. Technical Brochure, Industrial Hydrocarbons, Inc.
10a. Technical Brochure, "Hydrocarbon Aerosol Propellants," Phillips Petroleum Company.
11. J. R. Frauenheim, *Aerosol Age* **20,** 23 (December 1975).
12. G. F. Ford, *Aerosol Age* **21,** 36 (May 1976).
13. G. W. Fiero and M. A. Johnsen, *Aerosol Age*, Part I, **14,** 22 (September 1969); Part II, **14,** 59 (October 1969); Part III, **14,** 48 (November 1969).
14. M. A. Johnsen, *Aerosol Age* **9,** 18 (July 1964).
15. G. Parks, *Aerosol Age* **7,** 31 (May 1962).
16. L. C. Goodhue, J. H. Fales, and E. R. McGovern, *Soap Sanit. Chem.* **21,** 123 (April 1945).
17. W. H. Reed, *Soap Chem. Spec.* **32,** 197 (May 1956).

18. Union Carbide Chemicals Company Technical Bulletin, "New Ucon Paint Propellant."
19. R. J. Scott and R. R. Terrill, *Soap Chem. Spec.* **38**, 2 (February 1962); *Aerosol Age* **7**, 18 (January 1962).
20. R. J. Scott and R. R. Terrill, *Soap Chem. Spec.* **38**, 130 (February 1962); *Aerosol Age* **7**, 18 (January 1962).
21. R. J. Scott and R. C. Werner, *Soap Chem. Spec.* **42**, 190 (December 1966).
22. Freon® Products Laboratory Data.
23. Technical Bulletin, "Aerosols under Fire," Dow Chemical Company, 1975; Dow Advertisement, *Aerosol Age* **22** (February 1977).
24. C. A. Teter, *Aerosol Age* **1**, 16 (August 1956).
25. R. W. Moore, U. S. Patent 746,866 (1903).
26. L. K. Mobley, U.S. Patent 1,378,481 (1921).
27. R. M. L. Lemoine, D. R. Patent 532,194 (1926).
28. J. Kalish, *Drug Cosmet. Ind.* **81**, 441 (October 1957).
29. A. R. Marks, *Aerosol Age* **3**, 64 (December 1958).
30. F. A. Mina, *Drug Cosmet. Ind.* **82**, 321 (March 1958).
31. J. S. Hinn, *Proc. 47th Mid-Year Meet. Chem. Spec. Manuf. Assoc.*, 53 (May 1961).
32. F. T. Reed, "Fluorocarbon Propellants," in H. R. Shepherd (Ed.), "Aerosols: Science and Technology," Interscience Publishers, Inc., New York, 1961.
33. R. C. Webster, "Compressed Gases," in A. Herzka (Ed.), "International Encyclopaedia of Pressurized Packaging" (Aerosols), Pergamon Press, Inc., New York, 1966.
34. R. C. Webster, J. C. Hinn, and P. A. Lychalk, *Proc. 49th Mid-Year Meet. Chem. Spec. Manuf. Assoc.* (May 1963).
35. M. A. Johnsen and F. D. Haase, *Soap Chem. Spec.*, Part I, **41**, 73 (July 1965); Part II, **41**, 93 (August 1965).
36. E. M. Dantzler, F. C. Holler, and P. T. Smith, *Soap Chem. Spec.* **41**, 125 (January 1965).
37. T. Anthony, *Soap Chem. Spec.* **43**, 185 (December 1967).
38. T. Anthony, *Aerosol Age* **12**, 31 (September 1967).
39. H. Hsu, *Aerosol Age* **20**, 27 (September 1975).
40. W. Gregg, *Aerosol Age* **20**, 19 (July 1975).
41. R. Bucaro, *Household Pers. Prod. Ind.* **43** (September/October 1975).
42. J. S. Hinn, *Proc. 47th Mid-Year Meet. Chem. Spec. Manuf. Assoc.* (May 1961).
43. R. C. Webster, *Aerosol Age* **6**, 20 (June 1961).
44. H. Meuresch, *Aerosol Report* **15**, A50, No. 4 (1976).
45. R. R. Rus, *Aerosol Age* **20**, 16 (July 1975).
46. E. Murphy, Personal Communication. GAF Corp.
47. J. Sievers, *Aerosol Age* **20**, 14 (July 1975).
48. R. C. Webster, J. S. Hinn, and P. A. Lychalk, *Proc. 49th Mid-Year Meet. Chem. Spec. Manuf. Assoc.* (May 1963).

5

CONTAINERS

The first aerosol containers produced in quantity were the high-pressure, heavy steel cylinders developed during World War II for insecticides. It was realized at that time that if aerosols ultimately were to be marketed to the civilian population, a low-cost, lightweight, disposable container would be necessary.[1,2] The logical container was the low-pressure beer can already in use. In 1946, the Crown Cork & Seal Company modified their seamless drawn beer can, the Crowntainer, to have a flat top. The new container was called the Spra-tainer. It had a drawn body made of black iron with an interior coated with baked enamel. The coated concave bottom was seamed into place.[3]

When Goodhue moved from the Department of Agriculture to Airosol, Inc. after the war, he used the Crown Spra-tainer for his work on low-pressure aerosols. On November 21, 1946, Airosol, Inc. received 104,832 blank cans from Crown.[4] The tops were reworked to accommodate a valve invented by Greenwood.[5] Airosol, Inc. made the first commercial shipment of a low-pressure aerosol insecticide trademarked "Jet" on March 7, 1947 in the Crown Spra-tainers. Crown subsequently supplied Spra-tainers with a valve on top. The containers were cold-filled upside down and closed by double seaming on the bottom. The container was difficult to handle inverted because of its shape. In 1947, Crown introduced a modified Spra-tainer with a bottom in place and a curled 1-in. opening in the top for valves with a 1-in. mounting cup. The aerosol industry liked the new container because the double-seaming bottom operation was eliminated. Most major products were packaged in this container in the period between its introduction and 1953.[4]

The Continental Can Company started development of a disposable aerosol container in 1943. In September 1945, Peterson demonstrated a Continental can with a tire valve and push-button actuator to Goodhue.[2] Shortly thereafter, Continental Can improved their valve and incorporated it into the end of a double concave container. The can body was formed from a single sheet of tinplate soldered along the seam. The container was closed by double seaming the bottom after cold filling in the inverted position.

Continental Can was ahead of Crown Cork & Seal at this time in the field of disposable aerosol containers. Production of the first commercial low-pressure aerosol insecticide became a race between Airosol, Inc., using the Crown container, and Continental Filling Corp., using the Continental can. However, Continental Filling was delayed by a combination of factors, including receipt of cans and valves from the Continental Can Company, their building program, and a need to develop filling equipment.[6] Their first shipment of 20,000 "Fly Tox" aerosols was in March 1948, about one year after Airosol, Inc. had shipped "Jet."

The popularity of the 1-in. opening in the Crown containers for valves with a mounting cup became so great that it was apparent to Continental Can that their containers would have to be modified to have the same opening. The Continental dome with a 1-in. opening was marketed in 1951. The manufacture of containers with valves already in the top was discontinued in 1953.[4]

The American Can Company manufactured 12-oz aerosol containers during 1948, but did not actively enter the field until about 1950. The 1-in. curl opening was adopted by American Can in 1955, and this opening has since, with few exceptions, become standard for the industry. Other companies that now manufacture metal aerosol containers include the Apache Container Corp., Diamond International Corp., National Can Corp., Sherwin-Williams Company, and Southern Can Company.

TYPES OF AEROSOL CONTAINERS

There are essentially four main classes of aerosol containers—steel, aluminum, glass, and plastic. There are four types of steel containers. These are listed below with the approximate share of the market they hold (October 1975). The market penetrations were estimated from information supplied by various container companies.

Types of Aerosol Containers	Approximate Share of Market (%)
1. Steel containers	
a. Three-piece, soldered, side-seam containers	70–80
b. Drawn and drawn and ironed containers	3–4
c. Soudronic welded containers d. Conoweld containers	12–13
2. Aluminum containers	3–4
3. Glass bottles	7–8
4. Plastic containers	0.1

The use of welded containers may increase depending upon the ability of the can manufacturers to reinvest in the business. Thus, expansion of welded containers will be determined by the growth per se of the aerosol industry.[7]

Each type of container has advantages and disadvantages. The suitability of a particular container for an aerosol product depends upon the relative importance of a number of factors. These include container cost, application for which the aerosol was developed, aesthetic appeal, corrosivity of the formulation, pressure, etc.

Pressure can be a decisive factor in container choice. The U.S. Department of Transportation (DOT) requires containers of a specified strength for different pressure ranges. The DOT pressure and size regulations for containers will be covered in more detail at the end of this chapter. However, a brief definition of the container designations for various pressure limits will be helpful in the subsequent discussion. For aerosols with pressures not exceeding 140 psig at 130°F, any type of metal container can be used as long as the container can withstand a pressure 1.5 times that of the aerosol at 130°F and the capacity of the container does not exceed 27.7 fluid ounces (fl. oz). Containers that meet these specifications are commonly referred to as "standard containers." Specification DOT 2P containers can withstand pressures of 240 psig without bursting. These are for products whose pressures exceed 140 psig, but not 160 psig, at 130°F. Specification DOT 2Q containers can withstand pressures of 270 psig. They are used for products whose pressures at 130°F exceed 160 psig, but not 180 psig.

Steel Containers

The metal most commonly used for steel containers is tinplate. It is a carbon steel alloy available in various gages with different quantities of tin on the surface. Tin is first deposited on the steel by electrolytic methods and then melted so that it forms a continuous film bonded to the surface. Tinplate is available with different thicknesses of tin on the surface. The various thicknesses are indicated by the terms $\frac{1}{2}$ lb, 1 lb, $1\frac{1}{2}$ lb, etc. The weight designation of tinplate used in the industry has the following meaning. There is a standard area over which varying weights of tin are spread. This standard area is known as a base box and is the area of both sides of 112 sheets of steel with dimensions of 14 in. X 20 in. The thickness of the steel sheets is indicated by the weight per base box, which runs from about 85–112 lb—about 0.010 in. to a little over 0.012 in. thick. The layer of tin on the steel surface is not uniform. The thickness of tin on samples of 1-lb tinplate has been found to vary from about 0.00003 to 0.00010 in. The theoretical thickness, assuming 1 lb of tin distributed evenly over the surface, would be 0.000065 in.[8,9]

Tinplate is used for three-piece containers with a soldered side seam, the drawn and ironed (D/I) container, the Soudronic container with a welded side seam, and two-piece drawn containers. Another metal, tin-free steel, used for the body of Conoweld containers, has the same steel base as tinplate. However, instead of tin, chromium and chromium oxides are deposited on the steel.

In the container industry, can dimensions are expressed by three-digit numbers. The first digit is inches and the last two are sixteenths of an inch. The container diameter is given first, followed by height. For example, a 211 × 413 container is $2^{11}/_{16}$ in. in diameter and $4^{13}/_{16}$ in. in height, measured from bottom double seam to top double seam. This container also is commonly referred to in the aerosol industry as a 12-oz container.

Two-Piece, Drawn, Tinplate Containers (Spra-tainers)[10]

Spra-tainers, manufactured by the Crown Cork & Seal Company, have been commercially available since World War II. The Spra-tainer is one of the strongest conventional aerosol containers. It meets both the DOT 2P and DOT 2Q pressure specifications. When manufactured with a sprayed interior lacquer, the Spra-tainer is very corrosion resistant. Because the design of the body-makers does not permit easy changes in height or diameter, it is manufactured in only two sizes, a 202 × 411, 6-oz container with an overflow capacity of 225 cc, and a 214 × 411, 12-oz container with an overflow capacity of 375 cc. The extensive drawing process does not permit high-speed production. This results in higher unit costs than for most other aerosol cans. However, it is used for products such as refrigerants and ether starting fluids where a strong, safe container is the primary consideration and unit cost secondary.

Manufacture

Spra-tainers are manufactured by a cold, deep-draw process. The tinplate can body is formed in two operations. First, a metal dish approximately 5 in. in diameter and 1 in. in height is punched from tinplate and conveyed to a body-maker. The dish is transformed to an elongated cylinder with a final diameter of either 202 or 214 by a series of deep-draw operations. The bottom of the cylinder is trimmed and flanged. A considerable amount of lubricant must be applied during the operation. The lubricant must be completely removed by solvent degreasing before organic coatings can be applied. After degreasing, the 1-in. opening of the container is formed by a curling die machine. An interior lacquer is then applied to the container body by an automatic spray machine. The coating, of either a phenolic or epoxy type, is flash-cured by an electrostatic process. After the bottom end is applied by double seaming, the finished container is leak tested and conveyed to the exterior finishing station. First an exterior base coat, usually white, is applied, followed by lithography and application of a clear varnish. The base coat and varnish are cured by electrical means.

Three-Piece, Soldered, Side-Seam Containers[9-12]

The three-piece, soldered, side-seam container is by far the most widely used container in the aerosol industry. The containers are available in a variety of sizes and shapes, with and without interior coatings. Capacities vary from 2.0 to 24 fl. oz. Containers up through 20 oz are available in both the standard and DOT 2P specifications. The 24-oz container meets only the standard specifications. Typical dimensions and capacities of the containers from the American Can Company are listed in Table 5-1. Continental Can and Crown Cork & Seal have essentially an equivalent line of soldered side-seam containers.

Many types of interior organic coatings have been developed. Interior coatings function as barriers to prevent a reaction between product and container, and to increase the resistance of the container to corrosion. In addition, the soldered side seam may have residues of solder flux which could initiate decomposition reactions. Therefore, striping compositions have been developed for covering the side seam. This operation is carried out regardless of whether the

TABLE 5-1   TINPLATE AEROSOL CONTAINER CAPACITIES
(AMERICAN CAN COMPANY)

	Fluid Ounces		Cu. Inch Overflow[a]	Cu. Centimeter Overflow[a]
Size Description	Nominal	Overflow[a]		
202 x 200	2.0	3.4	6.0	98
202 x 204	2.5	3.8	7.1	120
202 x 214	3	4.9	8.8	145
202 x 314	4	6.7	11.5	189
202 x 406	6	7.7	13.9	227
202 x 412	6.5	8.3	14.9	245
202 x 509	8	9.8	17.7	289
202 x 608	9	11.6	20.9	343
202 x 700	10	12.4	22.3	367
202 x 708	12	13.4	24.1	393
207.5 x 413	10	11.6	20.9	343
207.5 x 509	12	13.4	24.1	393
207.5 x 605	14	15.1	27.4	450
207.5 x 701	16	16.9	30.5	500
207.5 x 708	17	18.0	32.4	531
211 x 413	12	13.7	24.8	405
211 x 604	16	17.8	32.2	527
211 x 612	18	19.2	34.7	568
211 x 713	20	22.3	40.3	659
300 x 709	24	26.9	48.5	795

[a]Inserting a valve decreases the capacity by appoximately 5 cc. These above values represent average capacities.

TABLE 5-2 TYPICAL COATINGS USED IN AEROSOL CANS (AMERICAN CAN COMPANY)

General Classification	Types	Representative Uses
Epoxy	Urea-formaldehyde, phenolic, modified and epon esters	Epoxies are good general-purpose enamels, well suited for a wide range of water base, water/alcohol, and mild solvent base products.
Phenolic	Vinyl modified	Phenolic body coatings provide excellent chemical resistance for a wide range of hard-to-hold products.
Vinyls	Solution type	Solution vinyls are used primarily as a top coat to provide superior enamel coverage for hard-to-hold products.
Vinyl organosols	Dispersions of vinyl resins; may contain pigments	Organosols can be applied in very heavy film weights, providing the extra resistance required by some products.
Double-coat systems	Epoxy plus vinyl; epoxy plus phenolic; and double organosol combinations	These double-coat systems combine the properties of their component parts as well as providing excellent coverage for hard-to-hold products.
Side-seam stripes— post-solder	Vinyls, phenolics, and epoxy-vinyls	These materials are sprayed over the side-seam after it is soldered to provide coverage over areas not protected by the regular inside body coatings.
Side-seam stripes— pre-solder	Epoxy-phenolic	These stripes are spray applied over the side-seam before it is soldered to provide coverage over areas not protected by the regular body coatings.

rest of the can interior is coated. Some typical interior coatings and side-seam striping compositions are summarized in Table 5-2.

Many anhydrous aerosols are sufficiently noncorrosive so that they can be packaged in uncoated containers. Johnsen[13] has estimated that about half of the tinplate containers used in the United States are uncoated. The container companies should be consulted first if there is any question about the type of container or coating required for a given aerosol formulation.

Manufacture

Can bodies are formed from tinplate blanks that have been slit and trimmed to the required size. The edges are then notched to provide a lock-and-lap seam construction. The lock portion is four layers thick and the lap portion is two layers thick. After treatment with flux to clean the tinplate, the notched body blank is wrapped around a mandrel that expands to interlock the folded edges. It is then hammered tight to form the side seam and passed over a roll that rotates in a bath of solder. The soldered lap-lock and lap-seam combination is strong enough so that the containers can meet DOT 2P specifications. The lap area, which can be found either inside or outside, is strongest when it is on the inside. The open ends are flanged after the cylinder has been formed. This keeps the cylinder round and also provides the hook for double seaming the ends. Double seaming is the operation in which the bottom and top ends are attached to the cylinder. It is usually carried out in the vertical position. The two operations involved in double seaming are shown in Figure 5-1. Before the seaming operation, the end units are treated with sealing compound to provide a seal when the double seam is compressed. Sealing compound is a rubber-based material normally composed of synthetic rubber, resin, and filler.

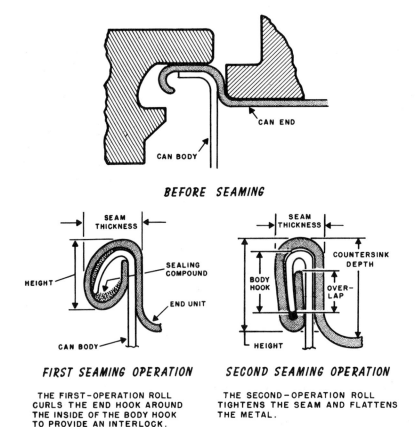

BEFORE SEAMING

FIRST SEAMING OPERATION

THE FIRST-OPERATION ROLL
CURLS THE END HOOK AROUND
THE INSIDE OF THE BODY HOOK
TO PROVIDE AN INTERLOCK.

SECOND SEAMING OPERATION

THE SECOND-OPERATION ROLL
TIGHTENS THE SEAM AND FLATTENS
THE METAL.

Figure 5-1   Double-seaming operations. (*Courtesy of the Continental Can Company*)

The solder is usually one of two types: lead or tin. Lead solder, the most commonly used, has a composition of 98% lead and 2% tin. Tin is added to increase the ability of the solder to wet the surface of the tinplate. Lead solder has fairly good creep resistance and is less expensive than tin solder. For high-pressure systems, lead solder is modified by replacing part of the tin with silver. This produces a solder with high resistance to creep.

Pure tin solder has poor creep resistance and is seldom used. It is alloyed either with .5% silver or 2% antimony for high-pressure systems. This produces side-seam strength equal to that obtained with high-pressure lead solder.

Three-Piece, Welded, Side-Seam Containers

The alternate method to soldering for seam bonding of aerosol containers is welding. Welding eliminates the soldering operation. This saves considerable metal because the area of the overlap is far less than that in soldered cans. The thinner stock cuts material costs. The cans are considered more attractive by designers. They permit almost full circle decoration and the advantages of embossed or sculptured surfaces. Aesthetically, a welded seam presents a better appearance to the consumer. The fact that the side seam contains no lead is another advantage.

Welded side-seam containers can hold a variety of products. Compared to soldered cans, welded side-seam containers provide superior shelf life for some products, but are inferior for others. Some corrosive products have greater stability in three-piece soldered containers than in welded side-seam containers because the solder in the side seam acts as a sacrificial anode in the electrochemical corrosion reactions. However, welded containers provide better enamel coverage at the side-seam area. This reduces the possibility of corrosion induced by solder flux residues. The storage stability of any aerosol should be determined in welded containers.

There are two commercial methods for welding: Soudronic and Conoweld. These are discussed in the following sections.

Soudronic Containers[10,13-18]

The Soudronic welded tinplate can is a companion package to the soldered side-seam container. It provides the marketer with a more varied selection for his particular needs. At the present time, Soudronic containers constitute only a small part of the aerosol container market. However, they are expected to show a rapid growth in the near future. They are manufactured by American Can, Crown Cork & Seal, and the Southern Can Company. Sizes currently available correspond to existing soldered cans. The diameters are 202, 207.5, 211, and 300 with heights ranging from 214 to 908.

Properties. Strength of the welded seam is equal to or exceeds that of any soldered aerosol can. In fact, the metal itself may shear before the weld breaks.

However, since conventional domes and bottoms are seamed on, the limiting factor is dome and bottom strength. Therefore, the overall strength of the Soudronic can approximates that of soldered side-seam containers. Soudronic cans are available in both the standard and DOT 2P specifications, but not in the DOT 2Q specifications.

Soldered cans are prone to solder fatigue after prolonged storage. This can result in stress fracture of the solder with possible product leakage or loss of pressure. High-pressure fatigue testing in the laboratory by Crown Cork & Seal has shown no stress cracking of the welded seam.[10]

In general, anhydrous-alcohol- and oil-based products are adaptable to Soudronic can packaging. Aqueous-based products should be thoroughly tested. The welded area presents more potential iron exposure than the soldered side seam. If a product is affected by base steel, packaging in Soudronic cans may result in problems. Conversely, some products, such as alkaline oven cleaners, may attack solder, resulting in side-seam leakage. The Soudronic can is superior for such products.

Manufacture. The manufacture of Soudronic welded cans is based upon an electronically controlled resistance welding principle using a copper wire as an intermediate electrode. A uniformly welded seam is obtained.

Tinplate is first sheared into blanks of the desired size. These are transported to a "flexor," where the blanks are passed through bending rolls to relieve sheet stress. This guarantees a uniform rounding of the can body. The sheets are then led to a "body roll-former," where the blanks are rounded with the edges slightly overlapping. The round blanks are delivered to a welding station, where welding is performed on the lap using twin copper wire electrodes, one inside and one outside. The electrical current is first applied gradually and then stepped up to full capacity. The tin from the sheet melted by welding is deposited on the copper wire electrodes, leaving a clean contact area. Line voltage and current density must be carefully controlled to assure a narrow, uniform, and leakproof seam. The melted surface tin binds the overlapped section of the welded seam. After the body is welded, it is fed into a conventional line where top and bottom ends are flanged. Domes and bottoms are then seamed on.

The speed of the line will depend on the particular models of the Soudronic machines used. Equipment is available from semiautomatic slow speeds of 20–50 cans/min to high-speed automatics of over 150 cans/min. In general, output from each Soudronic welder will approximate about 50% of output from a conventional fabricated soldered aerosol line using a solder bath. Accordingly, two Soudronic welders need to be placed in tandem to maintain the same output as a conventional line.

Conoweld Containers[19-20]

Conoweld containers are manufactured by Continental Can. Conoweld is a high-speed process (about 375 cans/min on a single line) for producing containers suitable for necking-in and stylizing. Sizes presently available are as follows: 202 X 314 (4 oz), 202 X 406 (6 oz), 202 X 509 (8 oz), 211 X 413 (12 oz), and 211 X 604 (16 oz). The containers are used for a variety of products, including oven cleaners, antiperspirants, underarm deodorants, starch sprays, etc.

Manufacture. The Conoweld process utilizes a 500-cycle modified ac power source that provides essentially a square wave rather than a conventional sine wave. The preformed body blank cylinder is passed through two rotating electrode rings—one inside and one outside. These rings also serve as a forging mechanism, producing a solid-state forged side seam. The body is formed from tin-free steel. After welding, the ends are attached using conventional double-seaming operations. The ends can be either tin-free steel or tinplate. Normally, tinplate is used. The cans are usually supplied with an interior coating.

Drawn and Ironed Containers[10,21,22]

The drawn and ironed (D/I) aerosol container is a relatively recent development in aerosol can manufacturing technology. It probably will have a profound effect upon aerosol packaging in the next decade. The first commercial D/I cans were aluminum. They were produced by the Apache Container Corp. in the fall of 1971. Shortly thereafter, Apache began manufacturing seamless tinplate steel aerosol containers. They have since discontinued the manufacture of aluminum containers.

Basically, the process is not really new. Cold-drawing of metal is an old art. Crown Spra-tainers have been manufactured using the cold-draw process since the late 1940's. The process of ironing metal has also been used for years to make ammunition. However, the combination of these two processes to produce aerosol cans at speeds approximating 500 (or more) cans/min was a considerable technical engineering accomplishment.

Tinplate containers currently available from Apache range from 202 X 509 (9 oz) to 202 X 700 (10 oz) and from 211 X 413 (12 oz) to 211 X 713 (20 oz). Crown is offering a necked-in 18-oz container designated 211/207.5 X 612.

Properties. The absence of bottom and side seams results in a significant improvement in can strength. In addition, since the top seam is not required to seal over the four thicknesses of metal found in conventional side seams, its strength is also improved. The usual D/I cans meet the standard and DOT 2P specifications. They also meet DOT 2Q pressure requirements. However, since the DOT 2Q specifications require a 0.008-in. minimum wall thickness, the normal wall thickness in the D/I cans, which is 0.007 in., must be increased to meet the letter of specification when 2Q certification is required.

Elimination of the side seam removes a major coating problem. Interior linings (Apache offers approximately a dozen single- and double-coat systems) are sprayed and cured after the body is formed. To date, plain tinplate containers appear to be satisfactory for anhydrous-alcohol- and oil-based products. Aqueous-based products probably will require an interior lacquer coating.

Manufacture. The big advantage of the D/I process is metal and labor efficiency. The ironing process reduces the amount of metal used by 30–40%. The process can be accurately controlled so that extra thickness is left in the can bottom for additional strength during shipment and at the top to insure a stronger double seam. Labor costs are lower, since D/I can-making is a continuous process. This eliminates the intermediate handling required on conventional three-piece, soldered, side-seam lines.

The D/I can body is generally formed in two distinct operations. Cup-drawing is the first and ironing the second. The cups are made from plain metal, fed into a cupping press that stamps out 4–8 cups/strike at a rate of 120 strikes/min, and draws them into a cup. The cups are then fed automatically into a special press known as an ironer. The ironer draws the tinplate cup to the final diameter. The ironer or body-maker is a machine in which a special plug forces the cup through a series of concentric ironing rings that stretch the cup to three or more times its original height. By this process, the side wall is reduced in thickness by 30–40% of the original thickness. The can body is trimmed to height and degreased. Exterior lithography is carried out over 360° of body surface using high-speed printing presses. One or two interior coatings are applied by spray application. The top of the body is necked-in and a conventional dome seamed on.

Aluminum Containers[4,11,13,23–26]

In 1954, an aluminum aerosol container with a curled 1-in. opening and a capacity of $2^{1}/_{4}$ fl. oz became available from the Peerless Tube Company. Two years later, another container with a 20-mm opening and a capacity of 16 cc was produced.[4] Regardless of this early start, aluminum containers still constitute less than 5% of the aerosol market at the present time. The primary reason for this low penetration is cost, which is significantly higher than that of steel containers. Aluminum containers are used when package appearance and container strength rather than cost are the main considerations. The feminine deodorant spray is an example of a product in which these two factors were predominant. Since it was a feminine cosmetic product, container appearance was particularly important. Also, most of the sprays were formulated with fluorocarbon 12 and had fairly high pressures. This required a container which could meet DOT 2Q specifications. Aluminum containers qualified in both respects.

Aluminum containers have other advantages. Aluminum is much lighter than steel. As a result, the product is easier to handle and shipping weights are reduced. Bursting strengths are high, particularly with one-piece containers. The latter meet both the DOT 2P and the DOT 2Q specifications. Aluminum containers are quite corrosion resistant because of the continuous film of aluminum oxide which forms in the presence of oxygen and water vapor. In tinplate containers, evacuation of air is often desirable to minimize corrosion. This may have the opposite effect in aluminum containers because of the necessity of maintaining the oxide coating. In addition, with one-piece containers there are no side seams where corrosion can occur and no bottom or top seams. The one-piece container with the drawn-in neck allows considerable latitude in the basic shape and permits continuous interior coatings. Aluminum containers can be supplied with enamel, vinyl phenolic, or epoxy-type coatings, if desired.

One disadvantage of aluminum containers is that they are nonmagnetic. Therefore, filling lines must be equipped with spring-loaded pucks to hold the cans in position.

One-piece, extruded, aluminum containers are available in sizes up to 35 oz. The Peerless Tube Company lists 21 different sizes. The three smallest, with diameters from $7/8$ to 1 in. and heights from $2^9/_{32}$ to $3^{13}/_{16}$ in., have a 20-mm opening. The overflow capacities range from 0.71 fl. oz (21 cc) to 1.55 fl. oz (46 cc). The remaining containers have the standard 1-in. curl opening. The largest container in this series is a 2.089 in. × 6.5 in. can with an overflow capacity of 11 fl. oz (325 cc). Sizes and capacities of aluminum containers available from American Can are listed in Table 5-3. The Impact Container Corp. lists seamless aluminum containers with the following nominal capacities: 12 fl. oz (360 cc), 16 fl. oz (475 cc), 18 fl. oz (527 cc), 21 fl. oz (630 cc), 32 fl. oz (930 cc), and 35 fl. oz (1,046 cc). These containers are equipped with the standard 1-in. curled

TABLE 5-3   ALUMINUM AEROSOL CONTAINER CAPACITIES
(AMERICAN CAN COMPANY)

Size Description	Fluid Ounces		Cu. Inch Overflow[a]	Cu. Centimeter Overflow[a]
	Nominal	Overflow[a]		
Miniature—One-Piece				
010 × 208	1/4	.35	.63	10.3
010 × 400	1/2	.57	1.03	16.9
014 × 304	3/4	.78	1.42	23.2
014 × 400	1	1.02	1.84	30.1
Travel Size—One-Piece				
106 × 208	1	1.80	3.20	53.0
106 × 313	2	2.80	5.00	83.0
108 × 403	3	3.60	6.50	107.0
108 × 505	4	4.70	8.40	138.0
MiraSpra—Two-Piece				
202 × 405	6	7.80	14.00	230.0

[a]The above values represent average capacities.

opening. The Cliff Manufacturing Company also has available a series of high-pressure aluminum containers with capacities ranging from 16 to 100 fl. oz.

Manufacture

The one-piece, seamless container is made by impact extrusion from aluminum slugs. The two-piece units are produced by drawing or extruding the shell and double seaming the shell to the bottom or top.

Glass Containers[13,27-29]

Glass as a packaging material is unsurpassed in its ability to resist corrosion, leakage, and permeation. Glass aerosol containers have an excellent safety record spanning more than 24 years. One of the first glass bottle products to appear on the market was "Larvex," an aqueous-based mothproofer formulated with Freon® 114. This product appeared in 1952 in a 16-oz uncoated bottle.[30] The first aerosol cologne in a glass bottle was Jasmin, marketed in 1953 by Zonite Products Corp.[31] The possibility of breakage and the potential hazard involved with products packaged under pressure in glass bottles resulted in a search for a suitable means for protecting glass bottles. In 1952, the Bristol-Myers Company announced the details of a successful method of coating aerosol bottles with plastic.[32] The Wheaton Glass Company acquired the rights to this process and made plastic-coated bottles available to the industry in 1953. The first product packaged in a plastic-coated bottle was a cologne by Parfums Corday, which appeared in 1954.[33]

The introduction of glass aerosol containers had a considerable impact upon the aerosol industry. Metal containers were not particularly attractive for packaging products such as colognes and perfumes. Glass bottles could be supplied in a variety of colors, sizes, and shapes. The aesthetic appeal of glass bottles stimulated the development of a whole series of personal products.

There was another important effect. Aerosol laboratories now had low-cost test tubes available. Prior to this time, visual observations of aerosol formulations had to be carried out in glass compatibility tubes.[34] These were expensive and most aerosol laboratories had few in stock. Formulation work was limited, since each tube had to be cleaned and reassembled after use. With the advent of glass bottles, innumerable formulations could be prepared inexpensively, observed over a period of time, and discarded after use. The availability of glass bottles for formulation work in the laboratory undoubtedly hastened the introduction of many new aerosol products.

The Foster-Forbes Glass Company, Wheaton Plasti-Cote Company, and Carr-Lowery are basically the sole producers of uncoated glass bottles. Foster-Forbes has the major share of the market.[28] Stock bottles they offer range in size from $5/16$ oz to and including 4 oz. Wheaton Plasti-Cote and Owens-Illinois produce coated bottles. Wheaton Plasti-Cote has the largest share of the coated-bottle market. Wheaton Plasti-Cote supplies bottles ranging from about $1/4$ to

10 oz (overflow capacities are 10–355 cc). The bottles are available in a variety of shapes and colors.

Pressure Limitations

There are no formal industry pressure limitations for glass aerosol containers. Those that do exist are self-imposed by loaders, marketers, and glass manufacturers.

In general, pressures in uncoated bottles are not allowed to exceed 18 psig at 70°F. Those in coated glass bottles usually do not exceed 25 psig at 70°F. These pressures include that contributed by air.

Quality Control Tests

Primary tests performed by glass companies are inspection for visual defects, impact resistance, pressure retention, vertical load, thermal shock, and glass distribution. *Impact testing* is used mostly to test out new container designs in the developmental stage. The impact tester is a weighted pendulum device available from American Glass Research. *Internal pressure retention* is a routine test applied to samples during any production run. Two devices are used: a minimum specification pneumatic tester which applies air pressure to the bottles for about 16 sec at 150 psi and a hydrostatic increment tester which provides 29 progressively higher test pressures to destruction. The *vertical load test* determines resistance to compressive or crushing forces. The minimum acceptance level is 500 psi. To determine *glass distribution*, samples are cut into four vertical sections with a silico-carbide saw or a diamond cutting wheel. Glass thickness at critical areas is measured. If wall thickness falls below a minimum value at these critical points, the mold is removed and bottles produced subsequent to the last acceptable test are rejected. Wheaton requires a minimum wall thickness of about 0.050 in. for coated bottles and 0.080 in. for uncoated bottles. Greater wall thicknesses may be required for unusual shapes.

As far as drop tests on uncoated bottles are concerned, Budzilek feels that they prove only one thing—glass will break if sufficient impact is delivered.[27] Drop tests enjoy a popularity far in excess of any significant data that can be obtained. Probably their greatest value is to determine design acceptability during developmental stages. There are two types of tests: One determines the mean break height and the other uses a fixed height and determines the incidence of breakage from that height to a steel plate.

Further details on the various tests carried out by the glass companies are available in References 12 and 28.

The Plastic Coating

The plastic coating has a number of functions. It retains glass fragments in the event of breakage, preserves glass surfaces from weathering, and protects glass surfaces from abrasion in case of bottle-to-bottle contact during loading or storage. It also increases impact resistance. Impacting forces four or five times greater than those required to break an uncoated bottle are needed to break a plastic-coated bottle.

The coating itself is a plasticized, high-molecular-weight polyvinyl chloride resin. It is applied by a hot-dip process. This is followed by a heat cycle to fuse the resin and develop tensile strength, tear resistance, abrasion resistance, and impact absorbing properties. The normal coating on an aerosol bottle varies from about 0.035 to 0.055 in., depending upon the size of the bottle and surface area to be covered.

There are three major coating systems in use.

Nonbonded Vented Coatings. These coatings require a bottle with a groove directly under the bottle lip. When plastic is applied, it fills the groove, forming a thick ring. The combination of the ring of plastic and the valve ferrule holds the plastic coating and glass together. Vent holes are added to the plastic to relieve pressure if the bottle breaks.

Bonded Coatings. An adhesive primer is applied to the glass before the plastic coating is applied. The coating adheres firmly to the glass after fusing.

Laminated Coatings. The advantage of the laminated coating (trade name "LamiSol") is that it resists higher pressures than either coating alone. Two coatings with different physical properties are applied to the bottle and fused together. Lamination provides greater tear strength.

Laminated coatings usually are not bonded to the glass. The bottle is manufactured with a groove directly under the bottle lip. This holds the bottle and plastic together. Vent holes are added to the coating to relieve pressure in the event of breakage.

Plastic Containers[13,33,35,36]

Plastic containers have been considered for aerosol products for many years.[33,37,38] Some of the resins tested were melamine resins, phenolic resins, nylon, acetal copolymers, polypropylene, polyethylene, polystyrene, polycarbonates, and polyacrylonitrile barrier resins.[39] Until recently, however, plastic containers had not been successful. They had a number of disadvantages. One was their high permeability. Permeation may occur in either direction. Moisture may escape from aqueous solutions or be absorbed from the atmosphere by anhydrous solvents such as ethyl alcohol. Propellant vapors may escape or solvents may diffuse through the plastic. Oxygen may permeate from the outside and attack oxygen-sensitive components such as perfumes. Some plastics had odor problems, while in others, migration of plasticizer from the plastic caused unacceptable changes in the product or discoloration on the outside of the container. The high cost of plastic containers was also a limiting factor. The majority of these containers have disappeared from the market. In the containers now available, most of these disadvantages have been overcome. Plastic containers now seem ready to gain a foothold in the aerosol container market.

Present Status

At the present time, two plastics appear to be satisfactory for a number of aerosol products. One is Celcon acetal copolymer resin, manufactured by the Celanese Plastics Company. Long-term performance data on this plastic are available and a number of products are packaged in containers made from Celcon. The other is a relative newcomer, General Electric's Valox, a thermoplastic polybutylene terephthalate resin. Celanese also has a thermoplastic polybutylene terephthalate resin, "Celanex." It will not be promoted for aerosol containers until additional long-term performance data are available and a market need has been established.[40]

General Properties

The strength of plastic containers permits a freedom in design and style not possible with glass containers. Plastic containers with a wall thickness of 0.04 in. can withstand a pressure of 700 psig without bursting. They have excellent creep and impact resistance. They are lightweight and resistant to corrosion and breakage. Disposal problems of used plastic containers are far less than those of other types. When plastic containers are incinerated, carbon dioxide and water are formed with very little ash. The explosion hazard with containers made of Celcon is considerably reduced. Celcon melts before explosive pressures build up.[39] Internal pressures are released when the container melts.

Containers Made of Celcon

In the early 1960's, Celcon had an odor which was traced to the amine stabilizers present.[43] The development of a new stabilizer system eliminated the odor.[44] According to Maczko, moisture permeation remains a minor problem.[40] In many cases, reformulations of the product to allow for loss of moisture during storage will minimize moisture permeation problems.

Aerosol products in containers molded from Celcon include "Air of Beauty 2000" (Dearling Associates), "Aqua Net" hair spray and "Body by Faberge," "Intimate" and "Moondrops" cologne spray mists (Revlon), and an anesthetic for operating rooms (Astra Chemical Company).[41] These are all packaged in 1-oz containers with the exception of "Air of Beauty 2000," which is in a 5-oz container. The Carry-On Division of Prosol Corp. markets a hair spray, skin lotion, breath spray, and feminine deodorant spray to the beauty salon and barbershop trade. They also market a "Take Along" line which includes an antiseptic, insect repellent, athlete's foot spray, first aid spray, and cologne line.[42]

Aerosol containers made from Celcon may be obtained from Charter Supply Corp. (5 oz), Cypro, Inc. ($\frac{1}{2}$ and 1 oz), Imco Container Company (2, 3, and 6 oz), and Wheaton Plasti-Cote (3 oz). Celanese has tooling available for containers with capacities of 2.5-6 fl. oz. Plastic containers are normally supplied with a 20-mm opening. Standard glass bottle valves with an aluminum ferrule can be crimped on using available equipment. Valves with a plastic skirt, referred to as ultrasonic valves, are also available. They are attached to the container by ultrasonic welding after the concentrate has been added. Suppliers of the valves include Precision Valve Corp., Seaquist Valve Company, and Dispenser Products Division of Ethyl Corp. (formerly VCA/ARC).

Prosol Corp. will supply a complete aerosol ready for marketing. They do not market containers alone. The Prosol unit consists of an all-plastic valve assembly that is snap-fitted to the container. The container, molded without a bottom, is filled in the inverted position with an undercap filler. The bottom is attached and welded ultrasonically.[42]

Valox Containers

American Can is now offering a 3-oz stock "wing" design for immediate use. They have the capability of manufacturing various custom shapes. The container can be decorated by hot-stamping or other methods of decoration. The containers are manufactured by extrusion blow molding.

According to Silverman, Platt, and Spencer,[45] Valox containers are equivalent to glass in their ability to safeguard fragrances. They are practically impermeable to water, ethanol, and fluorocarbon and hydrocarbon propellants. Valox containers are equivalent to acetal in creep resistance, do not stress-crack under a variety of test conditions, and are unsurpassed in drop impact strength.

Valox containers have been under evaluation by selected marketers since late 1973. Thus far, Valox containers appear to be satisfactory for packaging colognes, perfumes, antiperspirants, shaving creams, disinfectants, and drugs. Among the materials which are incompatible with Valox are chlorinated solvents such as methylene chloride, strong oxidizing agents, and concentrates with a very low or very high pH. Oven cleaners and depilatories should be avoided.

Manufacture

The two general methods for manufacturing plastic containers are extrusion and injection blow molding. They differ primarily in the preparation of the tube of molten resin (the "parison") from which the molded container is formed. At present, the technology of the extrusion process is the most advanced. It is the preferred and most widely used method. Injection blow molding is becoming increasingly popular, however. It is reported to orient the molecules of plastic, which results in stronger containers.

The extrusion blow molding process produces a parison which is slightly smaller than the diameter of the final container. The container mold closes over the hollow tube of plastic and pinches off the end. Air under pressure forces the tube to assume the shape of the mold. There are a variety of arrangements for making the parison, bringing the mold into position around the parison, and completing the blow molding and final cooling. In one type of system, parisons are extruded continuously while a mold comes into position, closes around the parison, and then moves out of the way while another parison is being extruded.

In the injection blow molding process, the parison is injection molded onto a core pin and then transferred while still molten to a separate cavity. Here it is blown to the desired shape. Details on the extrusion and injection blow molding processes may be obtained from Reference 46.

Barrier Containers

Barrier containers have a specialized construction designed to keep the product completely separated from the propellant at all times from the initial filling to complete discharge. Two different types of containers are used for this purpose. One is a bag-in-can container and the other uses a plastic piston. The bag-in-can system consists primarily of a metal container equipped with an inner plastic bag fastened at the 1-in. opening at the top of the container. The bag contains the product. The propellant, which is usually loaded through the bottom of the container, remains outside the bag. The propellant provides the pressure that forces the product out of the container when the valve is opened. The bag must be flexible enough to fold up as the product is discharged. Containers are usually supplied with the plastic bags already in place. Piston containers are constructed with a free-moving plastic piston. The product is loaded into the space above the piston. The propellant is loaded through a valve in the bottom of the container and remains below the piston. When the valve is opened, the piston forces the product out of the container.

The main advantages of barrier packages are as follows:

1. Products can be packaged without alteration of their composition. Reformulation of the product is not necessary.
2. Very viscous products can be packaged. Products with high viscosity usually cannot be formulated as aerosols using conventional systems.
3. The ratio of product to propellant is high, a consumer advantage.
4. Sterility, required for foods and certain pharmaceuticals, is easier to achieve with barrier containers than with conventional aerosol systems.
5. The package can be discharged in any position.

The disadvantage of barrier packages is higher cost. Barrier packages are not considered to be competitive with conventional aerosols, since the types of products packaged in the barrier systems cannot be formulated as typical aerosols.

A number of barrier containers have been introduced. Only two—Continental Can's "Sepro" can and American Can's "Miraflo" piston container—have achieved any significant commercial success.

Bag-in-Can Systems

The "Sepro" Can (Continental Can Company).[47,48]   The "Sepro" can is illustrated in Figure 5-2. The unit consists essentially of a three-piece can with a flexible plastic bag fastened to the dome, which takes the usual 1-in. valve cup. Originally, the bottom end of the can was perforated for a solid rubber charging grommet for introduction of propellant. This was changed a few years ago to a one-piece valve which allows the propellant to flow around the valve. After the propellant is charged, the valve is seated with a flange.

The bag is made either of polyethylene or Conaloy, a blend of nylon and polyolefins. The "Sepro" bag has an accordian pleated side wall to insure conrolled bag collapse during product discharge. This prevents the bag side wall from collapsing inward and pinching off product before the bag is emptied. Product–propellant compatibility is the major consideration in selection of material for the bag. All propellants and products have some degree of permeation. If permeation is excessive, product characteristics may be altered or cor-

Figure 5-2 The "Sepro" can. (*Courtesy of the Continental Can Company*)

rosion can occur. Conaloy has the best all-around properties and the best resistance to permeation by propellant. Polyethylene has better resistance to moisture permeation. It is easier to process than Conaloy.

Several methods can be used to fill the bag with product. Entrapment of air during filling must be avoided, since air can cause popping or sputter during discharge. Conventional bottom filling (filling from the bottom up with a retracting filling spout) using a piston-type positive-displacement filler is reported to be adequate for most products. If necessary, spin fillers, which rotate the cans while the product is being filled, can be used. Spin filling is particularly effective in minimizing air entrapment. Single- or four-head spin fillers are available from the Elgin Manufacturing Company.

Both types of propellants are used with the "Sepro" can—liquefied gases (fluorocarbons and hydrocarbons) and compressed gases (nitrogen and nitrous oxide). Propellant selection is determined by the type of product, permeation characteristics, pressure required, and economics. The only commercial fluorocarbon with FDA approval for use with foods is Freon® 115. Hydrocarbons are less expensive than fluorocarbons, but are flammable and may impart an off-flavor to food products. Food grade hydrocarbons should be specified. A major advantage of liquefied gases is that the pressure remains relatively constant throughout discharge. With compressed gases, the pressure drops continuously as the product is discharged. Consequently, the discharge rate decreases during use. Pressures of about 100 psig should be employed with compressed gases. The propellant chamber should constitute about 40–50% of the total package volume to assure at least 95% product discharge. The new Continental charging valve permits charging rates up to 30 cc/sec with minimum propellant loss.

Specific products that have been commercially packaged in the "Sepro" can include shaving cream gels, cream hair removers, greases, barbeque sprays, ceramic tile caulking compounds, contraceptive creme-gels, shampoos, body lotions, and toothpastes. A large variety of other products have been successfully test packaged. Fillers capable of handling commercial packs with the "Sepro" can include Diamond Chemical, Barr Company, Peterson/Puritan, Aerosol Services, and Aerosol Techniques.

Presspack System (Cebal, France).[49]   The Presspack was developed by Cebal, a Paris-based aluminum container manufacturer. It is similar to the "Sepro" can in that the flexible inner polyethylene bag contains the product and the propellant is injected through a hole in the base of the can. The hole is then sealed with a rubber bung. The main difference is in the design of the polyethylene bag. The inner flexible plastic bag is pleated longitudinally instead of horizontally. When the product is discharged, the Presspack bag collapses inward on itself. After product discharge is completed, the bag touches the can wall in four places at the top and bottom to minimize rattling. The system is designed for use with nitrogen or air as the propellant.

"Aeratron" Barrier Pack System (Dynatron Company).[50,51] The "Aeratron" is
a relatively new development. The collapsible liner, made of polyethylene, is
constructed with longitudinal pleats and fits inside conventional containers. The
"Aeratron" design differs from the "Sepro" and Presspack systems in that propellant is charged through the top of the container. The container does not have
to be spun during filling. Filling is reported to be relatively simple. The polyethylene bag is filled with product through a tube. The valve is placed on the
container which then moves to an under-cup gasser. This lifts the valve and
draws a vacuum. Since the inner liner is filled with product, the vacuum is drawn
from the area between the liner and the container walls. Propellant, either vapor
or liquid, is loaded into the free space. The valve is then crimped. Liquefied or
compressed gases can be used. The "Aeratron" is offered for licensing in two
sizes. The first is designed to fit the 202 X 509 (8 oz) container and has a capacity of about 5.5 fl. oz. It is reported to work well with compressed gases. The
second is under development and will also be used with the 202 X 509 container.
However, it will occupy 80–90% of the inside volume of the container.

Piston Containers

"MiraFlo" Piston Container (American Can Company)

The aluminum "MiraFlo" piston container is illustrated in Figure 5-3. According
to Boyne,[52] it is used for packaging hand lotions, hair treatments, cheese spreads,
cake icings, and caulking compounds. The unit consists of a free-moving piston
in a coated extruded aluminum container. The piston is constructed of a plastic
such as polyethylene and is essentially a hollow cylinder with the upper end
closed and the bottom end open. The upper end is fairly rigid, but the sides are
flexible in order to maintain a seal with the wall. The principle behind the piston
container is that the product itself forms the seal between the piston and wall.

The can is shipped completely assembled. The product is filled through the
1-in. opening in the piston can. In order to minimize entrapped air, addition of
as much product as possible is recommended. Ideally, only enough space for the
valve cup should be left.

The "MiraFlo" unit is pressurized through a center hole in the bottom of the
container with a special gassing and plugging unit called the "energizer." This
equipment is available as a single-can gassing unit or a multihead machine. After
the propellant has been loaded into the container by the "energizer," a rubber
plug is pushed into the opening while the container is still under pressure.
According to Hoffman and Marchak,[53] the resistance of the plug to blow-out
exceeds the buckling strength of the bottom of the container. Compressed gases
are normally used for propellants at a pressure of about 100 psig. Nitrogen is
preferred because of low product solubility. Liquefied gases can also be employed, provided the plastic used for the piston is impervious to the liquefied
gases.

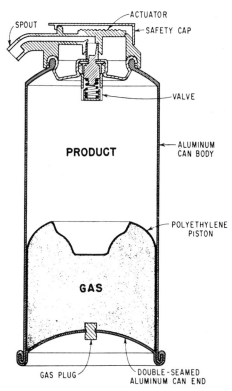

Figure 5-3 The aluminum "MiraFlo" piston container. (*Courtesy of the American Can Company*)

Since the piston can depends upon the product to form the seal between the piston and the container wall, the product must be quite viscous. Otherwise, the propellant will leak past the seal or the product will seep down into the propellant chamber. Even if the viscosity of the product is satisfactory at room temperature, higher temperatures may decrease the viscosity sufficiently so that the seal is broken. In order to overcome these disadvantages, smaller pistons with either three or four piston rings have been proposed.[54,55] It is reported that this type of construction permits products with low viscosity to be packaged without the danger of the propellant by-passing the seal. In addition, the rings maintain an effective seal even if the container is dented. An improved gassing and sealing device called the "Prestoplug closure" is reported to allow both compressed and liquefied gases to be loaded.

The American Can Company also has under development a piston container which utilizes a three-piece, Soudronic welded, side-seam container. Although there are no commercial applications as yet, the development is considered to be promising.[56]

THE DEPARTMENT OF TRANSPORTATION (DOT) REGULATIONS
FOR AEROSOLS

The Federal Government has been concerned with the safe transportation of hazardous materials by common carrier for many years. Regulations in effect for hazardous materials cover the packaging, marking, labeling, shipping paper preparation (bill of lading), filling, and handling of these materials. The regulations are published in Title 49 Code of Federal Regulations (CFR), parts 100-199, revised December 31, 1976. They are also reprinted in Agent R. M. Graziano's Tariff No. 31, "Hazardous Materials Regulations of the Department of Transportation by Air, Rail, Highway, and Water," issued March 1, 1977, effective March 21, 1977. A copy of Tariff No. 31 may be obtained by writing to R. M. Graziano, Agent, 1920 "L" St., N.W., Washington, D.C. 20036.

The DOT regulations pertaining primarily to the pressures of aerosol products are summarized in this chapter because the pressure in an aerosol determines the type of container that may be used. The regulations themselves are quite extensive. The following summary is not intended to be a substitute for legal counsel or advice from the DOT if there is any difficulty in application or interpretation of the regulations. Law firms especially competent in this field are also available.

Prior to 1966, the Interstate Commerce Commission (ICC) had jurisdiction over the shipment of compressed gases. Rules governing interstate shipment and the definition of a compressed gas had been developed with the aid of the Bureau of Explosives. At that time, a compressed gas was defined in Agent W. S. Topping's Freight Tariff No. 4 as "any material with a gas pressure exceeding 25 psig at 70°F, or any liquid flammable material with a Reid vapor pressure exceeding 40 psia at 100°F." Under these rules, any aerosol with a pressure over 25 psig at 70°F had to be shipped in a Specification ICC 4B 300 cylinder. Investigations in the middle 1940's had shown that pressures in the 35-40 psig range were necessary to produce satisfactory sprays for aerosol insecticides. Therefore, the ICC regulations needed to be modified to allow pressures up to 40 psig in the lightweight, disposable containers developed by Continental Can and Crown Cork & Seal. The regulations finally were amended on July 28, 1947 to permit pressures as high as 40 psig at 70°F in nonrefillable containers, providing the containers were filled with nontoxic and nonflammable materials and that each filled container was heated to 130°F without evidence of leakage, distortion, or other defect. During this same period, the refrigeration supply industry expressed a desire for a single-charge, lightweight container for refrigerants. A further exemption to the ICC regulations which permitted packaging refrigerant gases with pressures up to 70 psig in specification 2P containers was obtained. The revised regulations also allowed aerosol products with pressures up to and including 60 psig at 70°F to be packaged in specification 2P containers.

On October 15, 1966, a law passed by Congress established a new Department of Transportation (DOT). The DOT then assumed responsibility for safety laws relating to railroads, motor carriers, etc., that previously had been the function of the ICC. This law became effective April 1, 1967.

Summary

The first step with any aerosol product is to determine if it is classified by the DOT definition as a hazardous material. If it is, then it is regulated by the DOT. A nonhazardous material is not regulated. Whether a product is hazardous may be determined by consulting Title 49 CFR, parts 100-199, or Graziano's tariff. In any case, the paragraph numbers and words will be the same.

Paragraph 172.101, toward the front of Graziano's Tariff No. 31, lists hazardous materials. The most specific listing for any product must be found. If the product is a fire extinguisher or a liquefied petroleum gas, it must be considered in that category. Under the heading of "Aerosol Products," the reader is referred to "Compressed gases, n.o.s." The term n.o.s. means "not otherwise specified" (as a hazardous material). Compressed gases, n.o.s., is listed twice, both as a nonflammable compressed gas and as a flammable compressed gas. To the right, in the column headed "Exceptions," there is a paragraph reference, 173.306, to the compressed gases. This must be read, because it outlines the conditions under which a product can be shipped in low-pressure containers if it is a compressed gas.

The main reason most aerosols are regulated by the DOT is because they meet the definition of a compressed gas. The next step, therefore, is to determine if the aerosol product is classified as a compressed gas. The DOT definition is found in Paragraph 173.300(a), Tariff No. 31, and is as follows:

> "The term 'compressed gas' shall designate any material or mixture having in the container an absolute pressure exceeding 40 psi at 70°F, or regardless of the pressure at 70°F, having an absolute pressure exceeding 104 psi at 130°F, or any liquid flammable material having a vapor pressure exceeding 40 psi absolute at 100°F, as determined by ASTM Test D-323."

If the aerosol pressure does not exceed the limitations at either of the first two temperatures, it is not classified as a compressed gas and is not regulated by the DOT (unless it is classified as a flammable liquid or any of the other hazard classes).

Compressed gases normally are shipped in cylinders, tank cars, cargo tanks, or portable tank containers (see paragraphs 173.301-173.305, 173.314, and 173.315). If there were no exceptions to the regulations, aerosol products would have to be packaged in cylinders etc. However, exceptions are provided for aerosol products. Those meeting the required conditions are excepted from labeling (except when offered for transportation by air) and specification packaging, unless this is required as a condition of the exception. Before discussing the exceptions, it might be well to define the latter terms. Specification packaging means using a can or a carton meeting specific construction requirements as designated in the regulations with an identifying number. Labeling means placing the appropriate label, such as a red flammable or green nonflammable gas label, on the shipping carton. Unless there is an exception, it is necessary to use specification containers and label them with a hazard sticker when shipping a hazardous material. The exceptions for compressed gases concerned with pressure and size limitations are listed below.

"Paragraph 173.306—Limited Quantities of Compressed Gases.

(a) Limited quantities of compressed gases for which exceptions are permitted as noted by reference to this section in Paragraph 172.101 of this subchapter are excepted from labeling (except when offered by transportation by air) and unless required as a condition of the exception, specification packaging requirements of this subchapter when packed in accordance with the following paragraphs. In addition, shipments are not subject to Subpart F of Part 172 of this subchapter, to Part 174 of this subchapter except Paragraph 24, and to Part 177 of this subchapter except Paragraph 177.817.

(1) When in containers of not more than 4 fluid ounces water capacity (7.22 cubic inches or less) except cigarette lighters. Special exceptions for shipment of certain compressed gases in the ORM-D class are provided in Subpart N of this part.

(3) When in a metal container charged with a solution of materials and compressed gas or gases which is nonpoisonous provided all of the following conditions are met. Special exceptions for shipment of aerosols in the ORM-D class are provided in Subpart N of this part.

  (i) Capacity must not exceed 50 cubic inches (27.7 fluid ounces).

  (ii) Pressure in the container must not exceed 180 psig at 130°F. If the pressure exceeds 140 psig at 130°F but does not exceed 160 psig at 130°F, a specification DOT 2P (Paragraph 178.33 of this subchapter) inside metal container must be used. If the pressure exceeds 160 psig at 130°F, a specification DOT 2Q (Paragraph 178.33a of this subchapter) inside metal container must be used. In any event, the metal container must be capable of withstanding without bursting a pressure of one and one-half times the equilibrium pressure of the content at 130°F.

  (iii) Liquid content of the material and gas must not completely fill the container at 130°F.

  (iv) The container must be packed in strong, outside packaging.

  (v) Each completed container filled for shipment must have been heated until the pressure in the container is equivalent to the equilibrium pressure of the content at 130°F (55°C) without evidence of leakage, distortion, or other defect.

(vi) Each outside packaging must be marked 'INSIDE CONTAINERS COMPLY WITH PRESCRIBED REGULATIONS'."

Exception No. 1 includes all types of containers—metal, glass, and plastic. If the *overflow* capacity of the container does not exceed 4 fl. oz, the product is exempt from restrictions. If the capacity of a glass or plastic container exceeds 4 fl. oz., but the pressure is 25.3 psig or less at 70°F, the product is not subject to DOT regulations because it is not classified as a compressed gas. If, however, the capacity of the glass or plastic container exceeds 4 fl. oz and the pressure is greater than 25.3 psig at 70°F, then the product cannot be shipped interstate without a special exemption.

Exception No. 3 (ii) defines the types of metal containers that must be used. For products in which the pressure does not exceed 140 psig at 130°F, all of the standard metal containers may be used—side seam or drawn. The container must be capable of withstanding a pressure 1.5 times the equilibrium pressure at 130°F. The maximum would be 210 psig. If the pressure exceeds 140 psig at 130°F, but not at 160°F, specification DOT 2P containers are required. If the pressure exceeds 160 psig, but not 180 psig, at 130°F, then DOT 2Q containers must be used.

The DOT regulations with respect to container capacity are summarized below.

Paragraph 173.306(a)(1) and 173.306(a)(3)(i)

Container Capacity

0-4 fluid ounces (Paragraph 173.306(a)(1)

Exempt from specification packaging and labeling (except by air). The exemption from specification packaging indicates that there are no pressure restrictions on containers of 4 fluid ounces or less. The labeling exemptions refer to the outside of the cartons.

Greater than 4 fluid ounces—27.7 fluid ounces
(Paragraph 173.306(1)(3)(i)(ii))

Exempt from labeling. Only nonrefillable metal containers are permitted with these capacities. Glass or plastic containers with capacities greater than 4 fluid ounces with products defined as compressed gases require a special exemption. Pressures may not exceed 180 psig at 130°F. The pressure of the aerosol product dictates the type of container required, that is, standard, DOT 2P, or DOT 2Q.

The requirements for a container to be classified as 2P or 2Q are listed in Paragraphs 178.33 and 178.33a, respectively. These specifications include the type and size of the containers, material of construction, method of manufacture, wall thickness, tests, and marking. Some portions of the specifications are listed below.

Paragraph 178.33 Specification 2P

178.33-2 Type and Size (b)

The maximum capacity of containers in this class shall not exceed 50 cubic inches (27.7 fluid ounces). The maximum inside diameter shall not exceed 3 inches.

178.33-7 Wall Thickness (a)

The minimum wall thickness for any container shall be 0.007 inch.

178.33-8 Tests (a)

One out of each lot of 25,000 containers or less, successively produced each day, shall be pressure tested to destruction and must not burst below 240 pounds per square inch gauge pressure. The container tested shall be complete with end assembled.

178.33-9 Marking (a)

On each container by printing, lithographing, embossing, or stamping "DOT 2P" and manufacturer's name or symbol. If symbol is used, it must be registered with the Bureau of Explosives.

Paragraph 178.33a Specification 2Q

All of the specifications for DOT 2Q containers except the following are identical to those for DOT 2P containers:

178.33a-7 Wall Thickness (a)

The minimum wall thickness for any container shall be 0.008 inch.

178.33-8 Tests (a)

One out of each lot of 25,000 containers or less, successively produced per day, shall be pressure tested to destruction and must not burst below 270 pounds per square inch gauge pressure.

178.33a-9 Marking (a)

On each container by printing, lithographing, embossing, or stamping "DOT 2Q" and manufacturer's name or symbol. If symbol is used, it must be registered with the Bureau of Explosives.

The main differences between DOT 2P and DOT 2Q containers are the greater wall thickness and higher bursting pressure requirements for the 2Q containers.

Another point of interest concerning containers and the DOT regulations is given below:

1. Several companies market seamless aluminum containers with capacities greater than 27.7 fl. oz. These containers are classified as cylinders. They meet the requirements in DOT Specification 39 for nonreusable (nonrefillable) cylinders. (Paragraph 178.65.)

REFERENCES

1. L. D. Goodhue, *Aerosol Technicomment*, Vol. XII, No. 2.
2. L. D. Goodhue, *Aerosol Age* **15**, 16 (August 1970).

3. *Aerosol Age* **1**, 75 (December 1956).
4. R. A. Foresman, "The Metal Container," in H. R. Shepherd (Ed.), "Aerosols: Science and Technology," Interscience Publishers, Inc., New York, 1961.
5. C. W. Greenwood, U.S. Patent 2,506,449 (1950).
6. H. E. Peterson, Personal Communication, Peterson/Puritan Filling Corp.
7. J. S. McClelland, Personal Communication, Continental Can Company.
8. W. C. Beard, "Valves," in H. R. Shepherd (Ed.), "Aerosols: Science and Technology," Interscience Publishers, Inc., New York, 1961.
9. H. Obarski, "Steel Aerosol Containers," in J. J. Sciarra and L. Stoller (Eds.), "The Science and Technology of Aerosol Packaging," John Wiley & Sons, Inc., New York, 1974.
10. A. S. Glessner, Personal Communication, Crown Cork & Seal Company.
11. Technical Brochure, "Aerosol Containers, Materials and Capacities," American Can Company, November 1973.
12. H. Obarski, Personal Communication, Continental Can Company.
13. M. A. Johnsen, W. E. Dorland, and E. K. Dorland, "The Aerosol Handbook," Wayne E. Dorland Company, Caldwell, N.J., 1974.
14. L. M. Garton, Personal Communication, American Can Company.
15. G. Meister, Personal Communication, National Can Corp.
16. *Aerosol Report* **13**, 183 (1974).
17. J. Boatwright, Personal Communication, Southern Can Company.
18. Technical Brochure, Soudronic Ltd.
19. D. R. Terrian, Personal Communication, Continental Can Company.
20. R. B. Mesrobian, *Aerosol Age* **19**, 19 (July 1974).
21. D. M. House, Personal Communication, Apache Container Corp.
22. E. G. Maeder, *Aerosol Age* **19**, 12 (November 1974).
23. S. T. Craige, "Aluminum Containers," in J. J. Sciarra and L. Stoller (Eds.), "The Science and Technology of Aerosol Packaging," John Wiley & Sons, Inc., New York, 1974.
24. Technical Brochure, "A Glance at Peerless," Peerless Tube Company.
25. L. Messina, Personal Communication, American Can Company.
26. Technical Brochure, "Container Selection Guide 1067," Impact Container Corp.
27. E. Budzilek, Personal Communication, Wheaton Plasti-Cote Company.
28. T. R. Herst, Personal Communication, Foster-Forbes Glass Company.
29. T. R. Herst, "Glass Containers," in J. J. Sciarra and L. Stoller (Eds.), "The Science and Technology of Aerosol Packaging," John Wiley & Sons, Inc., New York, 1974.
30. *Modern Packaging* **26**, 90 (January 1953).
31. *Glass Packer* **21** (1953).
32. R. H. Thomas, *Proc. 38th Mid-Year Meet. Chem. Spec. Manuf. Assoc.* (June 1952).
33. R. H. Thomas, "Glass and Plastic Containers," in H. R. Shepherd (Ed.), "Aerosols: Science and Technology," Interscience Publishers, Inc., New York, 1961.
34. Freon® Aerosol Report FA-14, "Handling Freon® Fluorinated Hydrocarbon Compounds in the Laboratory."
35. M. J. Kakos, "Plastic Containers," in J. J. Sciarra and L. Stoller (Eds.), "The Science and Technology of Aerosol Packaging," John Wiley & Sons, Inc., New York, 1974.
36. *Aerosol Age* **18**, 15 (June 1973).
37. J. N. Owens, *Aerosol Age* **4**, 20, 53 (April/May 1959).
38. J. C. Pizzurro and R. Abplanalp, *Aerosol Age* **2**, 37 (June 1957).
39. J. Maczko, Oral Presentation to the Society of Plastics Engineers, Eastern Section, October 1974.
40. J. Maczko, Personal Communication, Celanese Plastics Company.
41. W. S. Pearson, Personal Communication, Cypro, Inc.
42. E. Sellers, Personal Communication, Pelorex Corp.

43. *Aerosol Age* 17 (June 1972).
44. *Modern Packaging* 45 (September 1972).
45. A. Silverman, J. R. Platt, and K. B. Spencer, *Aerosol Age* 20, 16 (March 1975).
46. Technical Bulletin, C3C/Blow Molding, Celcon Acetal Copolymer, Celanese Plastics Company, July 1973.
47. Technical Brochure, "Sepro" Technical Data, Continental Can Company.
48. L. F. Irland and J. W. Kinnavy, *Drug Cosmet. Ind.* 101, 42 (1967).
49. *Aerosol Age* 18, 38 (May 1973).
50. D. E. Casey, *Aerosol Age* 20, 30 (April 1975); 22, 27 (June 1977).
51. D. E. Casey, Personal Communication, Dynatron Company.
52. R. W. Boyne, *Detergent Age* 5, 151 (1968).
53. H. T. Hoffman and N. Marchak, *Modern Packaging* 34, 129 (1961).
54. *Aerosol Age* 13, 52 (1968).
55. T. C. Clark, *Aerosol Age* 11, 28 (1966).
56. L. M. Garton, Personal Communication, American Can Company.

6

VALVES
AND ACTUATORS

The first large-scale commercial aerosol valve was the oil burner nozzle used for World War II insecticides. During this period, however, work had already started on the development of other types of valves. A valve designed by Bridgeport Brass was tested on a Crown can in 1943 by Goodhue. During the same year, C. D. Thoms, Advanced Packaging Company, and Goodhue visited the Surgeon General of the Army to demonstrate a new valve. Unfortunately, the valve flew into pieces during the demonstration. Goodhue reported that Thoms was so embarrassed he walked out of the room without bothering to pick up the parts.[1] The Ansul Chemical Company and Armstrong Engineering Company also submitted lightweight containers with and without valves to Goodhue in 1943. Some of the containers were equipped with toggle valves. Sjokin and Peterson,[2] Continental Can Company, invented a reliable valve in 1945. It was supplied only in the end of a Continental can and could not be used for the containers with a 1-in. curled opening that became common several years later. The dependable all-purpose valve developed by Abplanalp that could be crimped into containers with a 1-in. curled opening ultimately became one of the most popular valves in the aerosol industry.[3]

AEROSOL VALVES

There is a tremendous variety of valves available. As a result, it is usually possible to obtain a valve which will be satisfactory for any aerosol product. Besides the common or general types of valves used for the majority of aerosol products, there are valves for special applications. These include metering valves for delivering specific quantities of product at each discharge, vapor tap valves, valves for use in either the upright or inverted position, tip-sealing valves, high-flow-rate valves, low-flow-rate valves for compressed gases, powder valves, and co-dispensing valves. Valves are available for metal, plastic, or glass containers.

The best source of detailed information on all the varieties of valves, actuators, and overcaps is the literature supplied by the various valve companies. A comprehensive survey of the different types of valves has been published by Johnsen.[4] Beard,[5,6] Harris and Platt,[7] and Gregg[8] have written excellent review articles on valves. Valve manufacturers include Clayton Corp., Emson Research, Inc., Newman-Green, Inc., Precision Valve Corp., Risdon Manufacturing Company, Seaquist Valve Company, Summit Packaging Systems, Inc. (formerly Aerosol Products Division of Scovill Manufacturing Company), and Dispenser Products Division of Ethyl Corp. (formerly VCA/ARC).

Common Spray Valves

The principles involved in the construction of valves for glass or plastic containers are essentially the same as those for metal containers. However, most glass and plastic containers have a 20-mm opening compared to the 1-in. curled opening in metal containers. Valves for glass bottles are provided with a metal ferrule which is crimped around the outside of the bottle lip. The valve stem gasket in the glass bottle valve rests on top of the bottle; therefore, the only metal that comes in contact with the aerosol formulation is the spring in the valve cup. Valves for metal containers are assembled with a mounting cup. This is crimped to the container by expanding the mounting cup inside the 1-in. opening in the container.

An exploded view of the major components of a typical valve for metal containers is shown in Figure 6-1. They consist of an actuator, valve stem or core, valve stem gasket, spring, housing or valve body, and dip tube. These components are assembled into a mounting cup. The construction at the bottom of the housing is called the tail piece. The dip tube is connected to the tail piece.

Types of Valve Construction

Rus[9] divides the major types of aerosol valves into four groups: (1) vertical action, "sliding" gasket, male; (2) vertical action, "deflecting" gasket, male; (3) tilt action; and (4) vertical action, "sliding" gasket, female. Although the spray valves differ somewhat in appearance and construction, the first three types operate basically in about the same way. The stem of the valve, which is open to the atmosphere, contains a small orifice (or orifices). In the closed position, the orifice is sealed off from the aerosol by an elastomeric gasket. This prevents the aerosol in the container from entering the stem. When the valve stem is pushed down or tilted to the side, the orifice is moved away from the gasket and down into the aerosol container. The aerosol product is then free to flow through the orifice and out into the atmosphere. The female valve has a slightly different construction, which will be described later. Details of the valves, given below, were supplied by Rus.[9]

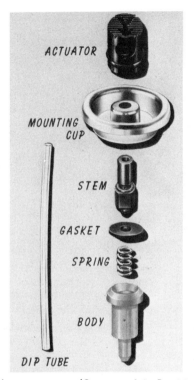

Figure 6-1   Valve components. (*Courtesy of the Precision Valve Corp.*)

Vertical Action, "Sliding" Gasket, Male. This group is representative of the largest number of valves, both by specific design and by total units sold. The Dispenser Products Division of Ethyl Corp. (formerly VCA/ARC) KN-38 and PARC-39, Seaquist NS-31, Risdon, and Precision NP valves are among those that use this basic design. In this type of valve, the valve stem, which has a uniform diameter, slides through the stem gasket, which remains stationary.

The operation can be seen in more detail in Figure 6-2, which illustrates a typical valve in both the open and closed positions. The stem is hollow from the top down to a point slightly below the inlet orifice and is sealed at the bottom. The valve housing is constructed so that a shoulder on the housing fits tightly against the valve stem gasket. The stem extends down into the spring as shown. When the valve is not actuated and is in the closed position, the spring inside the housing forces the shoulder on the stem up against the valve stem gasket. The stem contains the inlet orifice. This can be seen in Figure 6-2 in the middle portion of the stem. In the off position, the inlet orifice is closed by the valve

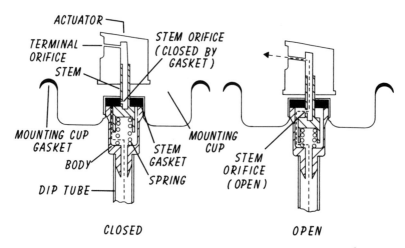

Figure 6-2 Vertical action, "sliding" gasket, male valve; from Rus.[9]

stem gasket. When the stem is depressed by pushing down on the actuator, the inlet orifice is moved down and away from the gasket so that the inlet orifice is open to the interior of the aerosol. The aerosol product can then leave the container. The pressure in the aerosol forces the product up the dip tube and into the valve housing. From here it enters the exposed inlet orifice and travels up the stem and out the orifice in the actuator.

Vertical Action, "Deflecting" Gasket, Male. In this valve, the stem is necked immediately before it passes through the gasket. This can be seen in Figures 6-1 and 6-3. The groove in the stem contains the inlet orifice. The gasket fits in and under the shoulder where the stem is necked. When the stem is depressed, the gasket is forced to "yaw" as it deflects, allowing material to pass between the gasket and stem and enter the orifice. In general, a shorter stroke is required to open and close the valve, since the stem orifice does not completely pass through the gasket before the product enters the orifice. The Precision Valve Corp. uses this design for many of its valves.

The operation of this type of valve in the open and closed positions is illustrated in Figure 6-3.

Tilt Action. Tilt action valves vary in their construction. In some cases, the stem orifices are sealed off from the aerosol in the closed position by an elastomeric gasket. Tilting the stem to one side breaks the seal and allows the aerosol to enter the stem.

Another type of construction is illustrated in Figure 6-4. In this case, the valve stem is molded to have a slight cup (not the valve cup) at the bottom of

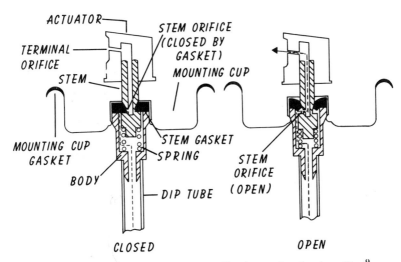

Figure 6-3 Vertical action, "deflecting" gasket, male valve; from Rus.[9]

the stem. The stem orifices (10 in this particular valve) are drilled vertically in a circle just inside the stem cup. In the closed position, the stem makes a seal with the valve gasket at two points: around the stem itself above the stem cup and on the rim of the stem cup. When the stem is tilted to one side, the seal between the stem cup rim and valve gasket is broken. This allows the aerosol to flow over the rim, down into the stem cup, through the orifices, and out the valve and actuator.

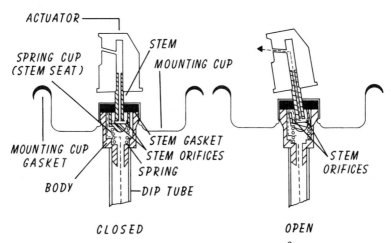

Figure 6-4 Tilt action valve; from Rus.[9]

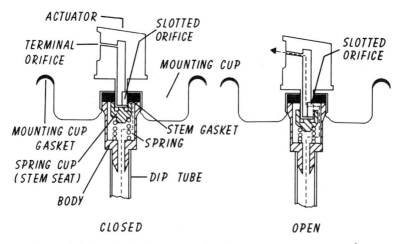

Figure 6-5   Vertical action, "sliding" gasket, female valve; from Rus.[9]

Vertical Action, "Sliding" Gasket, Female. The construction and operation of valves of this type in the open and closed positions is illustrated in Figure 6-5. Industry old-timers still call this the "paint valve" design, although it is used for a variety of other household, personal care, and industrial products. The unique feature of this design is that all valve orifices are incorporated in the stem, which is connected to the actuator instead of the valve. The stem can thus be removed for cleaning if clogging should occur.

In the closed position, the spring cup or stem seat forms a seal against the stem gasket. This closes off the aerosol from the slotted orifice in the stem. When pressure is applied to the actuator, the spring cup is depressed, breaking the seal between the cup and valve gasket. The aerosol then flows between the spring cup and gasket, into the slotted orifice in the stem, and out through the actuator.

Valves using this construction include the Dispenser Products Division of Ethyl Corp. (formerly VCA/ARC) AR-74 and the Newman-Green B-14, R-10, and R-70.

Valve Stems

Valve stems are usually constructed from nylon, acetal, or polyester resins. Metal stems are used for formulations that might soften plastic. Orifices are molded in plastic stems and drilled or pierced in metal stems.

The expansion chamber is formed by the hollow portion of the stem. Although the value of an expansion chamber in promoting small particle size in aerosol sprays was recognized as early as 1901, the first high-pressure insecticides developed during World War II were equipped with an oil-burner-type nozzle without an expansion chamber. Beard[5] has suggested that the reason for the absence of an expansion chamber is that cooling produced by vaporization of the high-pressure propellant precipitates the insecticide in an expansion chamber. Also, the nozzles on the insecticides were constructed to impart a swirling motion to the product. This caused some break-up of the spray by mechanical action.

When aerosol insecticides subsequently were packaged with low-pressure propellants, it was found that valves without expansion chambers did not produce fine enough sprays. Fulton and co-workers investigated the effect of variations in orifice, actuator, and expansion chamber dimensions upon particle size.[10] The best valve had an expansion chamber with a capacity of 0.008 cubic inches (cu. in.) and inlet and outlet orifice diameters of 0.015 in. and 0.021 in., respectively.   The most uniform sprays were obtained when the ratio of the inlet and outlet diameters was about $2:3$. The difference in pressure in the interior of the aerosol and expansion chamber during discharge was about 5 psi.

In present-day valves, the inlet orifice is normally the stem orifice and the outer orifice is in the actuator. The orifice in the stem usually determines the spray rate and is termed the metering orifice. Generally, only one orifice is present in the stem, but valves with stems containing a multiple number of orifices are available. Inlet orifices commonly range from about 0.012 to 0.035 in.,[14] but may reach almost $1/8$ in. in diameter.[8] In some valves, the stem is molded to the actuator (female-type valves). In this case, the actuator contains both orifices, since the end of the stem is slotted to produce a rectangular orifice.

Valve Stem Gaskets

An aerosol product can fail completely if the proper valve gasket is not used. Excessive weight loss during storage may result from leakage through the valve gasket. Staining or coloring of the formulation may occur from contact with the gasket. Odor characteristics are particularly important. Even if a gasket itself has no odor, it may affect the odor of the final product by selective absorption or reaction with some of the perfume components. The choice of the most suitable valve gasket for any product, therefore, can be critical. The valve companies are the best source of information and generally can recommend an appropriate gasket material. Ahrweiler[11] has written an excellent article describing the properties for valve gaskets. Flanner[12] determined the effect of various combinations of propellants, solvents, and gasketing materials upon valve leakage. Stanave, Ward, and Hutchins[13] investigated weight loss through gaskets of neoprene, Buna N, polyvinyl alcohol, and Teflon laminates using a high-solvency propellant system containing fluorocarbon 21. Factors affecting valve leakage are discussed in detail in Chapter 10, "Solvency."

Three types of elastomers are generally used for valve gaskets: Buna N with a Shore durometer hardness of 50 or 70, neoprene with a hardness of 70, and butyl rubber. The first two are the most common, but butyl may become more important in the future because of its resistance to chlorinated solvents. Buna N is the best for aromatic hydrocarbons[11] but has very poor resistance to chlorinated solvents. Neoprene has the best all-around resistance to solvents in general, but is inferior to butyl for chlorinated solvents.[42]

When the valve is in the closed position, the gasket is in contact with the vapor phase of the aerosol. In the open position, the liquid phase of the aerosol passes between the valve gasket and stem. The gasket must withstand both liquid and vapor phase contact without excessive permeation, swelling, distortion, or shrinkage. Poor sealing of the stem orifice by the gasket and permeation through the gasket are the two main causes of excessive valve leakage. Preliminary indications of the suitability of a gasket for an aerosol formulation can be obtained by a combination of a 48-h immersion test[15] and permeation studies with propellant alone (discussed in Chapter 10, "Solvency").

Immersion studies allow three effects to be evaluated: swelling or shrinkage, hardening or softening, and general deterioration. Swelling or shrinking can be determined by measuring the outside diameter of the gasket before and after immersion. Gasket hardening may be measured using a Shore durometer gage. Hardening indicates that stem leakage during intermittent spray may be a problem. Softening indicates that disintegration may occur at a later date. General deterioration after immersion tests is determined by rubbing the gasket across a sheet of white paper. Flaking off of pieces of black gasket on the paper indicates significant deterioration. Deterioration can occur without excessive swell, shrinkage, hardening, or softening. Excessive permeation through the gasket material suggests that leakage from the vapor phase will probably exceed allowable limits.

Neither shrinkage nor too much swelling is desirable. Shrinkage is sometimes accompanied by hardening resulting from extraction of plasticizer in the gasket. This can cause seeping or squirting of material around the stem when the valve is actuated. Too much swelling results in problems such as clogging of the actuator by particles from the gasket or failure of the valve to close completely. According to Green,[15] gasket swelling in the range of 5–15% offers optimum valve performance if the gasket has not changed significantly in hardness. If an aerosol product causes more than 15% swell, a gasket with an oversized inside diameter may be beneficial.

The results of short-term tests should be interpreted with care. Final choice of a gasket will depend upon room temperature and accelerated shelf tests, coupled with intermittent spray tests. Particular attention must be paid to leakage rates.

Low toxicity of the valve gasket is a necessity. During 1973, a problem arose with the neoprene used for mounting cup and stem gaskets. These gaskets were cured with NA-22, 2-mercaptoimidazoline, an accelerator which is suspected of being a carcinogen. On November 30, 1973, the FDA banned the use of NA-22 in neoprene for food container closures. On May 2, 1974, the FDA published in the Federal Register (pp. 15306-07) a notice of a proposed regulation banning the use of NA-22 in neoprene products for closures and product delivery systems for drugs and cosmetics. The standard 30-day comment period would have expired on June 3, 1974. However, an extension of 70 days to August 12, 1974 was granted. Both the Cosmetic, Toiletry, and Fragrance Association (CTFA) and Chemical Specialties Manufacturers Association (CSMA) have essentially requested a reasonable time to determine if there is any degree of human risk resulting from inhalation of or dermal exposure to NA-22 or any of its by-products when using aerosol products. At this time (March 1977), the FDA has not yet taken any action.

The potential problem does not exist with Buna N gaskets, which are not accelerated with NA-22. Meanwhile, alternate neoprene gaskets that do not contain NA-22 have been developed by gasket manufacturers. These are being extensively evaluated by fillers, valve manufacturers, and marketers. The American Gasket and Rubber Company, for example, has developed several new neoprene compounds—AGR-2369 and AGR-2371—that have been accepted by some of the valve companies.[16] Neoprene gaskets with NA-22 are still being manufactured for certain product areas not under the jurisdiction of the FDA.

The major suppliers of valve gaskets are the American Gasket and Rubber Company, Avon, Reeves Bros., Rubber Products, and Vernay.

Mounting Cups

Mounting cups for valves with the 1-in. curled opening are usually made from 1-lb tinplate. They may be obtained either unlined or coated with epoxy, vinyl, or phenolic resins. The sealing or flowed-in gasket in the mounting cup is added by applying liquid neoprene compound to the side and in the lip of the rotating cup. The liquid neoprene is dried and vulcanized in place. Cut gaskets are available for special applications.

Until the recent NA-22 controversy, the most widely used flowed-in gasket material was Dewey and Almy's neoprene GK-45, accelerated with NA-22. Dewey and Almy now supply three NA-22-free neoprene compounds: GK-45 NV, GK-45 NVH (same as GK-45 NV except for slightly higher solids and viscosity, and GK-70.[17] Recently, Gribens[18] reported the development of a new series of mounting cup sealing compounds based upon polyurethane technology. The specific compound that Dewey and Almy are promoting is "Dara-thane 100." This is a 100% total solids formulation consisting primarily of polyurethane prepolymers, inorganic fillers, and a curing system. It is claimed to be equal to the GK-45 family of compounds in performance and to be environmentally acceptable.

Chemical Products' "Chem-O-Seal" R-8835 also does not contain NA-22 and is used in limited quantities for mounting cups.[9]

Valve Springs

Valve springs are usually made of 302 or 316 stainless steel.

Housing or Valve Body

The housing or valve body serves to hold the spring and to provide a connection for the dip tube. The valve body may be made from metal or plastics such as nylon or acetal resin. The lower end of the valve body or tail piece usually has an inside diameter of about 0.060–0.080 in. Tail pieces with restrictions are available. These range in size from about 0.0135 to 0.040 in.[14] When the diameter of the tail piece is smaller than the orifice in the stem, the tail piece becomes the main metering orifice. The purpose of this is to obtain a further reduction in spray rate or add another expansion chamber. Clogging may be a problem with a small-diameter orifice in the tail piece because there is no wiping action of the valve components to clean the orifice. Some tail pieces have filters molded to the bottom to minimize clogging.

Dip Tubes

Dip tubes are normally constructed of polyethylene, but are available in nylon or other plastics. Dip tubing is available with inside diameters from 0.014 to almost 0.5 in. Oversize dip tubes with diameters of 0.25 in. or larger are referred to as jumbo dip tubes. They provide somewhat of a reservoir of material in case the product is used in the inverted position.

The behavior of dip tubes in aerosol formulations must be checked. If the aerosol causes stress cracking or the dip tube does not fit the tail piece properly, the dip tube may fall off during aging or pressure filling. Some formulations cause excessive swelling and growth of dip tubes. The swelling and growth of dip tubes may be determined by immersing a premeasured length of the dip tubing in the aerosol for 48–72 h. The linear elongation per inch is calculated after the aged dip tube is remeasured. The exact length of dip tube required can be obtained by cutting off the bottom of the container to be used and measuring the distance from the corner of the bottom to the valve tail piece. Knowing the linear elongation per inch caused by the aerosol, it is possible to determine the initial length of dip tube that should be used. This reduces the possibility of seal-off by the dip tube.[15]

Dip tubes are usually attached to the outside of the tail piece. However, capillary dip tubes are sometimes used, particularly for aqueous-based products. These may be inserted inside the tail piece. A detailed discussion of dip tubing, covering historical development, manufacture, materials of construction, and problems, has been published by Seckel.[19]

Identification of Common Valves

Quite often it is desirable to determine the manufacturer of an aerosol valve. Identification sometimes is not easy because of the variety of valves available. There are many valve assemblies that appear virtually the same at first glance. However, by observing a few outward physical characteristics and making some simple measurements with an inexpensive vernier caliper, it is usually possible to identify a valve. These operations involve measuring the diameter of the pedestal in the mounting cup and the stem while noting the type of crimp and the characteristics of the actuators.

The information necessary to identify valves manufactured by Dispenser Products Division of Ethyl Corp. (formerly VCA/ARC), Clayton Corp., Newman-Green, Precision, Risdon, Summit, and Seaquist is listed in Tables 6-1, 6-2, and 6-3. Table 6-1 lists the characteristics of vertical action, male valves; Table 6-2, lists the characteristics of tilt action, male valves; and Table 6-3 lists those of vertical action, female valves (female valves are those in which the valve stem is an integral part of the actuator). The information in Tables 6-1, 6-2, and 6-3 was supplied by the various valve companies.[9,20-25]

Special Valves

Metering Valves

Metering valves are designed to deliver specific quantities of a product each time the valve is actuated. Metered quantities are necessary for many pharmaceutical products where the dosage of the drug must be controlled. Metering valves are also useful for products such as the more expensive perfumes where it is desirable to limit the quantity of material sprayed in order to avoid waste. Metering valves that deliver quantities ranging from a fraction of 1 cc up to 15 cc with each depression of the valve are available.

Metering valves are generally used with glass bottles or small metal containers. The construction of one type of metering valve for glass bottles is illustrated in Figure 6-6. The valve body or housing is the metering chamber, and its size determines the quantity of product discharged. During discharge, the metering chamber is sealed off from the rest of the product in the container so that only the material in the metering chamber is released. When the product is not being used, the metering chamber is closed to the atmosphere and open to the interior of the container.

As shown in Figure 6-6, the valve stem is constructed with a shoulder, below which is a spring. An elastomeric gasket is located at the bottom of the spring and rests on the opening from the metering chamber to the tail piece. The inlet orifice in the valve stem is located in the upper portion of the stem. When the valve is in the off position, the orifice is above the valve stem gasket and outside the valve cup. The lower end of the valve stem is flattened and sealed at the

TABLE 6-1 VALVE IDENTIFICATION—VERTICAL ACTION, MALE

Valve	Mounting Cup		Stem Diameter (in.)	Actuator Characteristics	Comments
	Pedestal Diameter (in.)	Type of Crimp			
Dispenser Products Division of Ethyl Corp (formerly VCA/ARC)					Stem extensions are:
KN-38	0.442	All four ARC (now the Dispenser Products Division of Ethyl Corp.) valves are crimped with eight dimples, generally hemispherical in shape	0.130	The design on top of actuators for the KN-38, PARC-39, and KN-44 valves is: △	0.308 in.
PARC-39	0.442		0.130		0.334 in.
KN-44	0.442		0.158		0.307 in.
KN-37	0.442		0.125 (top) 0.130 (bottom)	See comments	0.308 in. The KN-37 is normally used with combination actuator-cap (dust cover) and not with standard size actuators
Summit	0.475 (largest of any valve)	Four-point	0.130	Letter "S" molded either directly below or above	Specialty actuators do not have letter "S"

Precision	0.400	Eight-point	0.158	molded arrow on standard actuators Some valves have stem outside diameter of 0.158 in. which are necked down to 0.125 in. about 1.16 in. from top
Seaquist NS-31/41	0.412	Four-stake square clinch	0.125 with a lock ring	Black and uncolored arrows are distinctive: Seaquist is the only valve with a lock ring on the stem
NS-32	0.412	Four-stake square clinch	0.158 with a lock ring	Actuators are distinctive. See Seaquist's catalog
Risdon 460/480	0.460	Valve assembly is staked to mounting cup with a complete ring, i.e., multifinger stakes	0.125	See Risdon's catalog
510 Powder	0.460	Four-point stake clinch	0.125 metal stem	Uses a larger than normal dip tube

TABLE 6-2 VALVE IDENTIFICATION–TILT ACTION, MALE

| Valve | Mounting Cup | | Stem Diameter (in.) | Actuator Characteristics | Comments |
	Pedestal Diameter (in.)	Type of Crimp			
Dispenser Products Division of Ethyl Corp. (formerly VCA/ARC)					
T-56	0.442	Eight dimples, generally hemispherical in shape and equally spaced	0.120		Design on top of actuators for T-56 and K-38 is:
K-38	0.407	No crimp—press fit assembly	0.121		
Seaquist					
ST-70	0.426	Four-stake square clinch	All stems contain a 0.125-in. stepped stem shank:		

Summit	0.475	Four-point	0.153	Letter "S" molded either directly below or directly above molded arrow on actuator	Specialty actuators do not have molded "S"
Clayton	0.435	Four-collett crimp	0.145		There are two flats on the valve stem directly above the pedestal top. This is the only valve with a vertical metering orifice. This can be seen by looking down the stem:

Front   Side

TABLE 6-3 VALVE IDENTIFICATION—VERTICAL ACTION, FEMALE

Valve	Mounting Cup		Actuator Characteristics	Comments
	Pedestal Diameter (in.)	Type of Crimp		
Newman–Green	0.39	Eight-point crimp. More of rectangular type than square	Design on top of actuator is: $\vee$	If valves are separate, the Newman–Green valve can be distinguished from the AR-74 by noting the dip tube. The Newman–Green valve has a swedged-in dip tube; the AR-74 is over the tail piece
Dispenser Products Division of Ethyl Corp. (formerly VCA/ARC)				
AR-74	0.382	The ARC (now the Dispenser Products Division of Ethyl Corp.) crimp consists essentially of four squares with two points in each square		
LA-10	0.416	No crimp—press fit		The LA-10 has a threaded pedestal. This valve is seldom used with its actuator—the DO-10. Generally used with special fitments which have their own metal actuating pin

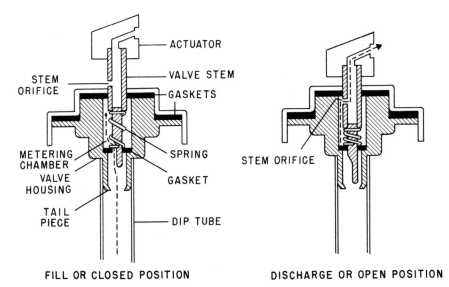

STEM ORIFICE

METERING CHAMBER
VALVE HOUSING
TAIL PIECE

ACTUATOR
VALVE STEM
GASKETS
SPRING
GASKET
DIP TUBE

STEM ORIFICE

FILL OR CLOSED POSITION   DISCHARGE OR OPEN POSITION

Figure 6-6 Metering valve.

bottom. In the off position, the flattened portion of the valve stem extends into the elastomeric gasket at the bottom of the metering chamber. Because the stem is flattened at this point, the gasket does not make a complete seal around the valve stem. The aerosol product fills the metering chamber through the opening between the gasket and the flattened portion of the valve stem.

When the valve is depressed, the rounded portion of the valve stem, located above the flattened section, ultimately passes through the gasket at the bottom of the metering chamber. When this occurs, the metering chamber is sealed off from the rest of the product in the aerosol container because the bottom gasket makes a complete seal around the rounded part of the valve stem. This takes place while the inner orifice is still located above the valve stem gasket in the upper part of the valve. Upon further depression of the valve stem, the part of the stem containing the inlet orifice is pushed down past the valve stem gasket and into the metering chamber. The product in the metering chamber is then discharged through the inlet orifice, into the hollow portion of the valve stem, and out through the outer orifice in the actuator. When the valve is released, the portion of the stem containing the inlet orifice rises through the valve stem gasket and closes off the metering chamber to the outside. This occurs before the lower flattened portion of the valve stem reaches the lower gasket at the bottom of the metering chamber. After the flattened stem section reaches the lower gasket, the product in the container fills the empty metering chamber through the opening between the gasket and the valve stem in preparation for the next discharge.

Vapor Tap Valves

Vapor tap valves have holes of varying sizes in the valve body. When the valves are actuated, propellant vapor passes through the hole in the valve body and mixes with the product passing up the dip tube. This produces a softer spray. It has also been reported that the sprays are less chilling than those from valves without vapor taps. A typical vapor tap can be observed in the valve body in Figure 6-1.

The diameters of the holes in the vapor tap and the tail piece can be adjusted so that aerosol products can be sprayed either upright or inverted. When the container is inverted, the product passes through the hole in the valve body and the propellant vapor passes through the end of the dip tube.

A disadvantage of the vapor tap valves is that loss of propellant through the vapor tap reduces the concentration of propellant in the aerosol. In products formulated with low concentrations of propellants, this may cause a noticeable change in spray characteristics during product use. In products that contain combinations of propellants, the composition of the propellants in the vapor phase is usually different from that in the liquid phase. Loss of the propellant from the vapor phase may change the composition of the remaining propellant sufficiently so that a change in spray characteristics occurs during use.

Valves for Both Upright and Inverted Discharge

Several valves are designed to spray in either the upright or inverted position. Two of these—Seaquist's "SA" valve and Risdon's 360° valve—use metal ball-check valve systems. The Seaquist "SA" valve is shown in Figure 6-7. The metal ball, which is in the vapor area, closes off the vapor tap when the valve is upright.

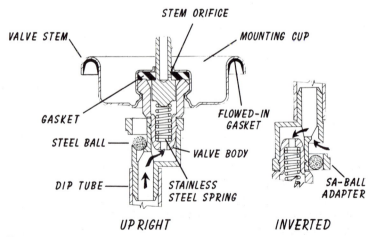

Figure 6-7   The Seaquist "SA" valve. (*Courtesy of the Seaquist Valve Company*)

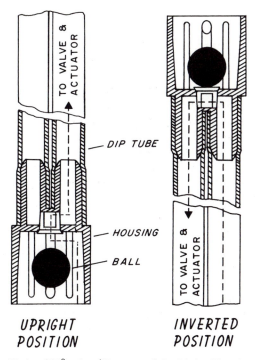

UPRIGHT
POSITION

INVERTED
POSITION

Figure 6-8 The Risdon 360° valve. (*Courtesy of the Risdon Manufacturing Company*)

When inverted, the vapor tap is open to the aerosol. The product can proceed directly through the vapor tap, into the valve housing, and out through the stem.

Risdon's 360° valve, shown in Figure 6-8, has two dip tubes, one of which is connected to the valve housing. The metal ball is in the liquid phase. In the normal or upright position, the aerosol flows unhindered up the dip tube and out through the valve as shown. In the inverted position, the metal ball closes off the bottom entrance as shown. The aerosol then follows the path indicated in Figure 6-8.

Some valves have a jumbo, or macro, dip tube.[14] The larger inner diameter of the jumbo dip tube provides a reservoir for the product which allows an inverted aerosol to spray for 4–8 sec. Spraying after the contents of the dip tube have been eliminated can cause excessive propellant loss.

Another type is the balanced orifice valve. In this system, a vapor orifice helps create a fine, warm spray in the upright position and acts as a body restriction in the inverted position. Without such a restrictive orifice, excessive propellant loss would occur during inverted spraying.[14]

Tip-Sealing Valves

Tip-sealing valves have been developed for use with viscous products in barrier packages such as the "MiraFlo" or "Sepro" can. The valves are particularly useful for food or pharmaceutical products where contamination must be avoided. Because these valves provide a sanitary seal at the tip, they may open an entirely new product area. There are no passageways or pockets where residual product can dry up, clog, or become contaminated.

The Beard Universal Seal-Tip valve[26] is illustrated in both the open and closed positions in Figure 6-9. The Beard valve is presently used on a sterile pharmaceutical lubricant packaged in the "Sepro" can.[27] Beard, Inc. and Bespak Industries, Ltd., have concluded a license agreement for the manufacture and sale of this valve throughout Europe.

Compressed Gas Valves

There has been considerable activity in the field of compressed gas aerosols during the past several years. Part of the reason for this is that three valve companies—Dispenser Products Division of Ethyl Corp. (formerly VCA/ARC), Precision, and Seaquist— have developed special valves for compressed gas products. These valves have two advantages—they produce aerosols with better spray characteristics and lower spray rates than had been possible with compressed gas aerosols in the past. They also permit higher filling speeds. Lower delivery rates are achieved by reducing the flow channels in the actuator or

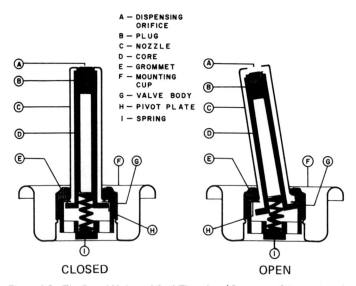

A — DISPENSING ORIFICE
B — PLUG
C — NOZZLE
D — CORE
E — GROMMET
F — MOUNTING CUP
G — VALVE BODY
H — PIVOT PLATE
I — SPRING

CLOSED   OPEN

Figure 6-9   The Beard Universal Seal-Tip valve. (*Courtesy of Aerosol Age*)

button, stem, valve body, or dip tube. A very small vapor tap can also be added to the valve body.

Precision's main efforts in reducing spray rate have been directed toward developing microcapillary dip tubes with diameters as small as .013 in. Production facilities are also available for valve stems with orifice diameters as small as .010–.013 in. Orifices with diameters as low as .008 in. have also been produced, but are too prone to clogging.

Precision's "hex" gasket valve for compressed gases was developed to aid in filling with compressed gases. The valve accomplishes two functions desirable for fast gas absorption: (1) The pressure in the container is raised almost instantly to a pressure where absorption is rapid and (2) it permits the proper ratio between gas flow around the valve stem directly into the head space and that through the dip tube. New one- and two-piece mechanical break-up actuators are now available with orifices as small as .008–.012 in. The one-piece actuator has been modified to a .013-in. orifice. These valve–actuator combinations produce a fine particle size with alcohol- or aqueous-based products.

Seaquist's developments have concentrated on microorifices and "Ultramist" actuators. In addition to their GP3 Ultramist actuator systems and valves with small orifices, Seaquist has developed efficient filter systems. One of these, a nylon mesh material attached to the valve body orifice, is illustrated in Figure 6-10.

The Dispenser Products Division of Ethyl Corp. (formerly VCA/ARC) has developed valves with relatively small stem orifices. The molded body orifice can be the usual round orifice, generally .013 in. in diameter, or a "filterite'. orifice, introduced around 1970. The filterite orifice is rectangular rather than round. The Dispenser Products Division of Ethyl Corp. (formerly VCA/ARC) found that particulate matter tends to clog on one side of the rectangle and often would drop down the dip tube after spraying. It is reported to be more effective than a single hole of the same relative size or a multitude of smaller holes. Jumbo or macro dip tubes can be used with the filterite system. They fit around the outside of the valve body.

More efficient mechanical break-up actuators which impart greater radial velocity to the spray as it goes through a tiny swirl chamber within the actuator have been developed. The Dispenser Products Division of Ethyl Corp. (formerly VCA/ARC) Super Marc actuator series has nine insert combinations with orifice diameters ranging from .008 to .031 in.

Foam Valves

Foam valves differ principally from spray valves in that foam valves have a large-diameter delivery spout. The delivery spout serves as an expansion chamber. The end of the delivery spout is the outer orifice. The inner orifice or orifices are usually large. Clayton Corp.'s "Nozzle-Down" standard foam valve

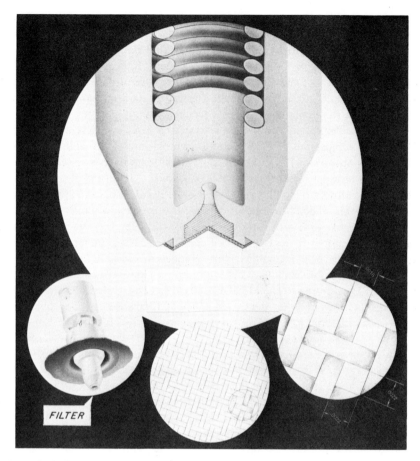

FILTER

Figure 6-10 Compressed gas valve with a nylon mesh filter. (*Courtesy of the Seaquist Valve Company*)

does not have a dip tube and is used in the inverted position only. It is available with two or four .030-in. stem orifices or four .060-in. stem orifices. The valve is illustrated in Figure 6-11. Most major valve manufacturers supply valves that can be used either for sprays or foams, depending upon the actuator. When a spray actuator is used for a foam aerosol, the product usually discharges as a foamy stream. With a foam actuator, expansion takes place in the delivery spout and the product discharges as a foam.

Co-Dispensing Valves

There has been considerable interest for many years in the possibilities for new aerosol products offered by co-dispensing valves.[29-36] Co-dispensing techniques allow two materials which react with each other to be packaged in the same

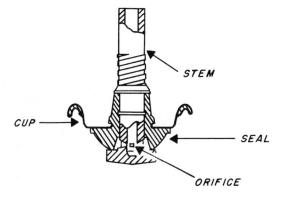

Figure 6-11   "Nozzle-Down" foam valve. (*Courtesy of the Clayton Corp.*)

container. This is achieved by using two-compartment (bag-in-can) containers. When the two components are discharged through a co-dispensing valve, they come into contact and react.

The first commercial co-dispensed product was Gillette's "Nine Flags Thermal Shaving Cream," marketed in 1967.[29] This was followed by Gillette's "The Hot One," another shaving lather.[33] Subsequently, the Toni Company, a division of Gillette at that time, marketed two co-dispensed hair dyes—"Magic Moment" and "Look of Nature." These products were equipped with a co-dispensing valve developed and manufactured by the Gillette Company. This valve is illustrated in Figure 6-12.

A number of other co-dispensing valves were developed during this period. Clayton Corp.'s "Clay-Twin" was designed for use in the inverted position.[32] It is constructed with three 0.30-in. orifices in the body and five 0.025-in. orifices in the base for discharge of material in the bag. A gasket seals off the holes in the base of the valve stem when the product is not in use. This prevents the components in the bag and the main portion of the container from mixing and reacting prematurely.[30] Products marketed with the "Clay-Twin" valve included Carter Products' shaving cream, "Rise-Hot," a hair conditioner, "Heats On," by Helene Curtis, and three Max Factor Geminesse aerosols, a thermal facial cleanser, moisturizer, and beauty oil.[34]

A co-dispensing valve developed by Valve Corp. of America (now the Dispenser Products Division of Ethyl Corp.) could be modified for use in either the inverted or upright position.[32] The ratio of the two reactants can be changed by varying the diameter of the dip tube in the upright model and the diameter of the tube chamber plug in the inverted model. The valve contains two intake ports which operate with a single actuation. Metering is controlled by a notch in the valve stem. Products using this valve included a deicer, "Hot-Melt," by Aeroseal, and an antiperspirant by Mitchum Thayer.[34]

Dart Industries also manufactured a co-dispensing valve, available with either

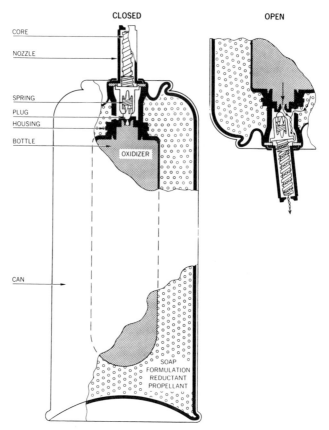

CLOSED   OPEN

CORE

NOZZLE

SPRING

PLUG

HOUSING

BOTTLE

OXIDIZER

CAN

SOAP
FORMULATION
REDUCTANT
PROPELLANT

Figure 6-12 Construction of "The Hot One." (*Courtesy of the Gillette Company*)

a 1-in. or 1.5 in. mounting cap. Rexall Drug's shaving cream and "Quiet Touch," a hair dye marketed by Clairol, Inc., used the Dart valve.[34]

Unfortunately, with one exception, the co-dispensed products failed to live up to initial expectations. At the present time, the only co-dispensed product still on the market is Gillette's shaving lather, "The Hot One." This product has sold steadily since its introduction and has maintained a significant portion of the shaving cream market. One of the main reasons other products failed was unsatisfactory operation of the co-dispensing valves. In co-dispensing, it is essential to maintain the same fixed ratio of the two components at all times during discharge. Most co-dispensing valves had the problem of "throttling." The ratio of the two components varied during discharge, depending upon the extent to which the valve was actuated. Product characteristics were not constant during discharge because it was difficult to maintain a constant pressure on the actuator.

The Dart valve was modified several years ago to reduce the throttling effect. The Precision Valve Corp. has the license to manufacture the valve. In 1972, Peterson/Puritan became the exclusive licensee of Dart Industries for packaging co-dispensed aerosol products with the Dart valve.[34] The Clayton and Valve Corp. of America (now the Dispenser Products Division of Ethyl Corp.) co-dispensing valves are still commercially available, but are not used on any product at present.

Powder Valves

Beard[37] has given an excellent summary of the requirements for a powder valve. The path from the valve seat to the spray orifice should be very short. This minimizes the space in which product can accumulate to collect moisture or fall back to cause clogging. The valve should have a sharp shut-off and a high valve seating pressure to minimize leakage. Beard suggests that the valve be operated wide open without any attempt to control spray rate by varying the pressure on the actuator. Otherwise, product will build up on the valve seat. Available powder valves are discussed in Chapter 21, "Aerosol Suspensions."

Miscellaneous Valves

A three-position spray valve, Variable V-8 Model Series, is available from Newman-Green. By rotating the actuator to any one of three positions, the spray rate may be varied from low to medium to high. This valve is presently used on Gillette's "Dry Look" hair spray.

Recently, Risdon has developed a new "High Flow" valve. This valve, used on Glamorene's "Drain Power," has a delivery rate of approximately 40 g/sec, compared to the conventional flow rate of 1–1.5 g/sec. The actuator, made of high-density plastic, is gripped by a large spring. This is slipped onto a plastic collar that snap-fits into the mounting cup of the valve.[38]

The Sprayon "Danvern" valve is a unique valve designed particularly for coating compositions. It was invented and first produced in 1951. The valve is available in this country only on aerosols custom-filled by Sprayon. Two models are available. One produces the conventional conical spray pattern and the other a fan-shaped spray. When the sprayhead is depressed, the actuating pin is pushed down, thus depressing the valve plug and creating an open position. This allows the aerosol to go up the stem, around the actuating pin, and through the metering orifice. There is sufficient clearance around the stem to allow the liquid to exit through the terminal orifice.

The "Danvern" valve is illustrated in Figure 6-13.

The Aquasol Dispenser System—Precision Valve Corp. The Aquasol dispenser system is designed primarily for use as a three-phase aerosol with water as the solvent and hydrocarbon propellants. The propellants must have a density less

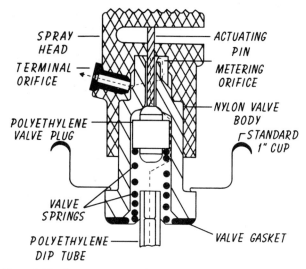

Figure 6-13 "Danvern" valve. (*Courtesy of Sprayon Products, Inc.*)

than that of water so that they will float on top of the aqueous phase. When the valve is actuated, product and propellant vapor discharge separately through twin ducts into the valve system. The ducts, or channels, converge in a swirl chamber where product and propellant are blended with violent force and emerge as a fine particle spray.

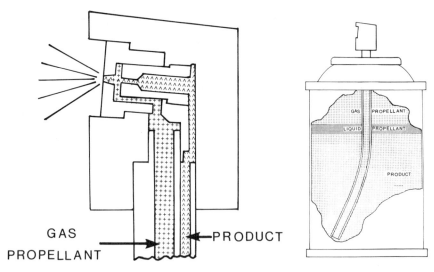

Figure 6-14 The Aquasol* dispenser system—basic principles. (*Courtesy of the Precision Valve Corp.*)

*Trademark owned by the Precision Valve Corp., Yonkers, New York.

The basic principles of the Aquasol system are illustrated in Figure 6-14, which shows the three-phase aerosol and the twin ducts for propellant and aqueous phase (product). The particular ducts that carry the aqueous phase and propellant depend upon the position of the container.[43] In the upright position the duct connected to the dip tube carries the product and the other duct provides passage for the propellant. When the container is inverted, with the dip tube protruding into the vapor phase, the system is reversed.

One of the main advantages claimed for the Aquasol system in comparison with conventional aerosol systems using fluorocarbon propellants is the lower product cost. The latter results from use of hydrocarbon propellants, lower propellant concentrations, and water as the solvent. The Aquasol system can be filled with standard equipment.

ACTUATORS OR BUTTONS

Actuators are available in a tremendous variety of sizes and shapes. Actuators which spray in either a horizontal or vertical position or at various angles may be obtained. The actuator is a major factor in determining whether the product will be dispensed as a foam, spray, or stream. The actuator contains the terminal orifice. These generally range from .012 to .060 in., but the most common have orifices of .016–.020 in. Actuators with orifices having a somewhat larger diameter than that of the metering orifice in the valve stem are often chosen to obtain a more uniform particle size in the spray. The shape of the orifices affects the spray. Those with a standard taper produce moderate-width sprays, while reverse taper actuators provide wider sprays. These two actuators are illustrated in Figure 6-15.

Many aerosol products in which coarse sprays are desired, for example, window cleaners or furniture polishes, are formulated either with low concentrations of liquefied gas propellants or with compressed gases. In these products, the concentration of propellant usually is not sufficient to break up the concentrate into a spray when a standard actuator is used. The result is a stream. In

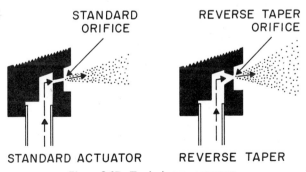

Figure 6-15 Typical spray actuators.

order to obtain a spray from these products, it is necessary to break up the stream by mechanical action. Most valve companies manufacture actuators that function by breaking up the stream mechanically. Although these actuators usually have specific trade names, they are generally referred to as mechanical break-up actuators. They function by imparting a swirling motion to the discharge so that it leaves the orifice in the form of a conical sheet. The film breaks up into droplets a short distance from the actuator. The swirling motion of the liquid stream is obtained by a tangential arrangement of the feed channels leading to the terminal or outlet orifice. A typical mechanical break-up actuator is shown in Figure 6-16.

Actuators are constructed either with one or two pieces. The two-piece actuators have a small disk or plug inserted into the face of the actuator. The terminal orifice is in the disk or plug. The chamber behind the insert may be constructed to impart a swirl to the discharge, thus providing mechanical break-up features, or it may have a straight-through design.

In products that contain sufficient liquefied gas propellant to give a fairly good spray with a standard actuator, a mechanical break-up actuator will generally produce a somewhat finer spray with a wider angle. Atomization of liquids by mechanical break-up actuators has been discussed in detail by Shay and Falkowski.[39]

Some actuators contain both inlet and outlet orifices. These actuators are manufactured with a stem-like tube. The tube fits into the valve and has a slit at the bottom which serves as the internal orifice. These valves are particularly useful for products such as aerosol paints because both the inlet and outlet orifices can be cleaned by removing the actuator. The orifices in the stem may be round or rectangular. They vary in diameter from about 0.0135 to 0.055 in.

In recent years, actuator caps or spray-through overcaps have become increasingly popular. Johnsen[34] estimates that in 1971 about 22% of the aerosol cans

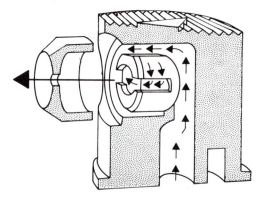

Figure 6-16 Mechanical break-up actuator.

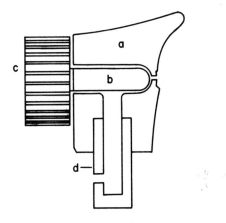

ACTUATOR BODY (a)

ACTUATOR PLUNGER OR PISTON (b)

ACTUATOR DIAL (c)

VALVE STEM (d)

Figure 6-17 "VariSeal" Aerosol Actuator. (*Courtesy of the Essex Chemical Corp.*)

were equipped with spray-through overcaps. Actuator caps are available in a wide variety of styles from most of the major valve companies. The caps have a number of advantages. Package appearance is improved, and the product is easier to operate. The construction of these actuators generally makes it simpler for the user to determine the direction of the spray. Some actuators have grooves for the finger which minimize fatigue during extended spraying. Actuator caps which fit around the valve button are available. Others take the place of the button.

A new actuator, the Essex "VariSeal" Aerosol Actuator, has been developed by the Essex Chemical Corp.[40] It can be used with liquefied and compressed gas systems as well as barrier packs. The actuator combines two features: a manually adjustable dial for gradually varying the discharge from a fine mist to a stream to a completely off position, and an automatic tip-sealing mechanism. The latter minimizes the possibility of clogging due to crystallization or film formation resulting from evaporation of liquid remaining in the actuator after discharge. The actuator is illustrated in Figure 6-17. The piston probe is held against the inner orifice wall by means of a spring and closes off the terminal orifice when the actuator is not in use. When the actuator is depressed, the internal pressure of the aerosol forces the piston probe back from the orifice and allows the product to escape into the atmosphere.

CHILD-RESISTANT CLOSURES

The Federal Poison Prevention Packaging Act of 1970 essentially requires that all aerosol products containing 2% or more of free or chemically unneutralized sodium and/or potassium hydroxide carry child-resistant closures. A number of aerosol oven cleaners contain over 2 wt.% sodium hydroxide and therefore come under these provisions of the Poison Prevention Packaging Act. Johnsen[34] estimates that oven cleaners to which the law applies comprise about 1% of the aerosol industry. Many industry observers believe that the government will eventually require child-resistant closures for aerosols such as insecticides, furniture polishes, medicinals, and other products that contain such solvents as kerosene and turpentine. This could amount to 5-15% of the aerosol volume.

There are three types of child-resistant closures available: the overcap, actuator overcap, and compound actuator button. A list of suppliers of these closures is available in Reference 41.

REFERENCES

1. L. D. Goodhue, *Aerosol Age* **15**, 16 (August 1970).
2. H. Peterson, Personal Communication, Peterson/Puritan Filling Corp.
3. R. Abplanalp, U.S. Patent 2,631,814 (1953).
4. M. A. Johnsen, W. E. Dorland, and E. K. Dorland, "The Aerosol Handbook," Chap. 6, Wayne E. Dorland Company, Caldwell, N.J., 1974.
5. W. C. Beard, "Valves," in H. R. Shepherd (Ed.), "Aerosols: Science and Technology," Interscience Publishers, Inc., New York, 1961.
6. W. C. Beard, *Aerosol Age* **11**, 35, 40 (April/May 1966).
7. R. C. Harris and N. E. Platt, "Valves," in A. Herzka (Ed.), "International Encyclopaedia of Pressurized Packaging" (Aerosols), Pergamon Press, Inc., New York, 1966.
8. W. A. Gregg, "Aerosol Valves," in J. J. Sciarra and L. Stoller (Eds.), "The Science and Technology of Aerosol Packaging," John Wiley & Sons, Inc., New York, 1974.
9. R. R. Rus, Personal Communication, Dispenser Products Division of Ethyl Corp. (formerly VCA/ARC).
10. R. A. Fulton, A. Yeomans, and E. Rogers, *Proc. 36th Mid-Year Meet. Chem. Spec. Manuf. Assoc.*, 51 (June 1950).
11. J. Ahrweiler, *Aerosol Age* **11**, 51 (June 1966).
12. L. T. Flanner, *Soap Chem. Spec.* **41**, 208 (December 1965).
13. F. J. Stanave, J. B. Ward, and H. H. Hutchins, *Am. Perfum. Cosmet.* **84** (October 1969).
14. R. R. Rus, *Soap, Cosmet. Chem. Spec.*, **48**, 107 (May 1972).
15. E. Green, *Aerosol Age* **19**, 12 (July 1974).
16. R. Angvall, Personal Communication, American Gasket and Rubber Company
17. J. A. Gribens, Personal Communication, W. R. Grace & Company; see also *Aerosol Age* **16**, 28 (July 1971).
18. J. A. Gribens, *Aerosol Age* **20**, 22 (September 1975).
19. P. H. Seckel, *Proc. 49th Ann. Meet. Chem. Spec. Manuf. Assoc.*, 34 (December 1962).
20. P. Kotuby, Personal Communication, Risdon Manufacturing Company.
21. R. J. Bucaro, Personal Communication, Seaquist Valve Company.
22. J. Mehta, Personal Communication, Newman-Green, Inc.

23. W. Gregg, Personal Communication, Precision Valve Corp.
24. B. Keys, Personal Communication, Summit Packaging Systems, Inc.
25. R. Chrishug, Personal Communication, Clayton Corp.
26. W. C. Beard, *Aerosol Age* **17**, 17 (April 1972).
27. W. C. Beard, Personal Communication, Walter C. Beard, Inc.
28. R. R. Rus, *Aerosol Age* **20**, 17 (July 1975).
29. *Aerosol Age* **12**, 124 (October 1967).
30. H. Boden, *Aerosol Age* **13**, 19 (May 1968); Freon® Aerosol Report A-74.
31. *Detergent Age* **5**, 122 (1968).
32. N. E. Platt, *Detergent Age* **5**, 137 (1968).
33. *Aerosol Age* **13**, 32 (April 1968).
34. M. A. Johnsen, W. E. Dorland, and E. K. Dorland, "The Aerosol Handbook," Chap. 8, Wayne E. Dorland Company, Caldwell, N.J. (1974).
35. R. E. Moses and P. Lucas, U.S. Patent 3,341,418 (1967).
36. R. M. Friedenberg, U.S. Patent 3,240,396 (1966).
37. W. C. Beard, *Proc. 41st Ann. Meet. Chem. Spec. Manuf. Assoc.*, 74 (December 1954); *Soap Chem. Spec.* **31**, 139 (1955); U.S. Patent 2,959,325 (1960).
38. H. T. Smith, *Household Pers. Prod. Ind.* **42** (August 1974).
39. J. J. Shay and J. R. Falkowski, *Aerosol Age* **20**, 52 (December 1975).
40. Soap, *Cosmet. Chem. Spec.*, **51**, 58 (November 1975).
41. Annual Buyers Guide, *Aerosol Age* **20** (October 1975).
42. P. A. Sanders, *Aerosol Age* **21**, 35, 41 (October/November 1976).
43. *Aerosol Age* **22**, 35 (June 1977).

7

FILLING METHODS

All aerosol propellants are gases at room temperature and atmospheric pressure. Regardless of whether they are liquefied or compressed gases, they are shipped and stored as liquids under their own vapor pressure. The compressed gases— nitrogen, carbon dioxide, and nitrous oxide—subsequently are filled into aerosol containers as gases. Liquefied gas propellants are added as liquids. They are maintained in the liquefied state either by cooling below their boiling points or by applying sufficient pressure.

There are essentially three methods for filling liquefied gas propellants into aerosol containers: cold or refrigeration filling, pressure filling, and undercap filling. Cold filling involves metering refrigerated liquefied propellants directly into open containers. The valve is then placed on the container and crimped. In pressure filling, the valve is crimped in the container before the propellant is added. The liquefied propellant, under high pressure, is filled into the container through or around the valve stem. Undercap filling utilizes a filling head that forms a tight seal on the container shoulder. The filling head holds the valve above the container while propellant under pressure is added through the 1-in. opening in the container. Both pressure filling and undercap filling are carried out at room temperature. It is estimated that 5% or less of the aerosol units in the United States are loaded by cold filling, about 20% by pressure filling, and 75% by undercap filling.[1]

The equipment and machinery designed for filling aerosol products vary from simple laboratory apparatus to extremely elaborate and sophisticated production lines. The major manufacturers of production equipment are J. G. Machine Works, Kartridg Pak Company, Nalbach Engineering, and Terco, Inc. Kartridg Pak is the largest and the only manufacturer of rotary equipment. Other suppliers include Aerosol Machinery Company and Aerosol Products International. Some manufacturers of commercial equipment also supply laboratory equipment. One of the best known suppliers for the small laboratory is Builder's Products.

LIQUEFIED GAS PROPELLANTS

Cold Filling

The arrangement of a more or less typical commercial cold-filling line is illustrated in Figure 7-1. The following steps are involved in the cold-filling process:

STEP   1. Empty aerosol containers are removed from their shipping cartons and placed in the unscrambler. The unscrambler separates the containers and sends them into the line at predetermined intervals.

STEP   2. After the cans leave the unscrambler, they are inverted and cleaned by an air blast which removes dirt and other foreign matter. In some cases, air cleaning is carried out in the presence of a continuous vacuum.

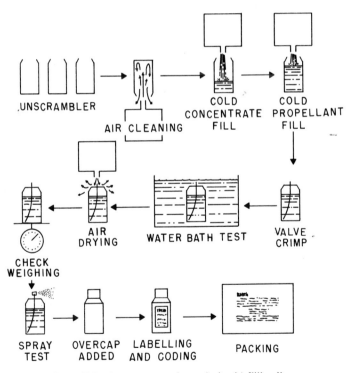

Figure 7-1   Arrangement of a typical cold-filling line.

STEP 3. The concentrate is cooled to a temperature of about $0°$ to $-10°F$, depending upon the product and propellant, and added to the open container. If the concentrate were not cooled, propellant losses would be excessive.

STEP 4. The propellant is cooled to a temperature somewhat below its boiling point, usually between $-20°$ to $-40°F$, and metered into the open container.

STEP 5. The valve is placed on the container and crimped. The can may be coded at this point for identification purposes.

STEP 6. Filled containers are heated to a temperature of $130°F$ in a water bath as required by the DOT regulations. The leaking or bursting of cans in the water bath is an indication of a faulty filling operation that must be corrected immediately. Two possible causes are improperly crimped valves or overfilled containers that become liquid full at $130°F$.

STEP 7. After passing through the water bath, the containers are air dried. This prepares the containers for labeling and coding and also removes water from areas where corrosion could occur, for example, the valve cup.

STEP 8. Containers may be weighed at this point to be certain that product fill is correct. In many plants, the check weighing operation is automatic.

STEP 9. Actuators are placed on valves if necessary and the product may be spray tested. Other tests, such as those for foam products, may be carried out at this point.

STEP 10. Protective caps or overcaps are placed on the containers.

STEP 11. For the purpose of identification, the cans are labeled and coded if this has not been done previously.

STEP 12. The complete aerosol products are packed in shipping cartons.

The cold-filling process is used mostly for perfumes and some pharmaceuticals. Certain perfumes initially contain impurities that are insoluble in the concentrate. These have an unsightly appearance and could cause valve clogging. Many impurities can be eliminated by cooling the perfume concentrate and filtering to remove precipitated material. The propellant is then cold-filled. Some impurities, however, are soluble in the perfume concentrate, but insoluble in the final aerosol because the propellant is a poor solvent. In such cases, it is advantageous to premix the perfume concentrate and propellant, cool the mixture to a predetermined temperature, and allow the mixture to stand several days. The mixture of propellant and concentrate is then filtered and cold-filled. This procedure gives products that are sparkling clear.

The first aerosol product packaged in low-pressure containers in the late 1940's was the insecticide. The formulations at that time averaged about 85 wt. % Freon® 12/Freon® 11 (50/50) and 15 wt. % concentrate. The cold-filling process was popular for insecticides because the production rate was fairly high. Large quantities of cold, liquefied propellant could be added rapidly through the 1-in. opening in the metal containers. A typical rate of fill was 60 cans/min. High-speed pressure-filling or undercap-filling equipment had not yet been developed. Another advantage of the cold-filling procedure was that enough liquefied propellant vaporized during filling to displace most of the air remaining in the container after addition of concentrate. Residual air increased the pressure in the container and could affect corrosion. However, with the development of high-speed pressure-filling equipment, the disadvantages of cold filling far outweigh its advantages, except for a few products. The disadvantages of cold filling are as follows:

1. Expensive refrigeration equipment is required for cooling the concentrate and propellant.
2. The method is suitable only for products whose concentrates can be cooled without freezing, precipitation of active ingredients, or excessive increase in viscosity. These points can be checked by first loading the concentrate into a glass bottle and observing the concentrate after cooling to the cold-filling temperature. If fluorocarbons 12 and 11 are used, precipitation of active ingredients or an excessive increase in viscosity of the concentrate during cooling sometimes may be avoided by premixing fluorocarbon 11 with the concentrate. Fluorocarbon 11 is a fairly good solvent.
3. Ice may form on the cold-filling nozzles and contaminate the product. Excessive moisture can cause corrosion.
4. Temperatures of the concentrate and propellant must be adjusted with care to obtain adequate removal of air from the container without incurring excessive propellant losses.

In the laboratory, cold filling is easily carried out by leading the propellant through copper coils cooled below the boiling point of the propellant. The coils may be enclosed in a deep-freeze unit or an insulated box. Suitable valves and connections are available for attaching the unit to a propellant cylinder. If an insulated box is used, it is advantageous to cool the coils with a mixture of dry ice and a nonflammable solvent such as fluorocarbon 11 for safety reasons. Laboratory cold filling is discussed in Reference 2.

Conventional Pressure Filling

Pressure filling was developed initially for products that could not be cold-filled. These included aqueous-based products that froze, for example, shaving lather, and concentrates in which the active ingredients were not soluble at cold-filling temperatures.

A typical conventional pressure-filling line is illustrated in Figure 7-2 and involves the following steps:

STEP 1. Unscrambling—same as cold filling.
STEP 2. Air cleaning—same as cold filling.
STEP 3. The concentrate, at room temperature, is added to the aerosol container.
STEP 4. Air is removed from the vapor phase in the container either by purging with propellant or by evacuation. If air is removed by evacuation, the valve is placed on the container and crimped in the same operation.
STEP 5. If air is removed by purging, the valve is placed on the container and crimped as soon as the purging operation is completed.
STEP 6. Propellant is pressure-filled through the valve. The total pressure in the system for filling propellant is generally about 50 psi higher than the vapor pressure of the propellant.
STEP 7. Filled containers are heated to a temperature of 130°F in the water bath.

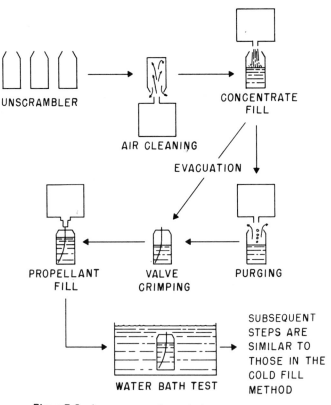

Figure 7-2 Arrangement of a typical pressure-filling line.

From here on, the steps are the same as those described for the cold-filling process.

Among the advantages of pressure filling are the following:

1. No refrigeration equipment is required.
2. The concentrate does not have to be cooled; therefore, the problems of insolubility of active ingredients in the cold concentrate or high viscosity at cold-filling temperatures are eliminated.
3. The procedure is more suitable for loading hydrocarbons or other flammable propellants.

Initially, when only single charging heads were available, a major disadvantage of pressure filling was the slow rate of fill (about 15 cans/min). The slow rate was the result of having to fill the propellant through the valve and thus through the small valve stem orifice. At present, the rate of fill averages about 20 cans/min.[2] In recent years, rotary pressure-filling units with multiple filling heads have been developed. These have high production rates. This eliminates one of the main disadvantages of conventional pressure-filling procedures. The Kartridg Pak Company manufactures units with 6, 12, or 15 heads. The latter has a production rate of at least 275 cans/min.[3]

Aerosol containers can now be pressure-filled through the valve stem (and thus through the metering orifice), through and around the valve stem simultaneously, or through the actuator and around the valve stem. The latter two procedures require special valves and filling head adapters. The valve companies should be consulted for information on valves which are designed for this type of filling. Filling equipment manufacturers will provide special adapters for the valves. The advantage of filling around the valve stem is that it is faster than filling through the stem.

Pressure-filling laboratory equipment is available from some commercial equipment manufacturers. One of the simplest and best for the small laboratory is the glass burette unit offered by Builder's Products.[4] This is described in References 2 and 4.

Undercap Filling

The latest filling technique is undercap filling. The propellant filling head performs three functions. It evacuates the container to remove air, fills the propellant, and crimps the valve. This is illustrated in Figure 7-3. After the concentrate has been added at room temperature, the valve is placed on the container as in conventional pressure-filling procedures. The undercap-filling head then contacts the top of the container, forming an airtight seal. Air is removed from the container by evacuation. The valve is held slightly above the 1-in. opening in the container by a combination of the initial suction resulting from the vacuum

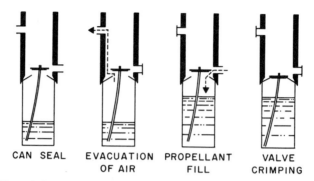

CAN SEAL   EVACUATION   PROPELLANT   VALVE
                OF AIR   FILL   CRIMPING

Figure 7-3 Undercap filling. (*Courtesy of the Kartridg Pak Company*)

system and the mechanical action of the collett in the filling head. After evacuation of air, the vacuum port closes and propellant is admitted through another port. It passes underneath the valve cup and into the container through the 1-in. opening. The valve is then seated and crimped. The undercap-filling technique thus combines the best features of both cold and pressure filling. The operation is carried out at room temperature and propellant is added through the 1-in. opening in the container.

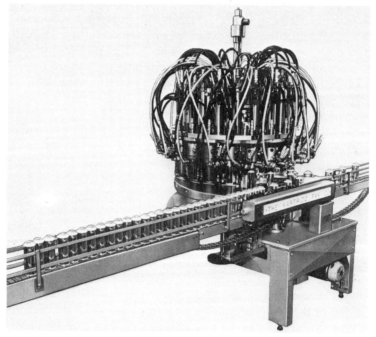

Figure 7-4 18-Head rotary undercap filler. (*Courtesy of the Kartridg Pak Company*)

Rotary undercap fillers with 3, 6, 9, or 18 heads are available. Production rates range from 55 to over 300 cans/min.[3] The model with 18 heads is illustrated in Figure 7-4. Kartridg Pak has developed a propellant reclaim system which practically eliminates propellant loss. The Kartridg Pak reclaim system is standard on the 18-head undercap filler. It is also available as an accessory for the 6- or 9-head undercap filler.

Laboratory model undercap fillers also are manufactured by Kartridg Pak.

COMPRESSED GASES

The four commercial methods for filling with compressed gases have been described by Rink,[1] Hayes,[5] Sievers,[6] and Bucaro[7] and in Reference 8. These consist of: (1) the saturator method; (2) impact gassing with pressure fillers; (3) impact gassing with undercap fillers; and (4) gasser/shakers. The saturator method is illustrated in Figure 7-5. The main function of the saturator, as the name implies, is to saturate a solvent such as ethyl alcohol with a compressed gas. The latter is usually carbon dioxide. The saturated solvent is then pumped to an undercap filler or a pressure filler and loaded into a container that has the concentrate already present. The type of product, amount of fill, and other factors will determine whether it is more practical to use the undercap or pressure filler as the filling equipment.[1] Few products filled using a saturator

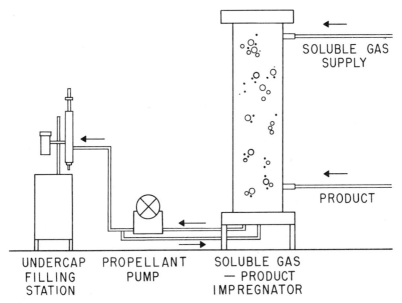

SOLUBLE GAS
SUPPLY

PRODUCT

UNDERCAP	PROPELLANT	SOLUBLE GAS
FILLING	PUMP	— PRODUCT
STATION		IMPREGNATOR

Figure 7-5 A typical saturator with an undercap filler. (*Courtesy of the Kartridg Pak Company*)

are single-stage fills. In most cases, a portion of the solvent is saturated and loaded into a container that has the active ingredients dissolved in the remainder of the solvent. The concentrate itself can be saturated, but this may result in difficulty in cleaning the saturator after the run is completed.

In carrying out the operation, compressed gas under high pressure (about 200 psi) enters the upper part of the saturator tower. The solvent is pumped into the saturator tower through a spray manifold located below the compressed gas inlet. The pressure of the solvent from the pump is approximately 300 psi, so that a nozzle differential of 100 psi is maintained. This is essential to obtain a good spray distribution and solvent droplet size. The solvent spray is directed toward the upper part of the tower, where it comes in contact with the descending compressed gas. The saturated solvent drops to the lower portion of the saturator. Sufficient pressure (maximum 200 psi) must be maintained in the equipment to prevent loss of gas from the solvent. The lower part of the saturator is connected to a booster pump which increases the pressure an additional 50 psi. The saturated liquid is fed to the propellant pump at 250 psi. The line between the saturator and booster pump should be as short as possible to prevent vaporization. The saturated liquid is them pumped to an undercap or pressure filler where it is loaded into the container. Production rates of 200 cans/min can be obtained with rotary undercap fillers.[8]

Impact gassing with a pressure filler is a technique of filling high-pressure compressed gas (approximately 588 psi) through a valve that has previously been crimped into place. The concentrate is added prior to crimping of the valve by conventional means. The main technology in impact gassing is the valve design, which allows a simultaneous injection of gas under high pressure directly into the head space and through the dip tube. Valves are selected to provide the proper ratio of gas flow around the valve stem into the head space and through the valve stem into the dip tube.

The resulting turbulence in the container breaks the concentrate into a fine mist that results in almost instant saturation. This essentially makes a miniature, efficient saturator out of each aerosol can.[9] The pressure in the can rises to about 120 psig as carbon dioxide is being introduced, but quickly settles to a pressure slightly over the final saturation pressure.[5] The method is limited to low-viscosity products that are easily saturated. Otherwise, high pressures could momentarily build up and cause bursting of the containers. Line speeds range from 100 to 200 cans/min.

Several companies have developed valves suitable for impact gassing. The Precision hex gasket valve has been used to impact pressure fill an estimated 25 million cans of product worldwide. Most of the Dispenser Products Division of Ethyl Corp. (formerly VCA/ARC) KN-38 valves can also be impact pressure filled at high rates.

Impact gassing with an undercap filler uses the same basic technique as that with pressure filling with one exception. High-pressure compressed gas is charged only into the head space above the concentrate already in the container. This creates turbulence on the surface of the concentrate and facilitates solution of the gas into the liquid concentrate. The rate at which the gas is absorbed is slower than impact gassing with pressure-filling equipment, where gas enters the container through the dip tube as well as directly into the head space. Products that can be impact gassed with an undercap filler are limited to those with high solubility for the compressed gas.

The fourth method of filling compressed gases—using a gasser/shaker—is probably the oldest method in the aerosol industry. The technique is just what the name implies. The liquid concentrate is loaded conventionally into containers and the valve crimped. The container is then moved to a gasser/shaker where compressed gas is forced through the valve at the desired pressure. The container is shaken vigorously during filling to establish equilibrium as rapidly as possible. The gasser/shaker is the most suitable equipment for filling compressed gas aerosols when the solubility of gas in the concentrate is low.

Autoproducts, Inc. (formerly Andora Automation, Inc.) manufactures a 12-head rotary gasser/shaker. The production rate depends upon the shaking time required to achieve saturation. The rate is 120 cans/min at a gassing/shaking time of 4 sec and decreases to 30 cans/min at a gassing/shaking time of 16 sec. Autoproducts also manufactures gas pressure checking machines for use in combination with their gasser/shakers. The multihead unit is capable of pressure checking 140 cans/min.[10] In-line-type gasser/shakers are manufactured by Nalbach Engineering and LeMay Machine Company.

Laboratory filling with compressed gases can be achieved using a gasser/shaker or by manual techniques involving the pressure gage on the pressure-filling burettes.[2]

REFERENCES

1. T. J. Rink, Personal Communication, Kartridg Pak Company, March 1976.
2. P. A. Sanders, "Aerosol Laboratory Manual," Industry Publications, Inc., Cedar Grove, N.J., 1975.
3. Technical Brochure, "A Complete Line of Filling Equipment," Kartridg Pak Company, 1973.
4. Technical Bulletin, "Aerosol Laboratory Equipment," Builder's Products.
5. C. S. Hayes, *Aerosol Age* **20**, 34 (June 1975).
6. J. Sievers, *Aerosol Age* **20**, 14 (July 1975).
7. R. J. Bucaro, *Household Pers. Prod. Ind.* **43**, 40 (October 1975).
8. Technical Bulletin, "CO_2 Aerosol Packaging," Cardox Products Division, Chemetron Corp., March 20, 1975.
9. R. R. Rus, *Aerosol Age* **20**, 17 (July 1975).
10. Technical Brochure, "The Andora Rotary Gasser/Shaker," Autoproducts, Inc.

8

VAPOR PRESSURE

The *pressure* in a container filled with a gas results from the impact of gas molecules against the walls of the container. The magnitude of the pressure is a function of the number of molecules striking the walls and their velocity. The *vapor pressure* of a liquid is defined as the pressure of the vapor (gas) in equilibrium with the liquid (see Chapter 2). The vapor pressure of any liquid is constant at a given temperature, regardless of the quantity of liquid present. If the liquid phase composition of an aerosol formulated with a liquefied gas *does not change* as the product is discharged, the vapor pressure will remain constant.

Vapor pressure is an important property of aerosol products. It must be known because of the limitations that the DOT regulations place upon the pressure of aerosol products. It is also useful for quality control. Pressure measurements on a product during filling serve as a check on such factors as the ratio of propellant to concentrate and effectiveness of purging or evacuation in removing air.

RAOULT'S LAW

Practically all aerosol products consist of mixtures of propellants with other components such as solvents and active ingredients. Components that are soluble in the propellant lower the vapor pressure of the propellant. The reason for this is that the vapor pressure of the propellant is a function of the frequency with which the molecules escape from the surface of the liquid phase and enter the vapor phase. This determines the number of molecules in the vapor phase. When another material is dissolved in the propellant, the concentration of propellant molecules in the surface is decreased. Therefore, the rate of escape of the propellant molecules is diminished. The material dissolved in the propellant can be a solid, liquid, or gas.

The lowering of the vapor pressure of a liquid by the addition of another substance is known as Raoult's Law. Raoult's Law can be stated as follows, where the term "solvent" refers to the original liquid and the term "solute" refers to the material that is added to the solvent:

The depression of the vapor pressure of a solvent upon the addition of a solute is proportional to the mole fraction of solute molecules in the solution.

The vapor pressure is also proportional to the mole fraction of the solvent if the solute has essentially no vapor pressure itself. This can be expressed by the following equation, where P_s is the vapor pressure of the solution, P is the vapor pressure of the pure solvent, and x is the mole fraction of the solvent:

$$P_s = xP \qquad (8\text{-}1)$$

If the material added to the liquid has an appreciable vapor pressure itself, its vapor pressure will also be lowered as a result of dilution by the initial solvent. Thus, the addition of fluorocarbon 11 to fluorocarbon 12 lowers the vapor pressure of fluorocarbon 12; and conversely, the addition of the fluorocarbon 12 will decrease the vapor pressure of fluorocarbon 11. The lowering of the vapor pressure of the solute is known as Henry's Law. Raoult's Law applies to the solvent and Henry's Law to the solute.[1]

DALTON'S LAW OF PARTIAL PRESSURES

The total pressure in any system is equal to the sum of the individual or partial pressures of the various components. When two or more gases occupy the same space, neither of the gases interferes with the pressure of the other. In 1802, John Dalton observed that in any mixture of gases, each gas exerted the same pressure it would if it were alone in the volume occupied by the gases. Therefore, the total pressure of the mixture is equal to the sum of the partial pressures of the gases in the mixture. This is known as Dalton's Law.

CALCULATION OF VAPOR PRESSURE

When combined, Raoult's Law and Dalton's Law provide a means of calculating the vapor pressure of any solution if the composition of the solution and the molecular weights and vapor pressures of the pure components are known. In carrying out the calculation, partial pressures are obtained first by multiplying the mole fraction of each component by its vapor pressure. The total pressure of the solution is obtained by adding together the partial pressures of the components.

For example, the vapor pressure of a mixture of 30 wt. % fluorocarbon 12 and 70 wt. % fluorocarbon 11 at 70°F can be calculated as shown below. The vapor pressures and molecular weights of fluorocarbons 12 and 11 are as follows:[2]

Fluorocarbon 12

$$\text{Vapor pressure at } 70°F = \;\; 84.9 \text{ psia}$$
$$\text{Molecular weight} \qquad = 120.9$$

Fluorocarbon 11

$$\text{Vapor pressure at } 70°F = \;\; 13.4 \text{ psia}$$
$$\text{Molecular weight} \qquad = 137.4$$

1. Assume there is 100 g of solution. The moles of the two components in 100 g of the propellant blend are obtained by dividing the weights of the components by their molecular weights:

$$\text{Moles of fluorocarbon } 12 = \frac{30}{120.9} = 0.248$$

$$\text{Moles of fluorocarbon } 11 = \frac{70}{137.4} = 0.509$$

$$\text{Total moles in mixture} \qquad = 0.757$$

2. The mole fraction of each component is obtained by dividing the moles of each component by the total moles in the mixture:

$$\text{Mole fraction of fluorocarbon } 12 = \frac{0.248}{0.757} = 0.33$$

$$\text{Mole fraction of fluorocarbon } 11 = \frac{0.509}{0.757} = 0.67$$

3. The partial pressure of each component at a specific temperature is obtained by multiplying the absolute vapor pressure of the pure component by its mole fraction in the liquid phase:

Partial pressure of fluorocarbon 12 = 0.33 X 84.9 psia = 28.0 psia
Partial pressure of fluorocarbon 11 = 0.67 X 13.4 psia = 9.0 psia

4. The total pressure is obtained by adding the partial pressures of the components:

Total pressure of fluorocarbon 12/fluorocarbon 11 (30/70) mixture
= 28.0 + 9.0 = 37.0 psia or 22.3 psig at 70°F

The same method of calculating vapor pressures is used regardless of whether the added component is a solid, liquid, or gas.

CALCULATION OF THE COMPOSITIONS OF BLENDS WITH SPECIFIC VAPOR PRESSURES

Sometimes a propellant blend with a specific vapor pressure is desired. If the composition of the blend is not known, it can be calculated if the vapor pressures and the molecular weights of the individual components are known.

Two-Component Blends

The composition of a two-component blend with a specified vapor pressure can be calculated using the following equation:

$$P_1(1 - X_2) + P_2 X_2 = P \qquad (8\text{-}2)$$

where

P_1 = Vapor pressure (psia) of component 1
P_2 = Vapor pressure (psia) of component 2
P = The specified vapor pressure of the blend that is desired
X_1 = Mole fraction of component 1 in the liquid phase of the blend
X_2 = Mole fraction of component 2 in the liquid phase of the blend

Since P_1, P_2, and P are known, X_2 can be calculated; X_1 is obtained by subtracting X_2 from 1, since the sum of X_1 and X_2 must equal 1.

The preceding equation is derived as follows:

Let Pa_1 = Partial pressure of component 1 in the blend
Pa_2 = Partial pressure of component 2 in the blend

1. $Pa_1 + Pa_2 = P$
This follows from Dalton's Law of Partial Pressures.

2. $P_1 X_1 = Pa_1$
$P_2 X_2 = Pa_2$
This follows from Raoult's Law.

3. Therefore,

$$P_1 X_1 + P_2 X_2 = P \qquad (8\text{-}3)$$

4. $$X_1 + X_2 = 1, \text{ or } X_1 = 1 - X_2 \qquad (8\text{-}4)$$

5. Substituting the value of X_1 from Equation 8-4 into Equation 8-3 gives

$$P_1(1 - X_2) + P_2 X_2 = P \qquad (8\text{-}2)$$

The composition of the blend in weight percent can be obtained by multiplying the mole fractions by the molecular weights. This gives the weight in grams of each component in 1 mole of the blend. Thus, if Mwt_1 and Mwt_2 are the molecular weights, respectively, of components 1 and 2, then

$$X_1 Mwt_1 = \text{Grams of component 1 in 1 mole of blend}$$
$$X_2 Mwt_2 = \text{Grams of component 2 in 1 mole of blend}$$

$$\text{Wt. \% of component 1 in blend} = \frac{\text{Grams of component 1}}{\text{Total grams}} \times 100$$

$$\text{Wt. \% of component 2 in blend} = \frac{\text{Grams of component 2}}{\text{Total grams}} \times 100.$$

As a specific example, assume that the composition of a blend of fluorocarbon 12 and isobutane with a vapor pressure of 40 psig (54.7 psia) at 70°F is desired.

1. Let X_1 = Mole fraction of fluorocarbon 12 in the liquid phase
   X_2 = Mole fraction of isobutane in the liquid phase
   P_1 = Vapor pressure of fluorocarbon 12 at 70°F = 84.9 psia
   P_2 = Vapor pressure of isobutane at 70°F = 45.7 psia
   P = Desired vapor pressure of blend = 54.7 psia

2. Substituting in Equation 8-2 gives

$$84.9(1 - X_2) + 45.7 X_2 = 54.7$$

3. Solving for X_2 gives a value of 0.77 for the mole fraction of isobutane. The mole fraction of fluorocarbon 12 is, therefore, 0.23.

4. The composition of the blend in weight percent is determined as follows:

   Wt. in grams of fluorocarbon 12 in 1 mole = 0.23 X 120.9 = 27.8 g
   Wt. in grams of isobutane in 1 mole = 0.77 X 58.0 = 44.6 g
   Total weight of 1 mole in grams = 72.4 g

$$\text{Wt. \% of fluorocarbon 12 in blend} = \frac{27.8}{72.4} \times 100 = 38.4 \text{ wt. \%}$$

$$\text{Wt. \% of isobutane in blend} = \frac{44.6}{72.4} \times 100 = 61.6 \text{ wt. \%}$$

Therefore, the composition of the fluorocarbon 12/isobutane blend that has a calculated vapor pressure of 54.7 psia at 70°F is 38.4 wt. % fluorocarbon 12 and 61.6 wt. % isobutane.

Three-Component Blends

In any ternary system, there are many different combinations of the three components that have the same vapor pressure. These can best be shown on a triangular coordinate chart. In order to construct a vapor pressure plot on a triangular coordinate chart, it is not necessary to calculate the composition of three-component blends. The two points at each end of the vapor pressure plot fall on the base lines for two-component blends. Therefore, in order to draw the vapor pressure line, it is only necessary to know the composition of the two different two-component blends that have the same vapor pressure as that specified for the three-component blends. These two compositions are plotted on the triangular coordinate chart and a straight line drawn between the points. All possible combinations of the three components with the same vapor pressure fall on this line. The points at the end of the line give the composition of the two-component blends with the specified vapor pressure; all other points indicate combinations of three components with the same specified vapor pressure.

As an example, assume that the various compositions of fluorocarbon 12, fluorocarbon 11, and isobutane that have a vapor pressure of 54.7 psia (40 psig) at 70°F are desired. Vapor pressures of the three components are as follows:

Vapor pressure of fluorocarbon 12 at 70°F = 84.9 psia
Vapor pressure of fluorocarbon 11 at 70°F = 13.4 psia
Vapor pressure of isobutane at 70°F   = 45.7 psia

Using these values, the compositions of fluorocarbon 12/fluorocarbon 11 and fluorocarbon 12/isobutane blends having a vapor pressure of 54.7 psia can be calculated using Equation 8-2. The fluorocarbon 12/fluorocarbon 11 blend with the specified vapor pressure of 54.7 psia at 70°F has a composition of 54.8 wt. % fluorocarbon 12 and 45.2 wt. % fluorocarbon 11. The fluorocarbon 12/ isobutane blend with a vapor pressure of 54.7 psia has a composition of 38.4 wt. % fluorocarbon 12 and 61.6 wt. % isobutane. These two compositions are plotted on a triangular coordinate chart and a line drawn between them as shown in Figure 8-1. The compositions of all three-component blends with a

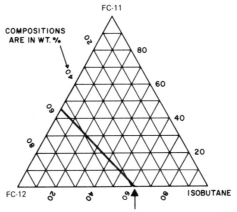

THE COMPOSITIONS OF ALL MIXTURES OF FLUOROCARBON 12, FLUOROCARBON 11, AND ISOBUTANE
WITH A VAPOR PRESSURE OF 54.7 psia AT 70°F FALL ON THIS LINE.

Figure 8-1 Compositions of fluorocarbon 12/fluorocarbon 11/isobutane blends having a vapor pressure of 54.7 psia at 70°F.

vapor pressure of 54.7 psia will fall on this line. In constructing the triangular coordinate charts it is assumed that there is no deviation from Raoult's Law.

DEVIATION FROM RAOULT'S LAW

In many cases, the vapor pressure of a solution calculated by Raoult's Law differs from the pressure actually measured. There are several reasons for this. One is that either the solvent or solute itself may be associated in the liquid state and thus is polymeric. The molecular weight of the polymeric system is higher than that of the monomer. Ethyl alcohol, for example, is highly associated. This is why ethyl alcohol has a much higher boiling point (78°C) than its isomer dimethyl ether (-25°C), which is not associated. The association can be illustrated as follows:

$$nC_2H_5OH \longrightarrow (C_2H_5OH)_n$$

The experimentally determined vapor pressures of fluorocarbon 12/ethyl alcohol mixtures are significantly higher than those calculated from Raoult's Law because the alcohol molecules are associated. The calculation of the vapor pressure first involves the calculation of the mole fractions of the individual components. The calculation assumes that ethyl alcohol has a molecular weight of 46. In reality, the effective mole fraction of ethyl alcohol in the fluorocarbon 12/ethyl alcohol mixture is much lower than the calculated value because of the presence of the higher molecular weight association complexes of ethyl alcohol. Consequently, the actual mole fraction of fluorocarbon 12 is higher than calculated and the vapor pressure of the mixture is higher. When the propellant contains no hydrogen (such as fluorocarbon 12), the ethyl alcohol is much more likely to associate with itself because of the possibility of hydrogen bonding between the alcohol molecules, in which fluorocarbon 12 cannot take part. Therefore, the major reason for the deviation is the self-association of ethyl alcohol molecules.

A comparison of the vapor pressures of fluorocarbon 12/ethyl alcohol mixtures, both calculated and experimental, is shown in Figure 8-2.

A somewhat different situation exists with a propellant that contains hydrogen, for example, fluorocarbon 22. In this case, not only can the individual components in a fluorocarbon 22/ethyl alcohol mixture associate, but they can also associate with each other. This is illustrated below:

$$xCHClF_2 + yC_2H_5OH \longrightarrow [(CHClF_2)_x \cdot (C_2H_5OH)_y]_n$$

The higher molecular weight of the association complex between fluorocarbon 22 and ethyl alcohol reduces the mole fractions of both fluorocarbon 22 and ethyl alcohol in the mixtures. The vapor pressures of the mixtures, therefore, are lower than if only the ethyl alcohol were associated with itself. As a result of

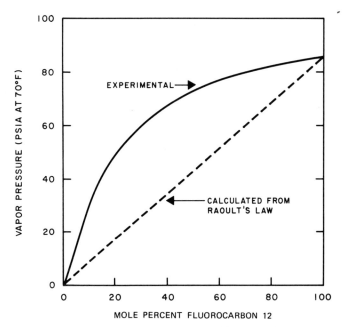

Figure 8-2 Vapor pressures of fluorocarbon 12/ethyl alcohol mixtures.

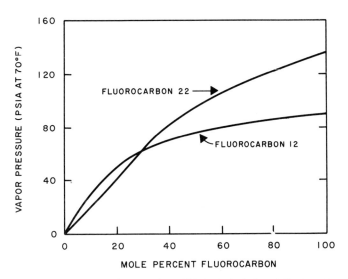

Figure 8-3 Vapor pressures of fluorocarbon 22/ethyl alcohol and fluorocarbon 12/ethyl alcohol mixtures.

the fluorocarbon 22/ethyl alcohol interaction, the vapor pressures of fluorocarbon 22/ethyl alcohol mixtures containing less than 30 mole % of fluorocarbon 22 are lower than those of corresponding fluorocarbon 12/ethyl alcohol mixtures, even though the vapor pressure of fluorocarbon 22 alone is considerably higher than that of fluorocarbon 12. This is illustrated in Figure 8-3.

Thus, in mixtures where a deviation from Raoult's Law occurs, it is usually not because Raoult's Law has failed, but only because the solutions are not truly homogeneous. Therefore, this is only an apparent deviation. There are some cases where the deviation is considered to be real.[3]

VAPOR PRESSURES OF PROPELLANT/ETHYL ALCOHOL/WATER SOLUTIONS[15]

The effect of increasing concentrations of water upon the vapor pressures of fluorocarbon 152a/ethyl alcohol solutions is shown in Table 8-1. At the same fluorocarbon concentration, substitution of water for alcohol causes a marked increase in the vapor pressure. The data are also illustrated on the triangular

TABLE 8-1 EFFECT OF WATER AT CONSTANT FLUOROCARBON
CONCENTRATIONS UPON VAPOR PRESSURES OF
FLUOROCARBON 152a/ETHYL ALCOHOL/WATER
SOLUTIONS

Composition (wt.%)			
FC-152a	Water	Ethyl Alcohol	Vapor Pressure (psig) at 70° F
72	—	28	50
45	—	55	40
45	9	46	50
30	—	70	30
30	10	60	40
30	20	50	50
21.5	—	78.5	20
21.5	11.5	67	30
21.5	19.5	59	40
21.5	29	49.5	50
10	25	65	20
10	34	56	30
10	40	50	40
5	38.5	56.5	20
5	47	48	30
5	52	43	40

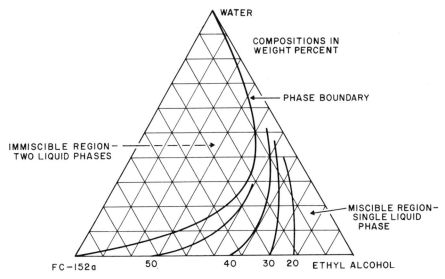

Figure 8-4 Vapor pressures (psig) of fluorocarbon 152a/ethyl alcohol/water solutions at 70°F.

coordinate chart in Figure 8-4, which also gives the compositions of ternary blends at constant vapor pressure.

The effect of water upon vapor pressure is opposite that which might have been expected. If both water and ethyl alcohol were monomolecular and did not associate with each other, the substitution of water for ethyl alcohol would cause a decrease in vapor pressure instead of an increase. Water has a lower molecular weight than ethyl alcohol. The substitution of water for ethyl alcohol at constant fluorocarbon concentration would decrease the mole fraction of fluorocarbon in the solution and, consequently, decrease the vapor pressure. The observed increase in vapor pressure due to water results from its polymolecular structure and its strong association with ethyl alcohol. These two factors, particularly the association with ethyl alcohol, cause an increase in the effective molecular weight of the water/alcohol portion of the solution and a resultant decrease in the mole fraction. Therefore, the mole fraction of the fluorocarbon increases with the addition of water, and the vapor pressure also increases.

VAPOR PRESSURES OF MISCELLANEOUS PROPELLANT BLENDS[15]

The vapor pressures of a series of miscellaneous propellant blends at 70°F are illustrated in Figures 8-5 through 8-8.

The vapor pressures and flammabilities of various combinations of fluorocarbon 133a, fluorocarbon 152a, isobutane, isopentane, and propane are available in Chapter 14, "Flammability."

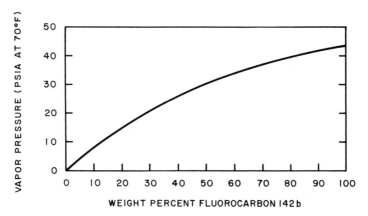

Figure 8-5 Vapor pressures of fluorocarbon 142b/ethyl alcohol mixtures as a function of composition.

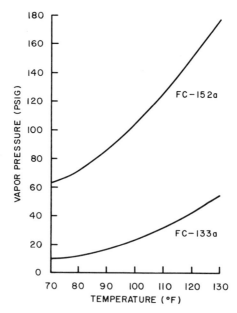

Figure 8-6 Vapor pressures of fluorocarbons 133a and 152a as a function of temperature.

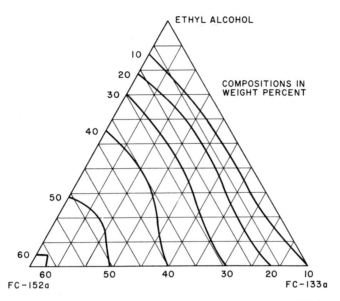

Figure 8-7 Vapor pressures (psig) of fluorocarbon 152a/fluorocarbon 133a/ethyl alcohol mixtures at 70°F.

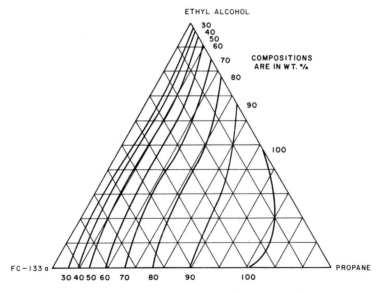

Figure 8-8 Vapor pressures (psig) of fluorocarbon 133a/propane/ethyl alcohol mixtures at 70°F.

PRESSURES OF AEROSOL FORMULATIONS

The vapor pressure of any aerosol formulation (excluding the effect of air) depends upon the vapor pressure of the propellant and the effect that other components in the formulation have upon the vapor pressure of the propellant. If concentrations of solvents and active ingredients are sufficiently high so that the mole fraction of the propellant in the aerosol is reduced, the vapor pressure will be lowered. In many products, concentrations of active ingredients are so low that the effect upon vapor pressure is negligible. Usually, the solvents in the concentrate have the most noticeable effect in depressing the vapor pressure of the propellant. In some cases, the components themselves have appreciable vapor pressures, and this will contribute to the total vapor pressure.

Effect of Air

Up to the present time, only the vapor pressure of the aerosol formulation itself, that is, the vapor pressure of the solution of liquefied propellant and concentrate, has been considered. This vapor pressure will be referred to in subsequent discussions as the vapor pressure of the aerosol formulation (P_f). Another factor that has an effect upon the pressure in the container is the concentration of air.

When air is present, it increases the pressure in the container. This follows from Dalton's Law of Partial Pressures: When two or more gases occupy the same space, each gas exerts the same pressure as if it were alone. If the pressure of air in the vapor phase of an aerosol is 10 psi, the total pressure will be equal to the vapor pressure of the aerosol formulation (P_f) plus the pressure of the air (10 psi). This can be expressed as follows:

$$P_t = P_f + P_a, \text{ or } P_t = P_f + 10 \qquad (8\text{-}5)$$

where

  P_t = Total pressure in the aerosol container
  P_f = Pressure of the aerosol formulation (propellant and concentrate)
  P_a = Pressure due to air

The total pressure in the aerosol container (P_t) will be referred to as the pressure in the container, the pressure of the aerosol product, or the total pressure in order to differentiate it from the vapor pressure of the aerosol formulation (P_f) alone.

All aerosol products contain air because initially all empty aerosol containers are full of air at atmospheric pressure, that is, about 14.7 psia. What happens to this air during the loading of the containers depends upon the method of filling, type of purging, etc. Air may also be present in the solvents used for the preparation of the aerosols.

Sometimes it is desirable to know how much pressure air contributes to the total pressure in the container. This is usually difficult to determine from a pressure measurement alone, since the pressure of the combination of propellant and concentrate (the aerosol formulation) may not be known. In these cases, a determination of the air concentration in the vapor phase of the product by gas chromatography coupled with a measurement of the pressure of the product will give both the pressure due to air alone and the vapor pressure of the aerosol formulation.

For example, assume that at 70°F the pressure of an aerosol product is 50.3 psig (65 psia) and the concentration of air in the vapor phase (as determined by gas chromatography) is 20 vol. % (this is numerically the same as mole %). Therefore, the pressure due to air is 0.20×65.0 psia = 13 psia. The vapor pressure of the aerosol formulation is $65 - 13 = 52$ psia or 37.3 psig at 70°F.

The concentration of air in the vapor phase by gas chromatography should be determined at the same temperature at which the pressure was measured. The solubility of air in the liquid phase of the aerosol is a function of the temperature, as is the pressure of the formulation. The solubility of air in the liquid phase of the aerosol determines the concentration of air in the vapor phase.

If the vapor pressure of the aerosol formulation (P_f) is known, the measurement of pressure in the product after loading will indicate how much air is present. The actual pressure in the container minus the pressure of the aerosol formulation without air will give the pressure due to air.

The effect of air upon the pressure of aerosol products is discussed in detail in Reference 4. The vapor pressures of combinations of fluorocarbons 11, 12, and 114 are given in Reference 2. Vapor pressures of mixtures of propellants and various solvents are given in References 5 and 6.

Pressure Measurement

Pressure measurements of aerosol products that do not contain air present no problem. A pressure gage is connected to the vapor phase of the aerosol with a shut-off valve between the gage and the aerosol. After the aerosol has been brought to the desired temperature, the valve between the gage and the aerosol is opened and the pressure read on the gage. When the valve between the aerosol and gage is opened, the vapor phase of the aerosol expands into the gage. This occurs when the vapor pressure of the aerosol is higher than that in the gage, which is at atmospheric pressure. There is no decrease in the pressure in the vapor phase as a result of loss of propellant vapor to the gage because sufficient liquefied propellant vaporizes to replace that lost to the gage. This maintains a constant pressure in the aerosol.

However, if air is present in the aerosol, as it is in practically all aerosols, a different situation exists. Consider a closed aerosol container filled with nothing except air at a pressure of 50 psi. If the valve of this container is opened so that air escapes, the pressure in the container will decrease just as the pressure in a tire decreases as a result of loss of air from a puncture.

If an attempt is made to obtain the pressure of the container with 50 psi of air, using a gage at atmospheric pressure, some of the air in the container enters the gage. This continues until the pressures in the container and gage are equal. However, since air left the container, the pressure in the container decreases because there is less air present than originally.

The pressure recorded by the gage is less than that in the aerosol container initially and is not an accurate value for the original pressure in the container. The extent to which the pressure in the container decreases depends upon the initial pressures in the container and the gage and the volumes of the container and gage.[8]

A reverse situation occurs if the pressure in the gage is higher than that in the container. If the air pressure in the gage were 100 psi and that in the container 50 psi, then when the valve between the gage and the container was opened, some of the air in the gage would enter the container and increase the pressure in the container. The final pressure recorded on the gage would be higher than the initial pressure in the container. Again, the gage reading would not give an accurate value for the original pressure in the container.

An accurate measurement is obtained if, before the valve between the aerosol and gage is opened, the pressure in the gage is adjusted so that it is the same as that in the aerosol container, that is, 50 psi. When the gage containing 50 psi of air is attached to the aerosol container and the valve between the two opened, there is no loss of air from the aerosol container into the gage nor any transfer of air from the gage into the aerosol container. The pressure recorded on the gage is a true indication of the pressure in the container.

The previous considerations can be applied to the pressure measurement of an aerosol product containing air. The total pressure of an aerosol product is equal to the sum of the vapor pressure of the aerosol formulation and the pressure of the air in the vapor phase (Dalton's Law):

$$P_t = P_f + P_a$$

If the pressure of the aerosol product is measured with a gage at atmospheric pressure, the vapor phase of the aerosol product will expand into the gage until the final pressure in the gage and the aerosol are equal. During the expansion of the vapor phase into the gage, the pressure of the formulation (P_f) remains relatively constant because additional liquefied propellant vaporizes to replace that lost to the gage. The final pressure recorded on the gage is lower than the initial pressure in the aerosol product because P_a decreases during the measurement.

The possibility that taking the pressure of an aerosol product with a gage at atmospheric pressure may give a lower reading than the original pressure was recognized early in the aerosol industry. A committee was established in the Chemical Specialties Manufacturers Association to develop a method for measuring the pressure of an aerosol which would take into account the air in the product. A procedure, which achieves essentially the ideal situation where the adjusted gage pressure is about equal to that in the aerosol,[7] was adopted. It is also described in detail in Reference 14. This method involves prepressurizing the gage before the pressure is measured. It is carried out with a modified gage with additional valves for adding air to the gage or releasing pressure from the gage while it is attached to the aerosol container.

In an aerosol product, part of the air is present in the vapor phase and part is dissolved in the liquid phase. The distribution ratio of air between the liquid and vapor phases is constant when the system is at equilibrium and depends upon such factors as the volume fill in the container, the solubility of air in the concentrate, and the temperature of the aerosol. For example, in a container filled 70% volume full with a fluorocarbon 12/fluorocarbon 11 (50/50) blend alone, at 70°F, 56% of the air is in the liquid phase and 44% is in the vapor phase. In a container filled only 50% volume full with a fluorocarbon 12/fluorocarbon 11 (50/50) blend, 35% of the air is in the liquid phase and 65% is in the vapor phase.[3] Because of the distribution of air between the liquid and vapor phases, it is necessary to shake an aerosol container at intervals before a pressure measurement. The agitation is necessary in order to establish equilibrium between the air in vapor phase and that in the liquid phase at the temperature at which the pressure measurement is carried out.

There are a number of processes that theoretically could take place during a pressure measurement with a nonprepressurized gage. These are discussed in detail in Reference 8. For example, one question that arises is whether the expanding vapor phase of the aerosol mixes with the air in the gage or merely compresses it. The two different processes would give different pressures. The data indicate that the expanding vapor phase of the aerosol merely compresses the air in the gage. The air in the gage, therefore, acts as a medium for transmittal of pressure.

The use of a prepressurized gage is necessary to obtain an accurate pressure. There is also another advantage to prepressurizing the gage. If vapor pressures are taken through the dip tube with a nonprepressurized gage, the aerosol product will enter the gage. Many aerosol products, such as aerosol paints, could contaminate the gage. Prepressurizing the gage will avoid this.

In the laboratory, the simplest method for taking pressures vapor phase is to prepare samples without dip tubes. Pressures can then be taken directly through the valve. Another procedure is to use a can-puncturing device that will puncture either the bottom or the side of the container. Can-puncturing devices allow pressures to be taken vapor phase on cans equipped with dip tube valves.

Removal of Air from Aerosol Containers

The presence of an excessive concentration of air in an aerosol product is undesirable for several reasons. Air may increase the pressure sufficiently so that

DOT limits are exceeded. The increased pressure due to air will also increase the spray rate. Air can cause deterioration of aerosol products by reacting with oxygen-sensitive components, such as perfumes. In addition, oxygen is a factor in many cases where corrosion occurs.[9, 10] For these reasons, it is advisable to reduce the air content to a minimum during the filling operation.

The method of filling has an influence on the amount of air trapped in aerosol containers. Products that are cold-filled normally contain relatively little air. The pressure due to air usually does not exceed 3–4 psi. In this procedure, the concentrate and propellant physically displace most of the air in the container when they are added. The combined liquid volumes of the concentrate and propellant can amount to as high as 85% of the volume of the container. Also, the propellant, which is added to the cold concentrate, generally vaporizes sufficiently so that most of the air remaining after addition of concentrate and propellant is displaced by the vaporizing propellant.

The problem of excess air occurs with pressure filling if no provision is made to eliminate the air. In this method, the concentrate added to the container displaces the amount of air corresponding to the volume of the concentrate. If the container is then capped, the air remaining in the container is trapped. When the propellant is pressure-filled, the propellant will compress the air already in the container and increase the pressure even further.

Most modern commercial pressure-filling equipment is capable of removing air from the containers either by evacuation or purging before the valve is crimped.[11] Purging of the vapor phase with propellant vapor or liquid propellant before the valve is crimped can be quite effective, but certain precautions must be observed. If liquid propellant is dropped into the container, the temperature of the concentrate must be high enough so that the propellant will vaporize immediately after it contacts the concentrate surface. If the propellant is easily soluble in the concentrate, enough propellant may dissolve in the concentrate so that purging is ineffective.

The technique of purging with propellant vapor is important. In one effective procedure, a slow stream of propellant vapor is started through the purging tube and the tube lowered into the container until it almost touches the surface of the concentrate. The purging tube is then raised slowly out of the container, allowing the propellant vapor to displace the air in the vapor phase. Purging with too vigorous a flow of vapor may create turbulence and result in ineffective purging. Regardless of whether purging is carried out with liquid or vapor, the container should be capped with the valve as soon as possible after purging has been completed. If the interval between purging and capping is too long, the propellant in the vapor phase may dissolve in the concentrate, allowing air to re-enter the container.

Attempts have been made to eliminate air by first purging the empty container with propellant and then adding concentrate. Usually this is not a very effective procedure because the turbulence caused by the addition of the concentrate results in some air remaining in the aerosol container. Propellant vapor purging with the purging tube outside and above the container has been tried with little success.

Another method for reducing air content is to premix some of the propellant with the concentrate before the concentrate is added to the container.[12] The concentration of propellant in the concentrate should be such that the mixture has a slight positive pressure. The concentrate then will be self-purging after it has been added to the container.

One of the first methods considered for removing air from pressure-filled products was to invert the container and spray enough material so that the vapor phase was discharged. This procedure removes the air initially present in the vapor phase and thus reduces the air concentration. However, it will not remove all the air in the aerosol because a proportion of the air is dissolved in the liquid phase. If the container is shaken after the vapor phase has been removed by venting, air will leave the liquid phase and enter the vapor phase in order to re-establish equilibrium between the liquid and vapor phases. Repeating this procedure will continue to reduce the air concentration. However, the air will never be eliminated entirely because of the air that remains in the liquid phase after each venting. This procedure is impractical because of the loss of propellant that occurs during the venting.

In laboratory work, it is sometimes desirable to reduce the air to a concentration below that obtained with either cold filling or purging. This may be accomplished by adding a slight excess of the propellant to the concentrate and boiling off the excess.[13] For example, if fluorocarbon 12/fluorocarbon 11/ethyl alcohol mixtures with a very low air content are desired, ethyl alcohol is loaded into the container, a slight excess of fluorocarbon 11 is added, the excess propellant is boiled off, and the container is capped while fluorocarbon 11 vapors are still overflowing the container. Fluorocarbon 12 is then pressure-filled.

REFERENCES

1. F. H. Getman and F. Daniels, "Outlines of Theoretical Chemistry," John Wiley & Sons, Inc., New York, 1931, p. 163.
2. Freon® Aerosol Report FA-22, "Vapor Pressure and Liquid Density of "Freon" Propellants."
3. F. H. MacDougall, "Physical Chemistry," The Macmillan Company, New York, 1936, p. 244.
4. Freon® Aerosol Report FA-15, "The Effect of Air on Pressure in Aerosol Containers."
5. L. Flanner, *Aerosol Age* 9, 27 (March 1964).
6. P. A. Sanders, *Aerosol Age* 11, 30, 31 (January/February 1966).
7. "Aerosol Guide," Aerosol Division, Chemical Specialties Manufacturers Association (CSMA), Inc., 6th ed., March 1971.

8. P. A. Sanders, *Aerosol Age* **11,** 17, 32 (August/September 1966) (Freon® Aerosol Report A-68).
9. F. L. LaQue and R. R. Copson, "Corrosion Resistance of Metals and Alloys," 2nd ed., Reinhold Publishing Corp., New York, 1963.
10. P. A. Sanders, *Soap Chem. Spec.* **42,** 74, 135 (July/August 1966) (Freon® Aerosol Report A-66).
11. Technical Bulletin, "Undercap Filling," Kartridg Pak Company.
12. F. A. Bower and R. G. Appenzeller, *Proc. 45th Ann. Meet. Chem. Spec. Manuf. Assoc.* (December 1958) (Freon® Aerosol Report A-45).
13. Freon® Aerosol Report A-51, "Freon® 11 S Propellant."
14. P. A. Sanders, "Aerosol Laboratory Manual," Industry Publications, Inc., Cedar Grove, N.J., 1975.
15. Freon® Products Laboratory Data.

9

SPRAY CHARACTERISTICS

One of the most important objectives in developing an aerosol formulation is to obtain a product with a satisfactory spray. Aerosols for inhalation therapy must produce a spray with sufficiently small particles ($10\,\mu$ or less) so that the product can reach the lungs. For space applications such as room deodorants or space insecticides, a fine spray with a small particle size (droplet size) is necessary so that the particles will remain suspended in the air long enough to be effective.

At the other extreme are products for residual applications such as window cleaners or insecticides for crawling insects. In this case, a coarse, wet spray, or even a stream, is required so that the product remains on the surface to which it is applied. This is desirable for product efficiency. In between these classes of products with very wet or very fine sprays is a whole series of aerosols in which sprays with medium particle size are most suitable. These include personal deodorants, hair sprays, colognes, coating compositions, etc.

MECHANISM OF SPRAY FORMATION

The overall process involved in the formation of a spray when an aerosol is discharged is the change in the propellant from liquid to gas. When the propellant vaporizes, it breaks up the concentrate into droplets. However, a common misconception is that all of the propellant flashes into vapor immediately as soon as the product leaves the outer orifice in the actuator. Investigations by York[1] and Wiener[2] showed that only a minor proportion of the propellant vaporizes at this point. York calculated that if a fluorocarbon 12/fluorocarbon 11 (50/50) blend were discharged from an aerosol container, approximately 14% of the propellant would flash into vapor. Wiener carried out a similar series of calculations for mixtures of fluorocarbon 12 with fluorocarbon 11 and fluorocarbon 114. The results are presented in a series of graphs showing percent flashing as a function of temperature for each propellant or propellant blend. For example, Wiener calculated that approximately 19% of a fluorocarbon 12/fluorocarbon 11 (50/50) blend would flash into vapor when sprayed into a room at 70°F. Thus, Wiener's calculations correlate with those of York.

The process of flashing when propellant is discharged is so rapid that the energy required for vaporization has to be drawn from the propellant itself and not from the surrounding air. The withdrawal of energy from the system cools the remaining liquid propellant droplets. Flash vaporization and resultant cooling continue until the spray droplets reach their boiling point at atmospheric pressure. This occurs almost immediately after the propellant leaves the outer orifice.

Some flash vaporization occurs initially in the expansion chamber of the valve before an aerosol leaves the actuator. When the aerosol enters the expansion chamber, a portion of the propellant vaporizes because the expansion chamber is at a lower pressure than the container. This vaporization cools the aerosol to a temperature corresponding to its boiling point at the expansion chamber pressure. When the resulting vapor–liquid mixture leaves the outer orifice, additional flashing of propellant occurs until the droplets are cooled to a temperature corresponding to the boiling point of the mixture at atmospheric pressure. After these two flash vaporizations, slower evaporation of the liquid droplets occurs with the vapor pressure essentially equal to atmospheric pressure. The rate of evaporation depends upon the rate of heat transfer from the atmosphere to the droplets.

The flashing of propellant is primarily responsible for the initial break-up of the liquid aerosol stream into droplets. The higher the proportion of propellant that flashes, the smaller the particle size of the droplets in the spray. Some of the factors that contribute to a high degree of flashing of propellant are a high specific heat, a low heat of vaporization, and a low boiling point.[2] Propellants with a high specific heat have a comparatively high reservoir of energy readily available for vaporization; those with a low heat of vaporization require less energy for vaporization. Propellants with a low boiling point (a high vapor pressure) have to be cooled to a greater extent than higher boiling propellants to reach the temperature at which the vapor pressure of the spray equals atmospheric pressure. This requires a greater degree of flashing.

Other factors such as surface tension, drag forces on the droplets, coalescence of the droplets, etc., are involved in the formation of a spray. However, relatively little is known about their importance.[1]

After the flash vaporizations in the expansion chamber and outer actuator orifice, the droplets continue to decrease in size as they travel through the air owing to secondary evaporation at atmospheric pressure. Therefore, the particle size depends upon the distance between the droplets and the valve. The secondary evaporation can be fairly rapid. When a solution of fluorocarbon 12/fluorocarbon 11 (50/50) is sprayed against a hand held at a distance of 1 in. from the valve, a considerable amount of the cold liquid propellant is deposited on the hand. At a distance of 2 ft, all the propellant has evaporated and there is no sensation of cold.

When high-boiling solvents with a low rate of evaporation are present, the change in droplet size after evaporation of propellant will be less than that in aerosols formulated with lower boiling solvents such as ethyl alcohol. Rapid evaporation of alcohol occurs, and droplets may evaporate almost completely when only a few feet from the valve.

FACTORS AFFECTING SPRAY CHARACTERISTICS

Spray characteristics may be varied by changing the concentration or type of propellant, valve, actuator, or solvent. The two most important propellant variables that affect the spray are vapor pressure and concentration in the product.

In the following section, the spray characteristics of various systems are described as very fine, fine, medium, coarse, etc. This classification is based on visual observation of the sprays. The terms have the following meaning, as used in this section:[3]

1. *Very fine.* The spray disappears at a distance of approximately 2-3 ft after it leaves the actuator. The spray does not wet paper at a distance of about 1 ft.

2. *Fine.* The spray travels for a distance of at least 5-6 ft before disappearing. Slight wetting of paper occurs at a distance of about 1 ft.

3. *Medium.* The spray tends to travel in a horizontal path and the particle size is noticeably larger than that in a fine spray. Wetting of paper occurs at a distance of about 1 ft.

4. *Coarse.* Fallout of large droplets is apparent. Heavy wetting of paper occurs at a distance of 2 ft.

5. *Streamy.* A broken stream consisting of a mixture of spray and stream.

6. *Stream.* A stream with essentially no spray.

Sprays with properties intermediate between the groups listed above are classified as fine-very fine, medium-fine, medium–coarse, etc.

Variation in Propellant Concentration and Type

An increase in propellant concentration decreases particle size. Energy is required to break up the liquid aerosol stream into particles or droplets. Since the propellant is the initial source of energy, an increase in propellant concentration results in a decrease in droplet size. At high propellant concentrations, a comparatively large quantity of energy is available to break up a relatively low amount of concentrate. This produces small particles. At lower propellant concentrations, proportionately less energy is available, and this has to be distributed

TABLE 9-1   EFFECT OF VARIATION IN FLUOROCARBON
152a/ETHYL ALCOHOL RATIO UPON
SPRAY CHARACTERISTICS[19]

Composition (wt. %)		Vapor Pressure (psig) at 70°F	Spray Characteristics[a]
FC-152a	Ethyl Alcohol		
90	10	58.0	Very fine
80	20	53.5	Very fine
60	40	45.5	Fine
50	50	42.0	Medium–fine
40	60	37.5	Medium
30	70	29.5	Medium–coarse
20	80	18	Coarse
15	85	11	Streamy
10	90	3.5	Stream

[a]Precision valve, 0.18-in. stem orifice; standard actuator, .018-in. orifice.

throughout a larger quantity of concentrate. The effect of varying concentrations of fluorocarbon 152a (62.5 psig at 70°F) and fluorocarbon 142b (29.1 psig at 70°F) in ethyl alcohol solutions upon spray characteristics is illustrated in Tables 9-1 and 9-2.

As a general rule, low-boiling propellants with high vapor pressures give finer sprays at the same concentration in an aerosol than higher boiling propellants with lower vapor pressures. A propellant with a high vapor pressure will flash to a greater extent than a propellant with a lower vapor pressure and give a finer spray. This can be seen from the data in Table 9-3, which shows the effect of

TABLE 9-2   EFFECT OF VARIATION IN FLUOROCARBON
142b/ETHYL ALCOHOL RATIO UPON
SPRAY CHARACTERISTICS[19]

Composition (wt. %)		Vapor Pressure (psig) at 70°F	Spray Characteristics[a]
FC-142b	Ethyl Alcohol		
80	20	25	Very fine
60	40	20	Fine
50	50	17	Medium
40	60	13	Coarse
30	70	7.5	Streamy
20	80	2.0	Stream

[a]Precision valve, .018-in stem orifice; standard actuator, .018-in orifice.

TABLE 9-3   SPRAY CHARACTERISTICS OF FLUOROCARBON
152a/FLUOROCARBON 133a/ETHYL ALCOHOL MIXTURES[19]

Composition (wt. %)			Vapor Pressure (psig) at 70°F	Spray Characteristics[a]
FC-152a	FC-133a	Ethyl Alcohol		
–	50	50	3	Stream
10	40	50	16.5	Coarse
20	30	50	27.0	Medium
30	20	50	33.5	Medium
40	10	50	39.0	Fine
50	–	50	42.5	Fine

[a]Precision valve, .018-in. stem orifice; standard actuator, .018-in. orifice.

changing the boiling point and vapor pressure of the propellant upon spray characteristics. These data illustrate how varying the ratio of fluorocarbon 152a to fluorocarbon 133a at a constant composition of 50 wt. % propellant and 50 wt. % ethyl alcohol affects the spray. Fluorocarbon 152a has a comparatively high vapor pressure and low boiling point. It gives a fine spray. Fluorocarbon 133a boils at about room temperature and has a low vapor pressure (11.0 psig at 70°F). The formulation with fluorocarbon 133a gives no spray. Blends of fluorocarbon 152a and fluorocarbon 133a give sprays in between these two extremes, with the particle sizes becoming smaller as the proportion of fluorocarbon 152a increases.

There is an exception to the rule that propellants with high vapor pressures give finer sprays at the same concentration in an aerosol than propellants with low vapor pressures. This occurs when the propellant with the high vapor pressure forms association complexes with the solvent, for example, fluorocarbon 22 and ethyl alcohol. Although fluorocarbon 22 itself has a higher vapor pressure than fluorocarbon 12 (121.4 psig at 70°F versus 70.2 psig), solutions of fluorocarbon 22 with ethyl alcohol at low propellant concentrations have lower vapor pressures than corresponding solutions with fluorocarbon 12. Ultimately, the vapor pressure/concentration curves cross each other. These curves are useful in clarifying the roles played by propellant vapor pressure and concentration in determining spray characteristics.

By selecting the appropriate compositions from these curves, the spray characteristics of fluorocarbon 22/ethyl alcohol and fluorocarbon 12/ethyl alcohol mixtures can be compared under the following conditions: (1) same fluorocarbon concentration–different vapor pressures; (2) different fluorocarbon concentrations–same vapor pressures; and (3) same fluorocarbon concentrations–same vapor pressures. These comparisons indicate how fluorocarbon concentration and vapor pressure of the aerosol affect spray properties.

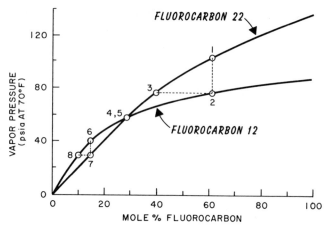

Figure 9-1 Vapor pressures of fluorocarbon 22/ethyl alcohol and fluorocarbon 12/ethyl alcohol mixtures as a function of mole percent concentration of fluorocarbon.

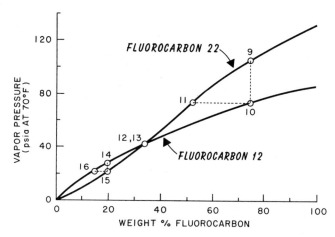

Figure 9-2 Vapor pressures of fluorocarbon 22/ethyl alcohol and fluorocarbon 12/ethyl alcohol mixtures as a function of weight percent concentration of fluorocarbon.

The curves showing the vapor pressures of fluorocarbon 22/ethyl alcohol and fluorocarbon 12/ethyl alcohol mixtures as a function of the mole percent concentration of the fluorocarbons are presented in Figure 9-1; those illustrating the variation of vapor pressure with weight percent concentration of the two fluorocarbons are shown in Figure 9-2. The numbers on Figure 9-1 and Figure 9-2 indicate the compositions of the various formulations whose spray properties were compared.[19] They are as follows:

Figure 9.1.

Formulations 1 and 2. Same mole percent concentrations of fluorocarbons 22 and 12, but vapor pressures are different.

Formulations 2 and 3. Different mole percent concentrations of fluorocarbons 22 and 12, but vapor pressures are the same.

Formulations 4 and 5. Mole percent concentrations and vapor pressures of fluorocarbons 22 and 12 are the same.

Formulations 6 and 7. Similar to formulations 1 and 2—same mole percent concentrations of fluorocarbons 22 and 12, but vapor pressures are different.

Formulations 7 and 8. Similar to formulations 2 and 3—different mole percent concentrations of fluorocarbons 22 and 12, but vapor pressures are the same.

The mole percent and weight percent concentrations of the formulations, vapor pressures, and results of the comparison of the spray characteristics of the samples are given in Table 9-4. Conclusions from the data in Table 9-4 are as follows:

1. At the same vapor pressure and mole percent concentration, different propellants give similar spray characteristics.
2. At the same vapor pressure but different mole percent concentrations, the formulation with the higher mole percent concentration of fluorocarbon produces a finer spray.
3. At the same mole percent concentration of fluorocarbon but different vapor pressures, the formulation with the higher vapor pressure produces a finer spray.

These results show that both vapor pressure and mole percent concentration affect spray characteristics. However, they do not indicate whether vapor pressure is more important than mole percent concentration, or vice versa.

Figure 9-2.

Formulations 9 and 10. Same weight percent concentrations but different vapor pressures.

Formulations 10 and 11. Different weight percent concentrations but same vapor pressures.

Formulations 12 and 13. Same weight percent concentrations and vapor pressures.

Formulations 14 and 15. Same weight percent concentrations but different vapor pressures.

Formulations 15 and 16. Different weight percent concentrations but same vapor pressures.

The weight percent and mole percent concentrations of the fluorocarbons in the formulations, vapor pressures, and results of the comparison of the spray

TABLE 9-4   SPRAY CHARACTERISTICS OF FLUOROCARBON 22/ETHYL ALCOHOL AND FLUOROCARBON 12/ETHYL ALCOHOL MIXTURES

COMPARISONS AT EQUAL MOLE PERCENT CONCENTRATIONS AND/OR VAPOR PRESSURES

Formulation No.	Fluorocarbon	Fluorocarbon/Ethyl Alcohol Ratio Mole %	Wt. %	Vapor Pressure (psia) at 70°F	Spray Comparison	Spray Characteristics
1	22	62/38	75/25	105	1 = 2	1 = 2. Both are very fine; no fallout; both are too fine to detect differences
2	12	62/38	81/19	78		
2	12	62/38	81/19	78	2 > 3 finer	2 is finer than 3. 2 is very fine; 3 is fine with slight fallout
3	22	39/61	55/45	78		
4	12	29/71	51/49	60	4 = 5	4 = 5. Medium–fine sprays; slight fallout
5	22	29/71	43/57	60		
6	12	15/85	32/68	42	6 > 7 finer	6 is finer than 7.6 is medium-coarse and wet; 7 is coarse and streamy with more fallout
7	22	15/85	25/75	30		
7	22	15/85	25/75	30	7 = 8	7 = 8. Coarse and streamy; fallout
8	12	9/91	20/80	30		

characteristics of the samples are given in Table 9-5. Conclusions from the data in Table 9-5 are as follows:

1. At the same vapor pressure, formulations with higher weight percent concentrations of fluorocarbon produce finer sprays. However, the formulations with higher weight percent concentrations also have higher mole percent concentrations.

2. At the same vapor pressure and weight percent concentration, formulations with higher mole percent concentrations produce finer sprays. This shows that mole percent concentration is more important than weight percent concentration.

3. At the same weight percent concentration and different vapor pressures, formulations with higher vapor pressures produce slightly finer sprays. The mole percent concentration of propellant in the latter was somewhat lower, indicating that in this particular instance, vapor pressure was more important than mole percent concentration.

4. At the same vapor pressure, formulations with higher mole percent and weight percent concentrations of propellant produce finer sprays.

As in the previous case, the data indicate that both the vapor pressure and the concentration of propellants affect spray characteristics. Again, the data are not sufficient to determine if one is more important than the other. It is probable that there is an optimum propellant concentration and vapor pressure for the best spray characteristics. Propellants with a high vapor pressure but low molar concentration give a poor spray, for example, the compressed gases. Propellants with a high molar concentration but low vapor pressure also produce a poor spray, for example, fluorocarbon 11.

Spray Characteristics of Fluorocarbon 152a/Ethyl Alcohol/Water Solutions[19]

At constant fluorocarbon concentrations, the substitution of water for ethyl alcohol in fluorocarbon 152a/ethyl alcohol mixtures has little effect upon spray characteristics, even though the pressure increases. At constant vapor pressures, increasing concentrations of water in fluorocarbon 152a/ethyl alcohol mixtures cause the spray to become increasingly coarser. In these systems, it is necessary to lower the concentration of fluorocarbon 152a as the concentration of water is increased in order to maintain a constant pressure. This indicates that in the fluorocarbon 152a/ethyl alcohol/water systems, spray characteristics are more dependent upon fluorocarbon concentration than vapor pressure.

Curves for compositions of fluorocarbon 152a/ethyl alcohol/water solutions having vapor pressures of 50, 40, 30, and 20 psig at 70°F are shown on the triangular coordinate chart of Figure 9-3. By selecting the appropriate compositions from these curves, the spray characteristics of fluorocarbon 152a/ethyl alcohol/water formulations can be compared under the following conditions:

TABLE 9-5 SPRAY CHARACTERISTICS OF FLUOROCARBON 22/ETHYL ALCOHOL AND FLUOROCARBON 12/ETHYL ALCOHOL MIXTURES

COMPARISONS AT EQUAL WEIGHT PERCENT CONCENTRATIONS AND/OR VAPOR PRESSURES

Formulation No.	Fluorocarbon	Fluorocarbon/Ethyl Alcohol Ratio Wt. %	Mole %	Vapor Pressure (psia) at 70°F	Spray Comparison	Spray Characteristics
9	22	75/25	62/38	105	9 = 10	9 = 10. Very fine; no fallout; both are too fine to detect differences
10	12	75/25	53/47	75		
10	12	75/25	53/47	75	10 > 11 finer	10 is finer than 11. 10 is very fine; 11 is medium–fine with slight fallout
11	22	53/47	40/60	75		
12	12	34/66	16/84	44	13 > 12 finer	13 is slightly finer than 12. Medium sprays with fallout
13	22	34/66	21/79	44		
14	12	20/80	9/91	30	14 > 15 finer	14 is finer than 15. 14 is coarse and streamy; 15 is very coarse and streamier than 14
15	22	20/80	12/82	24		
15	22	20/80	12/82	24	15 > 16 finer	15 is slightly finer than 16. 16 is almost a stream with heavy fallout
16	12	15/85	6/94	24		

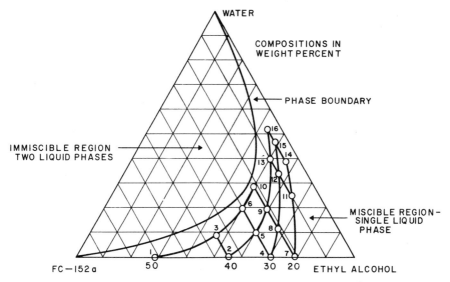

Figure 9-3 Vapor pressures (psig) of fluorocarbon 152a/ethyl alcohol/water solutions at 70°F.

(1) same fluorocarbon concentrations–different vapor pressures; (2) different fluorocarbon concentrations–same vapor pressures. The numbers on the triangular coordinate chart of Figure 9-3 indicate the compositions of the various formulations whose spray characteristics were compared.

Comparative spray characteristics of formulations with the same fluorocarbon 152a concentration but increasing concentrations of water are given in Table 9-6. Spray properties of formulations with the same vapor pressures but different concentrations of water are listed in Table 9-7.

Effect of Discharge

In any propellant blend where the components have widely differing boiling points and vapor pressures, for example, fluorocarbon 22 and isopentane, the composition of the vapor phase will be significantly different from that of the liquid phase. Thus, the vapor phase over a fluorocarbon 22/isopentane (35/65) blend has a composition of approximately 86.5 wt.% fluorocarbon 22 and 13.5 wt.% isopentane. As the product is discharged, the volume of the vapor phase increases. More propellant vaporizes to maintain equilibrium between the liquid and vapor phases. Since the propellant that vaporizes is predominantly fluorocarbon 22, the concentration of fluorocarbon 22 in the liquid phase decreases. The composition of the liquid phase therefore changes, with the concentration of fluorocarbon 22 relative to that of isopentane becoming correspondingly lower. The vapor pressure and spray rate of such a blend de-

TABLE 9-6 SPRAY CHARACTERISTICS OF FLUOROCARBON
152a/ETHYL ALCOHOL/WATER SOLUTIONS

COMPARISONS AT CONSTANT FLUOROCARBON 152a CONCENTRATIONS

Formulation No.	Composition (wt.%)			Vapor Pressure (psig) at 70°F	Spray Characteristics[a]
	FC-152a	Water	Ethyl Alcohol		
1	72	–	28	50	Very fine
2	45	–	55	40	Fine
3	45	9	46	50	Fine
4	30	–	70	30	Medium
5	30	10	60	40	Medium
6	30	20	50	50	Medium
7	21.5	–	78.5	20	Coarse
8	21.5	11.5	67	30	Coarse
9	21.5	19.5	59	40	Coarse
10	21.5	29	49.5	50	Coarse–streamy
11	10	25	65	20	Streamy
12	10	34	56	30	Streamy
13	10	40	50	40	Streamy
14	5	38.5	56.5	20	Stream
15	5	47	48	30	Stream
16	5	52	43	40	Stream

[a]Precision valve, .018-in. stem orifice; standard actuator, .018-in. orifice.

crease with discharge. The decrease may be sufficient to cause a change in spray characteristics.

When the product is equipped with a vapor tap valve, the decrease in vapor pressure and spray rate is aggravated. The vapor phase, which is rich in fluorocarbon 22, is lost continually through the vapor tap during discharge. This decreases the concentration of fluorocarbon 22 in the liquid phase correspondingly. This is illustrated in Table 9-8, which shows how the properties and composition of a fluorocarbon 22/isopentane (35/65) blend change with discharge. Data were obtained using five different valves with vapor taps having diameters from 0 (no vapor tap) to .040 in. Vapor pressures of the formulations varied from 52 to 56 psig at 70°F and were normalized to 50 psig. Compositions of the liquid phase after discharge were obtained from their vapor pressures using a composition/vapor pressure curve for fluorocarbon 22/isopentane blends. Compositions of the vapor phase were calculated.

The decrease in vapor pressure and spray rate with discharge became more

TABLE 9-7   SPRAY CHARACTERISTICS OF FLUOROCARBON
152a/ETHYL ALCOHOL/WATER SOLUTIONS

COMPARISONS AT CONSTANT VAPOR PRESSURE

Formulation No.	Composition (wt.%)			Vapor Pressure (psig) at 70°F	Spray Characteristics[a]
	FC-152a	Water	Ethyl Alcohol		
1	72	–	28	50	Very fine
3	45	9	46	50	Fine
6	30	20	50	50	Medium
10	21.5	29	49.5	50	Coarse–streamy
2	45	–	55	40	Fine
5	30	10	60	40	Medium
9	21.5	19.5	59	40	Coarse
13	10	40	50	40	Streamy
16	5	52	43	40	Stream
4	30	–	70	30	Medium
8	21.5	11.5	67	30	Coarse
12	10	34	56	30	Streamy
15	5	47	48	30	Stream
7	21.5	–	78.5	20	Coarse
11	10	25	65	20	Streamy
14	5	38.5	56.5	20	Stream

[a]Precision valve, .018-in. stem orifice; standard actuator, .018-in. orifice.

pronounced as the diameter of the vapor tap increased. The change in spray characteristics was less than would be expected from the change in liquid phase composition, vapor pressure, and spray rate when the vapor tap diameter did not exceed .020 in. This may be because as the vapor tap diameter becomes larger, the contribution of the vapor phase to the spray characteristics becomes correspondingly greater. Spray characteristics are a function of two factors: the liquid phase composition and the quantity of the vapor phase which is mixed with the liquid phase. As the diameter of the vapor tap is increased, the concentration of fluorocarbon 22 in the liquid phase decreases to a greater extent during discharge. This would tend to make the spray coarser. However, this is compensated to some extent by the increased flow from the larger vapor tap. Therefore, the change in spray characteristics which might be predicted from the change in the liquid phase composition during discharge is not as large as would be expected because of the opposing effect of the vapor tap. When the diameter of the vapor tap reached .030 in., the discharge was sputtery and difficult to classify.

TABLE 9-8   EFFECT OF DISCHARGE UPON PROPERTIES OF A FLUOROCARBON 22/ISOPENTANE (35/65) BLEND[19]

Property		Vapor Tap Diameter			
	No Vapor Tap	.013 in.	.020 in.	.030 in.	.040 in.
Vapor Pressure (psig) at 70°F					
Initial	50	50	50	50	50
85% Discharge	47	44	39	29	14
Spray Rate (g/sec) at 70°F					
Initial	.90	.89	.85	.67	.57
85% Discharge	.88	.85	.62	.38	.16
Spray Characteristics					
Initial	Very fine	Very fine	Very fine	Very fine	Very fine
85% Discharge	Very fine	Fine	Fine	Sputters	Sputters
Composition (wt.% ratio of fluorocarbon 22 to isopentane)					
Liquid Phase					
Initial	35/65	35/65	35/65	35/65	35/65
85% Discharge	31/69	29/71	25/75	19/81	11/89
Vapor Phase					
Initial	86.5/13.5	86.5/13.5	86.5/13.5	86.5/13.5	86.5/13.5
85% Discharge	84/16	82.5/17.5	80/20	73/17	59/41

In fluorocarbon 22/ethyl alcohol mixtures, the change in properties during discharge is greater than that with fluorocarbon 22/isopentane blends because alcohol has a lower vapor pressure than isopentane. The vapor phase of a fluorocarbon 22/ethyl alcohol mixture is essentially 100% fluorocarbon 22. It is lost continually through a vapor tap during discharge.

The change in vapor pressure, spray rate, spray characteristics, and liquid phase composition of a fluorocarbon 22/ethyl alcohol (35/65) mixture with discharge is illustrated in Table 9-9. Data were obtained using five different valves with vapor taps having diameters from 0 (no vapor tap) to .040 in. Experimental vapor pressures of the formulations varied from 32 to 34 psig at 70°F and were normalized to 30 psig. Compositions of the liquid phase after discharge were obtained from their vapor pressures by using a fluorocarbon 22/ethyl alcohol composition/vapor pressure curve. The change in spray characteristics became noticeable when the vapor tap diameter reached .020 in., but the product was still usable. When the vapor tap diameter reached .030 in., the change in vapor pressure, spray rate, and spray characteristics was so large that the product was useless after 85% discharge.

Effect of Valves and Actuators

Valves and actuators have a considerable effect upon the spray characteristics and spray rates of aerosol formulations. This has been discussed in Chapter 4, "Hydrocarbon Propellants, Compressed Gases, and Propellant Blends," and in Chapter 6, "Valves and Actuators." The effect of vapor tap valves has been covered in the previous section. Mechanical break-up actuators generally have a marked effect upon spray properties in comparison with the standard actuators: They generally give a finer spray with a wider cone.

The differences in spray characteristics of fluorocarbon 152a/fluorocarbon 133a/ethyl alcohol mixtures packaged with a standard valve and actuator and with a valve having a mechanical break-up actuator are shown in Table 9-10. The mechanical break-up actuator produces a much finer spray, particularly at lower concentrations of fluorocarbon 152a.

Because of the wide variety of valves and actuators available, no attempt will be made to correlate valve design and particle size. Whenever there is a question about the valve most suitable for any given product, valve manufacturers should be consulted.

MEASUREMENT OF PARTICLE SIZE (DROPLET SIZE)

A tentative method for measuring particle sizes of insecticides formulated with high-boiling solvents was developed by the U.S. Department of Agriculture.[4] This method involves spraying the aerosol into a wind tunnel. Droplets in the

TABLE 9-9   EFFECT OF DISCHARGE UPON PROPERTIES OF A FLUOROCARBON 22/ETHYL ALCOHOL (35/65) MIXTURE[19]

Property		Vapor Tap Diameter			
	No Vapor Tap	.013 in.	.020 in.	.030 in.	.040 in.
Vapor Pressure (psig) at 70°F					
Initial	30	30	30	30	30
85% Discharge	27	24	19	3	1
Spray Rate (g/sec) at 70°F					
Initial	.93	.89	.79	.52	.41
85% Discharge	.89	.76	.55	.06	.03
Spray Characteristics					
Initial	Medium	Medium	Medium	Medium	Medium
85% Discharge	Medium	Medium	Medium–coarse	Sputters	Sputters
Composition (wt.% ratio of fluorocarbon 22 to ethyl alcohol)					
Liquid Phase					
Initial	35/65	35/65	35/65	35/65	35/65
85% Discharge	32/68	31/69	27/73	16/84	14/86
Vapor Phase					
Initial	—Essentially 100% fluorocarbon 22 at all times				
85% Discharge	—Essentially 100% fluorocarbon 22 at all times				

TABLE 9-10   SPRAY CHARACTERISTICS OF FLUOROCARBON
152a/FLUOROCARBON 133a (10/90)/ETHYL
ALCOHOL MIXTURES[19]

Composition (wt.%)				
FC-152a/FC-133a (10/90)	Ethyl Alcohol	Vapor Pressure (psig) at 70°F	Valve A[a]	Valve B[b]
90	10	20.0	Medium	Very fine
80	20	18.5	Medium	Fine
70	30	16.5	Medium–coarse	Medium
60	40	14.5	Coarse	Medium
50	50	13.0	Streamy	Medium
40	60	11.0	Streamy	Coarse
30	70	7.0	Stream	Coarse
20	80	3.0	Stream	Very coarse

[a] Valve A—.080-in. body/.018-in. stem and .018-in. standard actuator.
[b] Valve B—.040-in. body/.016-in. stem and .016-in. mechanical break-up actuator.

spray are collected on a silicone-coated microscope slide and the diameters of 200 droplets are determined with a calibrated microscope. Particle size distribution is indicated by a cumulative weight percent method that shows the weight percent of the spray with a particle size smaller than or equal to a given diameter in microns. From the data, it is possible to determine the mass median diameter. This is the droplet size that divides the spray in half, so that 50 wt.% of the particles have a diameter less than the mass median diameter, while the remainder consist of particles with equal or larger diameters. The results may also be plotted as a typical frequency distribution curve.

The wind tunnel has sufficient length so that practically all the propellant flashes off and evaporates from the droplets before they reach the microscope slide. The droplets that reach the slide, therefore, have decreased to essentially their ultimate size. This method is suitable only for aerosols formulated with high-boiling solvents and a high enough concentration of propellant to produce a fine spray. The method is not satisfactory for aerosols with lower boiling solvents such as water or ethyl alcohol. These solvents evaporate completely in the wind tunnel and there is no deposition of droplets on the slide. Also, the particle size of sprays with very coarse particles cannot be measured satisfactorily because many of the droplets fall to the bottom of the tunnel before reaching the slide. Since only the finer particles are deposited on the slide, it is not possible to obtain a correct picture of the particle size distribution.

Lefebvre and Tregan[5] have described a method for determining particle size distribution in which aerosol droplets are caught on a thin film of magnesium oxide deposited on a glass slide. According to the authors, a perfectly circular hole is obtained where the droplet impacts the slide. The diameter of the hole is essentially the same as that of the droplet. Examination of the various holes on the slide indicates the particle size distribution of the spray. Lefebvre and Tregan also discuss the use and limitations of high-speed microphotography in investigations of particle sizes in aerosol sprays. Several other methods of measuring particle size of aerosols are discussed in References 6-12. These methods include collecting and weighing particles on filters of known pore size or precipitating the particles with an electrical precipitator in a tube of known weight. These methods are not generally used in the aerosol industry.

A detailed discussion of recent methods for measuring the size of aerosol particles is available in a collection of papers presented in a seminar and workshop sponsored by the National Bureau of Standards and the Food and Drug Administration in 1974.[13] The publication is entitled, "Aerosol Measurements." The principles discussed in particle size measurement include laser light scattering, optical imaging, Doppler shift, electromobility, piezoelectric effect, and beta-ray absorption. The seminar was organized so that instrument manufacturers, research groups, and government agencies could evaluate the most recent developments in instrumentation for measuring the particle size of aerosols. Some comments from the first paper[14] are summarized below.

The major factors to be considered in a method for measuring the particle size in aerosols are as follows: (1) size ranges; (2) speed of analysis; and (3) cost. For consumer aerosols, a size range of $1-300 \mu$ is considered wide enough for most aerosol droplet size distributions. For particles in this range having the high volatility of aerosol droplets, a maximum imaging and sensing time of 2 sec is desirable. Judging by these criteria, laser holographic methods are the best for most aerosol products. An article on the particle size measurement of an aerosol deodorant using laser holographic methods has been published by Hathaway.[15] The method is rapid and is independent of droplet distribution. Coupling with image analysis equipment improves the method. The cost of this system was estimated to be $60,000 in 1974. With a manual image analysis method, the capital cost was estimated at about $25,000. A light-scattering system provides rapid analysis, but the aerosol has to be discharged in a confined space and must be dilute. The cost of this type of system is estimated at about $10,000. If cost is a major factor, photographic imaging and impaction should be investigated. The photographic equipment is the only investment if measurement of droplet size on the photograph is carried out manually. Droplet size can also be measured with impactors by determining the amount of deposit on each stage of the impactor. The concentration of solids in the aerosol must be known. Sensitized paper can also be employed as a deposition surface with use of a Lundgren-type impactor. Droplet size is estimated by measuring the spots on the paper. The advantage of the impact method is that the collection time is fast; however, the analysis is long and expensive. These devices are estimated to cost about $1,000.

Most aerosol laboratories are not equipped with apparatus for measuring the actual particle size of aerosols. Root[16] has described a procedure for evaluating spray patterns which involves spraying the aerosol against a piece of paper treated with a dye–talc mixture. An advantage of this method is that the paper can be photographed after it has been sprayed, thus providing a permanent record of the spray pattern. Generally, spray patterns are described merely by visual observation. For consistency in visual classification of sprays, it is helpful to prepare a series of standard aerosols. For example, a series of fluorocarbon 152a/ethyl alcohol mixtures in which the fluorocarbon 152a/ethyl alcohol ratio varies from 90/10 to 10/90 in 10% increments can be prepared. The sprays from this series of control samples vary from very fine to stream and can be classified by visual observation as very fine, fine, medium–fine, etc. Whenever it is necessary to determine the type of spray of an aerosol, the product can be compared with samples in the control series. The aerosol product is then given the same classification as the control sample it most closely resembles.

For residual products with very coarse sprays, measurement of the actual particle size is considered to be impractical and difficult.[17] An indication of the particle size of residual products may be obtained, however, by determining the pick-up efficiency of the product. Pick-up efficiency is defined as the percentage of low volatile material deposited on a surface and is considered to be a function of particle size.[18] Pick-up efficiency is determined by spraying an aerosol product against a blotter target which is clamped to an aluminum sheet suspended from a balance. The amount of low volatile material sprayed from the container can be calculated from the weight of aerosol sprayed, assuming that the composition of the aerosol is known. The weight of low volatile material that reaches the target can be determined by the increase in weight of the blotter.

Sometimes it is desirable to replace the original propellant in an aerosol product with a different propellant while maintaining the original spray properties of the aerosol. An empirical method has been developed[10] for determining approximately what concentrations of different propellants will produce about the same spray properties with a given concentrate. The method is based upon a modification of the method for determining pick-up efficiencies of residual products, but it may also be used for comparing the spray properties of aerosols with finer sprays. It assumes that within certain limits, different propellants appear to give approximately the same spray characteristics with the same solvent (or concentrate) when the propellant/solvent ratios are adjusted to give the same pick-up efficiency. As an example, it was determined that mixtures of various propellants with odorless mineral spirits that gave a pick-up efficiency of 70% had the compositions given in Table 9-11.

As an illustration of the use of this type of data, assume that an aerosol product has been formulated with odorless mineral spirits as the solvent and a fluorocarbon 12/fluorocarbon 11 (50/50) blend. For reasons of economy, it may be desirable to use fluorocarbon 12 instead of the fluorocarbon 12/ fluorocarbon 11 (50/50) blend. The data in Table 9-11 indicate that the ratio

TABLE 9-11   PROPELLANT/ODORLESS MINERAL SPIRITS MIXTURES
WITH PICK-UP EFFICIENCIES OF 70%

Propellant	Propellant/Odorless Mineral Spirits Ratio (wt.%)
Fluorocarbon 12	63/37
Fluorocarbon 12/fluorocarbon 11 (70/30)	71/29
Fluorocarbon 12/fluorocarbon 11 (50/50)	81/19

of propellant to concentrate could be reduced from 81/19 about 63/37 while maintaining about the same spray pattern.

Data that illustrate the effect of variation in pick-up efficiencies of fluorocarbon 12/fluorocarbon 14 and fluorocarbon 11/fluorocarbon 114 combinations with both odorless mineral spirits and ethyl alcohol are available in Reference 10.

REFERENCES

1. J. L. York, *J. Soc. Cosmet. Chem.* **7**, 204 (1956).
2. M. V. Wiener, *J. Soc. Cosmet. Chem.* **9**, 289 (1958).
3. P. A. Sanders, "Aerosol Laboratory Manual," Industry Publications, Inc., Cedar Grove, N.J., 1975.
4. "A Tentative Method for Determination of the Particle Size Distribution of Space Insecticide Aerosols," *Proc. 43rd Ann. Meet. Chem. Spec. Manuf. Assoc.* (December 1956).
5. M. Lefebvre and R. Tregan, *Aerosol Age* **10**, 31, 32 (July/August 1965).
6. R. A. Fulton, *Aerosol Age* **4**, 22 (July 1959).
7. K. R. May, *J. Sci. Instr.* **22**, 187 (1945).
8. J. M. Polcher, R. I. Mitchell, and R. E. Thomas, *Proc. 44th Ann. Meet. Chem. Spec. Manuf. Assoc.* (December 1957).
9. W. B. Tarpley, *Aerosol Age* **2**, 38 (December 1957).
10. P. A. Sanders, *Aerosol Age* **11**, 30, 31 (January/February 1966).
11. R. R. Irani and C. F. Callis, "Particle Size, Measurement, Interpretation, and Application," John Wiley & Sons, Inc., New York, 1963; see also *Drug Cosmet. Ind.* **93**, 567, (October 1963).
12. J. H. Burson et al., *Rev. Sci. Instr.* **34**, 1023 (1963).
13. NBS Special Publication 412, "Aerosol Measurements," U.S. Dept. of Commerce, National Bureau of Standards, October 1974.
14. R. Davies and J. D. Stockham, "A Review of the Methods for the Particle Size Analysis of Aerosol Spray Can Droplets," in NBS Special Publication 412, U.S. Department of Commerce, National Bureau of Standards, October 1974.
15. D. Hathaway, *Aerosol Age* **18**, 28 (September 1973).
16. M. J. Root, *Aerosol Age* **2**, 21 (August 1957).
17. A. H. Yeomans, *Proc. 40th Ann. Meet. Chem. Spec. Manuf. Assoc.* (December 1953).
18. "A Tentative Method for Determining Pick-up Efficiency of Residual Aerosol Insecticides," *Proc. 43rd Mid-Year Meet. Chem. Spec. Manuf. Assoc.* (May 1957).
19. Freon® Products Laboratory Data.

10

SOLVENCY

In formulation studies where a homogeneous system (that is, one in which all of the components are mutually soluble) is desired, unnecessary experiments can often be avoided if the solvent properties of aerosol propellants and common aerosol solvents are known. When an aerosol propellant is added to a concentrate (which usually consists of active ingredients dissolved in a solvent or mixture of solvents), addition of the propellant results in the formation of a new solvent system. The new solvent system consists of propellant plus solvent (or solvents) in the concentrate. The new solvent system may be a better or a poorer solvent for the active ingredients than that in the original concentrate. It is possible for an active ingredient to be quite soluble in a solvent such as ethyl alcohol, but relatively insoluble in a mixture of propellants and ethyl alcohol. In some cases, it is necessary to change the propellant in order to obtain a homogeneous system.

Fluorocarbon propellants are soluble in a wide variety of organic solvents, and they are also solvents for a wide variety of compounds. A considerable amount of miscellaneous data has been accumulated on the solubility of propellants in a number of compounds and is available in Reference 1. Fluorocarbon propellants are not miscible with water nor with many glycols.[2]

A number of different physical properties of propellants and solvents are useful for comparing the solvent properties of a compound with other compounds or in predicting the miscibility or solvent properties of the compound itself. Two of these—Kauri-Butanol values and solubility parameters—will be discussed in the following sections. Kauri-Butanol values provide an indication of the relative solvent properties of a liquid in comparison with other liquids. Solubility parameters are useful for predicting whether two liquids will be miscible, an amorphous solid or polymer will dissolve in a solvent, or a propellant will attack a valve gasket elastomer.

KAURI-BUTANOL (K-B) VALUES

Kauri-Butanol values indicate the relative solvent powers of liquids. The test initially was designed for evaluating hydrocarbon solvents used in paint and lacquer formulations and was restricted to solvents with a boiling point above 104°F (40°C).[3] However, its use has been extended to include the lower boiling hydrocarbon propellants as well as the fluorinated propellants, for which it also appears to be applicable. Although the test has been criticized because of its artificial nature,[4] there appears to be a correlation between the Kauri-Butanol values and the more fundamental solubility parameters.

The Kauri-Butanol value of a solvent is the number of milliliters required to produce a certain degree of turbidity when added to 20 g of a standard solution of Kauri resin in n-butyl alcohol at 25°C. The concentration of Kauri resin in n-butyl alcohol has been standardized so that 20 g of the Kauri-Butanol solution requires the addition of 105 ml of toluene to reach the end point. Toluene, with a K-B value of 105, is the standard for solvents with Kauri-Butanol values over 60, while an n-heptane/toluene (75/25) blend, with a K-B value of 40, is the standard for solvents with K-B values under 60.

Because of the higher vapor pressures of liquefied gas propellants, the test procedure had to be modified to include the use of pressure equipment. In carrying out the test, 20 g of standard Kauri-Butanol solution is placed in a suitable pressure vessel, for example, a 4-oz aerosol bottle. The bottle is capped with a standard valve not having a dip tube and placed over a sheet of No. 10 print (normal newspaper print is satisfactory). The solution is then titrated with propellant using a pressure burette. The end point is reached when the sharp outlines of the 10-point print, as viewed through the liquid in the bottle, become obscured or blurred due to precipitation of the Kauri resin from solution. Details of the Kauri-Butanol test are given in References 3 and 24.

Kauri-Butanol values for fluorocarbons are listed in Table 10-1.[5] Fluorocarbon 21 has the best solvent properties, followed by fluorocarbons 132b, 11, 123, and 141b. The rest of the fluorocarbons have relatively poor solvent properties.

n-Butane, isobutane, propane, n-pentane, and isopentane all have Kauri-Butanol values of about 25 and are relatively poor solvents. Methylene chloride has a Kauri-Butanol value of 136. It is an excellent solvent. Kauri-Butanol values of a wide variety of organic solvents are available in References 4 and 6.

SOLUBILITY PARAMETERS

Solubility parameters provide a means for predicting whether mixtures of liquids will be miscible, an amorphous solid will dissolve in a solvent or mixture of solvents, or an elastomeric valve gasket will be affected by an aerosol propellant.

TABLE 10-1   SOLVENT PROPERTIES OF THE FLUOROCARBONS[5,6,17]

Fluorocarbon Number	Formula	Kauri–Butanol Value	Solubility Parameter
METHANE SERIES			
11	CCl_3F	60	7.5
12	CCl_2F_2	18	6.1
21	$CHCl_2F$	102	8.0
22	$CHClF_2$	25	6.5
31	CH_2ClF	35	9.0
32	CH_2F_2	–	7.5
ETHANE SERIES			
113	CCl_2FCClF_2	31	7.2
114	$CClF_2CClF_2$	12	6.2
123	$CHCl_2CF_3$	60	7.3
124	$CHClFCF_3$	22	6.8
132b	CH_2ClCCl_2F	72	7.9
133a	CH_2ClCF_3	22	7.7
134a	CH_2FCF_3	–	6.6
141b	CH_3CCl_2F	58	7.6
142b	CH_3CClF_2	20	6.8
143a	CH_3CF_3	–	5.8
152a	CH_3CHF_2	11	7.0
BUTANE SERIES			
C-318	C_4F_8 (cyclic)	10	5.0

Like boiling points and vapor pressures, solubility parameters are physical constants for compounds. Solubility parameters have been determined for a large number of organic solvents, propellants, and polymers. If two solvents have approximately the same solubility parameters, the chances are that they will be miscible. Therefore, by finding the solubility parameter for any liquid in a table and comparing it with the solubility parameter of another liquid, it is possible to predict the miscibility of the two liquids. If a resin has a solubility parameter similar to that of a solvent, the resin usually will dissolve in the solvent. The method fails in a number of cases, as will be discussed later, but a comprehension of solubility parameter theory makes possible a better understanding of much of the solubility phenomena encountered in the aerosol field.[7] Burrell[8,9] and Crowley, Teague, and Lowe[10,11] have written excellent articles on solubility parameters. The solubility parameters of a large variety of compounds are available in their papers.

The concept of solubility parameter was originated by Hildebrand.[12] In 1916, he suggested that a molecule will be most attracted by other molecules that have the same internal pressure. Thus, if solute and solvent molecules have the same internal pressures, an ideal solution will be formed. But if the internal pressures differ sufficiently, the molecules with the greater internal pressure will associate only with themselves and exclude those with lesser internal pressures. In this case, immiscibility will occur. In between these two extremes are mixtures of two components with internal pressures not sufficiently different to cause immiscibility. In these systems, a lessening of the attractive forces between the solute and solvent molecules occurs, and the partial pressures are greater than calculated by Raoult's Law.

The derivation of solubility parameters starts with the free-energy equation from thermodynamics, which is written as follows:

$$\Delta F = \Delta H - T \Delta S \tag{10-1}$$

This equation states that the change in free energy (ΔF) in any process is equal to the heat of mixing (ΔH) minus the product of the absolute temperature (T) and the change in entropy (ΔS). The process involved could be the mixing of two liquids, the dissolving of a solid in a liquid, etc. Burrell explains entropy as a measure of disorder or randomness, or it may be considered as a measure of the freedom of movement. As an example, a molecule tightly bound in a crystal lattice has low entropy, while a gas molecule which is essentially unrestricted in its movement has high entropy. Processes tend to proceed so that the system will increase in entropy.

One of the most important applications of the equation is that it can be used to determine whether any process will take place. If the change in the free energy (ΔF) is negative, then the process will occur. For example, assume that it is desirable to know whether or not a given material is soluble in one of the fluorocarbon propellants and that it is not possible to determine this experimentally. If the heat of mixing (ΔH), the temperature (T), and the change in entropy (ΔS) of the system were known, then the change in free energy (ΔF) could be calculated. If the heat of mixing (ΔH) were less than $T \Delta S$, then the change in free energy (ΔF) would be negative. Therefore, the process would take place, that is, the material would dissolve in the fluorocarbon.

In order to arrive at the point where solubility parameters come into the picture, it is necessary to rearrange Equation 10-1 and substitute a term containing solubility parameters for the heat of mixing. Hildebrand has shown that when two components are involved in a process, the heat of mixing (ΔH) is as follows:

$$\Delta H_m = V_m \left[\left(\frac{\Delta E_1}{V_1} \right)^{1/2} - \left(\frac{\Delta E_2}{V_2} \right)^{1/2} \right]^2 \phi_1 \phi_2 \tag{10-2}$$

where

ΔH_m = Overall heat of mixing (cal)
V_m = Total volume of the mixture (cc)

ΔE = Energy of vaporization of component 1 or 2 (cal)
V = Molar volume of component 1 or 2 (cc)
ϕ = Volume fraction of component 1 or 2

The term $\Delta E/V$ is the energy of vaporization per cubic centimeter and has been described as the cohesive energy density. It indicates the amount of energy that has to be absorbed by 1 cc of a liquid to overcome the intermolecular forces which hold the molecules together and represents the internal pressure of each component.

Equation 10-2 is now rearranged as follows:

$$\frac{\Delta H_m}{V_m \phi_1 \phi_2} = \left[\left(\frac{\Delta E_1}{V_1} \right)^{1/2} - \left(\frac{\Delta E_2}{V_2} \right)^{1/2} \right]^2 \tag{10-3}$$

At a given concentration, the heat of mixing per cubic centimeter is equal to the square of the difference between the square roots of the cohesive energy densities of the two components. The symbol δ, which Hildebrand termed the solubility parameter, was assigned to this quantity, so that the following equation results:

$$\delta = \left(\frac{\Delta E}{V} \right)^{1/2} \tag{10-4}$$

If solubility parameters are substituted for cohesive energy densities, then Equation 10-3 becomes

$$\frac{\Delta H_m}{V_m \phi_1 \phi_2} = (\delta_1 - \delta_2)^2 \tag{10-5}$$

The term $(\delta_1 - \delta_2)^2$ is proportional to ΔH. When ΔH in the original free-energy equation is replaced with the term $(\delta_1 - \delta_2)^2$, the following equation is obtained:

$$\Delta F \propto (\delta_1 - \delta_2)^2 - T\Delta S \tag{10-6}$$

As the solubility parameters (δ_1 and δ_2) approach each other, the term $(\delta_1 - \delta_2)^2$ approaches zero:

$$\text{as } \delta_1 \longrightarrow \delta_2, (\delta_1 - \delta_2)^2 \longrightarrow 0$$

Therefore, this term drops out of Equation 10-6, leaving it essentially as follows:

$$\Delta F = 0 - T\Delta S \tag{10-7}$$

If the solubility parameters approach each other, ΔF is usually negative, and the process under consideration will take place. This is why two solvents with about the same solubility parameters will usually be miscible, or why an amorphous solid will dissolve in a liquid having about the same solubility parameter as the solid. To repeat, when the solubility parameters are about the same, the term $(\delta_1 - \delta_2)^2$ becomes essentially zero and ΔF is usually negative.

During Hildebrand's work, it became known that the van der Waals forces of attraction between molecules contain three components. These are known as Keesom forces (dipole–dipole interactions), Debye forces (dipole–induced dipole interactions), and London– van der Waals forces (instantaneous dipole–induced dipole interactions). (See Chapter 15, "Surfaces and Interfaces," for a discussion of these forces.) Hydrogen bonds between molecules result from dipole–dipole interactions, but have special properties and are treated separately.[13,14] Hydrogen-bonding forces are by far the most important. The solubility parameter is an overall measure of all of these forces. Since these forces have different interactions, a similarity of solubility parameters does not necessarily assure that the system will be miscible.[10]

The solubility parameter concept explains many of the puzzling features of solvency. For example, there are known instances where a mixture of two solvents will dissolve a third component, although neither solvent by itself will dissolve the compound. Burrell[8] lists a number of examples of this phenomenon. The explanation for this is that one of the liquids has too high a solubility parameter to dissolve the compound, while the other liquid has too low a solubility parameter. The average solubility parameter of the mixture of liquids, however, is close enough to that of the compound so that the mixture is a solvent for the compound. The solubility parameter concept was useful in explaining some of the phenomena observed with the formation of foams from aqueous ethyl alcohol foam systems.[7]

Solubility parameter theory was also helpful in the selection of solvents for polymers. Small[15] increased the usefulness of the concept by developing an empirical procedure for calculating the solubility parameter of a polymer which had a known structure. However, as mentioned previously, even though a solvent and polymer may have the same solubility parameter, there is no assurance that the solvent will dissolve the polymer because the proportions of the London– van der Waals, Keesom, Debye, and hydrogen-bonding forces may not be the same in the solvent and polymer. There were many cases known where a solvent did not dissolve a polymer even though they had similar solubility parameters. Burrell improved solubility parameter theory by classifying solvents into three groups, depending upon their hydrogen-bonding capacity: weakly hydrogen-bonded, moderately hydrogen-bonded, and strongly hydrogen-bonded.[8] Lieberman[16] further improved the predictability of polymer–solvent behavior by establishing a more quantitative scale for the classification of compounds according to the sum of their hydrogen-bonding and polar forces. He devised two-dimensional plots of hydrogen-bonding numbers versus solubility parameters for various solvents and showed that solvents whose combination of solubility parameter and hydrogen-bonding number fell within the specified zone would dissolve the polymers on which he worked. However, there still remained a number of exceptions, that is, polymer–solvent systems which did not interact according to the Lieberman plots. The exceptions included cellulose polymers, vinyl chloride–acetate copolymers, and polymethyl methacrylate polymers. In an excellent series of articles, Crowley, Teague, and Lowe[10,11] reported that by using three-dimensional models based on combinations of solubility parameter, hydrogen-bonding numbers, and dipole moments, the solubility characteristics of solvents for these polymers could be predicted with much greater accuracy than had previously been possible. Their papers provide tables of solvents listing their solubility parameters, hydrogen-bonding numbers, and dipole moments.

The solubility parameters of the fluorocarbons are listed in Table 10-1.[17]

CHLOROCARBON AND FLUOROCARBON STRUCTURES, KAURI–BUTANOL VALUES, AND SOLUBILITY PARAMETERS

There is a definite relationship between the molecular structures of chloro- and fluorocarbons of the methane series and their Kauri–Butanol values and solubility parameters. This is shown in Figure 10-1.

The following conclusions can be drawn from the data:

1. Substitution of fluorine for chlorine decreases both the Kauri–Butanol value and the solubility parameter.
2. Substitution of one hydrogen atom for a chlorine atom results in an increase in the Kauri–Butanol value. Further substitution causes the Kauri–Butanol value to decrease. Kauri–Butanol values reach a maximum with one hydrogen atom in the molecule.
3. Substitution of hydrogen for chlorine increases the solubility parameter.

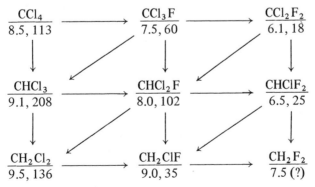

Figure 10-1 Relationship of structure to Kauri–Butanol values and solubility parameters.*
*The first figure under each compound is its solubility parameter and the second is its Kauri–Butanol value.

Additional hydrogen substitution further increases the solubility parameter. Solubility parameters do not reach a maximum with one hydrogen atom in the molecule as do the Kauri–Butanol values.

4. Substitution of hydrogen for fluorine increases both the solubility parameter and the Kauri–Butanol value. Substitution of a second hydrogen atom for fluorine further increases the solubility parameter and Kauri–Butanol value.

Similar relationships may exist for the ethane series of fluorocarbons, but sufficient data are not available for a detailed analysis. The large number of isomers makes the analysis much more complicated than in the methane series.

EFFECT OF PROPELLANTS UPON VALVE ELASTOMERS

Selection of the proper valve gasket elastomer for a formulation is highly important. The wrong gasket material can lead to excessive leakage or valve malfunction. The major elastomers used for aerosol valve gaskets are Buna N, a copolymer of butadiene and acrylonitrile; neoprene, a polymer of 2-chloro-butadiene-1,3; and butyl rubber, a copolymer of isobutylene and a diene, usually butadiene or isoprene.

A considerable amount of work on leakage of propellants through valve gaskets has been carried out at the Freon® Products Laboratory. This was published in Reference 18. A propellant loss not exceeding 3 g/year at room temperature is acceptable for most aerosol products packaged in containers with capacities of 6 fl. oz or larger. A fluorocarbon 12/fluorocarbon 11 (30/70) blend has a leakage rate no greater than 3 g/year through a neoprene gasket. This was used as a control. The study also showed that leakage rates per year generally could be extrapolated from 8-week weight-loss data. The weight loss of the fluorocarbon 12/fluorocarbon 11 (30/70) blend through a neoprene valve gasket after 8 weeks at room temperature was about 0.49 g. Therefore, weight losses not exceeding 0.49 g after 8 weeks were considered acceptable. The leakage rates reported in the following sections include that through the mounting cup gasket and container seams. However, leakage through the mounting cup gasket and container seams was negligible.

Leakage through Valve Gaskets

The weight loss of various chloro- and fluorocarbons from containers with valves having Buna N, neoprene, and butyl rubber gaskets is shown in Table 10-2. The compounds are listed in order of decreasing weight loss with a Buna N gasket.

Leakage through a Buna N gasket is unacceptably high with propellants 30, 21, 123, 22, 133a, and 124. It is satisfactory with propellants 152a, 142b, 12, isobutane, and propane. With a neoprene or butyl gasket, only propellants 30,

TABLE 10-2 PROPELLANT LEAKAGE THROUGH VALVE GASKETS

| Propellant | Leakage after 8 Weeks at Room Temperature (g)[a] | | |
	Buna N	Neoprene	Butyl
30 (CH_2Cl_2)	Excessive[b]	6.8	Erratic[c]
21	15.5	6.3	1.95
123	5.4	.23	.20
22	4.8	.29	.12
133a	2.6	.13	.08
124	1.0	.13	.05
11	.23	.80	1.80
152a	.20	.08	.04
142b	.06	.35	.02
12	.04	.11	—
Isobutane	.03	.12	—
Propane	.03	.09	—
12/11 (30/70) Control	.18	.49	—

[a]Standard precision valve with .018-in. stem orifice.
[b]Gasket swelled out of valve.
[c]High leakage, results variable.

21, and 11 have unsatisfactory leakage rates. Other propellants have acceptable leakage rates (that is, no greater than 0.49 g after 8 weeks at room temperature). The weight losses listed in Table 10-2 occurred with 100% propellant. In aerosol formulations where the propellant concentration is reduced by other solvents, for example, in alcohol-based products, leakage rates will be much less than those shown in Table 10-2. For example, in alcohol-based products with fluorocarbon 22, leakage rates with Buna N are satisfactory if the concentration of fluorocarbon does not exceed 35 wt. %.[18]

Methods for Predicting Valve Leakage

Solubility parameters are useful for predicting valve leakage if the solubility parameters and hydrogen-bonding capacities of both the fluorocarbons and the elastomers are known. For elastomers such as neoprene and butyl which have a low hydrogen-bonding capacity, solubility parameters alone provide an indication of the valve leakage to be expected. When the elastomer itself has a high hydrogen-bonding capacity, for example, Buna N, then it is necessary to know the solubility parameters and hydrogen-bonding capacities of both the fluorocarbon and the elastomer.

Elastomer swelling by the liquid phase of the propellant is generally an effective method for judging potential leakage rates with fluorocarbons. However, other effects can occur during immersion of the polymer in the propellant (such as extraction of compounding ingredients etc.). As would be expected, the correlation between solubility parameters and elastomer swelling is fairly good.

Kauri–Butanol values, in conjunction with hydrogen-bonding capacities, provide an indication of the effect of fluorocarbons on valve elastomers, but the results are less reliable than those using solubility parameters or elastomer swelling.

Solubility Parameters and Kauri–Butanol Values

Solubility parameters, Kauri–Butanol values, leakage through valve gaskets, and a qualitative estimation of the hydrogen-bonding capacities of the fluorocarbons[20] are listed in Table 10-3.

The compounds are divided into three groups, depending upon their hydrogen-bonding capacities. Propellants with high hydrogen-bonding capacities are listed

TABLE 10-3   RELATIONSHIP BETWEEN SOLUBILITY PARAMETERS, KAURI-BUTANOL VALUES, HYDROGEN-BONDING CAPACITIES, AND PROPELLANT LEAKAGE THROUGH VALVE GASKETS

Propellant	Solubility Parameter	Kauri–Butanol Value	Leakage after 8 Weeks at Room Temperature (g)[a]		
			Buna N	Neoprene	Butyl
Group I—Propellants with High Hydrogen-Bonding Capacities					
21	8.0	102	15.6	6.3	1.95
123	7.3	60	5.1	.23	.20
124	6.8	22	1.0	.13	.05
22	6.5	25	4.8	.29	.12
Group II—Propellants with Low Hydrogen-Bonding Capacities					
30 (CH$_2$Cl$_2$)	9.5	136	Excessive[b]	6.8	Erratic[c]
133a	7.2	22	2.6	.13	.08
152a	7.0	11	.20	.16	.04
142b	6.8	20	.06	.35	.04
Group III—Propellants with No Hydrogen-Bonding Capacity					
11	7.5	60	.23	.80	1.80
Isobutane	7.1	25	.03	.12	—
Propane	7.1	25	.03	.09	—
12	6.1	18	.04	.11	—

[a]Buna N—solubility parameter 9.4, hydrogen-bonding capacity high; neoprene—solubility parameter 9.2, hydrogen-bonding capacity low; butyl—solubility parameter 8.1, hydrogen bonding capacity low.
[b]Gasket swelled out of valve.
[c]Results too variable to establish a definite value.

in Group I, those with low hydrogen-bonding capacities in Group II, and the remaining, with essentially no hydrogen-bonding capacity, in Group III. The compounds are not listed in order of their hydrogen-bonding capacities in any of the individual groups, since quantitative data are lacking in many cases. Solubility parameters and hydrogen-bonding capacities of Buna N, neoprene, and butyl are also given.

Solubility parameters and hydrogen-bonding capacities are both important factors in valve gasket leakage. When an elastomer has a high bonding capacity, for example, Buna N, then the hydrogen-bonding capacities of the propellants are particularly important. If the elastomer has a low hydrogen-bonding capacity, solubility parameters become the most important factor. All of the compounds with high hydrogen-bonding capacities (Group I) caused unacceptable leakage with Buna N, regardless of their solubility parameters, which ranged from 6.5 to 8.0. Fluorocarbon 22 has a higher leakage rate with Buna N than fluorocarbon 124, although its solubility parameter is lower. This is probably because fluorocarbon 22 has a higher hydrogen-bonding capacity than fluorocarbon 124.[20]

In general, however, propellants with the highest solubility parameters had the highest leakage rates. If the solubility parameters of the propellant and elastomer are close enough, significant leakage will occur regardless of the hydrogen-bonding capacities. Fluorocarbon 21, with the highest solubility parameter (8.0), is the only propellant in Group I that causes unacceptable leakage with neoprene and butyl. In Group II, propellants 30 (solubility parameter 9.5) and 133a (solubility parameter 7.2) both have unacceptable leakage with Buna N. Only propellant 30 affects neoprene or butyl. None of the compounds in Group III had a significant effect upon Buna N. However, fluorocarbon 11, with a solubility parameter of 7.5, did affect neoprene and, to a greater extent, butyl. The solubility parameter of fluorocarbon 11 is closest to that of butyl (8.0) and next closest to that of neoprene (9.2).

From the data in Table 10-3, it should be possible to predict the behavior of other propellants with the three elastomers if the solubility parameters and hydrogen-bonding capacities of the propellants are known.

Kauri–Butanol values are less reliable for predicting valve leakage than solubility parameters, even if the hydrogen-bonding capacities of the propellants are available.

It seems probable that if quantitative data on hydrogen-bonding numbers were available for the propellants and valve gasket elastomers, a Lieberman-type two-dimensional plot of hydrogen-bonding numbers versus solubility parameters could be established for the fluorocarbons. Compounds whose combination of solubility parameters and hydrogen-bonding numbers fell within the specified zones would have a predictable high leakage rate with the particular polymer for which the zone was established. Fluorocarbon 22, although it has a low solubility parameter, has a strong hydrogen-bonding capacity, as evidenced by the negative deviation of its vapor pressure curves with ethyl alcohol from Raoult's

Law. For this reason, it would fall in the established zone. Methylene chloride has a low hydrogen-bonding capacity, but a comparatively high solubility parameter. Therefore, its combination of solubility parameter and hydrogen-bonding number evidently would also place it in the specified zone of solvents.

The hydrogen-bonding capacities of propellants 12, 11, 142b, isobutane, and propane are either zero or very low. The combination of solubility parameter and hydrogen-bonding capacity for these compounds apparently would fall outside the specified zone for strong interaction with Buna N.

Solubility Parameters, Kauri–Butanol Values, and Elastomer Swelling

The extent to which an elastomer will swell during immersion in the liquid phase of a propellant is often a valuable guide in predicting whether a particular elastomeric gasket can be used with a propellant without excessive leakage or malfunction. Elastomer swell is usually expressed as percent linear increase. However, other effects, such as extraction of plasticizers or fillers and changes in mechanical strength and hardness of the elastomer, can also occur during immersion in the propellants. Extraction of plasticizers may cause shrinkage, so that appreciable swelling may be covered up by extraction.

Literature values for the swelling of Buna N, neoprene GN, and butyl rubber by various propellants[19] are listed in Table 10-4, along with their solubility parameters, Kauri–Butanol values, and hydrogen-bonding capacities. Swelling of an elastomer is an indication of the combined effect of the solubility parameter and hydrogen-bonding capacity of a liquefied gas propellant. The data in Table

TABLE 10-4   SWELLING OF ELASTOMERS BY PROPELLANTS[5,19]

Propellant	Solubility Parameter	Kauri–Butanol Value	Hydrogen-Bonding Capacity	Percent Increase in Length at Room Temperature[a]		
				Buna N	Neoprene GN	Butyl
30 (CH_2Cl_2)	9.5	136	Low	52	37	23
21	8.0	102	High	48	28	24
11	7.5	60	Low	6	17	41
113	7.2	31	Low	0.5	3	21
152a	7.0	11	Low	1	2	1
142b	6.8	20	Low	3	3	3
22	6.5	25	High	26	2	1
114	6.2	12	Low	0	0	2
12	6.1	18	Low	2	0	6
C-318	5.6	10	Low	0	0	0

[a]Buna N–solubility parameter 9.4, hydrogen-bonding capacity high; neoprene GN–solubility parameter 9.2, hydrogen-bonding capacity low; butyl–solubility parameter 8.1, hydrogen-bonding capacity low.

10-4 show that swelling data can be very useful. The percent swell for any specific polymer correlates quite well with the solubility parameters and Kauri-Butanol values of the propellants if the hydrogen-bonding capacities are also considered. The one outstanding exception to the relationship between solubility parameter and swelling in this group of compounds is fluorocarbon 22 and Buna N. As previously discussed, this is the result of the high hydrogen-bonding capacities of both Buna N and fluorocarbon 22. Except for the latter, propellants with solubility parameters of 7.5 or lower have little effect on Buna N; those with solubility parameters of 7.2 or lower have little effect on neoprene GN; and those with solubility parameters of 7.0 or lower cause little swelling of butyl. This trend could be predicted, since Buna N has the highest solubility parameter and butyl the lowest.

Elastomer Swelling

The data in Table 10-5 show the relationship between leakage of propellants through valve gaskets and swelling of the elastomer by the propellants. The correlation appears to be quite good. Discussion in a later section will show that it is not an exact relationship, however. After 8 weeks fluorocarbon 11 has an unacceptable leakage of 0.80 g at a neoprene swell of 17%. Judging from this, an acceptable swell will fall somewhat below 17%.

Factors Affecting Valve Gasket Leakage[18]

The two main factors affecting leakage of propellant through valve gaskets are: (1) permeability of the elastomer by the propellant and (2) partial pressure of the propellant in the vapor phase. With any given elastomer and propellant, the leakage rate is proportional to the partial pressure of the propellant. The partial

TABLE 10-5   COMPARISON OF LEAKAGE AND SWELLING DATA FOR PROPELLANTS

Propellant	Buna N		Neoprene		Butyl	
	Swell (%)[a]	Leakage (g)[b]	Swell (%)[a]	Leakage (g)[b]	Swell (%)[a]	Leakage (g)[b]
30 (CH_2Cl_2)	52	Excessive	37	6.8	23	Erratic
21	48	15.5	28	6.3	20	1.95
11	6	.23	17	.80[c]	41	1.80
142b	3	.06	3	.35	3	.02
22	26	4.8	2	.29	1	.12
12	2	.04	0	.11	6	—

[a] Percent linear swell in liquid propellant.
[b] Leakage through valve gasket after 8 weeks at room temperature.
[c] Unacceptable leakage rate.

pressure is determined by the mole fraction of propellant in the liquid phase. Leakage is not proportional to weight percent concentration of propellant in the liquid phase except to the extent that it affects molar concentration.

Partial Pressure and Leakage of Propellants

Fluorocarbon 21 has a high leakage rate through both Buna N and neoprene; fluorocarbon 22 has a high leakage rate through Buna N, but a low rate through neoprene; and fluorocarbon 12, isobutane, and propane have low leakage rates through both Buna N and neoprene. Therefore, the leakage of blends of fluoro-carbons 22 and 21 through Buna N would be expected to be proportional to the sum of the partial pressures of both fluorocarbons, since they both have high leakage rates through Buna N. With neoprene, the leakage rate would be expected to be proportional only to the partial pressure of fluorocarbon 21, since the leakage rate of fluorocarbon 22 through neoprene is low. Likewise, fluorocarbon 12, isobutane, and propane have very low leakage rates through both Buna N and neoprene. The leakage rates of blends of fluorocarbon 21 with fluorocarbon 12, isobutane, or propane should be proportional only to the partial pressure of fluorocarbon 21.

The above conclusions were confirmed by leakage rate studies with blends of 85 wt. % fluorocarbon 21 with 15 wt. % of fluorocarbon 22, fluorocarbon 21, isobutane, and propane. The weight percent of fluorocarbon 21 remained the same, that is, 85 wt. % in the blends, but the mole fraction varied considerably as a result of the different molecular weights of the other propellants.

Liquid phase compositions, mole fractions, calculated vapor pressures, partial pressures, and leakage rates of the propellant blends are given in Table 10-6. The leakage rate through Buna N is highest with the fluorocarbon 22/fluorocarbon 21 blend and is proportional to the sum of the partial pressures of the two. The leakage rates of combinations of fluorocarbon 21 with fluorocarbon 12, isobu-tane, and propane are proportional only to the partial pressure of fluorocarbon 21. Fluorocarbon 21 is the only propellant in the group to have an appreciable leakage rate through neoprene. As predicted, the leakage rates of combinations of fluorocarbon 21 with fluorocarbon 22, and fluorocarbon 12, isobutane, and propane are essentially proportional only to the partial pressure of fluorocarbon 21.

Elastomer Permeability and Swelling

Loss of fluorocarbon through a gasket occurs only after the gasket has been completely permeated by fluorocarbon. Therefore, leakage should be propor-tional to the permeability of the elastomer by fluorocarbon. This was verified experimentally as described in the following section. Milled elastomers supplied commercially for the manufacture of valve gaskets were used. Buna N and butyl were obtained from Reeves Brothers; neoprene was obtained from the American

TABLE 10-6 RELATIONSHIP BETWEEN PARTIAL PRESSURE AND LEAKAGE

Composition of Liquid Phase (wt. %)					Mole Fraction Liquid Phase		Vapor Pressure (psia)	Partial Pressure (psia)		Leakage after 8 weeks at Room Temperature (g)	Gasket
FC-21	FC-12	FC-22	Isobutane	Propane	FC-21	Other		FC-21	Other		
85	15	–	–	–	.87	.13	31.1	20.1	11.0	10.7	Buna N
85	–	15	–	–	.83	.17	42.1	19.1	23.0	13.2	Buna N
85	–	–	15	–	.76	.24	28.6	17.6	11.0	9.5	Buna N
85	–	–	–	15	.71	.29	52.7	16.4	36.3	6.3	Buna N
85	15	–	–	–	.87	.13	31.1	20.1	11.0	4.8	Neoprene
85	–	15	–	–	.83	.17	42.1	19.1	23.0	3.0	Neoprene
85	–	–	15	–	.76	.24	28.6	17.6	11.0	2.7	Neoprene
85	–	–	–	15	.71	.29	52.7	16.4	36.3	2.3	Neoprene

From *Aerosol Age*.[18]

TABLE 10-7   VALVE LEAKAGE, PERMEABILITY, SWELLING, AND EXTRACTION WITH FLUOROCARBON 21

	Buna N	Neoprene	Butyl
Valve leakage (g) after 8 weeks at room temperature	4.54	1.4	.49
Permeability coefficient ((ml at STP) (mil)/(day) (100 in.^{2}) (atm))	1,950,000	138,000	57,500
Linear swell (%)[a]	38	22	22
Percent extracted[a]	13.3	7.3	3.8

[a]Samples cut crosswise to direction of milling. From *Aerosol Age*.[18]

Gasket and Rubber Company. The apparatus and procedure for determining permeability coefficients are described in References 18 and 21. The higher the permeability coefficient (K), the greater the rate of permeation. Permeability coefficients have the dimensions (ml at STP) (mil)/(day) (100 in.^{2}) (atm).

Data on valve leakage, permeability, swelling, and extraction with fluorocarbon 21 for the three elastomers are given in Table 10-7. Permeability correlates well with observed leakage through valve gaskets and is highest through Buna N and lowest through butyl. Percent linear swelling undoubtedly is a major factor, since it indicates the extent to which the fluorocarbon dissolves in the elastomer. However, correlation of leakage with swelling is not as exact as with permeation. Both neoprene and butyl were swelled to the same extent by fluorocarbon 21, but leakage and permeation rates with the two elastomers were different. The difference in extractables between the polymers may be a factor in the permeability.

The data with fluorocarbon 22 and the three elastomers are shown in Table 10-8. Again, there is a better correlation between valve leakage and permeability

TABLE 10-8   VALVE LEAKAGE, PERMEABILITY, SWELLING, AND EXTRACTION WITH FLUOROCARBON 22

	Buna N	Neoprene	Butyl
Valve leakage (g) after 8 weeks at room temperature	4.8	.29	.12
Permeability coefficient ((ml at STP) (mil)/(day) (100 in.^{2}) (atm))	7250	2700	550
Linear swell (%)[a]	21	1.1	4.4
Percent extracted[a]	<0.1	<0.1	<0.1

[a]Samples cut crosswise to direction of milling. From *Aerosol Age*.[18]

TABLE 10-9   ELASTOMER LINEAR SWELL (%) BY FLUOROCARBON 21

| Elastomer | Linear Swell (%) | | Value from Reference 3 |
| | Direction of Cut | | |
	Lengthwise	Crosswise	
Buna N	31	38	48
Neoprene	18	22	11,[a] 28[b]
Butyl	19	22	24

[a] Neoprene W.
[b] Neoprene GN. From *Aerosol Age*.[18]

than between valve leakage and linear swell or percent extractables. However, the linear swell and extractables are low for both neoprene and butyl.

Literature values of linear swell of elastomers by propellants are generally quite useful as an indication of the suitability of an elastomer for valve gaskets. However, literature values were not necessarily obtained with the compounded polymers supplied for valve gaskets. Factors such as the type of compounding and extent of cure of polymers will affect percent linear swell. In most cases, the compounded polymers used for valve gaskets are proprietary materials. The relationship between the percent linear swell of the elastomers used in this study and the values reported in the literature are given in Table 10-9. The agreement between the values for Buna N and neoprene is rather poor; those for butyl are much better. In addition, it was observed that the manner in which the samples were cut from the milled stock affected swelling. Samples cut crosswise to the direction of milling swelled more than those cut lengthwise.

Leakage of Propellant Blends

Propellant blends are commonly used in the aerosol industry. If the blend under consideration consists of a propellant with a low leakage rate through valve gaskets (for example, fluorocarbon 142b) and one with a high leakage rate (for example, methylene chloride), it may be necessary to determine the maximum concentration of the high leakage component that can be used without producing excessive leakage.

The maximum concentration of methylene chloride in a blend with fluorocarbon 142b can be determined as follows: A series of mixtures of the two components is prepared in which the composition of the blends is varied from 100/0 to 0/100 wt. % at 10% intervals. Leakage of these blends is then measured after storage. Any blend would have an acceptable leakage rate if the loss did not exceed 0.49 g after 8 weeks at room temperature. From a plot of the compositions of the test blends versus leakage, the composition of the blend that would produce a maximum leakage of 0.49 g is determined.

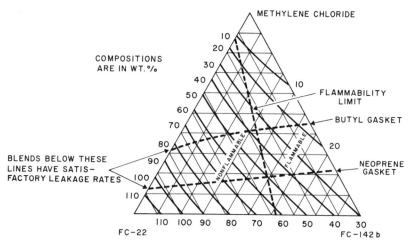

Figure 10-2 Vapor pressures (psig) at 70°F, flammability, and leakage of fluorocarbon 22/ fluorocarbon 142b/methylene chloride blends. (*Courtesy of Aerosol Age*)[18]

By evaluating other binary blends, a triangular coordinate chart which shows the compositions that give satisfactory leakage rates can be constructed for three-component blends. Vapor pressures and flammability data can also be plotted on the chart. Such a chart makes possible the selection of three-component blends having any desired vapor pressure and an acceptable leakage rate. A triangular coordinate chart of this type is illustrated in Figure 10-2 for ternary blends whose components are fluorocarbon 22, fluorocarbon 142b, and methylene chloride. Compositions that have satisfactory leakage rates with butyl or neoprene gaskets are shown on the chart. The superiority of the butyl gasket for these blends is evident.

EFFECT OF FLUOROCARBONS ON PLASTICS

The effect of fluorocarbons upon plastics has always been of interest to the aerosol industry, since some components of aerosol valves are constructed from plastic materials. Moreover, plastic aerosol bottles are presently being used.

A considerable amount of information on the effect of fluorocarbons on plastics is available in References 22 and 23. A brief summary on a limited number of plastics is given below.

Teflon® Tetrafluoroethylene Resin

No swelling was observed when submerged in most fluorocarbons, but some diffusion with fluorocarbons 12 and 22 was found. Fluorocarbon 21 caused swelling.

Polyvinyl Alcohol

Polyvinyl alcohol is not affected by fluorocarbons and is suitable for use with these compounds.

Vinyl Polymers

There is so much variation in the types of polymers and their plasticizers that generalizations are impractical. Specific materials should be tested before use.

Nylon

Nylon is generally suitable for use with the fluorocarbons. Many valve components are constructed from nylon.

Polyethylene

Polyethylene is suitable for many applications. Dip tubes are usually constructed of this plastic. However, it can be affected by strong solvents such as fluorocarbon 21.

Lucite® Acrylic Resins (Methacrylate Polymers)

This type of polymer is affected by many fluorocarbon propellants, particularly those that are the better solvents, for example, fluorocarbons 21 and 11. Aerosol coating compositions have been formulated with these polymers, indicating the solvent action of the fluorocarbons.

Polystyrene

Polystyrene is not suitable for use with most fluorocarbons, since it is generally affected by these compounds.

TABLE 10-10 PERMEABILITY AND LINEAR SWELL OF PLASTIC FILMS

Plastic	Fluorocarbon 21		Fluorocarbon 22
	Permeability Coefficient	Linear Swell (%)	Permeability Coefficient
Alathon® 3120 polyethylene	310,950	12.5	–
Mylar® polyester	412	2.5	<6.5
Teflon® fluorocarbon resin	247	2.5	<4.3
Kapton® polyimide	341	0.35	–
Polyvinyl alcohol	<6.6	–	–
Zytel® nylon	<9.1	–	<0.6
Surlyn® ionomer resin	6,100	–	–

From *Aerosol Age*.[18]

Phenolic Resins

As a general rule, phenolic resins are not affected by the fluorocarbons. However, there is enough variety in the resins so that any special application should be tested. Some linings in aerosol containers are prepared from phenolic resins.

Additional information on the permeability and percent linear swell of various plastics by fluorocarbons 21 and 22 is presented in Table 10-10. Nylon and polyvinyl alcohol were the least permeable of the plastics tested.

SOLUBILITY WITH WATER

The solubility of various fluorocarbons in water and that of water in fluorocarbons is illustrated in Table 10-11. Several conclusions can be drawn from the data:

1. Solubility of a fluorocarbon in water decreases when chlorine is replaced by fluorine in the molecule.
2. Solubility of a fluorocarbon in water increases when chlorine is replaced by hydrogen.

Judging by the limited information available, the solubility of water in fluorocarbons is not affected nearly as much by replacement of chlorine by fluorine or hydrogen as that of fluorocarbons in water.

TABLE 10-11 SOLUBILITY OF THE FLUOROCARBONS WITH WATER[1,6]

Fluorocarbon Number	Formula	Solubility with Water (wt. %) at 77°F	
		Fluorocarbon in Water (1 atm)	Water in Fluorocarbon
11	CCl_3F	0.11	0.011
12	CCl_2F_2	0.028	0.009
21	$CHCl_2F$	0.95	0.13
22	$CHClF_2$	0.30	0.13
31	CH_2ClF	1.27	0.12
32	CH_2F_2	0.44	–
113	CCl_2FCClF_2	0.017	0.011
114	$CClF_2CClF_2$	0.013	0.009
115	$CClF_2CF_3$	0.006	–
123	$CHCl_2CF_3$	0.46	–
124	$CHClFCF_3$	0.15	–
125	CHF_2CF_3	0.11	–
132b	$CH_2ClCClF_2$	0.49 (75°F)	0.085 (75°F)
133a	CH_2ClCF_3	0.43	–
142b	CH_3CClF_2	0.054 (70°F)	–
152a	CH_3CHF_2	0.28	–

MISCIBILITY OF THREE-COMPONENT SYSTEMS

A number of three-component solvent systems in which only certain proportions of the three components are miscible and form clear solutions are used in the aerosol industry. Examples of systems of this type are fluorocarbon/glycol/alcohol mixtures and fluorocarbon/water/ethyl alcohol mixtures.

An effective way of showing which particular compositions are miscible and form a single liquid phase, and which compositions are not miscible and form two liquid layers, is to determine the miscibility of various compositions in these mixtures and plot the data on a triangular coordinate chart. Extensive use of this method for illustrating the miscibility of fluorocarbon/glycol/ethyl alcohol mixtures is shown in Reference 2 and for fluorocarbon/water/ethyl alcohol mixtures in Reference 7.

A typical miscibility chart for a fluorocarbon 12/propylene glycol/ethyl alcohol system is shown in Figure 10-3. Compositions to the left of the phase boundary are not miscible and form two layers. Compositions to the right of the boundary are completely miscible and form a single liquid phase.

Several procedures can be followed to obtain the data required for drawing a phase boundary such as that illustrated in Figure 10-3.[24] Two miscible components such as ethyl alcohol and propylene glycol can be weighed into glass bottles and fluorocarbon 12 added until the solubility limit of fluorocarbon 12 in the mixture is exceeded. At this point, the solution either becomes cloudy or immiscible propellant droplets are noticeable. Since the weights of the original

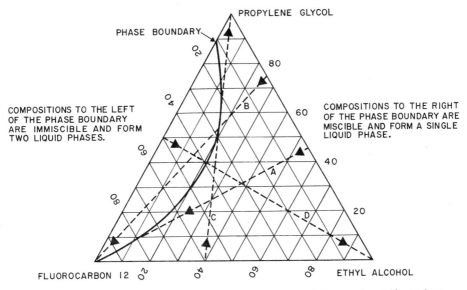

Figure 10-3 Determination of the miscibility characteristics of fluorocarbon 12/propylene glycol/ethyl alcohol mixtures by titration.

mixture are known, as is the weight of the fluorocarbon 12 added, the composition at the point where immiscibility occurred can be calculated. An example of this procedure is shown in Figure 10-3 using Composition B, consisting of 75 wt. % propylene glycol and 25 wt. % ethyl alcohol as the original mixture. As fluorocarbon 12 is added, the composition of the three-component mixture will change continually along Line B as shown by the arrows. Immiscibility occurs where Line B intersects the phase boundary.

Another procedure is to start with an immiscible mixture such as Composition D, which consists of 50 wt. % fluorocarbon 12 and 50 wt. % propylene glycol. Ethyl alcohol is added to the mixture by means of air pressure until the initially cloudy, immiscible mixture forms a clear solution. This end point is reached where Line D crosses the phase boundary. Since Line D crosses the phase boundary at about a right angle, the end point is quite sharp.

A problem that may arise in obtaining the data is illustrated in Figure 10-3 with a mixture such as Composition A, which consists of 46 wt. % propylene glycol and 54 wt. % ethyl alcohol. When fluorocarbon 12 is added to this mixture, the solution remains clear until Line A contacts the phase boundary. However, Line A is essentially tangential to the phase boundary at the point of contact, and the end point is not clear-cut. A similar situation occurs if a mixture such as Composition C, which consists of 61 wt. % fluorocarbon 12 and 39 wt. % ethyl alcohol, is titrated with propylene glycol. As propylene glycol is added, the composition will change along Line C in the direction shown by the arrows. Ultimately, Line C contacts the phase boundary, but again the end point will not be definite. If additional propylene glycol is added, a homogeneous system will result because the compositions will now fall to the right of the phase boundary.

REFERENCES

1. Freon® Technical Bulletin B-7, "Solubility Relationship of the Freon® Fluorocarbon Compounds."
2. P. A. Sanders, *Aerosol Age* **5**, 26 (February 1960).
3. 1966 Book of ASTM Standards, Part 20 (January 1966).
4. Freon® Aerosol Report FA-3, "Kauri–Butanol Numbers of Freon® Propellants and Other Solvents."
5. P. A. Sanders, "Principles of Aerosol Technology," Van Nostrand Reinhold Company, New York, 1970.
6. Freon® Aerosol Report FA-26, "Solvent Properties Comparison Chart."
7. P. A. Sanders, *Drug Cosmet. Ind.* **99**, 56 (August 1966).
8. H. Burrell, *Official Digest* **27**, 726 (1955).
9. H. Burrell, *Official Digest* **29**, 1159 (1957).
10. J. D. Crowley, G. S. Teague, and J. W. Lowe, *J. Paint Technol.* **38**, 269 (May 1966).
11. J. D. Crowley, G. S. Teague, and J. W. Lowe, *J. Paint Technol.* **39**, 19 (January 1967).
12. J. Hildebrand, *J. Am. Chem. Soc.* **38**, 1452 (1916); see also J. Hildebrand and R. Scott, "The Solubility of Nonelectrolytes," 3rd ed., Reinhold Publishing Corp., New York, 1949.

13. D. K. Sebera, "Electronic Structure and Chemical Bonding," Chap. 11, Blaisdell Publishing Company, Waltham, Mass., 1964.
14. L. N. Ferguson, "Electron Structures of Organic Molecules," Chap. 2, Prentice-Hall, Inc., New York, 1952.
15. P. A. Small, *J. Appl. Chem.* **3**, 71 (1953).
16. F. P. Lieberman, *Official Digest* **34**, 30 (1962).
17. Freon® Technical Bulletin D-73, "Solubility Parameters"; see also Reference 5.
18. P. A. Sanders, *Aerosol Age* **21**, 35, 41 (October/November 1976).
19. Freon® Product Bulletin B-12A, "Effect of Fluorocarbon on Elastomers."
20. S. Temple, Personal Communication, Freon® Products Laboratory.
21. H. M. Parmelee, *Refrig. Eng.* **66**, 35 (February 1958).
22. Freon® Technical Bulletin B-41, "Effect of Freon® Compounds on Plastics."
23. J. A. Brown, *Proc. 46th Ann. Meet. Chem. Spec. Manuf. Assoc.* (December 1959).
24. P. A. Sanders, "Aerosol Laboratory Manual," Industry Publications, Inc., Cedar Grove, N.J., 1975.

11

VISCOSITY

Viscosity is defined as the resistance of a liquid to flow or as the resistance experienced by one portion of a liquid when it moves over another portion.[1] In spite of the vast amount of experimental and theoretical work that has been carried out on viscosity, the theory of liquid viscosity is still incomplete.[2] Several theories on the structure of liquids[3] have been advanced to explain viscosity. These attribute a pseudocrystalline or a microcrystalline structure to liquids. The liquids thus possess some properties analogous to those of solids. These structures result from the strong intermolecular forces and close packing of the molecules in a liquid.

In aerosol formulations, viscosity problems occur mostly with products designed for surface-coating applications, for example, paints, lacquers, adhesives, etc. The essential ingredient in these products is a polymeric material, such as a resin or elastomer. Polymers consist of chains of molecules linked together and have molecular weights ranging from relatively low to very high. For any homologous series of polymers with the same molecular structure, the viscosity of solutions of the polymer increases as the molecular weight of the polymer increases.

The effect of molecular weight upon viscosity can be visualized by considering polymer chains in solution to exist as strands of barbed wire. The longer and larger the number of strands, the more entangled they will be in solution. The difficulty of moving one layer of the solution over another layer due to strand positions is also increased by entrapment of solvent molecules by the coils and segments of the polymer chains. This is a partial reason for the abnormally high viscosity of polymer solutions.

Such solutions are difficult to formulate as a fine-spray aerosol. When propellent is added to a polymer solution of high viscosity, the product discharges as a stream and not as a spray. If the solution of polymer is diluted by addition of more solvent, the polymer chains will start to separate, since the chains are in constant motion in the solution. If the solution is dilute enough, the polymer chains will untangle sufficiently so that the polymer molecules separate. The product may then be formulated as an aerosol.

The viscosity of any polymer solution is a function of the concentration of the polymer in the solution and increases as the concentration of polymer is increased. Therefore, dilution of the polymer solution by addition of solvent will decrease the viscosity because the concentration of the polymer in the solution has decreased. The extent to which viscosity is reduced by addition of solvent depends not only upon the particular polymer involved, but also upon the type of solvent used for dilution[4-6] If the added liquid is a relatively poor solvent for the polymer, the polymer chains will tend to curl up and the viscosity will decrease rapidly.

VISCOSITY OF THE CONCENTRATE

In many cases the viscosity of the polymer solution, that is, the concentrate, is a good indication of whether the concentrate can be formulated directly as an aerosol or whether it will have to be diluted with solvent. An effective and simple method for determining the relative viscosity of a polymer solution is to measure the viscosity with a #4 Ford cup. The #4 Ford cup consists essentially of a stainless steel cup with a small orifice in the bottom. The time required for the polymer solution to flow through the orifice and drain out of the cup is measured with a stopwatch. The viscosity is expressed as the number of seconds required to empty the cup. It has been observed experimentally with certain enamels[7] that a concentrate with a #4 Ford cup viscosity of about 20 sec may be formulated as an aerosol with a satisfactory spray. Nonaerosol paints and lacquers sold for brush application have viscosities considerably higher than 20 sec. They have to be diluted with solvents until the viscosity is low enough to produce an acceptable spray.

The experimental work involved in obtaining a concentrate with the correct viscosity for aerosol packaging can often be minimized by using the following procedure: First determine the #4 Ford cup viscosity of the concentrate; then dilute the concentrate with successive portions of a suitable solvent and determine the #4 Ford cup viscosity after each addition of solvent. The viscosity can then be plotted as a function of the concentration of the polymer in the concentrate or as a function of the percent of solvent added to the concentrate. The results obtained with a hypothetical Polymer X might appear as shown in Table 11-1. Assume that the concentration of Polymer X in the initial concentrate was 40 wt.%.

These data can be illustrated graphically as shown in Figure 11-1. From Figure 11-1 it is possible to determine what concentration of polymer in the concentrate will have any given viscosity. If the concentration of the polymer in the

TABLE 11-1   EFFECT OF VARIATION IN
                     CONCENTRATION UPON VISCOSITY

Concentration (wt.%) of Polymer X in Concentrate	#4 Ford Cup Viscosity (sec)
40	120
30	82
20	48
10	20
5	10

concentrate is not known, then the graph will show what concentration of added solvent is necessary to produce any desired viscosity.

As far as the proportion of propellant required is concerned, it has been found that about 40–50 wt.% fluorocarbon 12 in a coating composition will usually give a satisfactory spray if the viscosity of the concentrate is low enough.

Besides viscosity, another limiting factor in the formulation of an aerosol coating composition is the compatibility of concentrate with propellant. Fluorocarbon 12 is a relatively poor solvent. The addition of fluorocarbon 12 to the

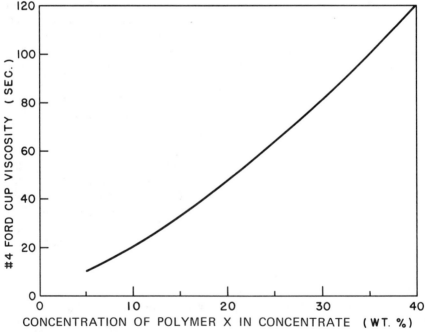

Figure 11-1   Relationship between solution viscosity and Polymer X concentration.

TABLE 11-2   COMPATIBILITY OF FLUOROCARBON 12
WITH SOLUTIONS OF POLYMER X

Concentration (wt.%) of Polymer X in Concentrate	Concentration (wt.%) of Fluorocarbon 12 in Mixture at End Point
40	2
30	11
20	26
10	50
5	63

concentrate may create a solvent system in which the polymer is not soluble. In order to determine the concentrations of polymer in the solvent that have sufficient compatibility with the propellant, the various solutions listed in Table 11-1 can be titrated with fluorocarbon 12 until precipitation of the polymer occurs. The precipitation of polymer is the end point of the titration. The composition of the solution at the end point can be calculated if the composition of the initial concentrate and the weight of fluorocarbon 12 added are known. The results that might be obtained with the hypothetical Polymer X are shown in Table 11-2.

The data in Table 11-1 and Table 11-2 can be combined on the same graph as illustrated in Figure 11-2. It is then possible to select reasonable compositions to evaluate. For example, a mixture containing 7.5 wt.% Polymer X in the concentrate has a #4 Ford cup viscosity of about 14 sec. This particular concentrate can accept up to 56 wt.% fluorocarbon 12 before precipitation of polymer occurs. Therefore, a solution containing 7.5 wt.% Polymer X would be a logical concentrate to test initially for spray properties.

Besides viscosity of the concentrate, other factors are involved in obtaining a satisfactory spray. With some high-molecular-weight polymers, a product with a satisfactory spray cannot be obtained even with #4 Ford cup viscosities below 20 sec.

VISCOSITY OF THE AEROSOL PRODUCT

Up to now, only viscosity of the polymer solution (the concentrate) has been discussed; however, one of the most important factors is the viscosity of the aerosol product itself. The measurement of aerosol viscosity is more complicated than the measurement of concentrate viscosity because of the problems introduced by the pressure of the aerosol. An apparatus and procedure for measuring the viscosity of aerosol compositions have been described by Morrow and Palmer.[4] The viscometer (consistometer) used the falling-ball principle to measure the viscosity of aerosol products under pressure. By using this apparatus, Morrow and

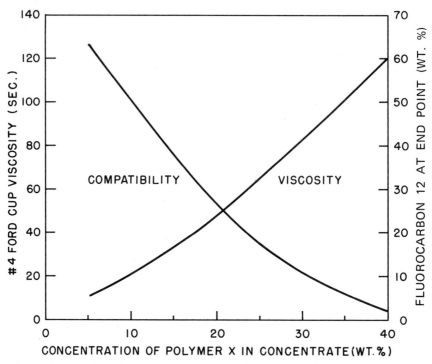

Figure 11-2 Relationship between concentrate viscosity, propellant compatibility, and Polymer X concentration.

Palmer were able to determine the relationship between the viscosity of an aerosol product and the pressure required to obtain a satisfactory spray. As would be expected, the higher the viscosity, the higher the pressure required.

The effect of adding additional solvent to the polymer solution upon viscosity has been discussed. The same effect is obtained when the propellants are added to the concentrate. The addition of the propellant reduces the viscosity of the concentrate. In this respect, the propellants, within the limits of their compatibility, act like the standard thinners used in the paint industry.

The extent to which the viscosity is reduced by the addition of the propellant is a function of the particular propellant used.[5,6] For example, a 50/50 solution of fluorocarbon 22/fluorocarbon 12 is a more effective thinner than fluorocarbon 12 alone.

EFFECT OF PROPELLANT CONCENTRATION

For many years it has been known that there is a minimum concentration of propellant necessary to provide a satisfactory spray for any polymer solution. There also is an upper limiting concentration of propellant at a given concentration of the polymer.[8] Thus, it was observed that with an acrylic ester resin dissolved in toluene, 25 wt.% fluorocarbon 12 was necessary in order to obtain a coarse, wet spray when the concentration of resin in the aerosol was 12 wt.%. When the concentration of resin was maintained at 12 wt.% and the concentration of fluorocarbon 12 increased above 25 wt.%, the spray became increasingly finer. Ultimately, the spray changed back to a stream when the concentration of fluorocarbon 12 was increased above 50 wt.%. This effect occurs because as the propellant increases at a constant resin concentration in the aerosol, the solvent concentration decreases proportionately, and the concentration of resin in the concentrate increases. Consequently, the viscosity of the concentrate increases. Also, as the proportion of propellant in the aerosol is increased, the extent to which the propellant flashes during discharge of the product increases. The cooling resulting from the increased flashing increases the viscosity of the product so that at higher propellant concentrations the product is so viscous that a stream rather than a spray is obtained.

TABLE 11-3   VISCOSITIES OF CHLOROCARBONS, FLUOROCARBONS, AND HYDROCARBONS

Compound	Formula	Liquid Viscosity (centipoise)	
Chlorocarbon			
Carbon tetrachloride	CCl_4	0.904	(77°F)
Chloroform	$CHCl_3$	0.542	"
Methylene chloride	CH_2Cl_2	0.416	"
Fluorocarbon			
11	CCl_3F	0.415	"
12	CCl_2F_2	0.214	"
21	$CHCl_2F$	0.313	"
22	$CHClF_2$	0.198	"
113	CCl_2FCClF_2	0.680	"
114	$CClF_2CClF_2$	0.360	"
123	$CHCl_2CF_3$	0.450	"
124	$CHClFCF_3$	0.163[a]	"
125	CHF_2CF_3	0.104[a]	"
133a	CH_2ClCF_3	0.40[a]	
142b	CH_3CClF_2	0.309	(86°F)
152a	CH_3CHF_2	0.227	"
Hydrocarbon			
Propane	C_3H_8	0.102	(68°F)
Isobutane	C_4H_{10}	0.156	"
n-Butane	C_4H_{10}	0.173	"
Isopentane	C_5H_{12}	0.224	"
n-Pentane	C_5H_{12}	0.234	"

[a]Estimated.

VISCOSITIES OF CHLOROCARBONS, FLUOROCARBONS, AND HYDROCARBONS[9, 10]

The viscosity of a propellant or solvent affects the sedimentation rate of a solid suspended in the liquid (Stokes' Law). This is of particular interest for aerosol formulations such as dry-type antiperspirants (see Chapter 21, "Aerosol Suspensions") where too rapid a sedimentation rate can result in consumer dissatisfaction.

The viscosities of various chlorocarbons, fluorocarbons, and hydrocarbons are listed in Table 11-3.

RELATIONSHIP OF STRUCTURE TO VISCOSITY

The relationship between structure and viscosity for the methane series of chloro- and fluorocarbons is illustrated in Figure 11-3. Several conclusions can be drawn from the data.

1. Substitution of hydrogen for chlorine in the molecule decreases viscosity. This is shown by comparing the viscosities of the chlorocarbons.
2. Substitution of fluorine for chlorine in the molecule decreases viscosity.
3. Substitution of hydrogen for fluorine increases viscosity.

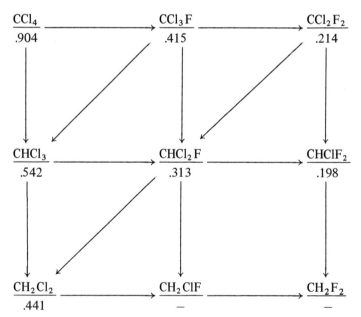

Figure 11-3 Relationship of viscosity to structure.*
*The number under each compound is the viscosity in centipoise.

The conclusions also hold for the ethane series of fluorocarbons, as shown by the data in Table 11-3 (for example, compare the viscosities of fluorocarbons 123, 124, and 125).

In the hydrocarbon series, viscosity increases with an increase in molecular weight, as would be predicted. Straight-chain hydrocarbons, that is n-butane and n-pentane, have slightly higher viscosities than branched-chain isomers.

REFERENCES

1. F. H. Getman and F. Daniels, "Outlines of Theoretical Chemistry," 5th ed., John Wiley & Sons, Inc., New York, 1931, p. 57.
2. H. Mark and A. V. Tobolsky, "Physical Chemistry of High Polymeric Systems," 2nd ed., Vol. 2, Interscience Publishers, Inc., New York, 1955, p. 282.
3. S. Glasstone, "Textbook of Physical Chemistry," 2nd ed., Macmillan Company, Ltd., London, England, 1956, p. 511.
4. R. W. Morrow and F. S. Palmer, *Proc. 39th Ann. Meet. Chem. Spec. Manuf. Assoc.* (December 1952) (Freon® Aerosol Report FA-13).
5. F. S. Palmer and R. W. Morrow, *Proc. 39th Mid-Year Meet. Chem. Spec. Manuf. Assoc.* (May 1953) (Freon® Aerosol Report FA-13).
6. F. S. Palmer, Freon® Aerosol Report FA-13, "The Formulation of Aerosol Coating Compositions with Freon® Fluorocarbon Propellants."
7. F. A. Bower, *Aerosol Age* 3, 21 (December 1958) (Freon® Aerosol Report FA-13).
8. P. A. Sanders, *Am. Paint J.* (October 1966) (Freon® Aerosol Report FA-13A).
9. Freon® Products Bulletin B-2, "Freon® Fluorocarbons, Properties and Applications."
10. "Handbook of Chemistry and Physics," 49th ed., The Chemical Rubber Company, Cleveland, Ohio, 1968–1969.

12

DENSITY

Aerosol propellant and formulation densities determine the weight of propellants that can be shipped in cylinders or tank cars, the weight of propellants that can be stored at filling plants, and the weight of aerosol products that can be packaged in an aerosol container. Quite often, there is some confusion concerning the terms density and specific gravity. The density of a substance is defined as the mass per unit volume at any given temperature. Thus, $d = m/v$, where d = density, m = mass, and v = volume. In the aerosol industry, density is usually expressed in g/cc or in lb/ft^3.

Specific gravity is the ratio of the weight of a given volume of a substance at one temperature (t_2) to the weight of the same volume of a reference material such as water. The temperature of the reference material (t_1) may be the same as the temperature of the substance whose specific gravity is under consideration, or it may be different. The specific gravity of any material is then reported using the symbol $d_{t_1}^{t_2}$, where t_1 is the temperature of the reference material and t_2 is the temperature of the substance. For example, iron has a specific gravity of 7.90 at 20°C compared to water at 4°C. Therefore, the specific gravity of iron is described as $d_{4°}^{20°} = 7.90$. In the metric system, the density of water is taken as one. Under these conditions, the density and specific gravity of a substance have the same numerical value. However, density has the dimensions of mass/volume, for example, g/cc, while specific gravity is a pure number and has no dimensions.

The numerical values of the densities of the various propellants were given previously in Chapters 3 and 4. The fluorocarbon propellants (with the exception of fluorocarbon 152a) have densities greater than one and the hydrocarbon propellants have densities less than one. Therefore, a greater weight of products formulated with the fluorocarbon propellants can be packaged in a given volume than products formulated with the hydrocarbon propellants, assuming that the weight percent concentration of the propellants is the same in both cases.

RELATIONSHIP OF DENSITY TO STRUCTURE

The relationship between density and structure for chloro- and fluorocarbons in the methane series is illustrated in Figure 12-1. Several conclusions can be drawn from the data:

1. Substitution of fluorine for chlorine decreases density.
2. Substitution of hydrogen for chlorine decreases density.
3. Substitution of hydrogen for fluorine has essentially no effect upon density.

CALCULATION OF DENSITIES OF MIXTURES

The densities of common blends of various propellants are usually available from the literature. However, quite often densities of various mixtures, particularly three-component blends, either have to be calculated or determined experimentally. The calculation of the density of a given mixture assumes that the volumes of the individual components are additive and that the total volume of the mixture is equal to the sum of the volumes of the individual components. The calculation of the density of a three-component mixture is illustrated by the following example.

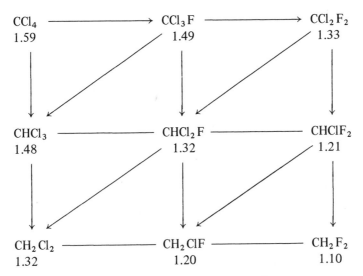

Figure 12-1 Relationship of density to structure.*

*The number under each compound is the density (g/cc) at $70°F$.

Assume that the density of a fluorocarbon 12/fluorocarbon 114/fluorocarbon 11 (50/30/20) mixture at 70°F is desired. The density is calculated as follows:

1. First, the densities of the three propellants (at 70°F) are obtained from a suitable source, such as Reference 1:

Fluorocarbon 12	1.32 g/cc
Fluorocarbon 114	1.47 g/cc
Fluorocarbon 11	1.48 g/cc

2. Assume that the weight of the three-component blend is 100 g. The volume (in cc) contributed by each of the components is calculated by dividing the weight of the component by its density:

$$\text{Volume of 50.0 g of fluorocarbon 12} = \frac{50}{1.32} = 37.9 \text{ cc}$$

$$\text{Volume of 30.0 g of fluorocarbon 114} = \frac{30}{1.47} = 20.4 \text{ cc}$$

$$\text{Volume of 20.0 g of fluorocarbon 11} = \frac{20}{1.48} = 13.5 \text{ cc}$$

3. The total volume of 100 g of the blend is obtained by adding together the volumes of the individual components. The total volume, therefore, is

$$37.9 \text{ cc} + 20.4 \text{ cc} + 13.5 \text{ cc} = 71.8 \text{ cc}$$

4. The density of the blend (at 70°F) is equal to m/v and is obtained by dividing the weight of the blend, 100 g, by the volume, 71.8 cc:

$$\text{Density of the blend} = \frac{100.0 \text{ g}}{71.8 \text{ cc}} = 1.39 \text{ g/cc}$$

A rather common error in calculating densities is to assume that the density of a mixture of liquids can be obtained merely by multiplying the weight percent concentration of each component by its density and adding up the resulting values for the various components. In some cases, where the densities of the various components are fairly close to each other, this procedure will give approximately the same result as that obtained by the correct method. However, when the densities differ appreciably, a considerable error will result when the density of the mixture is calculated by multiplying the weight percent concentration of each component by its density. The difference that results with the two methods is illustrated by the following example. Assume that the density of a fluorocarbon 12/propane (50/50) mixture is desired. The density of this mixture is calculated correctly as follows:

1. The densities of the two components (at $70°F$) are:

$$\text{Density of fluorocarbon 12} = 1.32 \text{ g/cc}$$
$$\text{Density of propane} = 0.50 \text{ g/cc}$$

2. The volume (in cc) contributed by each of the components in 100 g of the blend is:

$$\text{Volume of 50 g of fluorocarbon 12} = \frac{50.0}{1.32} = 37.9 \text{ cc}$$

$$\text{Volume of 50 g of propane} = \frac{50}{0.50} = 100.0 \text{ cc}$$

3. The total volume of 100 g of the fluorocarbon 12/propane (50/50) blend is

$$37.9 \text{ cc} + 100.0 \text{ cc} = 137.9 \text{ cc}$$

4. The calculated density of the blend (at $70°F$) is

$$\frac{100.0 \text{ g}}{137.9 \text{ cc}} = 0.72 \text{ g/cc}$$

Now, if the density of the blend is calculated erroneously by multiplying the weight percent concentration of each component by its density, the following result is obtained:

Contribution of fluorocarbon 12 = 0.50 × 1.32 g/cc = 0.66 g/cc
Contribution of propane   = 0.50 × 0.50 g/cc = 0.25 g/cc
Total = 0.91 g/cc

The density of 0.72 g/cc obtained by the first procedure differs considerably from the value of 0.91 g/cc obtained by the second method.

CALCULATIONS OF COMPOSITIONS OF BLENDS WITH A SPECIFIC DENSITY

It is sometimes desirable to determine the composition of a propellant blend that has a specific density. For example, in the formulation of emulsion systems, it is advantageous to use a propellant blend having the same density as that of water, since this decreases the rate of creaming.

Two-Component Blends

The composition of a two-component blend with a specific density can be calculated using the following equation:

$$X = \frac{100(d_1 d_2 - d_1 D)}{D(d_2 - d_1)} \qquad (12\text{-}1)$$

where

X = Weight (in g) of component 1 in 100 g of the propellant blend
d_1 = Density of component 1
d_2 = Density of component 2
D = Specified or required density of blend

Solving the equation for X gives the weight (in g) of component 1 in 100 g of the blend, which is numerically the same as weight percent concentration. Subtracting X from 100 gives the weight percent concentration of component 2. Equation 12-1 is derived as follows:

1. Assume that the total weight of the blend is 100 g. Then let X = the weight (in g) of component 1 in the blend. Therefore, $100-X$ is the weight (in g) of component 2 in the blend.
2. The total volume (in cc) of component 1 in 100 g of the blend is equal to the weight of component 1 in the blend divided by its density (d_1):

$$\text{Volume (in cc) of component 1} = \frac{X}{d_1}$$

3. The total volume (in cc) of component 2 in the blend is equal to the weight of component 2 in the blend divided by its density (d_2):

$$\text{Volume (in cc) of component 2} = \frac{100 - X}{d_2}$$

4. The total volume (in cc) of 100 g of the blend, therefore, is equal to the sum of the volume of component 1 and the volume of component 2:

$$\text{Total volume (in cc) of 100 g of blend} = \frac{X}{d_1} + \frac{100 - X}{d_2}$$

5. The density of the blend is equal to the weight divided by the volume.

Since the weight is 100 g and the total volume is

$$\frac{X}{d_1} + \frac{100 - X}{d_2} \text{ cc}$$

the density is as follows:

$$\text{Density of blend} = \frac{100}{\dfrac{X}{d_1} + \dfrac{100 - X}{d_2}}$$

6. Since the density desired is D g/cc, the equation then becomes

$$\frac{100}{\dfrac{X}{d_1} + \dfrac{100 - X}{d_2}} = D$$

7. Solving for X gives Equation 12-1.

As an example, assume that the composition of a fluorocarbon 12/isobutane blend with a density of 1.0 g/cc at 70°F is desired. The composition of the blend is obtained as follows:

1. The densities of fluorocarbon 12 and isobutane at 70°F are:

Density of fluorocarbon 12 = 1.32 g/cc
Density of isobutane   = 0.56 g/cc

2. Substituting these values in Equation 12-1, along with the value of 1.0 g/cc for D, gives

$$X = \frac{100(1.32 \times 0.56 - 1.32 \times 1.0)}{1.0(0.56 - 1.32)}$$

3. Solving the equation for X gives a value of 76.4 g of fluorocarbon 12. The weight (in g) of isobutane in the blend is 100 − 76.4 = 23.6 g. The composition of the blend with a density of 1.0 g/cc, therefore, is 76.4 wt. % fluorocarbon 12 and 23.6 wt. % isobutane.

Three-Component Blends

There are many combinations of three components that have the same density. The simplest approach is to calculate the compositions of the two binary blends that have the desired density and plot the data on a triangular coordinate chart. The two points at each end of the density plot fall on the base lines for the two-component blends. The compositions of these two blends are calculated using Equation 12-1 and plotted on the triangular coordinate chart. A straight line is drawn between the two points. All possible combinations of the three components that have the specified density fall on this line. In view of the number of different combinations that have the same density, the selection of the final blend will have to be based upon some other factor in addition to the density.

The procedure for preparing a triangular coordinate plot of compositions with a particular density can be illustrated as follows: Assume that the compositions of fluorocarbon 12/fluorocarbon 11/isobutane blends with a density of 1.0 g/cc are required. The fluorocarbon 12/isobutane blend with a density of 1.0 g/cc was shown in the previous section to have a composition of 76.4 wt.%

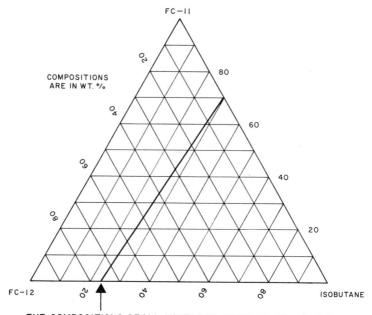

COMPOSITIONS
ARE IN WT. %

FC−11

FC−12

ISOBUTANE

THE COMPOSITIONS OF ALL MIXTURES OF FC−12, FC−11, AND
ISOBUTANE WITH A DENSITY OF 1.0 g/cc FALL ON THIS LINE.

Figure 12-2 Fluorocarbon 12/fluorocarbon 11/isobutane blends with a density of 1.0 g/cc at 70° F.

fluorocarbon 12 and 23.6 wt.% isobutane. Similarly, the composition of the fluorocarbon 11/isobutane blend with a density of 1.0 g/cc can be calculated using Equation 12-1. The composition of this blend is 70.6 wt.% fluorocarbon 11 and 29.4 wt.% isobutane. These two compositions are then plotted on a triangular coordinate chart and a straight line drawn between the points, as shown in Figure 12-2. All combinations of the three components that have a density of 1.0 g/cc fall on this line.

EXPERIMENTAL DETERMINATION OF DENSITY

In some instances, it is desirable to determine the density of a propellant blend experimentally. If precise results are not required, the density may be obtained by adding the propellant blend to a volumetrically calibrated pressure tube and thermostating the tube to the desired temperature. The weight of propellant added is obtained by weighing the glass tube, and the volume of the propellant is read directly from the tube. If more accurate results are desired, the densities can be obtained with a hydrometer. The procedure and apparatus for this method are described in Reference 2.

The densities of complete aerosol formulations can be obtained by using the same procedures as those described for the propellant blends. If the relative proportions of the propellant and concentrate and the densities of each are known, the density of the product can be calculated. In many cases, the concentrate consists of a complex mixture of ingredients, and it is difficult to calculate the density of the concentrate itself. In these cases, it is usually easier to obtain an approximate density of the concentrate experimentally by weighing a known volume of the concentrate. The density of the aerosol product is then calculated using the experimentally determined concentrate density.

EFFECT OF TEMPERATURE

The DOT regulations specify that an aerosol package must not become liquid full at 130°F. In order to determine the volume of any given weight of a product at 130°F, it is necessary to know the density. An indication of whether or not the container will be liquid full at 130°F can be obtained by assuming that the entire contents of the aerosol package consist only of propellant. Generally, the propellants expand more than any aerosol product with a rise in temperature. The volume fill of the container at 130°F can be calculated by using the propellant density data given in the literature, for example, that for the propellants in Reference 1. If the container will not be liquid full with propellant alone, it is unlikely that it will become liquid full with the aerosol product itself.

The change in density with temperature can also be obtained by measuring the densities at several temperatures. A curve illustrating the change in density with temperature can then be drawn. Densities at temperatures either higher or lower than those determined experimentally can be approximated by extrapolation.

REFERENCES

1. Freon® Aerosol Report FA-22, "Vapor Pressure and Liquid Density of Freon® Propellants."
2. "Tentative Hydrometric Determination of Aerosol Liquid Densities," Chemical Specialties Manufacturers Association (CSMA) Aerosol Guide, June 1957, p. 33.

13

STABILITY

There are many properties an aerosol package must have to be successful. One is adequate shelf life or storage stability. If an aerosol leaks as a result of container corrosion or the product itself deteriorates becasue of instability, the aerosol is not satisfactory regardless of its other properties. Premature marketing of an aerosol product with poor shelf life can be very costly if the product has to be withdrawn from the market. It is important, therefore, to take every precaution to be certain that an aerosol product has sufficient storage stability.

Poor shelf life can result from a number of causes. Leakage of the product may occur because of pinhole corrosion, even though the product appears to be stable. The odor may change without any noticeable container corrosion, or the product itself may deteriorate and lose its effectiveness or activity. A valve may fail to operate after a period of time either because the formulation has affected the valve gasket or because physical or chemical changes in the product result in the formation of particles that clog the valve orifice. Since the causes of poor storage stability are varied, an aerosol product must be subjected to a number of tests to insure that it has satisfactory shelf life.

In many cases, adequate information for making reasonable predictions already exists, and storage stability tests can be minimized or avoided. Aerosol fillers and suppliers, such as the propellant, valve, and container manufacturers, have accumulated extensive data on the storage stability of hundreds of products; also, information on the stability of aerosol propellants is readily available. In addition, an understanding of the systems and materials that cause corrosion can often be helpful in avoiding potentially unstable products.

CORROSION

Corrosion in metal containers is one of the major reasons for the poor shelf life of aerosol products. It can be slight as far as the effect on the material of the

204

container is concerned, but still sufficient to have an adverse effect upon aerosol components, such as perfumes. In extreme cases, corrosion can result in perforation of an aerosol container with a subsequent loss of product. Most corrosion is the result of electrochemical processes, and some of the fundamental concepts of electrochemical corrosion are discussed in the following section. The electrochemical aspects of corrosion have been summarized in an excellent booklet by LaQue, May, and Uhlig.[1]

Fundamentals of Corrosion

In order for electrochemical reactions to take place, certain conditions are necessary. An electrolyte for conducting electric current (assumed to move in the direction of positive charges), two electrodes, an anode and a cathode with a difference in potential, and a path for electrons to flow from the anode to the cathode must be present. An electrolyte is a liquid containing ions that conduct the current (positive ions). Water is an example of an electrolyte that contains hydrogen and hydroxyl ions. The two electrodes may consist of two different metals placed in an electrolyte solution. The electrochemical process with different metals is called galvanic action. The electrodes can also form at two different locations on the same piece of metal. A potential difference can arise from differences in the environmental conditions, differences in structure of the metal, concentrations of impurities, etc. The electrochemical processes are called local action in this case.

The path for the electrons can consist of a piece of wire or, if the two electrodes are located on the same piece of metal, the electrons can flow directly from the anode to the cathode through the metal. The complete system is called a corrosion cell and is illustrated in Figure 13-1a.

The potential difference between the two electrodes is the driving force that actuates the electrochemical process. The degree to which the potentials of the electrodes differ depends upon many factors, such as the types of metals, their environment, etc.[1] However, once conditions exist so that the electrochemical process can occur, a number of reactions take place simultaneously at both the anode and the cathode. Positively charged metal ions leave the anode and enter the electrolyte solution. This loss of metal from the anode, if allowed to continue, will ultimately result in its destruction. For this reason, most of the corrosion that occurs in electrochemical processes takes place at the anode. Meanwhile, the free electrons left in the anode as a result of the loss of the positively charged metal ions flow to the cathode through the metallic path. This anodic electrochemical process is illustrated in Figure 13-1b.

Meanwhile, a different series of reactions takes place at the cathode. Positively charged hydrogen ions migrate through the electrolyte solution to the cathode. Here the hydrogen ions meet the electrons that travel from the anode to the cathode and are neutralized to form hydrogen gas. This cathodic process is illustrated in Figure 13-1c.

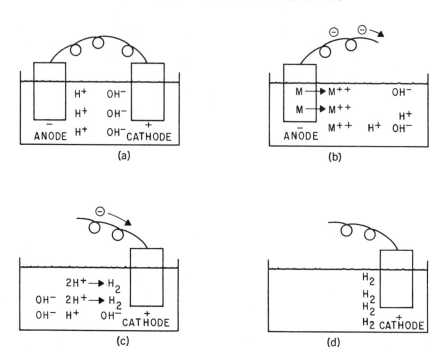

Figure 13-1 (a) Corrosion cell; (b) anodic reaction; (c) cathodic reaction; (d) cathodic polarization.

Polarization and Depolarization

The hydrogen formed at the cathode can either escape as hydrogen gas or remain and accumulate at the cathode. When sufficient hydrogen molecules collect at the cathode, they form a barrier and interfere with the neutralization of other hydrogen ions by the electrons in the cathode. This slows down the electrochemical process and reduces the rate of corrosion. This effect of hydrogen molecules on the cathode is called polarization, and the extent to which an electrode can be polarized depends upon such factors as the environmental conditions and the type of metal forming the electrode. Polarization is illustrated in Figure 13-1d.

Oxygen can have a pronounced effect upon polarization. If oxygen is dissolved in the electrolyte, it can react with the hydrogen at the cathode to form water or hydrogen peroxide and hydroxyl ions.[1] In any case, these reactions remove hydrogen molecules from the cathode and allow the neutralization of hydrogen ions by electrons, and thus corrosion, to continue. The reaction of oxygen with hydrogen is called depolarization, and oxygen is termed a cathodic depolarizer.

Relative Areas of the Anode and Cathode

The relationship of the area of the anode to that of the cathode can have a marked effect upon the rate of corrosion. When the area of the cathode is large compared to that of the anode, corrosion is favored. This is because the polarizing hydrogen formed at the cathode is spread out over a comparatively large area. Under these conditions, the hydrogen is easily accessible to attack by oxygen, which removes it from the cathode and allows corrosion to continue. When the area of the cathode is small compared to that of the anode, the rate of corrosion is much lower because the polarizing hydrogen molecules are able to concentrate on a smaller area and, consequently, are more difficult to remove. In order to minimize corrosion it is desirable to have the area of the cathode small relative to that of the anode.

The area relationship of the anode and cathode and its effect upon corrosion become very important when applied to aerosol tinplate containers. In these containers, the area of the tin coating is very large compared to that of the iron, which is usually exposed only through minute holes in the tin coating. If, in the tinplate container, iron is the anode and tin the cathode, then corrosion will be rapid because of the unfavorable relationship of the large area of the cathode and the small area of the anode. There is an additional reason why this particular situation is undesirable in aerosol tinplate containers. Since corrosion occurs mostly at the anode, it will be concentrated on the relatively small areas of iron available through the holes in the tin coating, and pinholing can result in a short time. The most desirable condition is obtained when the polarity is reversed: the tin is anodic and the iron cathodic. In this case, if corrosion does occur, the tin will be attacked initially instead of the iron. Generalized detinning occurs instead of pinholing, and the rate of corrosion will be much lower because of the favorable relationship of the large anode area and the small cathode area.

Effect of Oxygen upon Polarity of Tinplate Containers

Tin and iron are similar to each other in their electrochemical properties,[2] and comparatively small changes in the environment or the product may determine which metal will be the cathode and which the anode. One of the environmental factors that influences the polarity of the metals in the tinplate system is the concentration of oxygen. In the presence of low concentrations of oxygen, tin is anodic and iron cathodic. This produces a comparatively low rate of corrosion. It is one reason why vacuum-packed food products are stable in tinplate cans.[2] At higher concentrations of oxygen, the polarity of the system is reversed, with tin being cathodic and iron anodic. This condition promotes corrosion.

The effect of oxygen upon corrosion in tinplate containers can be summarized as follows: Low concentrations of oxygen retard corrosion because there is little depolarization of the cathode and because tin is anodic and iron cathodic; high concentrations of oxygen promote corrosion as a result of the depolarizing activity of oxygen and because tin is cathodic and iron anodic.

STABILITY OF THE FLUOROCARBONS

There are pronounced differences in the stability of the various fluorocarbons, depending upon the conditions to which the fluorocarbons are exposed. Hydrolytic stability and reactivity of the fluorocarbons with ethyl alcohol are summarized in Table 13-1 and are discussed below.

Fluorocarbon 11

Fluorocarbon 11 is one of the most commonly used propellants. As a result, its stability has been the subject of a number of investigations.[3,4,5,6] The stability of fluorocarbon 11 depends upon the particular system used. In some systems, fluorocarbon 11 is perfectly stable; in others, its use must be avoided.

Aqueous Systems

Fluorocarbons are not miscible with water. Therefore, aqueous systems containing fluorocarbons must be formulated as oil-in-water emulsions, water-in-oil emulsions, three-phase systems in which the water is layered over the propellant, or as three-component systems in which the third component is a solvent for water and the fluorocarbon.

In oil-in-water emulsions, or in three-phase systems with metal present, fluorocarbon 11 usually is unstable and should not be used. Metal appears to act as a catalyst for the reaction of fluorocarbon 11 with water, and reaction products that cause corrosion of the metal are formed.

The decomposition of fluorocarbon 11 by water in the presence of metals is not a simple hydrolysis reaction. Fluorocarbons 21 and 112 were identified among the reaction products and are considered to be formed by the following reactions (M denotes metal):[4]

$$2CCl_3F + M \longrightarrow MCl_2 + CCl_2FCCl_2F \qquad (13\text{-}1)$$

$$2CCl_3F + 2H_2O + 2M \longrightarrow 2CHCl_2F + MCl_2 + M(OH)_2 \qquad (13\text{-}2)$$

Acid is also formed, which indicates that some acid-forming reaction such as the one given below may take place (the specific reaction responsible for acid formation has not been identified):

$$CCl_3F + H_2O \longrightarrow \begin{array}{c} F \\ \diagup \\ C=O + 2HCl \\ \diagdown \\ Cl \\ \underset{(H_2O)}{\big\downarrow} CO_2 + HCl + HF \end{array} \qquad (13\text{-}3)$$

TABLE 13-1 FLUOROCARBON STABILITY[25]

Fluorocarbon Number	Formula	Aqueous Systems			Reactivity with Ethyl Alcohol
		Acidic or Neutral	Alkaline	Aqueous Alcohol	
11	CCl_3F	Stable in absence of metals. Decomposes rapidly in presence of metals such as iron and tin	Fairly stable in absence of metals. Decomposes rapidly in presence of metals such as iron and tin	Fairly stable in 95 vol. % alcohol with nitromethane as a stabilizer	Reacts via a free radical reaction. The reaction is inhibited by nitromethane
12	CCl_2F_2	Stable	Generally stable; slight decomposition at pH of 12	Stable	None
21	$CHCl_2F$	Stable	Unstable	Unknown	None
22	$CHClF_2$	Stable	Unstable	Stable	None
31	CH_2ClF	Unknown	Unstable	Unstable	Reacts to form a diether. Reaction cannot be inhibited
113	CCl_2FCClF_2	Stable	Unstable	Unstable	Reacts via a free radical reaction
114	$CClF_2CClF_2$	Stable	Stable	Stable	None
123	$CHCl_2CF_3$	Unknown	Unstable	Unknown	None
124	$CHClFCF_3$	Stable	Stable	Stable	None
133a	CH_2ClCF_3	Unknown	Unstable	Unknown	None
142b	CH_3CClF_2	Stable	Stable	Stable	None
152a	CH_3CHF_2	Stable	Stable	Stable	None

As a result of its instability in the presence of water and metals, fluorocarbon 11 is not suitable for use with oil-in-water emulsion products such as shaving lathers, starch sprays, and window cleaners. The use of lacquer-lined containers and lacquer-coated valve cups does not solve the problem of instability because there are always enough pinholes or other imperfections in the lacquer lining or coating on the valve cup to allow some contact of the formulation with the metal.

Nitromethane is a fairly effective inhibitor for retarding the decomposition of fluorocarbon 11 by water in metal containers. The evidence indicates that nitromethane functions as an inhibitor in this system by deactivating the metal surface so that it no longer acts as a catalyst for the fluorocarbon 11/water reaction.[4] The use of nitromethane in combination with other inhibitors for aqueous-based systems has been suggested.[7] However, nitromethane has not been sufficiently enough tested in aqueous-based aerosol products to allow determination of its effectiveness as an inhibitor for such systems.

In contrast to its instability in oil-in-water emulsions, fluorocarbon 11 can be fairly stable in water-in-oil emulsions.[8,9] In water-in-oil emulsions, fluoro-carbon 11 is part of the outer phase and water is the dispersed phase. The dispersed water droplets are surrounded by the oil-soluble emulsifying agent concentrated at the interface between the water and the fluorocarbon propellant. This prevents water from coming in contact with the metal, which is the catalyst for the decomposition. However, recent work indicates that the presence of nitromethane as a stabilizer is necessary to achieve adequate storage stability of water-in-oil emulsions with fluorocarbon 11.

Nonaqueous Systems

Fluorocarbon 11 is stable in most nonaqueous systems and has been widely used in products such as aerosol insecticides, room deodorants, hair sprays, etc. However, under certain conditions fluorocarbon 11 will react with ethyl alcohol via a free radical reaction mechanism to give acetaldehyde, hydrogen chloride, and fluorocarbon 21:[3]

$$CCl_3F + CH_3CH_2OH \longrightarrow CH_3CHO + HCl + CHCl_2F \qquad (13\text{-}4)$$

Secondary reactions then follow to give acetal and ethyl chloride:

$$CH_3CHO + 2CH_3CH_2OH \longrightarrow CH_3CH(OC_2H_5)_2 + H_2O \qquad (13\text{-}5)$$

$$CH_3CH_2OH + HCl \longrightarrow CH_3CH_2Cl + H_2O \qquad (13\text{-}6)$$

These reactions were of particular interest because of the corrosion occasionally observed with hair sprays formulated using ethyl alcohol and fluorocarbon 12/fluorocarbon 11 propellant mixtures. The free radical reaction between fluorocarbon 11 and ethyl alcohol is catalyzed by low concentrations of oxygen. Corrosion in hair spray containers was sometimes observed in products filled in the summer, when the warm conditions promoted effective purging of air during the filling of the samples. The products, therefore, had relatively low concentrations of oxygen.

High concentrations of oxygen inhibit the reaction between fluorocarbon 11 and ethyl alcohol. Products filled in the winter had higher concentrations of oxygen and were more stable. That oxygen has the property of functioning as a catalyst for free radical reactions at low concentrations and acting as an inhibitor at higher concentrations has been observed for other free radical reactions.[3]

Many compounds are inhibitors for free radical reactions. The most suitable and effective inhibitor for this reaction in aerosol products is nitromethane. Freon® 11 S, which contains nitromethane as a stabilizer for the reaction, is available commercially for use with aerosol products containing alcohols.[3]

Aqueous Ethyl Alcohol Systems

The use of nitromethane as a stabilizer either for aqueous systems or for alcoholic systems with fluorocarbon 11 has been discussed. On the basis of the stabilizing effect of nitromethane in these two systems, it might be expected that nitromethane would be equally effective in aqueous ethyl alcohol mixtures. This does not seem to be the case. Nitromethane is somewhat effective as a stabilizer in 95 vol. % ethyl alcohol/fluorocarbon 11 mixtures, but not in mixtures containing higher percentages of water.

Fluorocarbon 12

Fluorocarbon 12 is stable in both nonaqueous and aqueous aerosol systems, except under extremely alkaline conditions (pH 12 or higher). Therefore, it is suitable for use with almost any aerosol product. Fluorocarbon 12, in combination with fluorocarbon 114, has been used extensively as the propellant for aqueous-based foam aerosols.

Fluorocarbon 21

Fluorocarbon 21 is stable in acidic or neutral aqueous systems. It decomposes rapidly under alkaline conditions. It is stable in nonaqueous systems and does not react with ethyl alcohol.[10]

Fluorocarbon 22

Fluorocarbon 22 is stable in nonaqueous systems and in aqueous systems that are neutral or acidic. It decomposes rapidly in alkaline aqueous systems, but the reactions involved have not been determined.[6] Fluorocarbon 22 does not react with ethyl alcohol via a free radical reaction. In aqueous ethyl alcohol foam systems formulated with fluorocarbon 22, slight corrosion with pitting was observed around the can shoulder and on the valve cup after storage for two months at 120°F. The corrosion was not severe.

Fluorocarbon 31

Fluorocarbon 31 is unstable in either aqueous or aqueous ethyl alcohol systems. It reacts with anhydrous ethyl alcohol to form a diether as follows:

$$CH_2ClF + 2C_2H_5OH \longrightarrow C_2H_5OCH_2OC_2H_5 + HCl + HF$$

The reaction is not free radical in nature and cannot be inhibited.

Fluorocarbon 113

The stability of fluorocarbon 113 is similar to that of fluorocarbon 11. It is unstable in alkaline aqueous systems and in aqueous ethyl alcohol. It reacts with anhydrous ethyl alcohol via a free radical reaction. The reaction is inhibited by nitromethane.

Fluorocarbon 114

Fluorocarbon 114 is very stable in both nonaqueous and aqueous systems, and may be used with any aerosol product.

Fluorocarbon 123

Fluorocarbon 123 is unstable in alkaline aqueous systems. When tested with a shaving lather foam formulated with triethanolamine salts of fatty acids, fluorocarbon 123 caused considerable corrosion in the container, attacking the lacquer lining. Fluorocarbon 123 does not react with ethyl alcohol.

Fluorocarbon 124

Fluorocarbon 124 was stable when tested in an alkaline shaving lather formulation. Therefore, it can be used for aqueous aerosol foam products. Fluorocarbon 124 does not react with ethyl alcohol.

Fluorocarbon 133a

Fluorocarbon 133a is unstable in alkaline aqueous systems. When tested in a shaving lather formulated with triethanolamine salts of fatty acids, it caused mild corrosion in the containers. Fluorocarbon 133a appears to be somewhat more stable than fluorocarbon 123, but less stable than fluorocarbon 124. The stability of fluorocarbon 133a is not sufficient for use in alkaline aqueous foam products.

Fluorocarbon 142b

Fluorocarbon 142b is fairly stable in alkaline aqueous systems. When tested in an aqueous shaving lather formulated with triethanolamine salts of fatty acids, it caused essentially no corrosion after storage for two months at 120°F. Its hydrolysis rate is a function of temperature and is affected by the presence of metals. It probably is suitable as an aerosol propellant for most aqueous-based foam products, but should be adequately tested in these systems.

Fluorocarbon 142b has been used as the propellant for quick-breaking foams formulated with aqueous ethyl alcohol (hair setting foams). According to Johnsen,[11] fluorocarbon 142b has adequate resistance to hydrolysis in these systems.

Fluorocarbon 142b does not react with anhydrous ethyl alcohol via a free radical reaction.

Fluorocarbon 152a

Fluorocarbon 152a is extremely resistant to hydrolysis and is stable at a pH of 13.5.[10] Thus, it is more stable than fluorocarbon 12. Fluorocarbon 152a does not react with anhydrous ethyl alcohol and can be used in aqueous ethyl alcohol systems.

EFFECT OF OTHER AEROSOL COMPONENTS

Acids, Alkalies, and Salts

One component necessary for electrochemical reactions is an electrolyte solution. Inorganic acids, alkalies, and salts are electrolytes, and it is not surprising that these materials have a long history of causing corrosion. The hydrogen ion is one of the main factors involved in corrosion, since it participates in the cathodic reaction with electrons which forms free hydrogen.[2] The rate of corrosion depends upon the acidity of the solution, with a dividing line occurring in the neighborhood of pH 4.5. The rate of corrosion is fairly low above a pH of 4.5, but becomes quite rapid at lower pH values.

Alkalies are usually less corrosive than acids, but can produce anodic attack at localized areas, resulting in a pitting type of corrosion.[2]

Salts are good electrolytes and their solutions are highly conductive. As a result, they can be quite corrosive. The corrosiveness of a salt solution will depend upon both the concentration and the type of salt present.

Chlorinated Solvents

Chlorinated solvents such as methylene chloride, methyl chloroform, and trichloroethylene are used in aerosol formulations as solvents for active ingredients, vapor pressure depressants, or just low-cost substitutes for fluorocarbon 11. In some products, they have a more specific function, such as solvents for dry cleaning etc.

Chlorinated solvents usually are unstable in the presence of metals, heat, moisture, and oxygen; they may decompose to acidic decomposition products.

If this occurs during shipment in steel drums or in solvent applications involving the cleaning of metal components at elevated temperatures, corrosion problems may arise. As a result, many compounds have been tested as inhibitors for increasing the stability of the chlorinated solvents. A number of patents disclosing additives reported to be effective stabilizers have been issued.[12] Most common commercial chlorinated solvents contain stabilizers.

Methylene chloride is stable in most aerosol formulations and has been used in both oil-in-water and water-in-oil emulsion systems. Methyl chloroform appears to be less stable in aqueous systems. Any aqueous formulation containing a chlorinated solvent should be checked thoroughly for shelf life.

Surface-Active Agents

Many products on the market containing surface-active agents, for example, shaving lathers, upholstery cleaners, window cleaners, etc., have a perfectly satisfactory shelf life. However, some surface-active agents are very corrosive in metal containers. One of the best known classes of surfactants in this category is the sodium alkyl sulfates. A number of early attempts were made to package aerosol shampoos based upon detergents of the sodium alkyl sulfate type, but the products generally were too corrosive in the metal containers of that period to be marketed.[13,19] Some of these products appeared to have satisfactory storage stability judging from accelerated storage stability tests at elevated temperatures, but subsequently were found to cause perforation of containers in a relatively short time at room temperature.

There are a number of reasons why sodium alkyl sulfates are corrosive. They ionize in solution and also contain inorganic salts as impurities. Therefore, their aqueous solutions are good electrolytes and conductors of electricity. Another important factor, as shown by Root,[14] is that in these systems the tin is cathodic and the iron anodic. Corrosion is localized at the minute areas where iron is accessible, resulting in pinholing. This is an excellent example of the rapid corrosion that can occur when the system contains a cathode having an area much larger than that of the anode. Root also examined other formulations in which tin was found to be anodic and iron cathodic. These formulations appeared to have much better storage stability.

In contrast to the sodium alkyl sulfates, anionic triethanolamine salts of the fatty acids have been used extensively in aerosol foam products. The rate of corrosion is acceptably low. There are several reasons why these surfactants are less corrosive than sodium alkyl sulfates. Since triethanolamine fatty acid soaps are salts of a weak acid and a weak base, they tend to hydrolyze in solution rather than ionize. Also, they do not contain inorganic salts as impurities. Therefore, solutions of triethanolamine salts are much weaker electrolytes than solutions of sodium alkyl sulfates. West[15] has suggested that fatty acid soaps should be considered as corrosion inhibitors. Another factor may be the relative solubilities of the tin and iron salts of the fatty acids and lauryl sulfates. LaQue and Copson[2] have indicated that corrosion is reduced when corrosion products are insoluble. The tin and iron salts of the lauryl sulfates would be expected to be more soluble in the aqueous phase than the corresponding salts of the fatty acids.

Nonionic surfactants in general appear to cause much less corrosion than anionic agents. Since these compounds do not ionize, their solutions are less conductive. This is certainly one of the factors responsible for the lower rate of corrosion.

Cationic agents such as the quaternary ammonium compounds have been used at very low concentrations as germicidal agents in a number of products. It is likely that at much higher concentrations they could cause corrosion because their solutions are fairly good electrolytes.

Water

Water alone can cause considerable corrosion in metal aerosol containers. The corrosive effects of water are due largely to the presence of dissolved solids and gases.[2] The corrosive action of water will vary considerably depending upon the source of the water. Distilled water is not very corrosive, but seawater, with its relatively high concentration of salts, is quite corrosive. Water containing appreciable concentrations of carbon dioxide will cause significant corrosion of steel, but will not have much effect upon tin.

In view of the difference in corrosive activity observed with water from various sources, it is important in storage stability tests to use the same water as will be used in production. The stability of a test pack prepared in the laboratory with distilled water may be considerably different from that of samples prepared with deionized water. Nitromethane is a fairly good inhibitor for retarding the corrosion of tinplate containers by water.[4]

ALUMINUM CONTAINERS

Aluminum as a material of construction for aerosol containers has a number of advantages. It is highly resistant to corrosion because of a protective oxide film. The oxide film forms easily in the presence of oxidizing agents or oxygen and is reported to be stable under most conditions in the pH range from 4.5 to 8.5.[2]

Certain combinations of propellants with alcohols can corrode aluminum containers. Mixtures of anhydrous ethyl alcohol with fluorocarbon 114 were reported to cause perforation of aluminum containers at elevated temperatures. The corrosion was attributed to the formation of aluminum alcoholate and hydrogen. The increase in pressure resulting from the formation of hydrogen was also noted in other tests carried out with alcohols and propellants in the presence of aluminum strips.[6] The presence of 1.5 – 2.0 wt. % water had an inhibiting effect upon the formation of the alcoholate, which evidently is favored by anhydrous conditions.[16]

Combinations of fluorocarbon 11 and ethyl alcohol cause severe corrosion of aluminum containers manufactured in the United States. The addition of water does not sufficiently inhibit the corrosion-inducing processes. Since fluoro-carbon 11 can react with ethyl alcohol or water to form acidic reaction products, it is assumed that these products attack the protective oxide film on aluminum.

STORAGE STABILITY TESTS FOR AEROSOL PRODUCTS

There are no standard storage stability tests for aerosol products. Theoretically, an ideal test for an aerosol package would be one in which the product was tested under all the conditions to which it would be exposed during consumer use. This, of course, would be difficult to determine. A suggested product check that is helpful in avoiding premature marketing of an aerosol product has been published.[17] A tentative method for storage tests of aerosol insecticides which is valuable as a guide in setting up storage stability tests is also available.[18]

Most aerosol products are used at room temperature, and it would be most desirable to test the storage stability at room temperature. The disadvantage of this is that it requires long-term testing which sometimes extends into years. Many marketers are reluctant to wait for the results of room temperature tests because of the extended periods required. As a result, tests at elevated tempera-tures are generally used in conjunction with room temperature tests. Tests at elevated temperatures have been used in the rubber and petroleum industries for many years. It is well known in these industries that in many cases the storage stability of a product at elevated temperatures does not necessarily correlate with that at room temperature.

The same situation exists with respect to accelerated storage stability tests of aerosol products. In some cases, the lack of correlation between the results obtained from room temperature tests and those from tests at higher tempera-tures was striking. One well-known example is that of aerosol shampoos formu-lated with surfactants such as sodium lauryl sulfate. When these products were packaged in metal containers, they appeared to be essentially noncorrosive when tested at 130°F for two or three months. However, the containers perforated within a few weeks at room temperature.[13] According to Foresman,[13] no typical perforation corrosion of the type caused by shampoos has ever been observed at elevated temperatures. Foresman recommended a temperature of 100°F for accelerated tests because he considered it close enough to room temperature so that it was not unrealistic. In some cases, storage for three months at 100°F indicated the stability of a formulation for one year at room temperature.

Glessner[7] studied the storage stability of aqueous-based insecticides and reported that with some products an accelerated factor of 2–3.5 to 1 was observed with tests at 100°F. He also pointed out that the results of tests at 100°F or 130°F could be misleading, however, and might cause the rejection of a product with adequate stability at room temperature.

The results of tests at elevated temperatures must be viewed with a certain amount of caution. Many formulators carry out storage stability tests at both room temperature and elevated temperatures. If short-term tests at elevated temperatures produce results that look sufficiently promising, limited marketing of a product may be considered. If the results of tests at the elevated temperature indicate that the storage stability of the product may be doubtful, marketing may be delayed until some of the longer-term tests at room temperature have been completed.

Although there are no standard accelerated storage stability tests, there are some experimental conditions used by a considerable number of formulators for evaluating the stability of their products. These are listed below.

Products in Two-Piece Drawn Containers

Storage tests are carried out for one and two months at 130°F. Most formulators do not extend the tests at 130°F beyond two months. Very short-term tests (a few days) at 130°F are, in a sense, not really accelerated tests, since the product may reach this temperature for a short period of time during transport or storage.

Products in Three-Piece Side-Seam Containers

Storage tests in three-piece side-seam containers for 3, 6, and 12 months at temperatures from 100 to 110°F are common. In many cases, shorter-term tests at 120°F for 1 and 2 months are used.

One effect observed with some side-seam containers is side-seam solder fatigue during extended periods at elevated temperatures (130°F). Under these conditions, what often is being tested is the resistance of the soldered side seam to fatigue at 130°F rather than the resistance of the container to corrosion by the product.

In some cases, cycling temperature tests are used: Containers are stored for short periods of time at 120°F, for example, and then stored successively at temperatures of 98°F, 70°F, and 40°F. These tests are intended to subject the product to the extremes of temperature to which it might be exposed during shipment and storage before it reaches the consumer.

Glass Bottle Products

Glass bottle tests may be carried out at temperatures of 100°F or slightly higher.

Ovens and Sample Storage

It is advisable to check the temperature at the different locations in an oven to be certain that there are no areas where the differences in temperature may be significant. If samples of one product are stored in an oven at a location where the temperature is 130°F and others are stored in the same oven but at a location where the temperature is 120°F, comparisons of the two sets of samples after the same aging period may lead to false conclusions about the relative storage stabilities of the two sets of samples.

A convenient and effective way to store samples is to use boxes constructed of perforated iron sheets equipped with a hinged lid. The boxes can be made to hold any desired number of samples and can be stacked in an oven, thus utilizing the space available in the most effective way. If aerosol containers explode during storage, the metal boxes usually will retain the pieces. Of greatest importance is the fact that metal boxes provide protection for the technician removing the samples after storage.

The number of samples tested depends upon a variety of factors such as the desired reliability, available storage space, number of formulation variables involved, different types of containers and valves, etc. Some companies store samples in inverted, upright, and sideways positions in order to simulate the conditions to which the product might be exposed during transportation and storage.

Evaluation of Products after Storage

There are two main considerations in planning storage stability tests: the stability of the product itself and the stability of the complete aerosol package, which includes the container and the valve. As far as the product is concerned, it is the use for which the product is designed that determines the type of storage stability test. It must be determined if chemical reactions that reduce the effectiveness of the product occur during storage. If the product is a disinfectant, for example, then measurement of disinfectant activity before and after aging is one criterion of stability. Aerosol perfumes are checked for odor stability and aerosol foam products for foam stability after storage. The criterion of product stability is as follows: Does the product perform its function adequately after storage?

In determining the suitability of a container and valve for a product, there are many factors to consider. Corrosion can lead to off-odors, discoloration of the product, and leakage. Corrosion may be due to decomposition of the product, or the product itself may be corrosive. The extent of corrosion after a storage stability test can be determined by opening the container and removing the product. The interior of the container, the can bottom, the dip tube, and the valve cup are inspected for indications of attack, such as detinning, rusting, etc.

In many cases it is sufficient merely to record the extent of corrosion by describing the condition of the container in words. In other cases it is simpler to use a corrosion rating system such as the following:

CORROSION RATING SYSTEM

Corrosion Rating	Description
0	No corrosion
1	Generalized corrosion either liquid phase or vapor phase–spotty detinning
2	Complete detinning either liquid phase or vapor phase
3	Complete detinning both liquid phase and vapor phase
4	Complete detinning both liquid phase and vapor phase with spotty rusting
5	Complete rusting throughout container or leakage

The corrosion rating system can be made as simple or as detailed as desired and can be devised to fit the particular product being tested.

The effect of the product upon the valve during storage is of utmost importance. If the valve fails to operate properly, the reason for the failure can often be determined by taking the valve apart and examining the valve components. It is informative to compare valve components from the test sample with those from an unaged valve. This will usually show if excessive swelling or deterioration of the valve gaskets or plastic components has occurred. Sometimes valve failure can be eliminated by changing valves, but more often it is necessary to modify the formulation.

Quite often the first signs of corrosion in an aerosol container appear on the valve cup, which is usually uncoated. Coated valve cups may be obtained and may help in retarding corrosion. However, crimping the valve may cause strains in the coating so that the uniformity of the coating on the valve is impaired.

The frozen concentrate which has been removed from the container is placed in a beaker and allowed to warm up to room temperature. Inspection of the concentrate will determine if discoloration or some other indication of decomposition (for example, the formation of insoluble reaction products) has occurred. The measurement of properties such as pH can be carried out to determine if acidic reaction products have formed.

The subject of corrosion and storage stability is very complicated and the literature on corrosion and its causes is extensive. Additional sources of information regarding corrosion of aerosol products are given in References 19 through 24.

REFERENCES

1. F. L. LaQue, T. P. May, and H. H. Uhlig, "Corrosion in Action," The International Nickel Company, Inc., 1955.
2. F. L. LaQue and H. R. Copson, "Corrosion Resistance of Metals and Alloys," 2nd ed., Chap. 13, Reinhold Publishing Corp., New York, 1963.
3. P. A. Sanders, F. A. Bower, and L. Long, *Soap Chem. Spec.* **36**, 95 (July 1960) (Freon® Aerosol Report A-51); U.S. Patent 3,085,116.
4. P. A. Sanders, *Soap Chem. Spec.* **41**, 117 (December 1965) (Freon® Aerosol Report A-64).
5. Freon® Technical Bulletin B-2, "Properties and Applications of the Freon® Fluorinated Hydrocarbons."
6. H. M. Parmelee and R. C. Downing, *Soap Sanit. Chem.* **26** (1950), CSMA 2 (Freon® Aerosol Report A-19).
7. A. S. Glessner, *Aerosol Age* **9**, 98 (October 1964).
8. P. A. Sanders, *J. Soc. Cosmet. Chem.* **9**, 5 (1958) (Freon® Aerosol Report A-49).
9. P. A. Sanders, *Soap Chem. Spec.* **42**, 74, 135 (July/August 1966) (Freon® Aerosol Report A-66).
10. Freon® Aerosol Report FA-28, "Freon® 21, Aerosol, Solvent, and Propellant."
11. M. A. Johnsen, W. E. Dorland, and L. K. Dorland, "The Aerosol Handbook," Wayne E. Dorland Company, Caldwell, N.J., 1972.
12. British Patent 773,187; U.S. Patent 2,185,238; U.S. Patent 3,391,689; U.S. Patent 2,567,621; U.S. Patent 2,923,747; U.S. Patent 3,159,582.
13. R. A. Foresman, Jr., *Aerosol Age* **1**, 34 (September 1956).
14. M. J. Root, "Factors in Formulation Design," in H. R. Shepherd (Ed.), "Aerosols: Science and Technology," Interscience Publishers, Inc., New York, 1961.
15. C. W. West, *Aerosol Age* **6**, 20 (March 1961).
16. E. D. Giggard, *Aerosol Age* **6**, 20 (July 1961).
17. *Proc. 38th Mid-Year Meet. Chem. Spec. Manuf. Assoc.*, 53 (May 1952).
18. Chemical Specialties Manufacturers Association (CSMA) Aerosol Guide, 1971.
19. R. A. Foresman, Jr., *Proc. 39th Ann. Meet., Chem. Spec. Manuf. Assoc.*, 32 (December 1952).
20. M. J. Root, *Aerosol Age* **4**, 29 (June 1959).
21. M. F. Johnson, *Aerosol Age* **7**, 20 (June 1962).
22. L. M. Garton, *Aerosol Age* **7**, 108 (October 1962).
23. W. J. Pickett, *Proc. 40th Ann. Meet. Chem. Spec. Manuf. Assoc.*, 43 (December 1953).
24. M. J. Root, *Proc. 42nd Ann. Meet. Chem. Spec. Manuf. Assoc.*, 41 (December 1955).
25. Freon® Products Laboratory Data.

14

FLAMMABILITY

The flammability of an aerosol product is one of its most important properties. If an aerosol product is flammable, it may be potentially hazardous when shipped, stored in warehouses, or used in the home. The flammability of aerosol propellants is also important because of the influence the propellants have upon the properties of aerosol products themselves and, moreover, because of possible hazards involved in their manufacture, transportation, storage, and handling.

There are a number of tests that evaluate and differentiate various aspects of flammability. Therefore, it is not sufficient merely to report that a product or propellant is flammable; the particular test or tests used to determine the flammability must also be specified. As far as aerosol products are concerned, the tests of primary importance are flame extension, flashback, and flash point. The most important criteria for propellants or propellant blends is whether they form flammable mixtures with air when vaporized.

FLAMMABILITY OF AEROSOL PRODUCTS

Flammability Tests

Flammability tests for aerosol products were originally developed by the Bureau of Explosives (American Association of Railroads) for the Interstate Commerce Commission (ICC). These tests are listed in Agent R. M. Graziano's Tariff No. 31, Paragraph 73.300, effective March 31, 1977. Several of the tests were subsequently adopted by other government agencies and the New York City Fire Department.

Flame Extension Test

This test indicates how far the spray from an aerosol will extend a flame when the spray passes through the flame. The test equipment consists of a base 4 in.

221

wide and 2 ft long, marked at 6-in. intervals. A rule 30 in. long and marked in inches is supported horizontally on the side of the base and about 6 in. above it. A plumber's candle of such height that the top third of the flame is at the height of the horizontal rule is placed at the zero point in the base.

The aerosol container is placed 6 in. in back of the candle and sprayed so that the spray passes through the top third of the flame and at right angles to it. The height of the flame should be about 2 in. Two observers are required for the test. One observer notes how far the flame is extended, while the other operates the container. The test is positive if the flame projects more than 18 in. beyond the ignition source with the valve fully opened or if the flame flashes back and burns at the valve with any degree of valve opening.

Some laboratories use an asbestos backboard marked with vertical lines at 1-in. intervals instead of a ruler. In most cases, it is easier to read the flame extension with this equipment than with the ruler.

Drum Tests

These tests initially were developed for the ICC (now the Department of Transportation, DOT) to indicate the potential hazard if aerosol containers were sprayed or ruptured and leaked into a closed area where a flame or some other source of ignition was present. This situation could result if a fire occurred in warehouses or during transit in trucks or boxcars. The drum tests were never adopted by any other agency. They are now of only minor importance, since the DOT has eliminated flammability tests for aerosol products where the gross net weight of the shipping package does not exceed 65 lb and the aerosol container does not have a capacity exceeding 50 cu. in. (ORM-D Classification).

The apparatus consists of a 55-gal. open-head drum fitted with a hinged cover or plastic film over the open end. The closed end of the drum is equipped with a shuttered opening 1 in. in diameter for introduction of the spray. A plumber's candle is placed an equal distance from the ends on the bottom of the drum.

Open-Drum Test. The open-drum test is carried out with one end of the drum completely open. The spray from the aerosol with the valve fully opened is directed into the upper half of the open end and above the ignition source for 1 min. Any significant propagation of flame through the vapor/air mixture away from the ignition source is a positive result. Any minor or unsustained burning is a negative result. This test is hazardous to the operator and, therefore, was never very popular.

*Closed-Drum Test.** The closed-drum test is conducted either with the open end covered by a thin plastic film (containing a slit) or the hinged cover dropped

---

*A description of the flame projection apparatus, open-drum apparatus, closed-drum apparatus, and the methods of testing aerosol products may be obtained either from the

into place. The aerosol is sprayed into the drum with the valve fully opened for 1 min. After the drum atmosphere has been cleared from any previous tests, the procedure is repeated. Any explosion or rapid burning of the vapor/air mixture is a positive result.

Flash Point. The aerosol unit is cooled to about $-50°F$ and the cold aerosol product is transferred to the test apparatus. The aerosol is allowed to increase in temperature at a rate of about $2°F/min$. The test flame taper is passed across the sample at intervals of $2°F$ until the sample flashes, reaches a temperature of $20°F$, or has evaporated completely. The level of the liquid does not remain constant during the determination.

The method of determining the flash point of aerosol products is described in the CSMA Aerosol Guide. The ASTM designation for the flash point test is D 1510-55T, "Flash Point of Volatile Flammable Materials by Tag Open Cup Apparatus."

Consumer Hazard Tests

The potential flammability hazard that could result during consumer use of an aerosol is difficult to assess. At the present time, there are no standard flammability tests available to evaluate this hazard.

In the absence of any standard tests, some marketers use flame extension and flashback as an indication of consumer hazard. If the flame does not project more than 18 in. and the product does not flash back at any degree of valve opening, the consumer hazard potential is considered to be relatively low. This criterion is similar to that used by the Consumer Safety Products Commission to determine if aerosol products falling under their jurisdiction are flammable or nonflammable. Other tests that have been used involve spraying wigs or blotters. A lighted match is applied to the sprayed object after various time intervals to determine if ignition occurs and, if so, how readily the wig or blotter burns.

A more recent test for alcohol-based uncoated glass bottle aerosol products (such as colognes and perfumes) involves dropping the aerosols onto a steel plate in the presence of an ignition source. The height and diameter of any fireball that results are estimated or the fireball is photographed. This test is an attempt to simulate a situation that could occur in a bathroom if, for example, an aerosol cologne was accidentally dropped on a tile floor near an ignition source. Ignition sources used at the Freon® Products Laboratory include (1) three 3-in candles placed equidistant from each other at a distance of 1 ft from the point of contact of the bottle on the steel plate and (2) an electric arc obtained with a 15,000-V ignition transformer. This test distinguishes very effectively between

Bureau of Explosives, Association of American Railroads, American Railroads Building, Washington, D.C., or the Chemical Specialties Manufacturers Association, 1001 Connecticut Ave., N.W., Washington, D.C. The tests are also described in Reference 1.

colognes formulated with hydrocarbons and those formulated with nonflammable fluorocarbons.

The Electrostatic Problem with Powder Aerosols*

Many aerosols that were propelled with fluorocarbons have been reformulated with hydrocarbons as a result of the fluorocarbon/ozone controversy. The substitution of flammable hydrocarbons for nonflammable fluorocarbons in powder aerosol products (such as dry antiperspirants and fabric cleaners), however, created a problem: static ignition of punctured aerosol cans. During the past year, at least seven fires or explosions occurred with aerosol cans as a result of static ignition. Generally, the problem has occurred on aerosol filling lines or in aerosol laboratories where discarded or faulty cans have been punctured and thrown into a confined area, such as a bin. The vaporizing propellant formed flammable mixtures with air and the static electrical charge resulting from the rapidly escaping aerosol powder formulation sparked and ignited the flammable propellant/air mixture. Thus far, there is no evidence that this situation has occurred during consumer use, where the quantity of spray is controlled by a valve. However, this is being actively investigated at the present time, along with other aspects of the problem.

The opportunity for electrostatic charge separation exists wherever liquids or solid particles are conveyed or sprayed. It is not surprising, therefore, to find that charge separation can occur during the expulsion of contents from a punctured aerosol can. If the propellant is flammable, one must consider the possibility of ignition of the resulting vapor/air or gas/air mixture by static sparking.

In practice, the amount of charge separation is unlikely to be hazardous except in some instances where the product is dust-like and the can is punctured. The charging tendency of specific products cannot be predicted, and it must be determined experimentally. A method of evaluating the degree of electrostatic hazard is described below.

Flammability Considerations

To cause a fire, one must bring together all three legs of the well-known "fire triangle," that is,

1. Fuel.
2. Oxygen or air.†
3. Ignition source.

*This section on the electrostatic problem was contributed by John E. Owens, Senior Research Associate, Engineering Physics Laboratory, Du Pont Company, Wilmington, Delaware.
†For the present purpose, we shall disregard those few materials that can burn in the absence of oxygen.

A flammable atmosphere is one in which the fuel/air ratio is within the flammability limits, that is, above the lower flammability limit and below the upper flammability limit. These data have been published for many flammable materials (see M. G. Zabetakis, "Flammability Characteristics of Combustible Gases and Vapors," U.S. Bureau of Mines Bulletin 627 (1965)). The present discussion deals only with static sparking as a possible ignition source; other sources of ignition may exist and should be considered in hazard analyses.

To determine whether or not static sparking is likely to cause a fire as a result of puncturing an aerosol can, one needs to know the value of the minimum ignition energy (M.I.E.) for the particular propellant. These values, which have been published for many materials, correspond to the least amount of energy that is needed to ignite the most flammable fuel/air concentration. (See H. Haase, "Electrostatic Hazards; Their Evaluation and Control," English translation, Verlag Chemie, New York (1977)). For other fuel/air concentrations within the flammability limits, larger ignition energies are needed. The M.I.E. values for most hydrocarbon propellants are about 0.2 millijoule (mJ).

Electrostatic Charging Considerations

Although the exact nature of the charge-separation process remains poorly defined, a substantial accumulation of static charges can develop during loss of certain products from punctured aerosol cans. The presence of a charged cloud (of dust-like particles, for example) is, itself, not an ignition hazard. A sparking hazard may be present, however, if any of the following conditions exists:

1. Charged spray is deposited on an ungrounded conductor.
2. The aerosol can is not grounded.
3. A punctured can is held by an ungrounded person.

The magnitude of charging of the effluent from a punctured can is expressed in units of microcoulombs per gram. An experimental method of measuring this quantity, described below, is based on the fact that the charge on the can is equal to (but of opposite polarity to) the charge on the effluent. The total energy (W) available in a spark from can to ground is calculated from the total charge (Q) on the can and the capacitance (C) of the can to ground:

$$W = \frac{500Q^2}{C}$$

where

W = energy (mJ)
Q = total charge on can (coulomb, C)
C = can-to-ground capacitance (farad, F).

Energies of 0.2 mJ or larger would be considered hazardous for hydrocarbon propellants.

Measurement of Electrostatic Charge Separation

The aerosol can is held by an insulating clamp. A shielded lead with an alligator clip connects the can to an electrometer, such as the Keithley Instruments model 610C. The function switch is initially set to the 10^{-9} C range and the VOLTS-FULL-SCALE switch to 100 V. A ground wire is connected to the pin that will puncture the can. Then one proceeds as follows:

1. Puncture the can.
2. If the electrometer indicates downscale, change the POLARITY switch.
3. Set the VOLTS-FULL-SCALE switch to produce an on-scale indication. If the voltage exceeds 100 V, switch to the 10^{-8} or 10^{-7} C range. The voltage should continue to increase so long as the can continues to vent product, after which it will remain constant.
4. Total charge (Q) is read as the indicated voltage times the coulomb range. With this measurement and the number of grams of product vented, express the specific charge in microcoulombs per gram.
5. Press the ZERO button to reset the meter to zero.

Flammability Regulations

*The Department of Transportation (DOT) Regulations for Aerosols**

The following discussion of the DOT regulations is not intended to be a substitute for legal counsel or advice from the DOT concerning the application or interpretation of the regulations. Law firms especially competent in this field are also available.

The official text of the DOT regulations for aerosols appears in Title 49 Code of Federal Regulations (CFR) parts 100–199, revised December 31, 1976. The regulations are also reprinted in Agent R. M. Graziano's Tariff No. 31, "Hazardous Materials Regulations of the Department of Transportation," issued March 1, 1977, effective March 31, 1977. The paragraphs and numbers are the same in both publications.

If an aerosol product meets the DOT definition of a compressed gas, it is considered a hazardous material and is regulated by the DOT. The DOT definition is found in Paragraph 173.300(a), Tariff No. 31, and is as follows: "The term 'compressed gas' shall designate any material or mixture having in the container an absolute pressure exceeding 40 psi at 70°F, or regardless of the pressure at 70°F, having an absolute pressure exceeding 104 psig at 130°F, or any liquid flammable material having a vapor pressure exceeding 40 psi absolute at 100°F, as determined by ASTM Test D-323." If the pressure of the aerosol does not exceed the limitations at either of the first two temperatures, it is not classified as a compressed gas and is not regulated by the DOT (unless it is classified as a flammable liquid or any of the other hazard classes).

*Also see Chapter 5, p. 78.

A carrier may not transport an aerosol classified as a hazardous material unless it is accompanied with shipping papers which give a description of the material (see Paragraphs 172.200, 172.201, 172.202, and 177.817). In the past, the flammability of aerosols defined as compressed gases had to be known in order to prepare the shipping papers. Flammability was determined by the flame extension, flashback, closed-drum, and open-drum tests previously described. If any one of the tests was positive, the aerosol was classified as a flammable compressed gas.

Flammability properties still have to be known for many aerosol products. However, the Department of Transportation has recently amended its regulations concerning aerosols so that flammability tests have been eliminated for aerosol products which meet a new DOT classification, ORM-D (Other Regulated Material-D). The official text appears in the December 31 revision of the Code of Federal Regulations and in R. M. Graziano's Tariff No. 31.

A material may be shipped ORM-D if it is a consumer commodity, that is, a material packaged and sold for personal care or household use. Thus, it must be intended or suitable for consumer consumption. In order to qualify for the ORM-D classification, the gross weight of the package must not exceed 65 lb, and the capacity of the aerosol container must not exceed 50 cu. in. (27.7 fl. oz). Packaging must conform to sections 173.510, 173.1200, and 173.6. The outside marking must include the consignee's name and address and the proper shipping name followed by ORM-D or ORM-D AIR. No labeling or placarding is required and shipping papers are not necessary unless intended for air transportation.

An ORM-D material is defined as follows (Federal Register 49 CFR Parts 171-177, Monday, September 27, 1976, page 42567):

"Subpart J—Other Regulated Material: Definition and Preparation
Paragraph 173.500 Definitions
    (4) An ORM-D material is a material such as a consumer commodity which, although otherwise subject to the regulations of this subchapter, presents a limited hazard during transportation due to its form, quantity and packaging. They must be materials for which exceptions are provided in Para. 172.101 of this subchapter. A shipping description applicable to each ORM-D material or category of ORM-D materials is found in Para. 172.101 of this subchapter." (Para. 172.101 is a table listing hazardous materials.)

The qualifications that a material must meet in order to be classified as ORM-D and be shipped as a consumer commodity are described in the following section:

"Subpart N—Other Regulated Material: ORM-D
Paragraph 173.1200 Consumer Commodity
    (a) In order to be transported under the proper shipping name of "consumer commodity," a material must meet that definition. It may be

reclassed and offered for shipment as ORM-D material (see Para. 173.500) provided that an ORM-D exception is authorized in the specific section applicable to the material and that it is prepared in accordance with the following paragraphs. (The gross weight of each package must not exceed 65 pounds, and each package offered for transportation aboard aircraft must meet the requirements of Para. 173.6).

(8) Compressed gases must be:

(i) In inside containers, each having a water capacity of 4 fluid ounces or less (7.22 cu. in. or less), packed in strong outside packagings.

(ii) In inside containers charged with a solution of materials and compressed gas or gases which is nonpoisonous, meeting all of the following:

(A) Capacity may not exceed 50 cu. in. (27.7 fluid ounces).

(B) Pressure in the container may not exceed 180 psig at 130°F (55°C). If the pressure exceeds 140 psig at 130°F (55°C) but does not exceed 160 psig at 130°F (55°C), a specification DOT 2P (Para. 178.33 of this subchapter) inside metal container must be used. If the pressure exceeds 160 psig at 130°F (55°C), a specification DOT 2Q (Para. 178.33a of this subchapter) inside metal container must be used. In any event, the metal container must be capable of withstanding, without bursting, a pressure of one and one-half times the equilibrium pressure of the contents at 130°F (55°C).

(C) Liquid content of the material and gas must not completely fill the container at 130°F (55°C).

(D) The containers must be packaged in strong outside packagings.

(E) Each completed container filled for shipment must have been heated until the pressure in the container is equivalent to the equilibrium pressure of the contents at 130°F (55°C) without evidence of leakage, distortion or other defect."

Paragraph (8)(i) above refers to the transportation of both propellants alone and aerosols and limits the water capacity of the containers to 4 fl. oz or less. Paragraph (8)(ii) refers to aerosols where they are defined as a solution of materials and compressed gas or gases. If the material did not meet the conditions established for classification as ORM-D, that is, the gross weight of the package exceeded 65 lb or the container size was greater than 50 cu. in., then the flammability of the product would have to be determined using flame extension, flashback, and drum tests.

The Federal Hazardous Substances Act

The Federal Hazardous Substances Act first became effective in 1960 and was known as the Federal Hazardous Substances Labeling Act. The word "Labeling" has since been dropped. The purpose of the act was to minimize the possibility of injury to consumers by requiring that hazardous substances moving in inter-state commerce be properly packaged and labeled. The act applied only to products intended for household use and did not cover foods, drugs, cosmetics, pesticides. The act was administered at that time by the Food and Drug Administration.

The Consumer Products Safety Act became law on October 27, 1972. The act created an independent Federal agency, the Consumer Products Safety Commission (CPSC), which was assigned the task of guarding the public against "unreasonable risks of injury" from consumer products. The CPSC was given the responsibility for administering the Federal Hazardous Substances Act. According to the act, the contents of self-pressurized containers are extremely flammable if a flashback (a flame extending back to the dispenser) is obtained at any degree of valve opening and the flash point is less than 20°F. Contents of self-pressurized containers are flammable if a flame projection exceeding 18 in. is obtained at full valve opening or a flashback is obtained at any degree of valve opening (from CSMA Publication, "Law, Regulations and Agencies of Interest to the Aerosol Industry," 6th ed., July 1966). The flammability must be tested by methods in the regulations described in sections 1500.45 and 1500.46 (CFR 16).

Food and Drug Administration

At this time (March 1977) the Food and Drug Administration has not yet developed guidelines concerning definitions or regulations for flammability testing of aerosols. However, the FDA has indicated that the regulations issued by the Consumer Products Safety Commission on hazardous substances (16 CFR 1500.0 to 1500.46) would be acceptable for testing aerosol cosmetic products. The FDA expects to develop and issue proposed regulations for cosmetics which will, in essence, adopt the definitions and testing methods contained in the Consumer Products Safety Commission regulations.[2]

Environmental Protection Agency

The flammability regulations of the Environmental Protection Agency (EPA) for pesticides, as published in the Federal Register, Vol. 40, No. 129, Thursday, July 3, 1975, are as follows for pressurized containers:

Extremely Flammable—Flash point at or below 20°F; if there is a flashback at any valve opening.

Flammable—Flash point above 20°F and not over 80°F; or if the flame extension is more than 18 in. long at a distance of 6 in. from the flame.

The New York City Fire Department Regulations for Pressurized Products

The New York City Fire Department has promulgated regulations pertaining to aerosols or their propellants for products to be sold in New York City. The regulations describe the tests for aerosol products, their propellants, type of labeling for aerosols, and restrictions covering storage of certain aerosols. The New York City Fire Department was in litigation with the CSMA. This was settled in 1971 (see CSMA vs Lowery, 452 F.2D 431, '2D Circuit 1971'). The CTFA started litigation in the early 1960's. This case has not yet been decided. The New York City Fire Department is negotiating with industry at the present time in order to arrive at a viable regulation.[14] Excerpts from the proposed New York City Fire Department regulations are reproduced below:

DIVISION OF FIRE PREVENTION
FIRE DEPARTMENT

F.P.Directive 5-71 (Revised)   March 15, 1974
(Revokes F.P.Dir. 5-71 dated 2/15/71)

BY VIRTUE OF THE AUTHORITY VESTED IN ME AS FIRE COMMISSIONER under Sections 489 of 1105 of the New York City Charter, I hereby promulgate new Regulations for Pressurized Products. (Filed with the City Clerk on February 15, 1974).

REGULATIONS FOR PRESSURIZED PRODUCTS

SCOPE
1. NO EXTREMELY FLAMMABLE, FLAMMABLE or COMBUSTIBLE pressurized products shall be stored, sold or used in the City of New York without a Certificate of Approval and/or Permit as required by these regulations.
2. Products falling within the purview of the Federal Hazardous Substances Act (FHSA) and the Pesticide Act (FIFRA) are preempted from compliance with certain sections of these regulations, which are noted.

Note: No Certificates of Approval are required for pesticide products.

Section 1. DEFINITIONS
A. Pressurized products (consisting of essential ingredients and propellant) are products which are self-propelled from containers through valves, regardless of the size of the particles emitted, type of propellant or form in which the product is dispensed.
B. A combustible pressurized product is one which flashes between 100 degrees Fahrenheit and 300 degrees Fahrenheit, or where the flame projec-

tion is over three (3) inches but less than eighteen (18) inches at full valve opening.

C. A flammable pressurized product is one which flashes between 20 degrees Fahrenheit and 100 degrees Fahrenheit or where the flame projection is eighteen (18) inches or more but less than twenty-four (24) inches at full valve opening.

D. An extremely flammable pressurized product is one which flashes below 20 degrees Fahrenheit or where a flash back toward and burning within one (1) inch of the valve is obtained at any degree of the valve opening, or where the flame projection exceeds twenty-four (24) inches.

Section 8. PROHIBITIONS

A. Extremely Flammable, Flammable or Combustible De-icers for refrigerators or freezers.

B. Extremely Flammable, Flammable or Combustible Oven Cleaners, Oven Coatings or similar preparations.

C. No EXTREMELY FLAMMABLE or FLAMMABLE propellants shall be used for filling containers for pressurized products within the City of New York.

D. Carbon tetrachloride and mixtures containing carbon tetrachloride shall not be used in any pressurized products for household use manufactured, stored or used in the City of New York.

Section 10. MODIFICATIONS AND APPEALS

A. Whenever circumstances, conditions, limitations or surroundings are unusual, or such as to render it impracticable to enforce all the foregoing requirements, the Fire Commissioner may waive or modify such provisions to such extent as he may deem necessary consistent with public safety.

B. Any appeals from these regulations may be presented to the Board of Standards and Appeals pursuant to Section 666-6 of the New York City Charter.

JOHN T. O'HAGAN
Fire Commissioner
Chief of Department

FLAMMABILITY OF PROPELLANTS

The most important basis for judging the flammability of a propellant is whether it will form a flammable mixture with air. If a propellant is flammable, special precautions have to be observed in its manufacture, handling, shipping, and storage.

Some fluorocarbons are nonflammable, while others are flammable. Nonflammable fluorocarbons are listed in Table 14-1.

TABLE 14-1   NONFLAMMABLE FLUOROCARBONS[5]

Fluorocarbon	
Number	Formula
11	CCl_3F
12	CCl_2F_2
21	$CHCl_2F$
22	$CHClF_2$
114	$CClF_2CClF_2$
123	$CHCl_2CF_3$
124	$CHClFCF_3$
133a	CH_2ClCF_3
134a	CH_2FCF_3

Blends of nonflammable and flammable propellants are commonly used in the aerosol industry. The concentration of flammable propellant in the blend determines whether the blend will be flammable and form flammable mixtures with air. The maximum concentration of a flammable propellant that can be added to a nonflammable propellant without the resulting blend becoming flammable is reported as weight percent in the liquefied propellant blend. For example, a blend of 90 wt. % fluorocarbon 12 and 10 wt. % propane is nonflammable. A blend of 89 wt. % fluorocarbon 12 and 11 wt. % propane is flammable and forms flammable mixtures with air. The limit of propane in fluorocarbon 12 for nonflammable blends is, therefore, 10 wt. %.

Flammability Limits in Air

A gas is considered to be flammable if it forms flammable mixtures with air or if a flame is self-propagating in the mixture of gas and air.[3] Flammable mixtures liberate enough energy on combustion of any one layer of the gas to ignite the neighboring layers and each successive layer of unburned gas throughout the mixture.

There is only a certain range of concentrations in air at which a flammable gas is capable of forming flammable mixtures. The lowest concentration of gas that will form a flammable mixture with air is termed the lower flammability limit of the gas; the highest concentration is termed the upper flammability limit. A gas will form flammable mixtures at all concentrations between the lower and upper flammability limits. The reasons for these limits are as follows. A lower flammability limit exists because there has to be a sufficient concentration of the gas in air so that when the gas ignites, enough energy is liberated to ignite the neighboring layers of unburned gas. An upper flammability limit occurs because as the concentration of gas in air is increased, it replaces that amount of air corresponding to the volume of added gas. This, of course, decreases the supply of

TABLE 14-2   FLAMMABILITY LIMITS OF VARIOUS
             PROPELLANTS IN AIR

Propellant	Formula	Flammability Limits in Air (vol.%)	Reference
FC-31	CH_2ClF	10-27	5
FC-32	CH_2F_2	–	
FC-141b	CH_3CCl_2F	6.4-15.1	5
FC-142b	CH_3CClF_2	6.0-15.0	8, 5
FC-152a	CH_3CHF_2	3.9-16.9	5
Propane	C_3H_8	2.2-9.5	3, 6, 7
n-Butane	C_4H_{10}	1.9-8.5	3, 6, 7
Isobutane	C_4H_{10}	1.8-8.4	3, 6, 7
n-Pentane	C_5H_{12}	1.4-8.3	9
Isopentane	C_5H_{12}	1.4-8.3	9
Methylene chloride	CH_2Cl_2	13.4-19	5
		12-19	13
Methyl chloroform	CH_3CCl_3	6.8-10.5	10
		10.0-15.5	12
Dimethyl ether	CH_3OCH_3	3.4-18.0	6

available oxygen. When the amount of added gas reaches a certain limit, the amount of oxygen present in the mixture is insufficient to burn the gas. The flammability limits of gases in air are usually reported in volume percent. The flammability limits of various propellants in air are listed in Table 14-2.

Methods of Determining Flammability Limits

A standard procedure for determining the flammability limits of gases in air involves the use of an explosion burette or eudiometer tube, as described by Coward and Jones of the Bureau of Mines.[3] The eudiometer tube is a glass tube 5 cm in diameter and 150 cm in height, equipped at the bottom with an ignition source such as an electric arc or a match that can be ignited by a hot wire. Suitable inlets and mixing devices are provided so that mixtures of the flammable gas and air with known compositions can be prepared in the tube. Gas/air mixtures are generally considered to be flammable if, when the heat source is ignited, flame propagation in the tube occurs from bottom to top. The flammability limits of any flammable gas in air can be determined by preparing mixtures with increasing concentrations of the gas and noting whether or not the mixture propagates a flame.

The determination of the flammability limits of gases in air is not easy, as Coward and Jones have noted.[3] Differences may result because the type of apparatus and ignition source used to determine the flammability limits are different. There is no doubt that in some cases flammability limits obtained with the eudiometer tube do not represent the flammability limits in larger spaces. For example, Haenni, Fulton, and Yeomans[4] reported that mixtures of 84 wt. % Freon® 12 and 16 wt. % ethylene oxide were nonflammable when tested in a conventional eudiometer tube. However, when the tests were carried out in a 13.1-ft^3 autoclave, it was necessary to reduce the concentration of ethylene oxide to 12 wt. % in order to obtain a nonflammable mixture.

Flammability limits at the Freon® Products Laboratory are normally determined using a modified 55-gal. stainless steel drum (the explosion drum) or a 4 ft X 4 in. eudiometer tube. The explosion drum is used to minimize any error that might result with the smaller eudiometer tube. It gives the most accurate results. The open end of the drum is covered with polyethylene film so that the inside of the drum can be observed as the ignition source is activated. A copper tube is led through the top of the drum, almost equidistant from the front and back, and extended 4 in. below the top. It is equipped with a spray nozzle. A glass pressure burette is connected to the copper tube so that known quantities of the liquefied propellant mixture can be discharged into the drum. Stirring is provided by a fan blade attached to a rod that leads to the outside through a rubber stopper in the drum. The unit is rotated by a variable speed motor. Match ignition is used in the drum. A Nichrome wire spiral holding two kitchen matches is located at the mid-bottom of the drum, and the assembly is supported by a small frame. The wires are led through a rubber stopper in the drum. The explosion drum is illustrated in Figure 14-1.

Table 14-3 provides a comparison of the flammability limit of fluorocarbon 152a obtained using the larger size explosion drum with that obtained using the 4 ft X 4 in. eudiometer tube when both are equipped with match ignition. The lower flammability limits from both pieces of equipment agree very well, but the

Figure 14-1 The Freon® Products Laboratory explosion drum.

TABLE 14-3   DETERMINATION OF FLAMMABILITY LIMITS
IN AIR BY MATCH AND ELECTRIC ARC IGNITION[5]

Propellant	Equipment	Ignition Source	Flammability Limits in Air (vol.%)
Fluorocarbon	Explosion drum	Match	3.9-16.9
152a	4 ft × 4 in. Eudiometer tube	Match	4.0-13.2
	4 ft × 4 in. Eudiometer tube	Electric arc	5.5-12.5
Isobutane	4 ft × 4 in. Eudiometer tube	Match	1.6-7.2
	4 ft × 4 in. Eudiometer tube	Electric arc	2.4-6.4

upper flammability limit obtained with the 4 ft × 4 in. eudiometer tube was significantly lower than that obtained with the drum. In general, the lower flammability limits obtained with the eudiometer tube were more reliable than the upper flammability limits obtained with it.

Another reason that different flammability limits have been obtained for the same gas is that different ignition sources were used. Coward and Jones[3] have reported that if the ignition source is not sufficiently strong, some flammable mixtures will not be ignited, particularly if the compositions of the mixtures are such that they are near the flammability limit. They emphasized that the flammability of the gas mixture and not the capacity of the ignition source to initiate flame was of primary significance in testing.

Electric arc ignition, obtained with a 15,000-V ignition transformer, was found at the Freon® Products Laboratory to give erroneous flammability results. Evidently, it is a weaker ignition source than match ignition. This is shown in Table 14-3 by the flammability limits obtained for fluorocarbon 152a and isobutane using a 4 ft × 4 in. eudiometer tube equipped with both match and electric arc ignition. The lower flammability limit obtained with the electric arc is higher than that obtained with match ignition, while the reverse is true for the upper flammability limits. The flammability range with the electric arc, therefore, was smaller than that with match ignition.

Another example of the weaker strength of electric arc ignition compared to match ignition is provided by the data in Table 14-4. The table shows the results of flammability tests of various fluorocarbon 133a/isobutane ratios. With match ignition, a positive result was obtained when the isobutane concentration in the blend reached 6 wt. %. The concentration of isobutane had to be increased to 16 wt. % for a positive result with electric arc ignition.

Flammability of Two-Component Propellant Blends

Mixtures of nonflammable and flammable propellants have been of considerable interest to the aerosol industry for many years. Flammable hydrocarbons are

TABLE 14-4   DETERMINATION OF FLAMMABILITY OF
ISOBUTANE IN FLUOROCARBON 133a[a] BY
MATCH AND ELECTRIC ARC IGNITION[5]

4 ft × 4 in. Eudiometer Tube		
FC-133a/Isobutane Ratio (wt. %)	Ignition Source	Flammability
86/14	Electric arc	Negative
85/15	" "	Borderline
84/16	" "	Positive
96/4	Match	Negative
94/6	"	Positive

[a]Fluorocarbon 133a/isobutane blends tested at 10 vol. % in air.

low in cost and their blends with nonflammable fluorinated propellants are less
expensive than the nonflammable fluorinated propellants alone. As a general rule,
the mixtures were formulated to achieve maximum cost savings with as little increase in the flammability of the propellant as possible.

A hydrocarbon gas such as propane will form flammable mixtures with air
when its concentration in air falls within a specific range as defined by the lower
and upper flammability limits. When a nonflammable propellant like fluorocarbon 12 is added to air, the flammability range of propane is decreased, that is,
the propane is less flammable in the fluorocarbon 12/air mixture. As the concentration of fluorocarbon 12 in air increases, the flammability range of propane
continues to decrease until ultimately it reaches zero and the propane will not
form a flammable mixture with the fluorocarbon 12/air combination. This effect
is shown in Table 14-5.

The data in Table 14-5 are illustrated graphically in Figure 14-2. This is a typical
flammability limit curve and shows which compositions of propane/fluorocarbon

TABLE 14-5   EFFECT OF FLUOROCARBON 12 UPON THE
FLAMMABILITY OF PROPANE IN AIR

FC-12 in Air (vol.%)	Flammability Limits of Propane (vol.%)	
	Lower Limit	Upper Limit
0	2.2	9.5
6.0	3.0	7.0
12.0	3.7	4.5
13.0	Nonflammable	

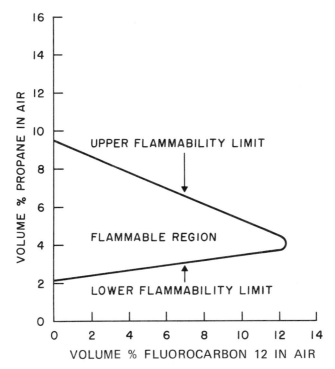

Figure 14-2 Flammability limits of propane in fluorocarbon 12/air mixtures.

12/air mixtures are flammable. The curve also shows how increasing the concentration of fluorocarbon 12 in air decreases the flammability range of propane.

The flammability limit curve in Figure 14-2 is plotted in volume percent, but it can be used to determine if any blend of fluorocarbon 12 and propane (whose composition is expressed in weight percent in the liquid phase) will form a flammable mixture with air when vaporized and what the flammability limits are. To do this, it is necessary to convert the weight percent ratios of the propellants in the liquid phase to volume percent ratios as gases after vaporization. Gas volume percent ratios (which are numerically the same as mole percent ratios) are obtained by dividing the weight percents by the molecular weights of the propellants.

The relationship between weight percent ratios in the liquid phase and volume percent ratios of the resultant gases when liquefied propellants are vaporized is shown in Table 14-6 for various fluorocarbon 12/propane blends. The data in Table 14-6 can be plotted on the flammability limit curve for fluorocarbon 12/propane blends in air as shown in Figure 14-3.

Lines A to E in Figure 14-3 show how the volume percent concentrations of fluorocarbon 12 and propane change in air as increasing amounts of any one of

TABLE 14-6   RELATIONSHIP BETWEEN LIQUID PHASE
COMPOSITION AND VAPORIZED COMPOSITION FOR
VARIOUS FLUOROCARBON 12/PROPANE BLENDS

Liquid Phase Composition *(wt. %)*		*Volume Percent Ratio of the Gases after Vaporization of the Liquefied Propellants*		
FC-12	*Propane*	*FC-12*	*Propane*	*Line*[a]
91.6	8.4	6.0	1.5	A
89.1	10.9	6.0	2.0	B
84.5	15.5	6.0	3.0	C
76.7	23.3	6.0	5.0	D
70.1	29.9	6.0	7.0	E

[a]Lines in Figure 14-3.

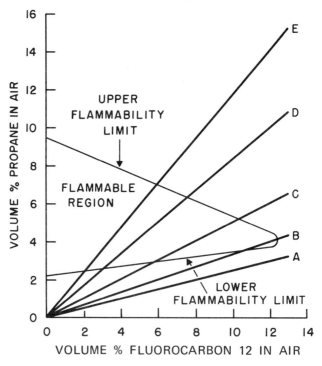

Figure 14-3 Liquid phase compositions and gas volume percent ratios for fluorocarbon 12/propane blends.

the liquid mixtures is vaporized. The resulting fluorocarbon 12/propane/air mixtures do not become flammable until the concentrations of the blends in air corresponding to their lower flammability limit are reached. This is the concentration at which the volume percent ratio line intersects the lower flammability limit curve. The fluorocarbon 12/propane/air mixtures remain flammable with increasing concentrations of fluorocarbon 12 and propane until the concentrations exceed the upper flammability limit.

Line A shows that a fluorocarbon 12/propane (91.6/8.4) blend will not form a flammable mixture with air at any concentration of the propellants, since line A does not cross the flammable region. The rest of the blends are flammable.

To determine experimentally if a given mixture of nonflammable and flammable propellants is flammable, increasing amounts of the test mixture are added liquid phase to the explosion drum and allowed to vaporize. The mixture of vaporized propellants in air is then tested with match ignition after each addition to determine if the mixture is flammable. The mixture was considered to be flammable if the flame extended to the top of the drum, the drum became warm, or there was an audible sound.

The explosion drum can be used to demonstrate if a propellant or a specific blend is flammable, as previously described. One of its most important uses, however, is to determine the maximum concentration of a flammable propellant that can be added to a nonflammable propellant without the blend becoming flammable. A typical example is the addition of flammable hydrocarbons to nonflammable fluorocarbons. The objective is to reduce the cost of the propellant blend below that of the fluorocarbon alone, while maintaining the nonflammability of the blend. For example, it was desirable to know the maximum concentration of isobutane that could be added to fluorocarbon 133a without the blend becoming flammable. A blend of the two was prepared with a concentration of isobutane (4 wt. %) estimated to be close to the limit of isobutane in fluorocarbon 133a for a nonflammable blend. This blend was found to be nonflammable. An additional blend with a higher concentration of isobutane, that is fluorocarbon 133a/isobutane (95/5), was then tested. This blend was found to be flammable. On the basis of these two test blends, the maximum concentration of isobutane that can be added to fluorocarbon 133a without the blend becoming flammable is 4 wt. %. The maximum concentration of fluorocarbons 142b and 152a for nonflammable blends with the nonflammable fluorocarbons are listed in Table 14-7. The maximum concentrations of various hydrocarbons for nonflammable blends with the nonflammable fluorocarbons are listed in Table 14-8.

The testing procedure just described provides the maximum concentration of a flammable component for a nonflammable blend with a nonflammable propellant. Increasing the concentration of the flammable component above this limit progressively increases the violence of the explosion when the blend is

TABLE 14-7 FLAMMABILITY OF FLUOROCARBON BLENDS

Nonflammable Fluorocarbon		Maximum Concentration (wt. %) of Flammable Fluorocarbon for Nonflammable Blend[5]	
Number	Formula	FC-142b	FC-152a
12	CCl_2F_2	74	34
21	$CHCl_2F$	62	20
22	$CHClF_2$	62	24
114	$CClF_2CClF_2$	–	30
123	$CHCl_2CF_3$	63	26
124	$CHClFCF_3$	67	29
133a	CH_2ClCF_3	65	21

TABLE 14-8 FLAMMABILITY OF FLUOROCARBON/HYDROCARBON BLENDS

Nonflammable Fluorocarbon		Maximum Concentration (wt. %) of Hydrocarbon for Nonflammable Blend[5]				
Number	Formula	Isobutane	n-Butane	Isopentane	n-Pentane	Propane
11	CCl_3F	10	–	–	–	8
12	CCl_2F_2	11	–	–	–	10
21	$CHCl_2F$	6	6	5	5	–
22	$CHClF_2$	6	6	5	5	–
114	$CClF_2CClF_2$	9	–	–	–	7
123	$CHCl_2CF_3$	9	–	8	–	8
124	$CHClFCF_3$	8	–	6	–	8
133a	CH_2ClCF_3	4	–	4	–	4

ignited. This is illustrated by the data in Table 14-9, obtained with various ratios of fluorocarbon 22 and isobutane in the explosion drum.

The above procedure indicates whether or not a given propellant or propellant blend will form a flammable mixture with air, but it does not show what the lower and upper flammability limits are. However, flammability limits may be obtained by taking a sample of the propellant blend/air mixture in the drum immediately before each test for analysis. Compositions of mixtures that are nonflammable may be obtained by this procedure, and a flammability curve can be drawn if enough mixtures are tested. In any of these tests, the incremental additions of propellant to the drum should be small enough so that one addition does not produce a composition with air that falls below the lower flammability

TABLE 14-9   FLAMMABILITIES OF FLUOROCARBON
22/ISOBUTANE BLENDS[5]

Composition of Blend (wt. %)		Flammability in Explosion Drum
FC-22	Isobutane	
94	6	Nonflammable
93	7	Flame hit top of drum and moved polyethylene cover—flammable
92	8	Explosion blew cover off of drum
91	9	Violent explosion blew cover off drum and flame extended out of drum

curve while the next addition produces a composition above the upper flamma-bility limit. This is most likely to occur when the flammability range in air is very small.

Flammability of Three-Component Propellant Blends

The flammability properties of three-component blends may be illustrated in a number of ways. For example, if a specific blend of two fluorocarbons, such as fluorocarbon 12/fluorocarbon 11 (30/70), with isobutane is under investigation, then a curve showing the flammability limits of the fluorocarbon blend/isobutane mixture can be constructed by obtaining the necessary data. A curve of this type, similar to that shown in Figure 14-2, will show not only the compositions of mixtures of fluorocarbon 12/fluorocarbon 11 (30/70) with isobutane that are flammable in air, but also the range of concentrations in air that are flammable. However, the data required for the construction of these curves is extensive and the experimental work is time consuming. Furthermore, such curves show only the flammability properties of mixtures of one specific blend, for example, fluorocarbon 12/fluorocarbon 11 (30/70), with the flammable component.

For most purposes, it is generally sufficient to prepare a triangular coordinate chart that shows the compositions of all combinations of the three components that form flammable mixtures with air. A chart of this type does not provide the flammability limits in air for any propellant mixture, but it does indicate if a particular blend will form a flammable mixture with air. The triangular coordi-nate chart can be constructed merely by determining the maximum concentra-tion of the flammable component (or components) that can be added to the nonflammable fluorocarbons without the blend becoming flammable. Data of this type are illustrated in Tables 14-7 and 14-8. The data for each of the two binary mixtures are then plotted on the appropriate base lines of a triangular coordinate chart and a straight line drawn between the two points. All compositions falling on the line or in the area bounded by the line and the apices of the nonflammable components are nonflammable.

As an example, a triangular coordinate chart showing the flammability of fluorocarbon 133a/propane/isobutane blends was desired. As shown in Table 14-8, the limiting concentration for both isobutane and propane in fluorocarbon 133a is 4 wt. %. These points were plotted on the fluorocarbon 133a/isobutane and fluorocarbon 133a/propane base lines of a triangular coordinate chart and a line drawn between the points. This line, referred to as the flammability limit, is shown in Figure 14-4. All compositions falling on the line or in the area to the left of the line are nonflammable. It is useful to include vapor pressure contours on the same triangular coordinate chart, as shown in Figure 14-4. A similar chart, showing the flammability and vapor pressures of fluorocarbon 133a/propane/ isopentane blends, is illustrated in Figure 14-5. The area of nonflammable compositions for both ternary blends is relatively small. However, nonflammable compositions with vapor pressures of 25 psig at 70°F are possible.

Other flammability data can also be added to the triangular coordinate chart. Figure 14-6 shows vapor pressures and flammability of fluorocarbon 133a/ fluorocarbon 152a/isobutane blends. In addition, compositions of the ternary blend that give flame extensions of 18 in. or less with a typical antiperspirant valve are also indicated.[5]

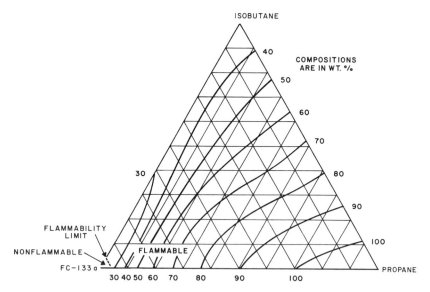

Figure 14-4 Flammability and vapor pressures (psig) at 70°F of fluorocarbon 133a/ propane/isobutane blends.

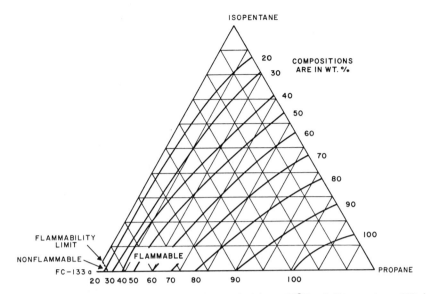

Figure 14-5 Flammability and vapor pressures (psig) at 70°F of fluorocarbon 133a/propane/isopentane blends.

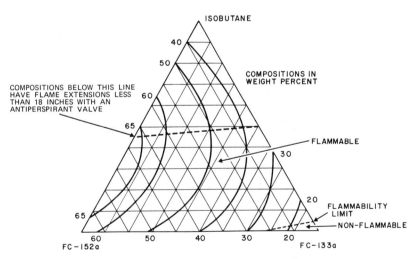

Figure 14-6 Flammability and vapor pressures (psig) at 70°F of fluorocarbon 152a/fluorocarbon 133a/isobutane blends.

This same technique can be used to prepare similar triangular coordinate charts for any three of the compounds in Tables 14-7 and 14-8 (one of the compounds must be nonflammable).

The possibility of determining the approximate lower flammability limit of propellant blends by spraying the mixtures into the closed drum with a candle as the ignition source has been proposed. The test would be carried out by spraying the blend into the drum until either an explosion occurred or the candle was extinguished. In a sense, this would be a modification of the DOT closed-drum test except that spraying would be continued beyond the 1-min limitation of the DOT test. Since the volume of the drum and the amount of blend sprayed into the drum could be determined, the composition of the mixture in the drum could be approximated by calculation.

Unfortunately, the method gives erroneous results. The reason for this can be seen by considering what takes place in the drum during the test. As the blend is sprayed, continuous combustion of the propellants occurs. The combustion of the blend not only diminishes the concentration of oxygen in the drum, but also forms nonflammable combustion products such as carbon dioxide and water. In most cases where the candle is extinguished during the test, the gas mixture that extinguishes the candle has an entirely different composition than the initial mixture of air and propellant blend. As a result, it is possible for a propellant blend that normally forms a flammable mixture with air when tested in a eudiometer tube or explosion drum to extinguish the candle when sprayed into a closed drum.

Flammability of Evaporating Mixtures

The flammability of various blends provides an indication of the potential hazard that might occur if the blends vaporized completely in air. This might result from a *liquid phase* leak from a cylinder, tank car, or storage tank or during filling.

However, another possibility could lead to a potentially hazardous situation. The composition of the vapor above any mixture of propellants with different boiling points is different from that of the liquid phase. The vapor over a fluorocarbon 12/fluorocarbon 11 (50/50) blend, for example, has a composition of approximately 86 wt. % fluorocarbon 12 and 14 wt. % fluorocarbon 11.[11] Whenever any mixture of propellants with different boiling points is allowed to evaporate, fractionation occurs and the composition of the vapor changes continually during the evaporation. Under these conditions, it is entirely possible that a mixture of nonflammable and flammable propellants, which itself is nonflammable in air when completely vaporized, will form flammable fractions at some point during evaporation. This could occur as a result of a spill or a *vapor phase* leak from a storage container. If a storage tank or some other container ruptured, flash vaporization of the propellant would occur until the liquid had cooled down to its boiling point at atmospheric pressure. Evaporation at atmospheric pressure would then take place.

If the flammable propellant is either the lowest or the highest boiling component in the blend, usually the blend will form flammable fractions during evaporation. If the flammable propellant is the lowest boiling component, the initial vapor phase fraction will contain the highest proportion of the flammable propellant and will be the most flammable (assuming that only one flammable component is present). This is the only fraction that needs to be analyzed or tested. If the flammable component is the highest boiling of the propellants, the last fraction will contain the highest proportion of the flammable propellant and will be the most flammable. This is the only fraction that needs to be analyzed and tested. If there are only two components in the blend, the flammable component will, of necessity, be either the lowest or highest boiling component (except for azeotrope formation).

When the blend consists of two nonflammable propellants and a third flammable propellant, then if the flammable propellant has a boiling point between those of the two nonflammable propellants, one of the intermediate fractions will contain the highest proportion of the flammable propellant and will usually be the most flammable.

If the compositions of a three-component blend that are flammable are available on a triangular coordinate chart, the flammability of a vapor phase fraction can be determined after analysis by locating the fraction on the triangular coordinate chart and noting whether it falls in the flammable or nonflammable region. If sufficient data are not available so that analysis of an evaporating sample is not enough to indicate whether the sample is flammable, a propellant blend having a composition corresponding to that of the vapor phase sample can be prepared and tested for flammability in the eudiometer tube or explosion drum.

Effect of Discharge upon the Flammability of a Fluorocarbon 22/Isopentane (35/65) Blend

For a limited time, the flammable fluorocarbon 22/isopentane (35/65) blend was of interest as a potential propellant for dry-type antiperspirants. The flammability properties (flame extension and flashback) of this blend before and after discharge are of interest because the blend is an example of a mixture of a high-pressure, nonflammable propellant and a low-pressure, flammable propellant. Blends of a nonflammable and a flammable propellant in which the vapor pressures of the two components differ appreciably are not uncommon in the aerosol industry. When such blends are used with vapor tap valves, the composition of the blend may change significantly during use. Such properties as vapor pressure, spray rate, and flammability can change as the product is discharged. The data obtained with this blend are useful for predicting the behavior of other similar blends.

The vapor phase in equilibrium with the liquid phase of the fluorocarbon 22/isopentane (35/65) blend has a composition of 86.5 wt. % fluorocarbon 22 and 13.5 wt. % isopentane. When the blend is discharged using a vapor tap valve, the vapor phase mixes with the liquid phase. Since the vapor phase has a higher concentration of fluorocarbon 22 than the liquid phase, the concentration of fluorocarbon 22 in the liquid phase will decrease. As a consequence, the liquid phase becomes more flammable. The change in vapor pressure, spray rate, flame extension, and flashback as the propellant is discharged is shown in Table 14-10. The blend was packaged with five different valves having vapor taps varying from 0 to .040 in. in diameter. There is little change in flame extension with vapor taps not exceeding .020 in. in diameter. However, when the vapor tap has dimensions of .030 in. or more, the product becomes more flammable with discharge.

The DOT Regulations (Propellants)

The definition of a compressed gas is the same for propellants alone as for aerosols, that is, "The term 'compressed gas' shall designate any material or mixture having in the container an absolute pressure exceeding 40 psi at 70°F or, regardless of the pressure at 70°F, having an absolute pressure exceeding 104 psig at 130°F, or any liquid flammable material having a vapor pressure exceeding 40 psi absolute at 100°F, as determined by ASTM Test D-323." Therefore, if a propellant is not classified as a compressed gas according to the above definition, it is not subject to the DOT regulations. If a propellant or propellant blend is classified as a compressed gas, its flammability must be determined (unless it falls under the ORM-D classification). In order to be classified as an ORM-D material, the propellant must meet the following conditions: It must be a consumer commodity and the gross weight of each package must not exceed 65 lb. In addition, the compressed gas (propellant only) must be "in inside containers, each having a water capacity of 4 fluid ounces or less (7.22 cubic inches or less) packed in strong outside packagings."

The flammability tests for a propellant alone are different than those for aerosols. The main concern with propellants is whether they form flammable mixtures with air. According to the DOT regulations (see Agent R. M. Graziano's Tariff No. 31, Paragraph 173.300), a compressed gas (propellant only) is classified as a flammable compressed gas "if either a mixture of 13% or less by volume with air forms a flammable mixture or the flammability range with air is greater than 12% regardless of the lower limit. These limits shall be determined at atmospheric pressure and temperature. The method of sampling and the test procedure must be acceptable to the Bureau of Explosives." The flammability range is defined as the difference between the minimum and maximum volume percentages of the material in air that forms a flammable compressed gas.

TABLE 14-10 VAPOR PRESSURE, SPRAY RATE, AND FLAMMABILITY PROPERTIES OF A FLUOROCARBON 22/ISOPENTANE (35/65) BLEND BEFORE AND AFTER DISCHARGE[a5]

Vapor Tap Diameter (in.)	Vapor Pressure (psig) at 70°F		Spray Rate (g/sec) at 70°F		Flame Extension (in.)		Flashback (in.)	
	Initial	85% Discharge	Initial	85% Discharge	Initial	85% Discharge	Initial	85% Discharge
0	50	47	.90	.90	18	18	0	0
.013	50	44	.90	.85	16	16	0	0
.020	50	39	.85	.62	16	16	0	0
.030	50	29	.67	.38	10	14	0	0
.040	50	19	.57	.16	0	14	0	0

[a]Precision valve, .018-in. orifice.

If any propellant or propellant blend is nonflammable in air, it is not classified as a flammable compressed gas. However, if it does form flammable mixtures with air, the lower and upper flammability limits of the propellant or propellant blend in air must be determined. If the complete flammability limit curve for mixtures of the blend with air is available, it is possible to determine from this curve if any propellant or propellant blend will be classified as a flammable compressed gas.

This is illustrated in Figure 14-7 using fluorocarbon 12/propane blends as the example. The flammability limit curve for fluorocarbon 12/propane blends in air is shown and a line indicating all compositions of fluorocarbon 12 and propane with a total concentration of 13 vol. % in air has been drawn.

The 13 vol. % line crosses the flammability limit curve, and any mixture of fluorocarbon 12 and propane that forms a flammable mixture with air at a concentration of 13% or less will be classified as a flammable compressed gas. Also shown in Figure 14-7 are the volume percent concentration lines of three fluorocarbon 12/propane blends. The weight percent ratios of the blends in the liquefied state are shown, and lines A, B, and C indicate the volume percent ratios of fluorocarbon 12 and propane after the liquids have been vaporized.

A fluorocarbon 12/propane (90/10) blend is nonflammable in air. This is illustrated by line A, which shows the volume percent ratios of the two components at various concentrations in air. Line A does not cross the flammability limit curve. The fluorocarbon 12/propane (89/11) blend forms flammable mixtures with air, since line B crosses the flammability limit curve, although the

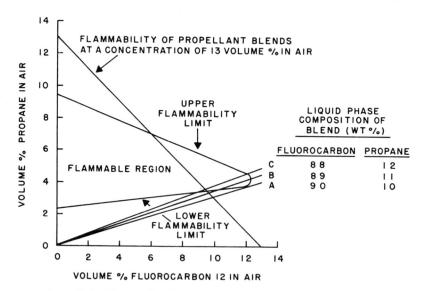

Figure 14-7   Flammability limits of fluorocarbon 12/propane blends in air.

TABLE 14-11   FLASH POINTS OF PROPELLANTS AND
              PROPELLANT BLENDS

Propellant or Propellant Blend	Flash Point (°F)	Reference
n-Butane	−101	7
Dimethyl ether	−42	6
Fluorocarbon 12/propane (91/9)	−67, −72	4
Isobutane	−117	7
Propane	−156	7
Propellant A (fluorocarbon 12/fluorocarbon 11/isobutane) (45/45/10)	No flash to dryness	4
Fluorocarbon 142b	No flash to dryness	4
Fluorocarbon 152a	>−58	4

lower flammability limit is greater than 13 vol. % and the range of concentrations where flammability occurs is very small. The fluorocarbon 12/propane (88/12) blend forms a flammable mixture with air at a concentration in air of slightly less than 13 vol. %. This mixture would, therefore, be classified as a flammable compressed gas and would have to be shipped in accordance with the specifications indicated by the DOT regulations.

Flash Points of Propellants and Propellant Blends

The flash points of propellants and propellant blends are often used as an indication of the flammability and potential hazard of the propellants. For certain regulations, such as those of the New York Fire Department, the flash points of the propellants are important. For single propellants, a flash point determination might indicate whether or not the propellant will form a flammable mixture with air. For blends of nonflammable and flammable propellants, the flash point determination might be considered as a rough indication of whether or not the propellant blend will form a flammable fraction during evaporation.

The flash points of a number of common flammable propellants and propellant blends are listed in Table 14-11.

REFERENCES

1. P. A. Sanders, "Aerosol Laboratory Manual," Industry Publications, Inc., Cedar Grove, N. J., 1975.
2. R. E. Newbury (Assistant to the Director of the Division of Regulatory Guidance, FDA), Personal Communication, March 1977.

3. H. F. Coward and C. W. Jones, Bulletin 503, Bureau of Mines, "Limits of Flammability of Gases and Vapors," 1952.
4. E. O. Haenni, R. A. Fulton, and A. H. Yeomans, *Ind. Eng. Chem.*, **51**, 685 (May 1959).
5. Freon® Products Laboratory Data.
6. Matheson Gas Data Book, The Matheson Company, Inc., 1965.
7. Phillips Hydrocarbon Aerosol Propellants, Technical Bulletin 519, Phillips Petroleum Company, 1961.
8. Freon® Technical Bulletin D-60, "Properties of K-142b, 1-Chloro-1,1-difluoroethane, CH_3CClF_2."
9. Technical Bulletin, Aeropres Corp.
10. J. M. Kuchta et al., *J. Chem. Eng. Data* **13**, 421 (July 1968).
11. Freon® Aerosol Report FA-22, "Vapor Pressure and Liquid Density of Freon® Propellants."
12. T. R. Torkelson, *J. Ind. Hygiene*, 353 (October 1958).
13. Dow Chemical Company Advertisement, *Aerosol Age* **22** (February 1977).
14. E. E. Kavanaugh, Personal Communication, Cosmetic, Toiletry, and Fragrance Association (CTFA), May 24, 1977.

II
EMULSIONS, FOAMS, AND SUSPENSIONS

15

SURFACES
AND
INTERFACES

Water has many properties that make it desirable for use in aerosol systems. It is an excellent solvent, nonflammable, readily available, inexpensive, and low in toxicity. However, its immiscibility with liquefied gas propellants is an obvious disadvantage. Fortunately, there are several ways in which the immiscibility can be overcome. One is to use the so-called three-phase system in which water is layered over the denser fluorocarbon propellants or below the lighter hydrocarbon propellants. Water can also be combined with propellants to a limited extent by addition of a co-solvent, such as ethyl alcohol, which is miscible with both water and propellants. Finally, water can be mixed with propellants in the presence of a suitable surface-active agent to give an emulsion system. The product can be discharged either as a foam or a spray. The latter procedure, that is, the formation of an emulsion system with the propellants, is the predominant method used at the present time.

The chemistry of emulsions and foams is one of surfaces. It is intimately involved with the realm of attractive and repulsive forces between molecules. These forces are responsible for surface and interfacial tension, strength of interfacial films, solubility properties of compounds, formation of molecular complexes, and the stability of emulsions and foams.

THE EIGHT COLLOIDAL SYSTEMS

Emulsions and foams belong to the class of substances designated as colloidal systems. A colloidal system consists of a dispersion of one material in another. Most definitions of colloidal systems have been based upon the particle size of the dispersed material. In the present discussion, a colloidal system is considered to be one in which the particle size of the dispersed phase is larger than molecular, but small enough so that interfacial forces are a significant factor in determining the properties of the system. This is similar to the definition proposed by Sennett and Oliver[1] and, as they state, it excludes true solutions and boulders in a lake, but includes about every dispersion in between.

Since colloidal systems consist of a dispersion of one substance in another, they are heterogeneous in nature. Colloidal systems are extremely commonplace.

There are eight colloidal systems, some examples of which are listed below.[2]

Solid-in-Solid

An example of this colloidal system can be observed in many churches. The ruby glass of the stained glass windows is a dispersion of metallic gold in glass. Some minerals are solid-in-solid dispersions.

Solid-in-Liquid

Suspensions are solid-in-liquid systems. Aerosol powders, which consist of powders suspended in the aerosol propellants, are examples of this system as long as they are confined in the aerosol container. Once they have been sprayed and the propellant evaporates, the result is a solid-in-gas (powder-in-air) system.

Solid-in-Gas

Dust and smoke are solid-in-gas systems. The haze of cigarette smoke or of a forest fire is due to small particles of carbon dispersed in air. The size of the carbon particles is often very small and may be as low as 0.01 μ. Aerosol powders, after discharge, produce a solid-in-gas system because the liquefied gas propellant vaporizes and leaves the solid powder particles suspended in the air.

Liquid-in-Solid

This colloidal system includes many precious gems. The opal is a dispersion of water in silicon dioxide, and the pearl is a dispersion of water in calcium carbonate. Gortner[2] has pointed out that valuable pearls have been destroyed by storage in a safety deposit box where the humidity was low.

Liquid-in-Liquid

Emulsions, which are dispersions of one immiscible liquid in another, fall in this class. Milk consists of a dispersion of fat droplets in water and is one of the most familiar examples of an emulsion. There are many examples of emulsions in the aerosol field, but these are generally emulsions only as long as they are in the container under pressure. After discharge, most of these products change to foams or sprays.

Liquid-in-Gas

Fogs and mists, dispersions of water droplets in air, are examples of this system. In the aerosol field, aqueous-based insecticides and deodorants produce water-in-air (liquid-in-gas) dispersions after they have been discharged.

Gas-in-Solid

The absorption of gases by materials such as carbon black may result in gas-in-solid systems. Some minerals are reported to contain finely dispersed gas bubbles.

Gas-in-Liquid

The familiar everyday foams belong in this class. Foams are dispersions of gases in liquids, for example, air bubbles in water. Aerosol foams, for example, shaving lathers and shampoos, are gas-in-liquid dispersions after discharge and consist of a vaporized propellant dispersed in an aqueous phase.

INTERFACIAL FORCES

Interfacial forces govern the properties of interfacial regions. These regions are present in all colloidal systems. They are the boundaries between two immiscible liquids, a liquid and a gas, a solid and a liquid, etc. Interfacial regions possess very special properties. These properties distinguish colloidal systems from non-colloidal systems, for example, true solutions.

Three of the forces active in the interfacial regions are the same attractive forces that exist between gas molecules. They are known collectively as the van der Waals forces of attraction. These forces[3-5,9] can result in weak bonds between molecules. These bonds in turn are involved in the affinity of one compound for another and thus in the formation of molecular complexes etc.

Dipolar Molecules

Dipoles in molecules are the basis for the van der Waals forces. Molecules possessing dipoles may be electrically neutral as a whole, but nevertheless possess localized charges resulting from a separation of the negative and positive charges in the molecule. Because of this separation, the center of negative charge does not coincide with the center of positive charge (Figure 15-1).

The separation of charges occurs for several reasons. In asymmetrical molecules such as ethyl alcohol and ethyl chloride, the various atoms that make up the molecule do not all have the same affinity for electrons. Thus, in the two

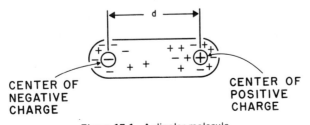

Figure 15-1 A dipolar molecule.

examples cited above, the chloride and oxygen atoms have a greater affinity for electrons than the other atoms in the molecule and tend to draw electrons toward themselves. They thus become the centers of negative charge in these molecules. This results in a deficiency of electrons in some other part of the molecule, which causes a local positive charge to develop. Because they have an affinity for electrons, chlorine and oxygen atoms are referred to as being electronegative.

The magnitude of the charge separation in dipolar molecules is indicated by the dipole moment of the molecule. The dipole moment u is equal to the product of the distance d between the centers of negative and positive charge and the charge e itself. Thus, the dipole moment of the molecule illustrated in Figure 15-1 is $u = de$, where d is the distance between the charges and e is the magnitude of the charge.

In asymmetrical molecules, the charge separation is permanent and the molecules have a permanent dipole. Molecules that are symmetrical do not have a permanent dipole. However, a temporary dipole can arise in these molecules when they are subjected to an electric field. This causes a displacement of the molecular electrons in the structure and an induced dipole results. One source of an electric field that can cause an induced dipole is a molecule with a permanent dipole.

Van der Waals Forces of Attraction

Attractive forces between molecules were postulated long before the time of van der Waals (see Reference 4 for an excellent article on the historical development in this field). Initially, the existence of attractive forces was suggested to explain surface tension. Later, in 1743, Clairault pointed out that these forces had to be applicable to all molecules. However, it remained for van der Waals in 1881 to lay the groundwork for the subsequent investigations which have led to the modern conception of these forces.

The well-known equation of state derived by van der Waals to account for the deviation of gases from the ideal gas law is

$$\left(P + \frac{a}{V^2}\right)(V - b) = RT \tag{15-1}$$

In this equation the constant a expresses the attractive force between molecules. Van der Waals did not specify or attempt to define the nature of these forces. Subsequent investigations have shown that the original van der Waals force actually contains three components and that all three are electrostatic in nature. These component forces have been given the names of the men who are most identified with them and are known, respectively, as Keesom, Debye, and London–van der Waals forces. The nature of these attractive forces will be considered in the following sections.

Keesom Forces (Dipole–Dipole Interactions)

Keesom, in an attempt to explain van der Waals forces in fundamental terms, proposed that in any group of dipolar molecules (molecules with permanent dipoles), there would be a certain number of collisions between the molecules.[6] As a result of their dipoles, some of the molecules would be oriented with respect to an adjacent molecule so that their dipoles would be aligned as shown in Figure 15-2.

Overall, there is a net attractive force which results from the interactions between the opposite charged ends of the dipoles. This is also referred to as the orientation effect, since it is the orientation of the dipoles that produces the attractive force. An examination of the equation which Keesom developed[3] to account for this effect shows that the dipole–dipole interactions are important for molecules with permanent dipoles at low or moderate temperatures, where the kinetic energy associated with the temperature is less than the energy of the dipole–dipole interaction.

Debye Forces (Dipole–Induced Dipole Interactions)

Although the orientation effect advanced by Keesom was a considerable step forward, it failed as a general explanation of the van der Waals forces for two reasons. First of all, many molecules that did not have dipole moments were known to possess attractive forces. Secondly, the orientation effect would be expected to disappear at high temperatures because the increased kinetic energy of the molecules would make orientation unlikely. However, it was known that the attractive forces did not disappear at higher temperatures.

Debye accounted for the presence of van der Waals forces at high temperatures by introducing another concept which did not depend upon temperature. Debye pointed out that the electron structure of a molecule is not fixed or rigid but consists of mobile clouds of electrons. When a molecule is approached by a dipolar molecule, the electric field associated with the dipolar molecule interacts with the mobile electron cloud of the first molecule and causes a separation of the charges. This effect is called polarization, and when it occurs, the polarized molecule temporarily becomes a dipole. The orientation of the two molecules— the dipolar molecule and the polarized molecule with an induced dipole—as they

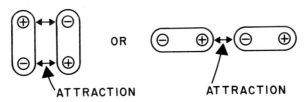

Figure 15-2 Dipole-dipole interaction.

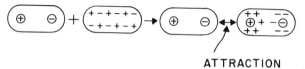

ATTRACTION

Figure 15-3 Dipole–induced dipole interaction.

approach each other is such that an attractive force exists between them.[7] The attractive force which results is called a dipole–induced dipole interaction because the dipolar molecule induces a dipole in the original molecule (Figure 15-3).

The Keesom orientation effect helped explain the attraction between the molecules of dipolar gases at moderate temperatures and the Debye induction effect explained the persistence of the van der Waals forces at high temperatures, but neither of these effects explained the attractive force between molecules that do not possess dipoles, for example, the rare gases.

London–van der Waals Forces (Instantaneous Dipole–Induced Dipole Interactions)

It became apparent at this point that there must be a third force present which was still unknown. It remained for London to clarify the nature of this force.[8,9] London suggested that even molecules without permanent dipoles, for example, the rare gases, have instantaneous dipoles. He pointed out that in any molecule, although the charge distribution may be perfectly symmetrical when averaged over a period of time, at any one instant there will be a region where the concentration of electrons is higher than average, thus producing a local negative charge. Consequently, there must be a corresponding region with a less-than-average concentration of electrons, producing a local positive charge.

Thus, each molecule is momentarily a dipole, and these instantaneous dipoles appear and disappear with great rapidity because of the high orbital speed of the electrons. These instantaneous dipoles act on the mobile electron clouds of adjacent molecules and induce in them a corresponding dipole. London showed that this always results in a net attractive force between the two molecules.

The London–van der Waals force (also called the London dispersion force) is the universal attractive force between molecules and in most cases accounts for a far greater portion of the attractive forces between molecules than the Keesom and Debye forces, even with dipolar gases. The London dispersion forces are additive for a group of molecules, and this is why the adhesion forces between macroscopic particles are often attributed to the London–van der Waals attractive forces.[10]

Hydrogen Bonds

Hydrogen bonds between molecules result from dipole–dipole interactions, but they have special properties and are usually treated separately.[3,5] They are mentioned at this point because they play an important role in determining the properties of interfacial regions. In some cases, they contribute more to the properties than the London–van der Waals forces.

As previously mentioned, an electronegative atom is one that has an affinity for electrons. Chlorine and oxygen are examples of atoms with a high electronegativity. Hydrogen has a relatively low electronegativity and when it is attached to an atom with a higher electronegativity, such as oxygen, the electrons are displaced toward the oxygen and the hydrogen becomes positive. The molecule then is dipolar and undergoes the electrostatic interactions characteristic of a dipolar molecule.

The formation of a dipole in an alcohol is illustrated in Figure 15-4a (R = alkyl). Since the hydrogen is positive, it will be attracted to a second electronegative atom. The hydrogen atom therefore becomes a bridge between two electronegative atoms, and hydrogen-bonded compounds are often referred to as hydrogen-bridge compounds. The attractive force between the positive hydrogen atom and the second electronegative atom is usually indicated by a dotted line. Alcohols are associated in the liquid state as a result of hydrogen bonding. A polymeric structure of an alcohol that results from hydrogen bonding is illustrated in Figure 15-4b.[11]

The hydrogen bond forms readily because hydrogen has a small radius, which allows it to approach closely to the second electronegative atom.

The water molecule itself is dipolar, $\overset{+}{H_2} - \bar{O}$, and water molecules have, therefore, a considerable attraction for each other. As a result, water molecules become linked to each other through hydrogen bonds and water exists in a polymeric state. A number of investigations on the structure of water have been carried out and are reported in References 12–14. Ferguson[5] has illustrated the structure of water with a planar representation as shown in Figure 15-5.

A considerable amount of energy is required to break hydrogen bonds. This is why water, in spite of its apparently small molecular size, has such a high boiling point. The hydrogen bonds must be broken before the individual molecules can be vaporized. This is also why ethyl alcohol has a much higher boiling point (78°C) than isomeric dimethyl ether (-25°C). The latter does not form hydrogen bonds.

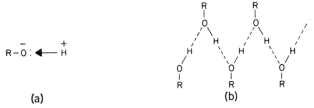

(a)

(b)

Figure 15-4 Hydrogen bonding in an alcohol. (a) Dipole formation. (b) Hydrogen bridge and polymer formation. From "Mechanism and Structure in Organic Chemistry" by Edwin S. Gould. Copyright © 1959 by Holt, Rinehart and Winston. Reproduced by permission of Holt, Rinehart and Winston.

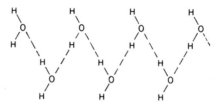

Figure 15-5 Polymeric structure of water. Lloyd N. Ferguson, "Electron Structures of Organic Molecules," © 1952, Prentice-Hall, Inc.

Hydrogen bonding is very important in surface and interfacial chemistry. The major part of the force causing the surface tension of water is due to hydrogen-bond formation. Hydrogen bonding undoubtedly has a considerable effect upon the properties of most interfacial films. The affinity of many compounds for water is the result of hydrogen bonding.

Ion-Dipole Interactions

There is another attractive force that has not been mentioned up to this point but is quite important. This is the interaction between an ion and a dipole, such as water. Ferguson[5] states that it appears likely that all ions are hydrated in aqueous solution. This interaction is of considerable importance in surface chemistry because many investigators believe that hydration accounts for much of the strength of interfacial films formed by surface-active agents. Ion–dipole interactions are also assumed to play a part in the formation of molecular complexes from combinations of anionic surface-active agents and long-chain polar alcohols or acids.

SURFACE TENSION

It has been known for many years that the surface of a liquid behaves as if it were covered with a skin having properties different from those of the bulk liquid underneath. Thus, if a small ring is lowered beneath the surface of the liquid and slowly raised, it is found that considerably more force is required to pull the ring through the surface and into the air than was necessary to move the ring up through the bulk of the liquid.

This effect (called surface tension) is due to certain attractive forces between the molecules. A molecule in the bulk of the liquid is attracted by the other molecules that surround it. Since the attractive forces are exerted on all sides of the molecule, the forces tend to cancel each other out. However, the molecules in the surface of the liquid are in a completely different environment. Here they are strongly attracted by molecules underneath them in the bulk of the liquid. However, there are essentially no molecules in the vapor phase above to provide a counter attractive force. Therefore, the attractive forces acting upon the molecules in the surface are unbalanced and tend to draw the molecules from the surface into the body of the liquid.

This force pulling on the surface molecules reduces the total number of molecules in the surface region and thus increases the intermolecular distance between those that remain. As Fowkes[15] has pointed out, the molecules in the surface resist this separation from each other because of the attractive forces between them. He pictures the molecules as being attached to each other by springs which represent the attractive forces, and it is then easy to visualize that it requires work to pull the molecules further apart. As a result, the surface molecules exist in a state of constant tension, and their natural inclination is to move closer to each other to reduce this tension. Therefore, surface tension causes the surface to contract, thus decreasing the total surface area.

The molecular structure of the liquid surface is, therefore, disordered or disoriented compared to that of the bulk phase. This effect is not confined to the surface molecules only, but extends some distance into the bulk of the liquid itself. There is some disagreement among various surface chemists as to the distance to which the disorder extends into the bulk phase, but the minimum distance is reported to be five to ten molecular layers.[16]

A simplified version of the molecular explanation of surface tension is illustrated in Figure 15-6. The attractive forces are shown as acting only in a vertical or horizontal direction.

Since it was necessary to expend energy in order to separate the molecules in the surface, this energy must still be present in the surface region. It can be released when the tension between the surface molecules is reduced by decreasing the total surface area. This is the meaning of surface free energy and the reason for its existence in the surface region.

Becher[17] has shown that surface tension can be defined on a physical basis as the work in ergs necessary to generate one square centimeter of surface. This is illustrated with a small rectangular wire frame containing a liquid film upon which work can be done. Assume that the small rectangular wire frame ABCD in Figure 15-7a is suspended vertically and contains a film of liquid as shown.

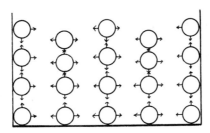

Figure 15-6   Molecular illustration of surface tension.

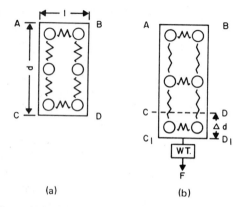

Figure 15-7 Physical interpretation of surface tension.

The film of liquid consists of molecules held together by attractive forces which are indicated by the springs suggested by Fowkes. The lower end of the frame, CD, is movable.

The surface area of the film, considering that it has two sides, is $2ld$. Now assume that a force F is exerted on the end of the frame CD by attaching some weights to the frame. This moves the end of the frame CD downward a distance Δd to a new position $C_1 D_1$. This is shown in Figure 15-7b.

The work done to bring the film to the new position is

$$W = F\Delta d \tag{15-2}$$

The force F exerted by the weights is opposed by the resistance of the forces of attraction between the molecules (as indicated by the springs). This counter opposing force exists along the length CD. If γ is the opposing force in dynes per centimeter resulting from the forces of attraction, then the force along the length CD is γl. Since there are two sides to the film, the force that opposes the expansion of the film is $2\gamma l$. Substituting this expression for the force F in Equation 15-2 gives

$$W = 2\gamma l\Delta d \tag{15-3}$$

The increase in surface (ΔS) that occurred when the film was expanded is $2l\Delta d$. Therefore, substituting ΔS for this expression in Equation 15-3 gives

$$W = \gamma\Delta S, \quad \text{or} \quad \gamma = \frac{W}{\Delta S} \tag{15-4}$$

Therefore, the surface tension γ is the work in ergs required to form one square centimeter of surface. Becher points out that since the surface tension is considered as the force acting along a 1-cm length, and since an erg is equal to a dyne-cm, it is appropriate to define surface tension in dynes/cm. This is the usual unit for surface tension.

Thus far, very little has been said about the specific intermolecular forces responsible for surface tension. In a symmetrical, nonpolar hydrocarbon liquid such as n-hexane, the attractive forces are due to the London dispersion forces, since the liquid is nonpolar and does not have a permanent dipole moment. The London dispersion forces can be represented by the symbol γ_{HC}^{d}. Therefore, the surface tension of the hydrocarbon, γ_{HC}, is due entirely to the London-van der Waals forces of attraction between the molecules.

This relationship can be represented as follows:

$$\gamma_{HC} = \gamma_{HC}^{d} \qquad (15\text{-}5)$$

In water, a polar liquid, the major intermolecular forces are hydrogen bonding (dipole–dipole interaction), which is symbolized by $\gamma_{H_2O}^{H}$, and the London-van der Waals forces, $\gamma_{H_2O}^{d}$. Therefore, the surface tension of water, γ_{H_2O}, is equal to the sum of the hydrogen bonding forces and the London-van der Waals forces:

$$\gamma_{H_2O} = \gamma_{H_2O}^{H} + \gamma_{H_2O}^{d} \qquad (15\text{-}6)$$

INTERFACIAL TENSION

In the previous section it was mentioned that the surface of a liquid acts as if a skin or membrane were present as a result of the surface tension. The same effect occurs at the interface between two immiscible liquids for essentially the same reason, that is, the unbalanced attractive forces exerted on the molecules in the interfacial region. Fowkes[15] indicated that at the interface between two liquids, for example, water and a hydrocarbon, there are actually two interfacial regions. One is the interfacial region associated with the aqueous phase, and the other is the interfacial region of the hydrocarbon phase (Figure 15-8).

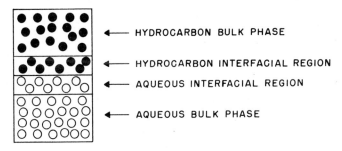

Figure 15-8   The two interfacial regions at a water–hydrocarbon interface.

At first glance, it might appear that the interfacial tension $\gamma_{H_2O/HC}$ between the water and hydrocarbon might be equal to the sum of the two individual surface tensions–that of water, γ_{H_2O}, and that of the hydrocarbon, γ_{HC}–because the molecules of each liquid in their own interfacial regions are pulled toward their own bulk phases by the same intermolecular forces that cause the surface tension. However, there is now an additional force at the interface between the two liquids that opposes the forces causing the surface tension of water and the surface tension of the hydrocarbon. This additional force is the London–van der Waals attractive force between the water molecules and the hydrocarbon molecules. Thus, in the aqueous interfacial region, the intermolecular forces exerted by the water molecules in the bulk phase tend to pull the interfacial water molecules into the bulk phase. However, this is resisted by the London dispersion forces between the water molecules in the interfacial region and the adjacent hydrocarbon molecules which tend to pull the interfacial water molecules in the opposite direction toward the hydrocarbon region. Therefore, the tension on the aqueous side is equal to the surface tension of water minus the attraction of the hydrocarbon molecules for the water molecules.

The same situation exists in the hydrocarbon interfacial region. The tension in the hydrocarbon interfacial region is equal to the surface tension of the hydrocarbon minus the attraction of the water molecules for the hydrocarbon molecules in the interfacial region.

It was shown previously that the surface tension of water arose from the two intermolecular forces: hydrogen bonding and the London–van der Waals attractive forces between the water molecules (Equation 15-6). The portion contributed by the London–van der Waals forces is called the dispersion component of the surface tension. The same condition applies to the surface tension of other liquids, although in the case of hydrocarbons, the intermolecular forces are practically all due to the London dispersion forces, since there is no hydrogen bonding.

Fowkes[15] has indicated that at an interface between two liquids, for example, water and a hydrocarbon, the magnitude of the London–van der Waals interaction between the water molecules and the hydrocarbon molecules, which tends to pull the molecules in an opposite direction from that of the surface tension forces, can be determined by the expression

$$(\gamma_{H_2O}^d \gamma_{HC}^d)^{1/2}$$

where $\gamma_{H_2O}^d$ and γ_{HC}^d are the dispersion components of the surface tensions of water and the hydrocarbon, respectively. Thus, the tension in the interfacial region of water would be

$$\gamma_{H_2O} - (\gamma_{H_2O}^d \gamma_{HC}^d)^{1/2}$$

and that in the interfacial region of the hydrocarbon would be

$$\gamma_{HC} - (\gamma_{H_2O}^d \gamma_{HC}^d)^{1/2}$$

Since the interfacial tension at the water-hydrocarbon interface is the sum of the tensions in the aqueous and hydrocarbon interfacial regions, it can be expressed as follows:

$$\gamma_{H_2O/HC} = \gamma_{H_2O} + \gamma_{HC} - 2(\gamma_{H_2O}^d \gamma_{HC})^{1/2} \qquad (15\text{-}7)$$

The values of the surface tension for water (72.8 dynes/cm at 20°C) and various hydrocarbons are known and, therefore, Fowkes was able to calculate the magnitude of the dispersion component for the surface tension of water. He reported that the dispersion component contributed 21.8 ± 0.7 dynes/cm at 20°C to the surface tension of water and, consequently, hydrogen bonding contributes 72.8 – 21.8 or about 51 dynes/cm. Hydrogen bonding, therefore, makes the larger contribution to the surface tension of water.

One practical aspect of interfacial tension is that it is an indication of the ease with which an emulsion can be prepared. The formation of an emulsion involves a considerable increase in the interfacial area of the dispersed liquid and, as in the case of surface tension, the increase in the interfacial area is opposed by the interfacial tension. Therefore, if the tension is high, emulsion formation is difficult and may require the expenditure of considerable mechanical energy. However, if the interfacial tension is low, emulsions may be formed with relative ease, and if the tension is sufficiently low, emulsions may form spontaneously.[18, 19] It should be emphasized that a low value of interfacial tension indicates only the relative ease of emulsification and not the stability of the emulsion once it is formed.

There are a number of ways of measuring the surface and interfacial tensions of conventional liquids, but special equipment is required for the liquefied gas propellants because they are under pressure. Kanig and Shin have described a

TABLE 15-1   SURFACE AND INTERFACIAL TENSIONS OF
             FLUOROCARBON PROPELLANTS

Compound	Surface Tension (Dynes/cm at 25°C)	Interfacial Tension (Dynes/cm at 25°C)
Propellant 12	8.4	56.7
Propellant 114	11.7	54.6
Propellant 11	18.9	49.7
Propellant 113	18.4	49.1

From Kanig and Shin, Proceedings of the Scientific Section of the Toilet Goods Association, Inc.[20]

pressure tensiometer constructed from a compatibility tube which utilizes the capillary rise principle.[20] The experimental values they reported compare very well with the theoretical calculated values available previously.[21]

The surface tensions of the liquefied gas propellants are low, as is expected of compounds in which the attractive forces between the molecules are comparatively low. The surface tensions of four fluorocarbon compounds and their interfacial tensions against water are listed in Table 15-1.

SURFACE-ACTIVE AGENTS

A surface-active agent is a compound attracted to and adsorbed at an interface. The interface can be the surface of a liquid or the boundaries between two immiscible liquids, a solid and a liquid, or two solids. The term surface-active agent is a general designation for all materials which are surface-active, but most agents have certain properties that make them desirable for specific uses. Thus, they may be classified primarily as emulsifying agents, wetting agents, foaming agents, suspending agents, detergents, etc., depending upon their particular function at an interface. A detailed listing of synthetic surface-active agents, along with their compositions, properties, and source, is given in Reference 22.

In order to possess the property of surface activity, a material must have a specific type of molecular structure. As a general rule, surface-active agents have two different parts: a polar part and a nonpolar part. The polar section is designed to have an affinity for water and is termed the hydrophilic (water-loving) portion. The other part of the molecule has an aversion for water and is called the lipophilic (oil-loving) portion. Thus, surface-active agents are constructed with two, almost opposite, parts (Figure 15-9).

There are many polar structures which confer affinity for water upon the molecule. These include ionizable groups, such as the sulfate, carboxylate, sulfonate, and quaternary ammonium groups, and nonionic structures, such as the polyoxyethylene chains that have an affinity for water as a result of hydrogen-bonding forces. Lipophilic properties are usually obtained with hydrocarbon chains.

It can be seen, therefore, that a molecule with this type of structure will seek a region, such as an oil–water interface, where it can satisfy its solubility ten-

Figure 15-9 Structure of a typical surface-active agent.

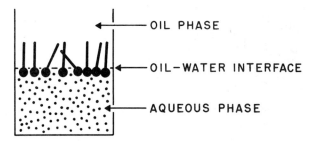

Figure 15-10 Orientation of a surface-active agent at an oil–water interface.

dencies by placing its lipophilic tail in the oil phase and its hydrophilic polar head in the aqueous phase (Figure 15-10).

At an aqueous surface, the surface-active agent can orient itself by immersing its polar head in the aqueous phase and allowing its nonpolar, lipophilic tail to extend into the air. This is shown in Figure 15-11.

Types of Surface-Active Agents

Surface-active agents are divided into three broad groups, depending upon their structure: anionic, nonionic, and cationic. For detailed description of the various types of agents within any particular group, see Reference 23.

Anionic Surface-Active Agents

Typical examples of compounds that fall into the anionic class are the salts of the fatty acids, such as sodium stearate and triethanolamine palmitate, and the various detergents such as sodium lauryl sulfate. These agents ionize in solution and the negative ion that results is adsorbed at the interface and imparts a negative charge to it.

In these compounds, the carboxylate or sulfate ionic head is the polar part that confers affinity for water upon the molecule, and the hydrocarbon chain

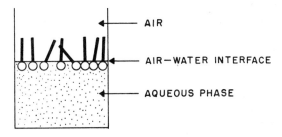

Figure 15-11 Orientation of a surface-active agent at an aqueous surface.

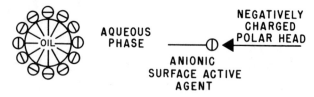

Figure 15-12 Orientation of an anionic surface-active agent at an oil–water interface.

is the part that has an affinity for oil. The anionic agents, therefore, orient themselves at an oil–water interface with the hydrocarbon chain extending into the oil phase and the ionized polar head into the aqueous phase. The orientation of an anionic agent at the interface of an oil droplet is illustrated in Figure 15-12.

Nonionic Surface-Active Agents

Nonionic surface-active agents do not ionize in solution. Esters of the fatty acids, such as glycerol stearate, diethanolamides of the fatty acids, and the ether alcohols produced by condensation of ethylene oxide with the fatty alcohols are typical nonionic surface-active agents. An example of the latter is polyoxyethylene (4) lauryl ether (the number in parentheses indicates the number of moles of ethylene oxide that were condensed with 1 mole of lauryl alcohol). The structure of this ether alcohol is $CH_3(CH_2)_{11}-O-(CH_2CH_2O-)_4H$.

In a molecule of this type, the lipophilic properties result from the presence of the hydrocarbon chain and the hydrophilic properties from the oxygen atoms in the ether linkages and the hydroxyl group. The affinity for water results from the hydrogen bonding of the oxygen atoms with water molecules as shown in Figure 15-13.

Cationic Surface-Active Agents

Cationic surface-active agents ionize in solution and impart a positive charge to the interface at which they are adsorbed. These agents constitute the smallest of the three classes of surface-active agents used in aerosols. The quaternary ammonium compounds are typical of cationic agents. The structure of one of these, benzyl dimethyl ammonium chloride, is as follows:

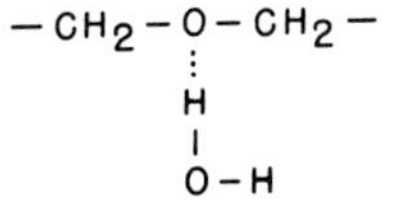

$$-CH_2-O-CH_2-$$
$$\vdots$$
$$H$$
$$|$$
$$O-H$$

Figure 15-13 Hydrogen bonding with the polyoxyethylene fatty ethers.

Ampholytic Surface-Active Agents

Ampholytic surface-active agents are a special class of compounds that contain both an acidic and a basic group in the same molecule. The molecule can be either negatively or positively charged in solution, depending upon the acidity of the medium. Thus, they are either anionic or cationic, depending upon the conditions. Thus far, they have not been used to any great extent in industry and only relatively few ampholytic surface-active agents are available commercially. Dodecyl-beta-alanine[23] is a typical ampholytic surface-active agent and has the structure $C_{12}H_{25}NHC_2H_4COOH$.

Selection of a Surface-Active Agent

Unfortunately, the selection of a surface-active agent or blend of surface-active agents for preparing a specific emulsion is still, to a considerable extent, a black art. The extraordinary large number of surface-active agents commercially available makes an empirical trial-and-error method of selection a costly and time-consuming process. There are, however, several general rules which are helpful. One is Bancroft's rule: The phase in which the surface-active agent is the most soluble will be the continuous phase.[24] Thus, a water-soluble agent tends to produce oil-in-water emulsions, and an oil-soluble agent tends to give water-in-oil emulsions. There are other factors involved that determine whether or not a system will be oil-in-water or water-in-oil, so that Bancroft's rule is not always applicable.

In most cases, a blend of several emulsifiers is more effective than a single emulsifier by itself.[25] Such blends usually contain both lipophilic and hydrophilic agents. In selecting blends, it should be realized that mixtures of cationic and anionic surface-active agents are incompatible as a result of their opposite charges and will react with each other. However, nonionic surface-active agents can be combined with either anionic or cationic agents.

One method for selecting an emulsifier or blend of emulsifiers is the HLB system devised by Griffin.[17,25] The letters HLB stand for hydrophile–lipophile balance, and the HLB number of an emulsifier shows the relationship between the hydrophilic and lipophilic portions of the emulsifier. A low HLB number indicates that the emulsifier is predominantly lipophilic and, therefore, will tend to form water-in-oil emulsions, while a high HLB number indicates that the emulsifier is predominantly hydrophilic and will tend to form oil-in-water emulsions. Thus, emulsifiers with HLB numbers in the range of 3–6 are useful for preparing water-in-oil emulsions, and those with HLB numbers of 8–18 are used for oil-in-water emulsions.

As Griffin has pointed out, the HLB number indicates the water or oil solubility of the emulsifier; thus, the HLB system is, in a sense, a refinement of Bancroft's rule that water-soluble emulsifiers should be chosen for oil-in-water emulsions and vice versa.

A simple method of shortening the list of potential surface-active agents is to ask that various manufacturers of surface-active agents recommend products for the particular application being considered.

Effect of Surface-Active Agents upon Surface and Interfacial Tension

Most surface-active agents lower the surface tension of water at relatively low concentrations, as illustrated in Figure 15-14.[17]

The reason that surface-active agents lower the surface tension of water is that after being adsorbed at a surface, they form a new surface in which the intermolecular forces of attraction are different from those in the original surface. The surface is now predominantly lipophilic, since it is composed essentially of the lipophilic tails of the surface-active agent. The lower surface tension of this lipophilic surface is due to the fact that it has a lower surface energy because the forces between the lipophilic tails are the lower magnitude London–van der Waals attractive forces.

The effect of surface-active agents upon interfacial tension can be explained in the same manner. The orientation of a surface-active agent at an oil–water interface results in the formation of a new interface composed primarily of the surface-active agent. The hydrophilic portion of the surface-active agent has an affinity for the water molecules in the aqueous phase, and the lipophilic hydrocarbon chain is attracted by the molecules in the oil phase. Since these forces oppose each other, the agent concentrates at the interface rather than dissolv-

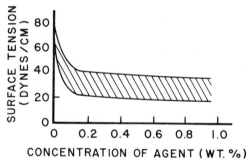

Figure 15-14 General effect of surface-active agents upon surface tension; from Becher.[17]

ing in either the oil or aqueous phase. Interfacial tension now decreases because of the lowered interfacial energy.

A high interfacial tension deters efforts of emulsifying one liquid in another, and the opposite is true for a lowered interfacial tension. A main function of an emulsifying agent is, therefore, to lower the interfacial tension at the oil-water interface so that the system can be emulsified with relative ease. It must be emphasized that a low interfacial tension does not increase the stability of the emulsion, but only makes it easier for emulsification to take place.

The range of interfacial tensions expected with solutions of most surface-active agents is illustrated in Figure 15-15.[17]

The interfacial tensions in water/propellant systems containing surface-active agents were determined by Kanig and Shin.[20] The data for systems containing 0.1 wt.% of a nonionic surface-active agent ("Igepal" CO-530) are shown in Table 15-2. "Igepal" CO-530 is a nonylphenoxypolyoxyethylene compound.

The interfacial tensions in Table 15-2 are similar to those illustrated in Figure 15-15 for a surface-active agent concentration of 0.1 wt.%.

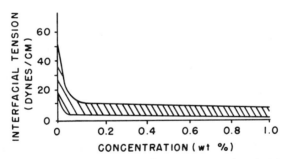

Figure 15-15 The effect of most surface-active agents upon interfacial tensions; from Becher.[17]

TABLE 15-2 INTERFACIAL TENSIONS IN WATER/
            PROPELLANT SYSTEMS

Propellant	Interfacial Tension (Dynes/cm at 25°C)
Propellant 11	12.3
Propellant 113	6.2
Propellant 12	3.5
Propellant 114	2.2

From Kanig and Shin, Proceedings of the Scientific Section of the Toilet Goods Association, Inc.[20]

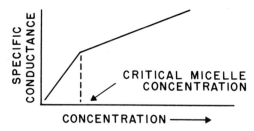

Figure 15-16   Effect of micelle formation upon electrical conductivity.

General Properties of Surface-Active Agents in the Bulk Phase—Micelle Formation

If the concentration in water of an ionizable surface-active agent such as sodium lauryl sulfate is gradually increased and the electrical conductivity of the solution measured during the addition of the agent, it will be found that initially there is a straight-line relationship between the concentration of the surface-active agent and the conductivity. Ultimately, a definite change in the slope of the conductivity versus concentration plot takes place. The concentration of the agent at which this change occurs is referred to as the critical micelle concentration (CMC) of the agent and the change in slope of the plot is due to the formation of aggregates of the surface-active agent called micelles. An idealized illustration of the change in slope of the conductivity versus concentration plot for an anionic surface-active agent is shown in Figure 15-16.

Not only does a change occur in the slope of the electrical conductivity curve, but similar changes have been reported for other properties of the solution, such as surface tension, osmotic pressure, refractive index, vapor pressure, and density.[17]

The structure of the micelles that are formed at the critical micelle concentration has been the subject of intense and continuing investigations. X-ray studies

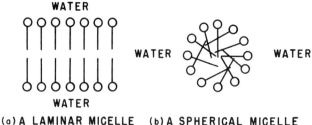

(a) A LAMINAR MICELLE   (b) A SPHERICAL MICELLE

Figure 15-17   Laminar and spherical micellar structures.

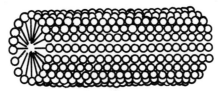

Figure 15-18   Sausage-shaped micellar structure; from Rosevear.[29]

have indicated that some micelles have a laminated structure with a definite spacing between the layers, as shown in Figure 15-17a. In this structure, the hydrocarbon tails of the surface-active agent are oriented toward the interior of the micelle, where they are probably held together by London–van der Waals attractive forces. The hydrophilic polar heads extend into the aqueous phase.

One difficulty with this structure, as Becher[17] has indicated, is that there is no simple way for the laminar micelle to end, and it might be expected to continue indefinitely from thermodynamic considerations. For this reason, Hartely[26] suggested a spherical structure for micelles, as shown in Figure 15-17b. The spherical micelle is much more disordered than the laminar structure.

Other workers have suggested disk- and sausage-shaped micelles to account for some of the properties that have been observed.[27-29] A typical sausage-shaped micelle is illustrated in Figure 15-18. Debye considered that the ends of the micelle were capped, so that the entire micelle had a hydrophilic surface.

Becher and Arai[30] have proposed a log boom rather than a radial sausage structure for the micelles in solutions of three polyoxyethylated lauryl alcohols. Their conclusion was based upon the results of light-scattering and hydrodynamic measurements. The log boom structure of the micelles is illustrated in Figure 15-19.

Micelles also form in nonaqueous solutions, and in this case the structure is inverted compared to that in aqueous solutions. In a spherical micelle, for example, the polar heads of the molecules are oriented toward the interior of the micelle with the nonpolar hydrocarbon tails extending in the oil phase[31] (Figure 15-20).

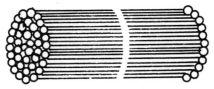

Figure 15-19   Log boom micellar structure; from Becher and Arai, *J. Colloid Science*, © Academic Press.[30]

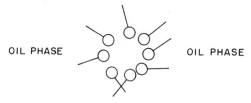

OIL PHASE   OIL PHASE

Figure 15-20 Spherical micellar structure in nonaqueous solutions.

Surface-Tension Lowering versus Micelle Formation

A plot of surface tension versus concentration for a typical surface-active agent shows that the initial addition of the agent causes a marked decrease in surface tension, which continues until the critical micelle concentration is reached. At this point, the surface tension tends to level off with continued addition of the surface-active agent (Figure 15-21).

Figure 15-21 can be interpreted as follows: Initially, increasing concentrations of the surface-active agent cause a continued decrease in the surface tension because the molecules first orient themselves at the air–water interface; this continues until the critical micelle concentration is reached, at which point further addition of the surface-active agent results in the formation of more micelles rather than a continued adsorption at the air–water interface.[31]

Addition of Hydrocarbon to an Aqueous Micellar System

When a hydrocarbon (or oil) is gradually added to an aqueous solution containing micelles of a surface-active agent, initially the hydrocarbon dissolves in the interior of the micelles because of the attraction between the hydrocarbon chains of the surface-active agent and the hydrocarbon molecules (illustrated in Figure 15-22 with a laminar micelle). This process is called solubilization, and the system exhibits the properties of a true solution because the vapor pressure of the hydrocarbon decreases in proportion to its mole fraction in solution in accordance with Raoult's Law.

The micelle continues to swell and increase in size as more hydrocarbon is

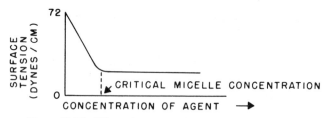

Figure 15-21 Effect of concentration upon surface tension.

WATER

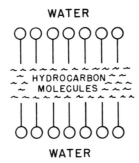

WATER

Figure 15-22   Solubilization of a hydrocarbon by micelles.

added until ultimately it ruptures. According to Ross and Fowkes,[31] this point is reached when the diameter of the micelle is about 500 Å.

REFERENCES

1. P. Sennett and J. P. Oliver, "Colloidal Dispersions," in S. Ross (Ed.), "Chemistry and Physics of Interfaces," Chap. 7, American Chemical Society Publications, Washington, D.C., 1965.
2. R. A. Gortner, "Outlines of Biochemistry," 3rd ed., Chap. 1, John Wiley & Sons, Inc., New York, 1949.
3. D. K. Sebera, "Electronic Structure and Chemical Bonding," Chap. 11, Blaisdell Publishing Company, Waltham, Mass., 1964.
4. H. Margenau, *Rev. Mod. Phys.* **11**, 1 (1939).
5. L. N. Ferguson, "Electron Structures of Organic Molecules," Chap. 2, Prentice-Hall, Inc., New York, 1952.
6. W. H. Keesom, *Physik Z.* **22**, 129 (1921).
7. P. Debye, *Physik Z.* **21**, 178 (1920).
8. F. London, *Physik Z.* **60**, 491: **63**, 245 (1930).
9. F. London, *Trans. Faraday Soc.* **33**, 8 (1937).
10. C. N. Davies, "Aerosol Science," Chap. 11, Academic Press, Inc., New York, 1966.
11. E. S. Gould, "Mechanism and Structure in Organic Chemistry," Holt, Rinehart and Winston, Inc., New York, 1959.
12. J. Bernal and R. H. Fowler, *J. Chem. Phys.* **1**, 520 (1933).
13. G. S. Forbes, *J. Chem. Educ.* **18**, 18 (1941).
14. N. D. Coggeshall, *J. Chem. Phys.* **18**, 978 (1950).
15. F. M. Fowkes, "Attractive Forces at Interfaces," in S. Ross (Ed.), "Chemistry and Physics of Interfaces," Chap. 1, American Chemical Society Publications, Washington, D.C., 1965.
16. W. Drost-Hansen, "Aqueous Interfaces," in S. Ross (Ed.), "Chemistry and Physics of Interfaces," Chap. 3, American Chemical Society Publications, Washington, D.C., 1965.
17. P. Becher, "Emulsions: Theory and Practice," 2nd ed., Chap. 2, Reinhold Publishing Corp., New York, 1965.
18. A. E. Alexander and J. H. Schulman, *Trans. Faraday Soc.* **36**, 960 (1940).
19. J. H. Schulman and E. G. Cockbain, *Trans. Faraday Soc.* **36**, 651 (1940).

20. J. L. Kanig and C. T. Shin, *Proc. Sci. Sec. Toilet Goods Assoc.* No. 38, 55 (1962).
21. "Properties and Applications of the Freon® Fluorocarbons," Freon® Technical Bulletin B-2.
22. McCutcheon's Detergents & Emulsifiers, 1977 North American Edition, MC Publishing Company, Ridgewood, N.J.
23. A. M. Schwartz, J. W. Perry, and J. Berch, "Surface Active Agents and Detergents," Vol. 2, Interscience Publishers, Inc., New York, 1958.
24. W. D. Bancroft, *J. Phys. Chem.* **17**, 591 (1913); **19**, 275 (1915).
25. W. C. Griffin, "Emulsions," in "Encyclopedia of Chemical Technology," 2nd ed., Vol. 8, John Wiley & Sons, Inc., New York, 1965, pp. 117–154.
26. G. S. Hartley, "Paraffin-Chain Salts," Hermann et Cie, Paris, 1936, p. 45.
27. P. Debye, *J. Phys. Chem.* **53**, 1 (1949).
28. P. Debye and W. E. W. Anacker, *J. Phys. Colloid Chem.* **51**, 18 (1947).
29. F. B. Rosevear, *J. Soc. Cosmet. Chem.* **19**, 581 (1968).
30. P. Becher and H. Arai, *J. Colloid Interface Sci.* **27**, 634 (1968).
31. S. Ross and F. M. Fowkes, "Emulsions and Dispersions," Course Booklet for ACS Short Course, 1968.

16

GENERAL
PROPERTIES
OF EMULSIONS

This chapter is concerned generally with conventional nonaerosol emulsions. Aerosol emulsions and foams consitute only a very small part of the general field of surface chemistry, and there is relatively little basic information available about aerosol systems. However, the properties of aerosol systems are governed by the same factors that determine the properties of conventional nonaerosol emulsions and foams. Therefore, in order to learn about aerosol emulsions and foams, it is helpful to understand the principles that apply to conventional systems. For this reason, some general properties of emulsions and foams arc considered before discussing the more specific and narrower field of aerosol systems.

There are several reasons for the lack of fundamental information about aerosol emulsion and foam systems. One is that such systems have appeared only recently, and there has not been sufficient time to develop much basic information. Also, much of the equipment needed for basic studies of systems under pressure has yet to be designed. Until recently, no suitable microscopic equipment for viewing emulsions under pressure had been developed. Finally, although aerosol emulsions are extremely useful and necessary for formulating many products, the emulsions themselves are only a means to an end, not the final form in which the product is applied or used. Thus, an aerosol may be present in the container as an emulsion, but after discharge it will be either a foam or a spray. This has tended to focus attention on the end product rather than on the intermediate emulsion.

DEFINITION OF AN EMULSION

Becher[1] has listed nine emulsion definitions suggested by various workers. Some of these include a condition of stability, while others impose a size limit upon the dispersed droplets. For the purposes of the present discussion, an emulsion is defined merely as a dispersion of one immiscible liquid in another. This is essentially the definition of a colloidal system used previously and thus includes aerosol emulsions where there is relatively little information about the droplet size of the dispersed phase.

TYPES OF EMULSIONS

In most emulsion systems, water is one of the immiscible liquids. The other liquid is usually organic in nature and is generally referred to as the oil phase, regardless of its structure.

There are two general types of emulsions. In one type, the oil is dispersed throughout water in the form of small droplets. This is called an oil-in-water emulsion and is abbreviated o/w. The oil is referred to as the dispersed or internal phase and water as the continuous or external phase. In the second type of emulsion, the situation is reversed and water is dispersed throughout the oil phase. This is called a water-in-oil emulsion and is abbreviated w/o. In this system, the oil is the continuous phase and water the dispersed phase.

This terminology applies to most aerosol emulsions where the liquefied gas propellant is considered to be the oil component of the emulsion. A dispersion of the propellant in an aqueous phase is called an oil-in-water emulsion, and a dispersion of water droplets in a liquefied gas propellant is a water-in-oil emulsion.

DETERMINATION OF EMULSION TYPE

A number of different methods are available to determine whether a conventional emulsion is oil-in-water or water-in-oil.[1,2] If an aerosol is formulated with a concentrate which itself is an emulsion system, then all of these methods are applicable to the concentrate, since it is not under pressure. Some of the procedures can also be used for an aerosol which contains propellant.

Drop Dilution

This method is useful for testing aerosol concentrates that are themselves an emulsion. In carrying out the test, a drop of the concentrate is allowed to fall upon a water surface. If the drop disappears or disperses throughout the water, the concentrate is an oil-in-water emulsion, since adding the concentrate to water merely dilutes the emulsion. If, however, the drop remains intact on the water surface, then the oil phase must be the external phase and the emulsion is a water-in-oil type.

If the concentrate is a water-in-oil emulsion, the aerosol prepared from the concentrate will also be a water-in-oil emulsion, since the addition of propellant to the concentrate will merely extend the oil phase. However, if the concentrate is an oil-in-water emulsion, it does not necessarily follow that the aerosol will also be an oil-in-water emulsion. In this case, it is possible that addition of propellant to the oil-in-water concentrate might cause phase inversion to a water-in-oil emulsion. Generally, it is possible to judge if this might occur by considering other factors, such as the proportion of the oil phase in the original concentrate, the quantity of propellant added, the type of emulsifying agent present, etc.

Because of technical difficulties, drop dilution is not a very satisfactory method for aerosol emulsions. Although it is possible to invert a glass bottle containing an aerosol emulsion (without dip tube) over a second glass bottle containing only water and transfer the emulsion to the second bottle by connecting the valve stems on the two bottles, the emulsion foams as it is being transferred, thus making any conclusions of doubtful significance. Cooling the bottles prior to transfer reduces the pressure and, in turn, the tendency to foam.

Electrical Conductivity

The conductivity of an emulsion depends essentially upon that of the continuous phase. Oil has a relatively low conductivity, while water has a high conductivity. Therefore, an oil-in-water emulsion will have a high conductivity, since water is the continuous phase. On the other hand, a water-in-oil emulsion will have a low conductivity, since oil is the continuous phase.

A simple apparatus for judging the conductivity of emulsions has been described by Griffin.[3] The apparatus contains a neon bulb that lights brightly if the emulsion is the oil-in-water type and only dimly (or not at all) for water-in-oil emulsions. The equipment was adapted for use with aerosol products by modifying a compatibility tube so that the conductivity of the aerosol could be determined in the tube.[4]

Creaming

Creaming occurs when the dispersed phase of an emulsion either settles to the bottom as a result of gravitation or rises to the top because it is less dense than the continuous phase. Becher[1] apparently was the first to point out that the manner in which an emulsion creams can be used to determine the emulsion type. It is an excellent method for aerosol emulsions because there is a fairly large difference in density between the liquefied gas propellants and water. Most of the fluorinated propellants are heavier than water. When they are dispersed in an aqueous phase to give an oil-in-water emulsion, the propellant droplets will settle in time and form a milky, dense layer at the bottom of the glass container. If the emulsion is the water-in-oil type, the creamed layer will appear at the top, since the dispersed water droplets, being less dense than the propellant, will rise (Figure 16-1).

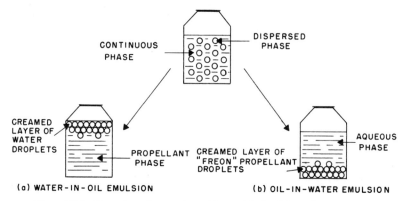

Figure 16-1 Creaming of aerosol emulsions with the "Freon" propellants.

Hydrocarbon propellants such as isobutane are less dense than water and, therefore, creaming occurs in a reverse manner from that of the fluorocarbon propellants. When an oil-in-water emulsion containing a dispersed hydrocarbon propellant creams, the emulsified hydrocarbon propellant droplets will rise to the top, since they are less dense than water. The creamed layer, therefore, appears at the top. In a water-in-oil emulsion, the dispersed water droplets settle to the bottom, since they are heavier than the continuous propellant phase.

If the composition of the propellant phase has been adjusted to have a density close to that of the aqueous phase (by use of propellant blends or propellant/solvent combinations), it is more difficult to determine emulsion type by creaming. If the two phases have approximately the same density, creaming may not occur for a considerable period of time. However, when creaming does occur, it is usually possible to determine the emulsion type by noting the proportion of the emulsion that creamed. For example, assume an emulsion is prepared with 20 wt.% propellant and 80 wt.% aqueous phase. If, after creaming, the milky creamed layer constitutes roughly 20 wt.% of the emulsion, then the creamed layer must consist of propellant droplets and the emulsion must be oil-in-water, regardless of whether the creamed layer appears at the top or bottom. If the creamed layer constitutes about 80 wt.% of the emulsion, then it must be a water-in-oil emulsion.

General Observations of Aerosols

Usually there is little difficulty in determining whether an aerosol emulsion is oil-in-water or water-in-oil merely from a knowledge of the emulsifying agents that are present and the concentration of propellant. Most aerosol foam products are formulated with water-soluble or water-dispersible surface-active agents and contain 10 wt.% propellant or less. Practically all of these emulsions have been shown to be oil-in-water and will foam when discharged unless significant quantities of a foam depressant, such as isopropyl alcohol, are present.

Aqueous-based systems designed to produce a fairly fine spray upon discharge are generally formulated with an oil-soluble, nonionic surface-active agent, such as polyglycerol oleate. These products normally contain a fairly high percentage of propellant, that is, 30 wt.% or more, and discharge as nonfoaming sprays. They are water-in-oil emulsions.

Bancroft's rule appears to be somewhat applicable to aerosol emulsions in that water-dispersible surface-active agents tend to produce oil-in-water emulsions and the oil-soluble agents, water-in-oil emulsions. This seems to be a more important factor than the concentration of propellant, although the latter undoubtedly plays a part in determining the type of emulsion. For example, an aerosol formulated with 30 wt.% of an aqueous shaving lather concentrate containing water-dispersible surface-active agents and 70 wt.% of a fluorocarbon propellant was an oil-in-water emulsion and discharged as a stream that foamed vigorously upon contact. On the other hand, a product formulated with 30 wt.% water and 70 wt.% of a fluorocarbon propellant containing the oil-soluble, nonionic polyglycerol oleate was a water-in-oil emulsion and discharged as a fine, nonfoaming spray.

Occasionally, an aerosol formulated as a water-in-oil emulsion will invert to an oil-in-water emulsion during storage. When this occurs, the spray changes from the initial fine, nonfoaming spray to a coarse, streamy, foamy spray. The change is usually unmistakable.

EMULSION STABILITY

Emulsion stability is generally considered to have two aspects: creaming, which leads to phase separation, and coalescence, which results in the breaking of an emulsion.[3,5] Coalescence results when two droplets come into contact and combine to form a large droplet. Coalescence may be preceded by the intermediate step of flocculation, where the droplets are held together in an aggregate or cluster, but still maintain their own individuality.

Some workers prefer to consider emulsion stability only from the viewpoint of coalescence, which leads to a change in the number of droplets and their size. They divorce creaming from emulsion stability on the basis that creaming merely separates the initial emulsion into two emulsions—a dilute emulsion and a concentrated emulsion—with the number of dispersed droplets remaining the same.[2,6] However, Becher[1] has pointed out that although creaming does not represent actual breaking of an emulsion, it is an indication of a situation that may lead to emulsion breaking. Creaming is favored by large droplet size, which may be an indication of coalescence. The closer packing of the droplets in the creamed layer also increases the tendency for coalescence. Schulman and Cockbain[7] considered phase separation (creaming) to be sufficiently valid for the comparison of relative emulsion stabilities when all the samples were prepared in the same way.

Creaming is quite important in aerosol products because a product that has creamed must be shaken before use. If significant creaming occurs while the product is being used, the discharge characteristics will change noticeably during use, and the consumer will end up with a product that is discharging either propellant or concentrate, depending upon whether the propellant is denser or lighter than the aqueous phase. For this reason, there is additional justification in the aerosol field for considering creaming as one aspect of emulsion stability.

The term phase separation is used in the present text to indicate any visible separation of the two phases of an emulsion, whether it occurs as a result of creaming or coalescence of the emulsified droplets to form a discrete separate phase. Complete phase separation is the term used to indicate that the emulsion has broken.

Creaming

Three factors have a major influence upon the rate of creaming of an emulsion. They are the relative densities of the dispersed phase (d_1) and the continuous phase (d_2), the droplet radius (r), and the viscosity of the continuous phase (η). This follows from Stokes' Law,[8] which gives the sedimentation rate (u) of a spherical particle in a liquid (g is the acceleration of gravity):

$$u = \frac{2gr^2(d_1 - d_2)}{9\eta} \tag{16-1}$$

If d_1 is larger than d_2, the dispersed phase will settle to the bottom. If d_1 is less than d_2, the rate of sedimentation u will have a minus sign, indicating that the dispersed phase will rise to the top (creaming).

Equation 16-1 shows that the rate of creaming is reduced by decreasing the radius of the droplets, decreasing the density difference between the dispersed and continuous phases, and increasing the viscosity of the continuous phase.

Little can be done to reduce the droplet size of an aerosol after the propellant has been added except mechanical agitation of the samples. Homogenization would be difficult because it would have to be carried out either in a closed system under pressure or at a sufficiently low temperature so that propellant losses would be minimized. The latter would be impractical with most propellants. If the concentrate is an emulsion itself, then homogenization of the concentrate prior to addition of the propellant might result in an aerosol emulsion in which the ultimate droplet size was smaller.

One method for minimizing creaming is to adjust the composition of the propellant phase so that it has approximately the same density as that of the aqueous phase. This can be done in several ways. If the aerosol contains very little organic material besides the propellant, then a propellant that has approximately the same density as that of the aqueous phase can be used. Methods for calculating the compositions of propellant blends with a specific density were described in Chapter 12. For oil-in-water emulsions, blends of fluorocarbon 12, fluorocarbon 114, and isobutane are generally satisfactory. The only problem with these blends is that they are usually flammable because of the quantity of hydrocarbon propellant that must be added to bring the density of the propellant blend close to that of water.

Sometimes the properties required for a particular aerosol product permit the presence of a higher-boiling hydrocarbon solvent in the concentrate, for example, odorless mineral spirits, VM&P naptha, etc. In this case, it is possible to adjust the proportions of the fluorocarbon propellant and hydrocarbon solvent so that the density of the resulting organic phase is about equal to that of water. This avoids the necessity of using a flammable propellant to obtain the desired density. In these cases, fluorocarbon 12 is normally used by itself as the propellant because of the vapor depressant effect of the hydrocarbon solvent.

The viscosity of the aqueous phase may be increased by addition of suitable thickening agents.[5] Combinations of surface-active agents with polar compounds, such as fatty alcohols, often produce highly viscous external aqueous phases in which the rate of creaming is very low.[9, 10]

Flocculation

Flocculation takes place when the droplets in an emulsion approach sufficiently close to each other so that they form clusters or aggregates. In these clusters, the droplets may remain as individual droplets, they may coalesce into a larger droplet, or they may redisperse as single droplets when the emulsion is agitated or shaken. These clusters will cream more rapidly than individual droplets because the radius of the cluster is larger than that of the droplets (Figure 16-2).

Coalescence

Coalescence occurs when emulsified droplets come into contact with each other and unite to form a larger droplet. This leads to phase separation and complete breaking of the emulsion. If this occurred to any significant degree while a product was being used, the spray or foam characteristics would change. A product

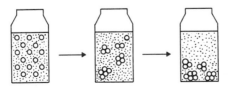

Figure 16-2 Flocculation of droplets in an emulsion.

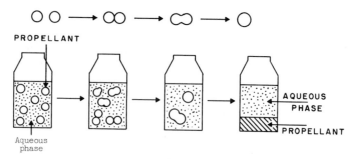

Figure 16-3 Coalescence and phase separation in an aerosol emulsion.

of this type would be of little value for the consumer market. Coalescence and complete phase separation are illustrated in Figure 16-3.

The reason emulsions coalesce is that they are thermodynamically unstable. Energy is required to form the emulsion initially (by shaking, homogenization, etc.). This energy is concentrated in the interfacial regions around each droplet, where it is manifested as interfacial tension. Therefore, the larger the interfacial area of an emulsion, that is, the smaller the droplet size and the greater the number of droplets, the greater the interfacial energy. Processes tend to proceed in a direction which will decrease the total energy of the system. The dispersed droplets can decrease interfacial energy by coalescing to form larger droplets, thus producing an overall decrease in interfacial area. Therefore, coalescence is a spontaneous process.

In an unstabilized aerosol system consisting of propellant and water, creaming, coalescence, and complete phase separation occur within a matter of seconds after the bottle has been shaken. Even if the density of the propellant is adjusted to match that of the aqueous phase, so that creaming is negligible, coalescence occurs so rapidly that complete phase separation again results in a few seconds. An emulsion that is to be marketed as a consumer product must have the droplets protected against coalescence while the consumer is using the product. The fact that an aerosol product creams during storage is not too important as long as a simple shake of the container redisperses the creamed layer and as long as creaming does not occur while the product is being used.

Coalescence of dispersed droplets can be delayed by addition of suitable surface-active materials to the system. Since the mechanism by which this stabilization is achieved depends to a considerable extent upon whether the emulsion is oil-in-water or water-in-oil, the stabilization of each of these two types of emulsions by surface-active agents is considered separately.

STABILIZATION OF OIL-IN-WATER EMULSIONS

Two major theories have been advanced to account for the stability of oil-in-water emulsions. One involves the presence of an electric charge on the droplets so that there is a mutual repulsion between the droplets when they approach each other. The second involves the formation of a protective interfacial film of surface-active agent around the droplets which prevents them from coalescing should they come into contact. There has been, and still is, a certain amount of controversy about the relative importance of these two mechanisms. However, each contributes to the stability of emulsions in a different way and they are both important.

The way in which the two mechanisms operate in stabilizing emulsions can be seen by considering a simplified picture of the processes that lead to the breaking of an emulsion. In the first place, the dispersed droplets must come into contact, either as a result of a collision or by flocculation due to an attraction between them. After contact, the droplets can either separate and redisperse again, remain flocculated but maintain their own individuality, or coalesce. If the rate of coalescence is sufficiently high, complete phase separation will occur and the emulsion will break. If the droplets possess an electric charge, this can decrease the rate and intensity of collisions, and it also will determine the extent to which the emulsion droplets flocculate. The electric charge, therefore, is the primary stabilizing factor. If, however, the droplets come into contact, then the strength and nature of the interfacial film determine if coalescence will take place. The interfacial film is the second and final stabilizing factor.

The degree to which these two mechanisms operate in an emulsion depends to a considerable extent upon the type of emulsifying agent present. In emulsions stabilized with either anionic or cationic agents, the droplets are charged and both the electric charge and the strength of the interfacial film are important. The droplets in emulsions stabilized with nonionic emulsifying agents can be charged, but the electric charge on the droplets is much smaller. The predominant stabilizing factor is the strength of the interfacial film. Emulsions can also be stabilized with gums. In these systems the strength of the interfacial film is the only important factor.

Stabilization by Electric Charge

The theory of the stabilization of dispersed particles by electric charge is based upon the fact that when two droplets with the same type of charge approach each other, there is always a repulsion between them (Figure 16-4). This repulsive force decreases the possibility of the particles approaching close enough so that flocculation or coalescence can occur.

The fact that the stability of certain colloidal dispersions was due to the elec-

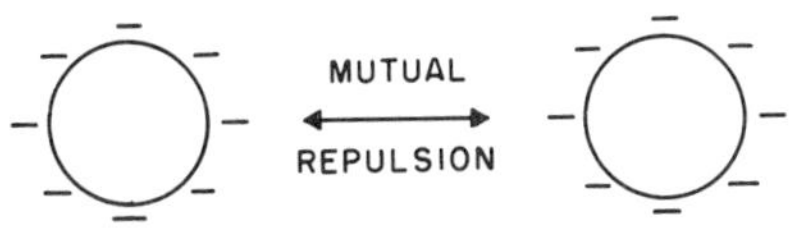

Figure 16-4 Mutual repulsion between two particles with the same charge.

tric charge on the particles was recognized around the turn of the century as a result of investigations of colloidal sols. These colloidal sols were dilute dispersions of submicroscopic solid particles. The small particle size of the solids was obtained by mechanical means, and there were no surface-active agents present. Schulze, as early as 1882,[11] reported that addition of electrolytes to colloidal dispersions caused them to become unstable and flocculate. Multivalent ions were found to have a much greater effect than monovalent ions. A few years later, Linder and Picton[12] discovered that the particles of a colloidal sol migrated under the influence of an electric field. This indicated that the particles were charged, and, from the direction in which the particles moved, they were able to determine the sign of the charge on the particles. This phenomenon is called electrophoresis. The velocity of migration was a function of the charge on the dispersed particles. Since the sign of the charge on the dispersed particles could be determined, it soon became apparent that the multivalent ions, which Schulze had found to have a disproportionate effect upon the flocculation of the particles, had an electric charge opposite in sign to that on the dispersed particles. The effect of electrolytes upon dispersions has since become known as the Schulze–Hardy rule and can be stated as follows: "The precipitating power of an electrolyte depends upon the valence of the ion whose charge is opposite to that on the colloidal particles."[13] Subsequently, Hardy[14] and Burton[15] demonstrated that the stability of the dispersions was related to their movement in an electric field. Since the movement was a function of the degree of charge on the particles, this showed a relationship between the charge on the particles and the stability of the dispersion.

The Electric Double Layer

An electric charge can arise on a particle by ionization, adsorption of other ions (for example, hydroxyl ions from solution), or by friction between the interfacial areas of the dispersed particles and the surrounding medium. In the case of emulsions, the charge can arise as a result of the orientation of an ionized surface-active agent such as sodium stearate at the oil–water interface. The molecule can be pictured with its hydrocarbon tail anchored in the oil droplet and the negatively charged polar head extending into the aqueous phase. In this case, the droplets would be negatively charged.

The actual structure and distribution of charges at the interface of a dispersed particle or an oil droplet have been the subject of many investigations, and the picture has not been completely clarified as yet. Helmholtz proposed in 1879[16] that the interfacial area consisted of a double layer of oppositely charged particles. According to this view, there is a layer of ions adsorbed at the surface of the particle with another layer of opposite charges located in the continuous phase adjacent to the surface charges so that a double layer of electrical charges is formed (Figure 16-5a). Helmholtz referred to the potential drop across the layer of charges as the electrokinetic potential.

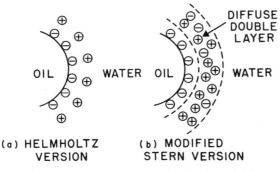

(a) HELMHOLTZ   (b) MODIFIED
    VERSION   STERN VERSION

Figure 16-5 Simplified version of the electric double layer.

The Helmholtz model of the double layer had the disadvantage that it required a sharp potential drop at the interface between the two series of charges, and it seemed unlikely, considering the general mobility of ions, that the outer layer of ions would remain oriented as required by the theory.[1] In order to avoid some of these difficulties, Gouy[17] assumed that the double layer was diffuse and the outer layer decreased in electrical density at increasing distances from the interface. Although this was an improvement, there were still some deficiencies in the theory. Stern[18] suggested a compromise between the two views which assumes that the double layer consists of two parts, one of which is approximately a single ion in thickness and remains attached to the surface of the particle, while the other (the Gouy layer) is diffuse and extends into the surrounding liquid. The potential drop from the first ionic layer is sharp, while that in the second diffuse layer gradually decreases as the distance from the interface increases. The modified Stern version of the double layer is illustrated in Figure 16-5b.

The Potential Energy Barrier (The D.L.V.O. Theory)

When electrically charged particles approach each other, there are two different forces which affect the behavior of the particles.[19] These forces are (1) the London-van der Waals force of attraction and (2) the electrostatic repulsive force resulting from the interaction of the electric double layers on the particles.

The theory concerned with the behavior of these charged particles is called the D.L.V.O. theory (from the initials of the scientists Derjaguin and Landau[20] and Verwey and Overbeek[21] who authored it).

If the London-van der Waals force of attraction between two particles is greater than the repulsive force regardless of the distance that separates the particles, the particles continue to approach until they come into contact. Flocculation then occurs with the formation of agglomerates. Under other conditions, the repulsive forces between the particles are larger than the London-van der Waals attractive force, and this tends to prevent contact between the particles or droplets, thus decreasing the possibility of flocculation and co-alescence. The repulsive force (potential energy barrier) between the particles increases in intensity as the droplets approach each other until it reaches a maximum. If the kinetic energy of the droplets is sufficiently large so that they can overcome this energy barrier, the London–van der Waals attractive force takes over and brings the particles or droplets together.

The London–van der Waals attractive force and the electrostatic repulsive forces between the particles oppose each other. Since they operate indepen-dently, the net interaction between the two particles is determined by the addi-tive effect of the forces. Therefore, the curve that shows how the force between the particles varies with distance is obtained by superimposing the curve for the London–van der Waals attractive force upon the curve that indicates how the re-pulsion varies with distance.

Many factors, including the radius of the particles, the electric charge, the electrolyte concentration, etc., affect the shape of the potential energy curve for two approaching electrically charged particles. A typical potential energy curve for two electrically charged particles is illustrated in Figure 16-6.

If the potential energy barrier X is much larger than the kinetic energy of the particles, then the droplets would be expected to bounce off each other without contact.[19]

The curve in Figure 16-6 shows a secondary minimum at a distance of about 100 Å. This is predicted by the D.L.V.O. theory and occurs because the Lon-don–van der Waals attractive force falls off more slowly with distance than the repulsion. This minimum becomes significant for large particles, such as those in emulsions, and it seems probable that weak aggregates in which the droplets are separated by a liquid film form at this distance. Because the forces that hold the droplets together in the aggregate are very weak, the droplets would be expected to redisperse under any slight mechanical shearing force. In highly concentrated emulsions, the droplets are crowded together. The high viscosity of these emul-

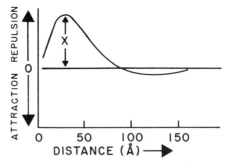

Figure 16-6 A typical potential energy diagram for two electrically charged particles; from Kitchener and Mussellwhite.[19]

sions is considered to be due to the formation of aggregates in the region of the secondary minimum.

It should be emphasized that the D.L.V.O. theory was developed to account for the stability of dispersions of submicroscopic inorganic solid particles. Attempts to extend the theory to include emulsions where the oil droplets are stabilized by interfacial films of an emulsifying agent should be viewed with caution.[19] The D.L.V.O. theory is useful in interpreting the effect of an electric charge in stabilizing emulsions against flocculation, but it does not apply to coalescence, which depends upon the disruption of the interfacial film of emulsifying agent.

Evidence that electrically charged emulsion droplets behave similarly to colloidal inorganic dispersions as far as flocculation is concerned is provided by the observation that the effect of added electrolytes upon the flocculation of emulsions follows the Schulze–Hardy rule. This was reported for emulsions of xylene in aqueous sodium oleate by Martin and Hermann[22] and by van den Tempel[23] for emulsions stabilized with the sodium salts of surface-active acids.

The Interfacial Film

There is also evidence which shows that the stability of emulsions to coalescence results from the presence of the interfacial film of emulsifying agent and not the electric charge. Thus, Limburg[24] found that addition of electrolytes to dilute oil-in-water emulsions without emulsifying agents present decreased electrophoretic velocity (thus the charge) and emulsion stability, as expected. However, when an emulsifying agent (saponin) was added to the emulsion, the addition of electrolyte to the emulsion caused the same decrease in electric charge as in the preceding case, but had no effect upon emulsion stability. Limburg concluded that the protective film of emulsifying agent was a more important factor in emulsion stability than electric charge.

King and Wrzeszinski[25] carried out a similar, but more extensive, series of experiments in which they determined the effect of electrolytes upon the stability and electrokinetic potential of oil-in-water emulsions stabilized with a number of surface-active agents, including sodium oleate and saponin. They came to essentially the same conclusion as Limburg and stated that there was no relationship between the electrokinetic potential on the droplets and the stability of the emulsions as long as sufficient emulsifying agent was present to cover the droplets with a monomolecular film of emulsifying agent. King and Wrzeszinski believed that the coherent protective film provided by the emulsifying agent was the major source of stability.

Recently, Kremnev, Nikishechkina, and Ravdel[26] measured the electric potential of emulsions of benzene in water stabilized with sodium oleate. They reported that although the addition of small amounts of sodium chloride caused the initially high electric potential to decrease sharply, there was no corresponding decrease in the stability of the emulsion. Kremnev and his co-workers stated that the coalescence of the emulsified droplets was prevented by a factor other than the repulsive force in the electric double layer.

Additional evidence was provided by Derjaguin and Titievskaya,[27] who found that when small amounts of sodium chloride were added to an aqueous sodium oleate solution in which two bubbles were pressed together, the salt caused the thickness of the water layer separating the bubbles to decrease until the electric double layers surrounding the bubbles overlapped. The bubbles did not coalesce, however.

Nature of the Interfacial Film

The investigations discussed in the previous section indicated that once the droplets were in contact, some factor other than electric charge was responsible for the stability of emulsions to coalescence. This factor is generally considered to be the nature and strength of the interfacial film of emulsifying agent that surrounds the droplet. The importance of the strength of the interfacial film can be seen when it is realized that in order for coalescence to occur, the molecules of the surface-active agent oriented at the interface of the two droplets have to be displaced and pushed aside so that the molecules of the oil droplets can come into contact. Therefore, resistance of the interfacial film to rupture and displacement is a necessary condition for stability of an emulsion to coalescence.

Besides the work discussed previously, there is a considerable amount of additional evidence for the theory that the stability of emulsions to coalescence is due to the strength of the interfacial film. Much of this evidence has resulted from investigations of molecular complexes that are formed between surface-active agents and long-chain polar compounds at interfacial regions. These complexes have been shown to have a considerable effect upon emulsion and foam properties, in both the aerosol and the nonaerosol fields. This effect is attributed to the strong interfacial film formed by these complexes.

An additional factor involved in the strength of the interfacial film is its hydration. There seems to be little doubt that hydration may be a very important factor in the strength of many interfacial films, but its role remains to be clarified.

Molecular Complexes

Molecular complexes, also known as association complexes or molecular addition compounds, have been postulated since 1823 (see Reference 28 for a list of the early papers in this field). The complexes formed between free fatty acids and their salts (the so-called acid soaps) were the first reported, but their existence was disputed until McBain and his co-workers demonstrated in 1933 that the acid soap consisting of equal molecular proportions of lauric acid and potassium laurate was a true chemical individual.[28-30] Many other acid soap complexes have been reported, and the pearlescence of some of the early lotions containing stearic acid was attributed to the formation of the acid soap complex.[31] The sodium stearate–stearic acid complex subsequently was investigated by Ryer[32] and more recently by Goddard and Kung,[33] who found that the ratio of stearic acid to sodium stearate in the complex was 1:1.

The real stimulus for the investigation of complex formation, however, was provided by the investigations of Schulman and his co-workers.[7,34,46] They showed that molecular complex formation was not confined to the acid soaps, but was a much broader phenomenon than had been suspected. Schulman and his co-workers also demonstrated that complex formation could have a considerable effect upon the properties of interfacial films and the stability of emulsions, and thus revealed the importance of complex formation in the field of surface chemistry.

Schulman and Rideal reported in 1937 that when a monomolecular film of a water-insoluble compound such as cetyl alcohol or cholesterol was spread over the surface of an aqueous solution containing sodium cetyl sulfate, the molecules of the surface-active agent penetrated the surface film of the cetyl alcohol and formed a complex with the alcohol.[34] Evidence for the existence of the complex was indicated by the fact that it was much more resistant to disruption by pressure than the film from either of the components by themselves. This is one reason why a complexed interfacial film around oil droplets is better able to retard coalescence than the film from the surface-active agent alone.

Schulman then extended his investigation to a study of the effect of complex formation upon the stability of oil-in-water emulsions prepared with mineral oil.[7] He and Cockbain observed that emulsions prepared with a surfactant alone, such as sodium cetyl sulfate, were unstable. However, the addition of a long-chain polar compound, such as cetyl alcohol, to the system resulted in the formation of excellent, stable emulsions. This effect of the alcohol upon emulsion stability was attributed to complex formation between the sodium cetyl sulfate and cetyl alcohol at the droplet interface. Thus, Schulman and Cockbain's results at the oil–water interface paralleled those at the air–water interface.

Oleyl alcohol was relatively ineffective in promoting complex formation. This was considered to be due to the steric hindrance resulting from the kinked structure of the double bond in oleyl alcohol, which prevented close association between oleyl alcohol and sodium cetyl sulfate.

The existence of the complexes formed between the alkyl sulfates and the long-chain alcohols was subsequently confirmed by Epstein and his co-workers[35] and Goddard and Kung[33] by analysis of the complexes. They prepared complexes from sodium alkyl sulfates and long-chain alcohols by two methods. One involved heating the two components to about 110°F for 10 min and then allowing the melt to cool overnight. In the other method, the complexes were prepared in either aqueous or aqueous alcohol solutions at elevated temperatures. The complexes precipitated from the cooled solution and were isolated by filtration. Analysis of the complexes was carried out by three methods: differential thermal analysis, infrared, and x-ray. In all cases, the ratio of alcohol to sulfate in the lauryl alcohol–sodium lauryl sulfate and myristyl alcohol–sodium myristyl sulfate complexes was found to be 1:2. This work established beyond a doubt the existence of the long-chain alcohol–alkyl sulfate complexes.

Miles and his co-workers[36,37] showed that films from aqueous solutions of sodium lauryl sulfate were fast-draining, but that the addition of lauryl alcohol resulted in films that were slow-draining. This effect was considered to be due to an increase in surface viscosity due to complex formation. The slow-draining films became fast-draining at a specific higher temperature called the "film drainage transition temperature (FDTT)." The change occurring at the transition temperature was due to thermal disruption of the complex.

The investigation of film drainage transitions was extended to nonionic systems by Becher and Del Vecchio,[38] who demonstrated by surface viscosity measurements that combinations of polyoxyethylene lauryl ethers with long-chain alcohols also exhibited film drainage transition phenomena.

Emulsion Stabilization by Molecular Complexes

The theory that the nature and strength of the interfacial film of emulsifying agent is the major factor controlling the stability of emulsions to coalescence has been mentioned, but the conditions that influence the nature of the interfacial film have not yet been discussed. Since the interfacial film is formed from the surface-active agent, one of the most important factors is the concentration of the agent at the interface. If relatively few molecules of the emulsifying agent are present at the interface, so that only a portion of the total interfacial area of the droplet is covered, then the interfacial film would be expected to be weak and provide little protection against coalescence. However, a condensed, strongly coherent film would be able to resist the displacement necessary before coalescence can occur. Molecular complexes can form strong, coherent films, and this undoubtedly is one way in which they stabilize emulsions.

Many anionic (and cationic) surface-active agents by themselves produce relatively poor, unstable emulsions regardless of the fact that they impart an electric charge to the droplets. The interesting aspect about these unstable emulsions is that the electric charge of the surface-active agent creates a situation that results in a sparse interfacial film and poor emulsion stability. This apparent paradox, in which the electric charge is both a help and a hindrance in stabilizing an emulsion, can be understood by referring to Figure 16-7a, which illustrates an oil droplet with an interfacial film of cetyl sulfate ions.

The oil droplets repel each other, since they have a similar charge, but the polar heads of the cetyl sulfate molecules oriented at the interface on the same

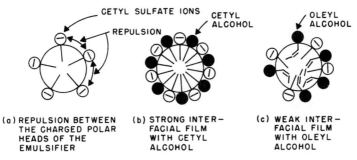

Figure 16-7 Molecular complex formation in emulsions.

droplet also repel each other because they too have the same charge. Therefore, the surface-active agent repels itself. The consequence of this self-repulsion is a low concentration of surface-active agent at the interface and a weak interfacial film. However, when cetyl alcohol is present, its molecules orient themselves between the sodium cetyl sulfate molecules and give a closely packed, condensed, strong interfacial film. A condensed film of this type, as pictured by Schulman, Cockbain, and Rideal, is illustrated in Figure 16-7b. The relatively weak complex with oleyl alcohol resulting from the steric hindrance of the double bond is shown in Figure 16-7c.

The formation of the complex between the sodium cetyl sulfate and cetyl alcohol molecules is due to the existence of attractive forces between the molecules. The attractive forces that account for the complex formation are probably the result of ion–dipole interactions between the ionized sulfate heads and cetyl alcohol molecules, hydrogen bonding, and possibly London–van der Waals forces of attraction between the hydrocarbon chains of the cetyl sulfate and cetyl alcohol molecules.[33,39] The cetyl alcohol molecules may function in part as insulators by screening the ionized cetyl sulfate molecules from each other, which would then reduce the repulsion between the electrically charged heads.[39,40]

There has been another theory advanced to explain the stabilization of emulsions by molecular complexes. This theory states that the high interfacial viscosity of the alkyl sulfate–cetyl alcohol complex prevents the interfacial film from being easily disrupted or pushed aside if the droplets do come into contact. The ineffectiveness of sodium lauryl sulfate alone as an emulsifying agent is considered to be due to the low interfacial viscosity of sodium lauryl sulfate interfacial films.[40] The high interfacial viscosity of the complex may be due to hydration. There seems to be little doubt that water in the immediate vicinity of interfaces is more viscous than that of the bulk phase. This has a considerable bearing upon emulsion stability, since it is possible that hydration layers have sufficient strength to prevent coalescence when droplets come into contact. This subject is discussed in Chapter 18 (see also References 20, 40, 41, and 42).

The attractive forces that account for hydrated layers include ion–dipole inter-actions and hydrogen bonding. Hydration may be one of the main factors in the stabilization of emulsions by nonionic surface-active agents. Polyoxyethyl-ene ethers can bind water molecules by hydrogen bonding with the oxygen atoms of the ether.[19]

Stabilization by Solids

It has been known for many years that emulsions can be stabilized by finely di-vided solids. A wide variety of solids have been reported to be effective for this purpose, including such materials as the basic sulfates of polyvalent metals, cal-cium carbonate, hydroxides of the polyvalent metals, etc.

In order for a solid to act as a suitable stabilizer, the particle size of the solid must be small compared to the size of the emulsified droplet, and the particle should have the proper wettability by the aqueous and oil phases.[19] If the solid stays suspended in either the aqueous or oil phase, then it obviously will not col-lect at the oil–water interface and will not act as a stabilizer. Pickering[43] has suggested that if the solid is more easily wetted by water than by oil, and oil-in-water emulsion will be formed. There seems to be some justification for this because Schlaepfer[44] prepared concentrated water-in-oil emulsions using soot, which is wetted more by oil than water.

Becher[1] has suggested that the concentration of solid stabilizers at an inter-face could produce an interfacial film of considerable strength that could ac-count for the stabilizing action of the solids. Verwey[45] has also indicated that emulsions increase in stability as the interfacial films around the droplets be-come more solid in character.

STABILIZATION OF WATER-IN-OIL EMULSIONS

The literature on water-in-oil emulsions is not nearly as extensive as that for oil-in-water emulsions. Schulman and Cockbain[46] investigated emulsions of water-in-mineral oil using various emulsifying agents and came to the conclusion that in order to obtain stable emulsions, the emulsifier should form uncharged inter-facial films that possessed considerable rigidity. They felt that the droplets could not be charged because an electric double layer would not be expected to exist in a nonionizing medium like mineral oil.

Albers and Overbeek,[47] however, found that there was a considerable electro-kinetic potential present in water-in-oil emulsions stabilized with the oleates of polyvalent metals. They reported that there was no correlation between the elec-trokinetic potential and stability of the emulsion to flocculation, and concluded that the stability of the emulsions was due to the formation of a thick film of the hydrolysis products of the oleates. The stabilizing activity of the hydrolysis products was explained on the basis that they could be considered as small particles, insoluble in both the oil and water phase. Because these products were mostly hydrocarbon in their structure, they were more easily wetted by the oil phase than the water phase and, therefore, formed water-in-oil emulsions. The particles collected at the interface and stabilized the emulsion mechanically.

Ford and Furmidge, in a recent paper,[48] studied the stabilization of water-in-oil emulsions using a variety of oil-soluble emulsifiers. They concluded that in order to obtain a stable water-in-oil emulsion, the interfacial film of the emulsifier should prevent the coalescence of water droplets while allowing the coalescence of oil droplets. The emulsifier should be hydrophobic in nature.

The water droplets in many water-in-oil emulsions have an irregular shape. This was observed as early as 1915 by Briggs and Schmidt[49] and has also been reported by a number of other investigators. Schulman and Cockbain[46] attributed the irregularity of the water droplets in their emulsions to the stiffness of the interfacial film. However, not all dispersed water droplets are irregular in shape.[1]

DETERMINATION OF AEROSOL EMULSION STABILITY

The emulsion stability of an aerosol can be determined by packaging the emulsions in clear glass bottles, shaking to emulsify the propellant, and noting the time until phase separation appears.

VISCOSITY OF EMULSIONS

Emulsions can vary in viscosity from very thin fluids to gels. There are many factors that affect the viscosity of emulsions and References 1, 2, and 50 should be consulted for details on this subject. Sherman[50] has listed a number of variables that influence the viscosity of emulsions; they include the following: (1) viscosity of the continuous phase; (2) volume concentration of the internal phase; (3) particle size distribution of the dispersed phase; (4) type of emulsifying agent; and (5) nature of the interfacial film.

Griffin[3] pointed out that the viscosity of emulsions is close to the viscosity of the external or continuous phase if the latter constitutes more than 50 vol.% of the emulsion. However, when the concentration of the dispersed phase is increased beyond 50 vol.%, the emulsion becomes increasingly more viscous and eventually becomes a gel.

The maximum volume that can be occupied by the internal phase is 74%, assuming that all droplets are spherical and the same size.[2,3] However, emulsions with as high as 99 vol.% internal phase can be prepared. This is possible when the droplets of the dispersed phase are nonuniform in diameter and the small droplets can be packed in between the larger droplets. In such high concentrations, the droplets become distorted and lose their spherical shape.

The type of surface-active agents present and the nature of the resulting interfacial film can have a profound effect upon the viscosity of the emulsion. An increase in viscosity has been observed in many cases with combinations of surface-active agents and long-chain alcohols that form molecular complexes at interfaces. The sodium alkyl sulfates, triethanolamine salts of the fatty acids, and the polyoxyethylene fatty ethers are examples of surface-active agents that have been observed to show this effect with the fatty alcohols.[7,9,10] These complexes form strong, condensed interfacial films, and hydration of these films may be a factor in the increased viscosity of the system.

The viscosity of the external phase may also be increased by the use of thickeners or gelling agents such as sodium carboxymethyl cellulose, methyl cellulose, and natural gums.

Determination of Viscosity of Aerosol Emulsions

No effective method for measuring the viscosity of aerosol emulsions quantitatively has yet been devised. Palmer and Morrow[51] reported a method for determining consistency (viscosity) of aerosol coating compositions which involved the use of a falling-ball viscometer. The apparatus would not be suitable for aqueous emulsion products, however, because of the foaming that would occur during transfer of the sample to the viscometer.

Crude comparisons of the viscosity of aerosols can be made by preparing the samples in clear glass bottles, inverting the bottles, and noting the flow characteristics of the emulsion.

DROPLET SIZE AND EMULSION APPEARANCE

Emulsions vary in appearance from milky, opaque systems to products that are transparent. The appearance of the emulsions is related to the size of the dispersed droplets. Griffin[3] reported the relationship between the appearance of an emulsion and its droplet size, as shown in Table 16-1.

TABLE 16-1 DROPLET SIZE AND EMULSION APPEARANCE

Droplet Size	Appearance of Emulsion
0.05 μ and smaller	Transparent
0.05–0.1 μ	Gray, semitransparent
0.1–1 μ	Blue-white
Greater than 1 μ	Milky white

From Ref. 3. Reproduced with permission.

The appearance of the emulsion can be best judged by shaking the bottle and observing a thin film of the emulsion as it drains down the side. Most aerosol emulsions are milky white, indicating a droplet size greater than 1 μ.

REFERENCES

1. P. Becher, "Emulsions: Theory and Practice," 2nd ed., Reinhold Publishing Corp., New York, 1965.
2. W. Clayton, "The Theory of Emulsions and Their Technical Treatment," 5th ed., The Blakiston Company, Inc., New York, 1954.
3. W. C. Griffin, "Emulsions," in "Encyclopedia of Chemical Technology," 2nd ed., Vol. 8, John Wiley & Sons, Inc., New York, 1965.
4. Freon® Aerosol Report FA-21, "Aerosol Emulsions with Freon® Propellants."
5. A. M. Schwartz, J. W. Perry, and J. Berch, "Surface Active Agents and Detergents," Vol. 2, Interscience Publishers, Inc., New York, 1958.
6. J. J. Bikerman, "Foams and Emulsions," in S. Ross (Ed.), "Chemistry and Physics of Interfaces," American Chemical Society Publications, Washington, D.C., 1965.
7. J. H. Schulman and E. G. Cockbain, *Trans. Faraday Soc.* **36**, 651 (1940).
8. G. G. Stokes, *Trans. Cambridge Phil. Soc.* **9**, 35 (1951).
9. P. A. Sanders, *J. Soc. Cosmet. Chem.* **17**, 801 (1966).
10. P. A. Sanders, *Soap Chem. Spec.* **43**, 68, 70 (July 1967).
11. H. Schulze, *Prakt. Chem.* **25**, 431 (1882); **27**, 320 (1883).
12. S. E. Linder and H. Picton, *J. Chem. Soc.* (London) **61**,, 148 (1892).
13. R. A. Gortner, "Outlines of Biochemistry," 3rd ed., Chap. 8, John Wiley & Sons, Inc., New York, 1949.
14. W. B. Hardy, *Proc. Roy. Soc.* (London) **66**, 110 (1900).
15. E. F. Burton, *Phil. Mag.* **11**, 425 (1906).
16. H. Helmholtz, *Wied Ann.* **7**, 537 (1879).
17. G. Gouy, *Compt. Rend.* **149**, 654 (1900).
18. O. Stern, *Z. Elektrochem.* **30**, 508 (1924).
19. J. A. Kitchener and P. R. Mussellwhite, "The Theory of Stability of Emulsions," in P. Sherman (Ed.), "Emulsion Science," Chap. 2, Academic Press, Inc., New York, 1968.
20. B. V. Derjaguin and L. Landau, *Acta Phys. Chim. USSR* **14**, 633 (1941).
21. E. J. W. Verwey and J. Th. G. Overbeek, "Theory of Stability of Lyophobic Colloids," Elsevier, Amsterdam, 1948.
22. A. R. Martin and R. N. Herman, *Trans. Faraday Soc.* **37**, 30 (1941).
23. M. van den Tempel, *Rec. Trav. Chim.* **72**, 419 (1953).
24. H. Limburg, *Rec. Trav. Chim.* **45**, 772, 854 (1926).
25. A. King and G. W. Wrzeszinski, *J. Chem. Soc.* **57**, 1513 (1953).
26. L. Ya Kremnev, L. A. Nikishechkina, and A. A. Ravdel, *Dokl. Akad. Nauk SSSR* **152**, 816 (1963).
27. B. V. Derjaguin and A. S. Titievskaya, *Kolloidn. Zh.* **15**, 416 (1953).
28. J. W. McBain and M. C. Field, *J. Phys. Chem.* **37**, 675 (1933).
29. J. W. McBain and M. C. Field, *J. Chem. Soc.* **37**, 920 (1933).
30. J. W. McBain and A. J. Stewart, *J. Chem. Soc.* **37**, 924 (1933).
31. F. Atkins, *Perfum. Essent. Oil Rec.* **332** (March 1934).
32. F. V. Ryer, *Oil Soap* **23**, 310 (1946).
33. E. D. Goddard and H. C. Kung, *Proc. 52nd Ann. Meet. Chem. Spec. Manuf. Assoc.*, 124 (December 1965).

34. J. H. Schulman and E. K. Rideal, *Proc. Roy. Soc.* (London) **B122**, 29 (1937).
35. M. B. Epstein, A. Wilson, C. W. Jakob, L. E. Conroy, and J. Ross, *J. Phys. Chem.* **58**, 860 (1954).
36. G. D. Miles, L. Shedlovsky, and J. Ross, *J. Phys. Chem.* **49**, 93 (1945).
37. G. D. Miles, J. Ross, and L. Shedlovsky, *J. Amer. Oil Chem. Soc.* **27**, 268 (1950).
38. P. Becher and A. J. Del Vecchio, *J. Phys. Chem.* **68**, 3511 (1964).
39. J. M. Prince, *J. Colloid Interface Sci.* **23**, 165 (1967).
40. J. T. Davies and E. K. Rideal, "Interfacial Phenomena," 2nd ed., Chap. 5, Academic Press, Inc., New York and London, 1963.
41. L. N. Ferguson, "Electron Structures of Organic Molecules," Prentice-Hall, Inc., New York, 1952.
42. J. W. McBain, "Colloid Science," D. C. Heath and Company, Boston, Mass., 1950.
43. S. U. Pickering, *J. Soc. Chem. Ind.* **13**, 129 (1910).
44. A. U. M. Schlaepfer, *J. Chem. Soc.* (London) **113**, 522 (1918).
45. E. J. W. Verwey, "Electric Double Layer and Stability of Emulsions," *Trans. Faraday Soc.* **36**, 192 (1940).
46. J. H. Schulman and E. G. Cockbain, *Trans. Faraday Soc.* **36**, 661 (1940).
47. W. Albers and J. Th. G. Overbeek, *J. Colloid Sci.* **14**, 501, 510 (1959).
48. R. E. Ford and C. G. L. Furmidge, *J. Colloid Interface Sci.* **22**, 331 (1966).
49. R. T. Briggs and H. F. Schmidt, *J. Phys. Chem.* **19**, 491 (1915).
50. P. Sherman, "Rheology of Emulsions," in P. Sherman (Ed.), "Emulsion Science," Chap. 3, Academic Press, Inc., New York, 1968.
51. F. S. Palmer and R. W. Morrow, *Soap Sanit. Chem.* **28**, 191 (December 1952).

17

GENERAL
PROPERTIES
OF FOAMS

A foam consists of a coarse dispersion of a gas in a liquid in which the volume of the gas is considerably larger than that of the liquid. The gas bubbles are separated by thin liquid films called lamellae. The foam in a washer or the foam that builds up while shampooing the hair results from mixing air with an aqueous detergent solution. In aerosol foams, the dispersed gas bubbles consist of vaporized propellant instead of air. One of the most common examples of an aerosol foam is shaving lather.

Aerosol foams are unique in that they are self-generating. An aerosol foam product is normally packaged in the container as an emulsion, with the liquefied gas propellant dispersed as droplets throughout the aqueous phase. When the product is discharged, the propellant vaporizes into a gas that is trapped by the aqueous solution. This forms a foam, as illustrated in Figure 17-1.

Since an aerosol foam is formed from an emulsion, the foam will inherit many of the properties of the original emulsion. The type of interfacial film around the dispersed propellant droplets in the emulsion, for example, influences the structure of the interfacial film around the vaporized propellant bubbles. The bubble size of the foam is determined in part by the size of the dispersed propellant droplets in the original emulsion. Generally, if an aerosol oil-in-water emulsion is fairly stable, the foam will also be stable.

There are many types of foams, and they can be formulated to possess a wide variety of properties. This is a considerable advantage in the development of

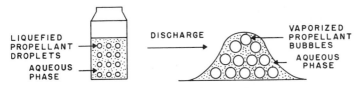

Figure 17-1 Formation of a foam from an aerosol emulsion.

products for different applications. Some foams are stable and show little wetting after many hours, others wet immediately and collapse after a few minutes, and some wet immediately but do not collapse for hours. All of these properties are manifestations of the stability of the foam.

FOAM STABILITY

Foams are thermodynamically unstable systems just like emulsions. The total surface area in a foam is fairly large. Therefore, there is a considerable amount of surface energy present. When the liquid films separating the gas bubbles in a foam break and the bubbles coalesce, the liquid films form droplets and the resulting surface area of the droplets is less than that of the original system. Since this process decreases the surface area, it also decreases the surface energy. As a consequence, film rupture and bubble coalescence are spontaneous processes. Foams from pure liquids are so unstable that they seldom have a lifetime longer than 1 sec. Nevertheless, like emulsions, foams can be stabilized by the addition of surface-active agents so that they last for considerable periods of time.

There are various definitions of foam stability just as there are for emulsion stability.[1,2] In the present discussion, a foam is considered to be stable for a given period only if there is no significant change in any of its properties during that time.

There are three main factors associated with foam stability: (1) drainage; (2) rupture of the liquid films followed by coalescence; and (3) change in bubble size.

Drainage causes the liquid films separating the gas bubbles to become thinner. This usually leads to film rupture. Rupture of the liquid films (lamellae) separating the bubbles leads to coalescence of the bubbles and complete collapse of the foam structure. Change in bubble size can lead to thinning of the lamellae and may cause mechanical shocks that result in film rupture.

Drainage

Drainage of a liquid in a foam is somewhat similar to the downward creaming in an emulsion except that in the foam the continuous liquid phase and not the dispersed gas phase settles to the bottom. This is illustrated in Figure 17-2.

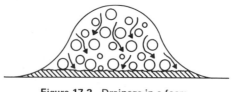

Figure 17-2 Drainage in a foam.

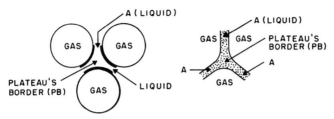

Figure 17-3   Location of a Plateau border in a foam.

Drainage in a foam occurs for two reasons: gravitation and capillary action resulting from surface tension. The drainage resulting from capillary action occurs at a location in the foam known as Plateau's border. The junction between three gas bubbles in a foam and the location of Plateau's border are illustrated in Figure 17-3.

Drainage occurs at the Plateau border because liquid flows from the lamellae at location A toward Plateau's border (PB). The reason for this is a difference in pressure at locations A and PB. The pressure depends upon the curvature of the interface between the liquid and the gas. At location A, the curvature is essentially zero, so that the pressure in the liquid is about the same as that in the gas. However, at location PB, the pressure is lower in the liquid than in the gas because of the negative curvature toward the gas. Therefore, the pressure in the liquid at PB is lower than that at A and liquid flows from A to PB.[1,2]

Rupture of the Liquid Films and Bubble Coalescence

Drainage of liquid weakens the foam structure because the liquid films separating the bubbles become thinner. This allows the gas bubbles to come closer together and may proceed to the point where the liquid films rupture and the gas bubbles coalesce. Davies and Rideal[2] reported that the mean thickness of the liquid films in a representative series of foams varied from about 3,300 to 30,000 Å, as calculated from air permeability data. The foam structure may remain fairly stable during drainage until the film thickness decreases to about 200 to 2,000 Å. At this point, the stability of the foam depends upon whether further thinning occurs or whether the foam is subjected to mechanical shocks. If further thinning reduces the thickness of the liquid films to about 50 to 100 Å, then molecular attractive forces cause the films to rupture. This is followed by bubble coalescence and the entire foam structure collapses. Bubble coalescence can also result from mechanical shocks, such as rubbing the foam between the hands.

Change in Bubble Size Distribution

Another factor that decreases the stability of foams is the change in size of the bubbles as the foam ages. In any foam, the small bubbles become smaller and the large bubbles grow larger. This occurs because the pressure in a small bubble is higher than that in a large bubble. This differential in pressure causes the molecules of the gas in the small bubble to diffuse through the liquid film and into the larger bubble.

The pressure in a gas bubble in a foam is greater than atmospheric pressure (P_a) by a factor, ΔP. The total pressure in the bubble (P_t) is

$$P_t = P_a + \Delta P \tag{17-1}$$

The magnitude of the factor ΔP is determined by the surface tension v and the radius of the bubble R:

$$\Delta P = \frac{2v}{R} \tag{17-2}$$

The total pressure in the gas bubble is

$$P_t = P_a + \frac{2v}{R} \tag{17-3}$$

Therefore, the pressure in a small bubble is higher than that in a large bubble because the radius of the former is smaller. As the small bubble becomes even smaller, its radius decreases further and the pressure increases; the opposite is true of larger bubbles. Therefore, the difference in pressure between the two bubbles increases until the smaller bubble disappears completely. Davies and Rideal[2] cite the example of a foam prepared from a solution of "Teepol" (a surface-active agent), where the total number of bubbles after 15 min was only 10% of the original number, although there had been no rupture of the liquid films. It has been suggested that the continual change in bubble size and the resulting rearrangement of the bubbles in the foam could lead to an increased possibility of mechanical shock followed by film rupture and coalescence.

Some of the factors that affect the rate at which the bubble size in a foam changes are the magnitude of the pressure difference between the bubbles, the permeability of the interfacial film to the gas, and the solubility of the gas in the liquid phase.

STABILIZATION OF FOAMS

Foams are inherently unstable, but they can be stabilized by the addition of surface-active agents. In order to stabilize a foam, it is necessary to prevent drainage, coalescence of the bubbles, and the change in bubble size. The major role of the surface-active agents is to increase the surface and bulk viscosities of the system, which reduces drainage, and to form a strong interfacial film around the bubbles, which retards coalescence if the bubbles do come into contact. A strong interfacial film may have a low permeability to gas molecules, and thus can decrease the rate of diffusion of the molecules from the smaller bubbles to the larger bubbles.

Electrical repulsion plays a comparatively minor part in the stabilization of foams. It does not stop drainage of the initial foam structure, but may prevent further thinning of the liquid film once the film thickness has decreased sufficiently so that a repulsion develops between the electric double layers.

The Electric Double Layers

Anionic and cationic surface-active agents are adsorbed at the gas-liquid interfaces in a foam structure in the same way that they are adsorbed at an oil-water interface in an emulsion system. This results in the formation of electric double layers similar to those postulated for emulsion systems[24] (Figure 17-4).

It can be seen from Figure 17-4 that there are two electric double layers in a single liquid film. As the liquid film thins as a result of drainage, the two opposing electric double layers come closer together, so that ultimately there is a repulsion between them. This is analogous to the situation in an emulsion when two electrically charged oil droplets approach each other. The energy barrier that develops in the film opposes further thinning. This can confer some stability to the foam structure, and it can also provide protection against shocks. This type of system, where stabilization does not occur until after an initial period of drainage and film thinning has taken place, is referred to as a pseudostable system.[2]

In the discussion on emulsions, it was pointed out that since the polar heads of the anionic surface-active agents have the same charge, they will repel each other when they are located on the same droplet. This self-repulsion results in a weak interfacial film because it reduces the concentration of the surface-active agent at the interface. The same phenomenon undoubtedly occurs at the gas-liquid interfaces in a foam, which is one reason why many surface-active agents by themselves produce relatively poor foams.

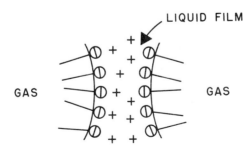

Figure 17-4 Electric double layers at gas–water interfaces.

Viscosity

Since drainage in a foam results from the flow of liquid between the gas bubbles, increasing the viscosity of the system decreases the rate of drainage. This has been demonstrated by a number of investigators. Miles, Shedlovsky, and Ross[3] studied the flow of liquid through various foams and found that the rate of flow decreased when the surface viscosity of the system increased. For example, the viscosity of a sodium lauryl sulfate solution was low and the rate of flow of liquids through the foam was high. When lauryl alcohol was added, the reverse situation occurred—the rate of flow of liquid through the foam was low and the viscosity of the system was high.

Subsequently, Miles and his co-workers[4] studied the drainage rates of aqueous films from glass frames that were suspended in a vertical position. The film from aqueous sodium lauryl sulfate drained rapidly, but the addition of lauryl alcohol resulted in films that drained slowly. This difference in the drainage rates was attributed to the increase in viscosity that resulted when lauryl alcohol was added.

In addition, it was noted that when the temperature of the slow-draining films from sodium lauryl sulfate–lauryl alcohol systems was raised, the films became fast-draining at a specific higher temperature. The temperature at which this occurred was called the film drainage transition temperature (FDTT). The change from slow to fast drainage took place because the sodium lauryl sulfate–lauryl alcohol complex, which was responsible for the higher viscosity of the system, dissociated at the higher temperature.

Epstein, Ross, and Jakob[5] devised an apparatus for studying the drainage transition temperature of foams and reported that the transition temperatures for foams from an aqueous sodium lauryl sulfate–lauryl alcohol system agreed with the drainage transition temperature for films from the same system.

In a further study of the sodium lauryl sulfate–lauryl alcohol system, Epstein and his co-workers came to the conclusion that the high surface viscosity and slow film drainage were due to the presence of a crystalline molecular complex formed from two molecules of sodium lauryl sulfate and one molecule of lauryl alcohol.[6]

Brown, Thuman, and McBain devised a sensitive rotational viscometer in 1953 and reported that foams with the highest stability were produced from solutions with appreciable surface viscosity.[7]

It seems likely that the increase in surface viscosity that occurs with the formation of molecular complexes, such as that from sodium lauryl sulfate and alcohol, is due to hydration of the interfacial film of the molecular complex. The effect of hydration and the formation of layers of "soft ice" upon the stability of emulsions is discussed in Chapter 18. This type of structure would increase the viscosity of the system, particularly if it extended into the bulk phase to any degree.

The Interfacial Film

A strong interfacial film serves at least two functions in a foam. There is evidence that the interfacial film formed from molecular complexes is less permeable to the passage of gases. This would decrease the diffusion of gases from the smaller bubbles to the larger bubbles and thus increase the stability of the foam. The low permeability of the interfacial film to the passage of gases is considered to be due to the presence of thin layers of soft ice formed by hydration.[2]

Finally, a strong condensed interfacial film would retard coalescence of gas bubbles when they came into contact. This would be quite important when the liquid films in the initial foam were very thin because of the high concentration of the gas phase relative to the liquid phase.

EVALUATION OF AEROSOL FOAMS

The ultimate test of any product is whether or not it is suitable for the application for which it was developed. If the product has been formulated as an aerosol shaving lather, for example, then the final test is how the product performs in shaving tests.

Usually, a considerable amount of exploratory investigation is necessary before a satisfactory product is developed. In order to arrive at this objective as rapidly as possible, it is necessary to know how certain variables and additives affect the properties of foams. Therefore, it is desirable to (1) characterize a foam as precisely as possible in order to have some basis for comparison with other foams and (2) have a permanent record of the properties of the foam. Properties of foams, such as density, viscosity, and stability, are very useful for this purpose. Not only do these properties serve to characterize a foam at a given time, but they are also useful as a means of determining to what extent a foam changes during aging (foam stability).

General Observations

A considerable amount of information can be obtained about an aerosol foam merely by discharging it onto a paper towel or onto the hand and observing it for a few minutes. The extent to which a foam wets a paper towel, retains its shape, or peaks when a glass rod or pencil is placed in it and then raised vertically all give an indication of foam stability and viscosity. The way the foam feels in the hand, the ease with which it spreads, and the degree to which it collapses when spread or rubbed are further indications of such properties as stability and density.

Bubble Size

Microscopic examination of foams can be very informative, and photomicrographs of foams can be quite useful, since they can be examined at leisure. Most professional photographers can suggest equipment suitable for taking photo-micrographs of foams.

At the Freon® Products Laboratory,[8] observations of foam structures are made with a Bausch and Lomb Sterozoom microscope, Model BVB-73, equipped with a 10X paired widefield eyepiece and a power pod magnification of 3X. Approximate bubble sizes of foams are determined using a micrometer disk (#31-16-08) which measures intervals of 0.001 in.

Photomicrographs of foams are taken at 30X magnification using a Spencer triocular single-stage microscope manufactured by the American Optical Company. It is equipped with a 15X eyepiece and a 3.5X objective lens. The camera is focused through 10X eyepieces with 3.5X objectives. The pictures are taken with an MP-3 Polaroid Multipurpose Industrial View camera with a 4×5 film adapter using surface illumination and a 1-sec exposure. Pictures are normally taken 20 sec after the foam has been discharged.

Density

The density of an aerosol foam can be obtained by filling a container of known volume with the foam and determining its weight. A large enough vessel should be used so that errors due to air entrapment in the foam are minimized. A crystallizing dish with a volume of 350 cc is used at the Freon® Products Laboratory.

In order to facilitate the discharge of the foam into the dish, a short piece of Tygon tubing about 3 in. long is attached to the foam actuator. The foam is layered carefully into the dish, keeping the end of the tubing slightly above the surface of the foam at all times. Enough foam is added so that its surface extends above the dish. After the dish is filled, an additional 1–2 min are allowed for any residual expansion. The excess foam is then removed with a spatula and the dish is weighed.

Several equations have been developed for calculating the density of an aerosol foam. Spitzer, Reich, and Fine[9] have published the following equation:

$$D = \frac{100M}{24,000X + 100M} \qquad (17\text{-}4)$$

where D is the density of the foam (g/cc), M is the molecular weight of the propellant, and X is the amount of propellant in the aerosol (wt.%).

Gorman and Hall[10] have reported the following equation for predicting foam density:

$$d_f = \frac{100}{\%PV_r} \qquad (17\text{-}5)$$

where d_f is the density of the foam (g/cc), $\%P$ is the amount of propellant in the initial emulsion (wt.%), and V_r is the volume occupied by 1 g of propellant vapor. The authors report fairly good agreement between published experimental values of various foams and calculated values.

The following equation was developed by Becher:[11]

$$D = \frac{W_a - W_p}{(22{,}400\,W_p/M) + V_a} \tag{17-6}$$

where W_a is the weight of the aqueous phase, W_p is the weight of the propellant, M is the molecular weight of the propellant, and V_a is the volume of the aqueous phase. Becher determined foam density by discharging approximately 50 cc of foam and drawing it up into a tared 10-cc hypodermic syringe (without needle). The syringe was then weighed. A second determination on the same sample was carried out by expelling half of the foam volume and reweighing the syringe. A large discrepancy between the two densities was assumed to indicate bubble occlusion and the results were rejected.

Stability

Drainage

The comparative rates of drainage of a series of foams can be obtained by discharging the foams into glass funnels located over graduates and measuring the quantities of liquid that collect in the graduates after various times. Since the initial weight of foam in the funnel is known, the percent drainage of the foams after any specified time can be calculated and the foams compared on this basis. If the drainage is determined for several different times, a curve for each foam, illustrating the percent drainage as a function of time, can be drawn. From this curve, the time required to produce any given percent of drainage, for example, 50%, can be obtained. The 50% drainage times can then be used for comparing the foams.

Wetting

Wetting is a consequence of drainage and, therefore, correlates with drainage. Wetting can be judged by discharging the foam onto a paper towel and determining the period of time that elapses before it is visibly wetted. DeNavarre and Lin[12] devised a simple electrical apparatus to measure wetting time. A piece of filter paper is placed over two electrical contacts in a circuit with a galvanometer. The foam is placed on the paper and the flow of current that occurs when the foam wets the paper is registered on the galvanometer. The time required to reach a predetermined galvanometer reading can be used as a basis for comparison with other foams.

Foam Collapse

Another indication of foam stability is the extent to which a foam will collapse during aging. This can be determined qualitatively for a series of foams by discharging them onto paper towels and noting their appearance after specified periods of time.

A somewhat more satisfactory procedure is to discharge the foams onto paper towels and observe them against a background of horizontal lines drawn at $1/4$- or $1/2$-in. intervals. If the height of the foam is noted immediately after discharge, any subsequent decrease or collapse can be determined by referring to the original height of the foam on the horizontal-line background.

Miscellaneous Tests

Any measurement that can be carried out on a foam before and after aging will give an indication of stability. Johnsen[13] uses a test that consists of attaching a light metal rod to a stainless steel disk and measuring the rate of fall of the disk and rod through a foam depth of 3 in. The difference in rate of fall between a fresh foam and one aged for 10 min is reported to give an excellent indication of foam stability.

Stiffness and viscosity measurements on fresh and aged samples of foam can be used as an indication of foam stability.

Overrun

Overrun is a measure of the change in volume that occurs when an emulsion of a liquefied gas propellant is allowed to expand into a foam. Overrun can be calculated as follows:[14]

$$\text{Percent overrun} = \frac{\text{Volume of foam} - \text{volume of liquid}}{\text{Volume of liquid}} \times 100 \qquad (17\text{-}7)$$

One method of determining overrun is to attach a piece of glass tubing to the foam actuator with rubber tubing of sufficient length so that the assembly reaches to the bottom of a 1000-cc graduate. The container is slowly discharged with the glass tube at the bottom of the graduate. As the foam builds up in the graduate, the glass tube is raised so that the end of the tube is always slightly above the surface of the rising foam. Overrun can be calculated from the volume of foam obtained and the weight of product discharged. The volume of the emulsion is calculated from its density.

The density of a foam changes during discharge, and the percent overrun is an average value for the quantity of product discharged. Therefore, the percentage of product discharged from the container for an overrun measurement should be specified.

Viscosity

The measurement of viscosity is an excellent way of characterizing foams. Carter and Truax[15] studied the effect of a number of variables upon the properties of aerosol shaving lathers. Viscosity measurements on the foams were carried out using a Brookfield Model HAT viscometer (20 rpm, T-type spindle A, Helipath stand).

Richman and Shangraw[16] have investigated the effect of variables upon the rheological properties of aerosol foams in considerable detail. Their articles give an excellent review of the historical background of the rheology of foams and the types of equipment that have been used for viscosity measurements. Richman and Shangraw determined viscosity with a Haake Rotovisco Viscometer (Gebruder Haake K.B., Berlin). They reported that the Rotovisco is a very versatile instrument that can measure apparent viscosities in the range of $5 \cdot 10^{-3} - 4 \cdot 10^{7}$ poise.

The resistance of a foam to penetration or deformation is a function of viscosity and can be used to characterize foams. In the Freon® Products Laboratory, the relative stiffness of aerosol foams is determined with a Curd Tension meter.[17] A crystallizing dish is filled with foam and placed on a scale platform located directly underneath a specially designed curd knife. The curd knife is then driven downward at a constant, controlled rate of speed by a synchronous motor. After the knife has contacted the foam, the resistance of the foam to penetration by the knife causes the scale platform to depress. The extent of the resistance of the foam is recorded on the scale dial in grams.

This is a relatively simple method for detecting changes in foams that occur during aging or variations that occur as the container is discharged. Stiffness measurements are also suitable for detecting changes in foam structure that occur with modifications in the aerosol formulation.

REFERENCES

1. J. J. Bikerman, "Foams: Theory and Industrial Applications," Reinhold Publishing Corp., New York, 1953.
2. J. T. Davies and E. K. Rideal, "Interfacial Phenomena," 2nd ed., Academic Press, Inc., New York and London, 1963.
3. G. D. Miles, L. Shedlovsky, and J. Ross, *J. Phys. Chem.* **49**, 93 (1945).
4. G. D. Miles, J. Ross, and L. Shedlovsky, *J. Amer. Oil Chem. Soc.* **27**, 268 (1950).
5. M. B. Epstein, J. Ross, and C. W. Jakob, *J. Colloid Sci.* **9**, 50 (1954).
6. M. B. Epstein, A. Wilson, C. W. Jakob, L. E. Conroy, and J. Ross, *J. Phys. Chem.* **58**, 860 (1954).
7. A. G. Brown, W. C. Thuman, and J. W. McBain, *J. Colloid Sci.* **8**, 491 (1953).
8. P. A. Sanders, *J. Soc. Cosmet. Chem.* **17**, 801 (1966).
9. J. G. Spitzer, I. Reich, and N. Fine, U.S. Patent 2,655,480 (October 1953).
10. W. G. Gorman and G. D. Hall, *Soap Chem. Spec.* **40**, 213 (May 1964).
11. P. Becher, Personal Communication.
12. M. G. DeNavarre and T. J. Lin, *Amer. Perfum. Cosmet.* **78**, 36 (July 1963).
13. M. A. Johnsen, "Laboratory Techniques," in H. R. Shepherd (Ed.), "Aerosols: Science and Technology," Interscience Publishers, Inc., New York (1961).
14. F. T. Reed, *Proc. 39th Ann. Meet. Chem. Spec. Manuf. Assoc.*, 30 (December 1952).
15. P. Carter and H. M. Traux, *Proc. Sci. Sec. Toilet Goods Assoc.* No. 35, 36 (1961).
16. M. D. Richman and R. F. Shangraw, *Aerosol Age* **11**, 36, 28 (May/November 1966).
17. P. A. Sanders, *Aerosol Age* **8**, 33 (July 1963).

18

AQUEOUS
AEROSOL
EMULSIONS
AND FOAMS

The use of foams for the application of products has a number of consumer advantages. Foams are usually formulated with small quantities of propellant. This provides a high proportion of concentrate in the container. Foams are easy to use, can be applied directly to the desired area with negligible waste, and are attractive in appearance. In addition, the properties of foams can be varied over a wide range, thus making them suitable for many different applications. Foams which collapse immediately after discharge or are so stable and coherent that they bounce can both be obtained. Some foams develop a sparkling surface as they age, while others make a crackling noise when rubbed.

Most aqueous foams are formulated in the container as oil-in-water emulsions in which propellant droplets, liquefied under pressure, are dispersed throughout the aqueous phase. When the emulsion is discharged, the dispersed liquefied propellant droplets vaporize upon reaching atmospheric pressure, thus producing a foam consisting of propellant gas bubbles dispersed in a continuous surfactant phase. Aerosols formulated as water-in-oil emulsions usually discharge as non-foaming sprays and can be used for products requiring a fairly fine droplet size, for example, room deodorants and space insecticides.

Although foams can be produced from systems other than oil-in-water emulsions, the latter account for the majority of foam products on the market. These products can be subdivided into two groups, depending upon whether they have been designed as sprayable foams or conventional foams. Sprayable foam products are equipped with spray valves having mechanical break-up actuators; these products discharge soft sprays that foam on contact. Conventional aerosol foam products are equipped with foam actuators and usually foam during discharge. These products are normally referred to as aerosol foams, with the understanding that sprayable foams are designated as such.

The relationship between aqueous emulsion systems and the type of products they produce is shown below:

310

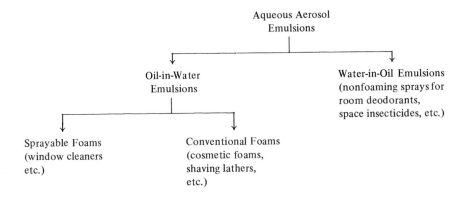

SPRAYABLE FOAMS

Window cleaners, starch sprays, oven cleaners, and bathtub cleaners are typical sprayable foam products. These aerosols are designed to give a soft, light spray that foams slightly on contact. The foam serves two purposes: It holds the product on a vertical surface until it can be wiped off with a cloth or paper towel and it shows where the surface has been sprayed. The foam should be stable enough to achieve the two preceding objectives, but not so stable that it is difficult to break down and remove by wiping.

Surface-active agents determine emulsion and foam stability and are essential components of sprayable foams. Many sprayable foams are designed for cleaning operations and may contain appreciable quantities of other additives, such as ammonia, isopropyl alcohol, etc. The surface-active agent should possess good wetting properties. Those for glass cleaners and similar products should preferably be liquid or solid at room temperature. Sodium dioctylsulfosuccinate is a commonly used surfactant for window cleaners. Waxy surfactants tend to leave films on glass that are difficult to remove. Where this property is not objectionable, surfactants such as the polyoxyethylene sorbitan esters, polyoxyethylene fatty ethers, and alkyl phenoxy polyethoxyethanols have proven to be effective.[1]

The low foam stability required usually necessitates surfactant concentrations of 1 wt.% or less. The low concentration may create a problem of emulsion stability, but this can often be overcome by combinations of emulsifiers or by the use of propellant blends with a density close to that of the aqueous phase. The latter minimizes creaming of the propellant. The calculation of the composition of blends with a specific density is described in Chapter 12.

Fluorocarbons, hydrocarbons, blends of the two, and compressed gases are utilized as propellants for sprayable foams. Hydrocarbons, such as isobutane, are used at concentrations of about 3.5–4.0 wt.%, while fluorocarbon blends are employed at higher concentrations of about 8–10 wt.%. Fluorocarbon propellants normally consist of blends of fluorocarbons 12 and 114 because these propellants are stable in aqueous systems. Fluorocarbons 124, 142b, and 152a could also be used. Fluorocarbon 152a would require the presence of a vapor pressure depressant such as isopropyl myristate, mineral oil, isopropyl alcohol, etc.

Fluorocarbon 11 is not satisfactory for use in oil-in-water emulsions in metal containers because it decomposes in the presence of water and metal.[2]

CONVENTIONAL AEROSOL FOAMS

These products constitute the largest group of aerosol foams on the market. All are equipped with foam actuators and either discharge as foams or as liquids that subsequently expand into foams. There are many cosmetic, pharmaceutical, and industrial products formulated to produce foams. Aerosol foams normally consist of an aqueous phase containing a surfactant or combination of surfactants, active ingredients, other additives (such as emollients, perfumes, etc.), and propellant.

Surface-Active Agents

Aerosol emulsions and foams with adequate stability for aerosol applications can be obtained with a wide variety of anionic, nonionic, and cationic surfactants. In some instances, a single agent is sufficient. Generally, however, a combination of surfactants or a combination of surfactants with other oil-soluble materials (for example, fatty alcohols or acids) has proved to be more effective than a single agent. Most surfactant combinations involve anionic surface-active agents. However, the use of nonionic agents has been increasing, particularly in cosmetic and pharmaceutical products. Cationic agents such as the quaternary ammonium compounds are generally used for their disinfectant or textile-softening properties rather than their emulsifying or foaming characteristics. However, some excellent foams can be obtained with certain cationic surfactants.[3]

Recent evidence indicates that in order for a surfactant to produce stable emulsions and foams, it should be insoluble in both the aqueous and organic phases. However, it must be wetted by the two phases so that it will collect at the interface between the dispersed propellant and the aqueous phase. The data upon which this conclusion is based are discussed in detail in a later section of this chapter. Surfactants soluble in the aqueous phase give unstable emulsions and foams. Those soluble in the organic phase generally produce water-in-oil emulsions or unstable oil-in-water emulsions.

A certain amount of experimental work is required to develop a satisfactory emulsion system for a specific product. However, there are several publications that can be of assistance. Reference 3 gives the results of an evaluation of over

400 surface-active agents in a simple aerosol system consisting of 86 wt.% water, 4 wt.% agent, and 10 wt.% fluorocarbon 12/fluorocarbon 114 (40/60) propellant. The properties of emulsions and foams obtained with these agents are tabulated and provide a starting point for formulating foam products. The properties of aerosol emulsions and foams prepared with triethanolamine salts of fatty acids are discussed in Reference 4, while those of systems formulated with nonionic polyoxyethylene fatty ethers are listed in Reference 5. In addition, there are a number of articles in the literature that disclose the types of surfactants used for various products.[6] The Spitzer, Reich, and Fine shaving lather patent[7] discloses a wide variety of surfactants useful in producing aerosol foams. Many reference books, both aerosol and nonaerosol, disclose the types of surface-active agents used for a variety of products.[6-13] McCutcheon's list of synthetic detergents and their uses, properties, structure, and source is useful for the formulation of emulsion products.[14]

Anionic Agents

Fatty Acid Salts. Salts of the fatty acids are the most commonly used anionic surfactants and are employed for a variety of products. These agents produce excellent emulsions and foams, are low in toxicity, and are essentially noncorrosive in metal containers. Although some products are prepared with sodium or potassium salts, the majority are formulated with triethanolamine salts. The salts are prepared by neutralizing fatty acids with the appropriate base. Most commercially available fatty acids are obtained from natural-occurring fats and oils and are composed of mixtures of fatty acids, regardless of how they are named. Triple-pressed stearic acid, commonly used in aerosols, has a composition of about 45 wt.% stearic acid and 55 wt.% palmitic acid. This mixture is obtained from tallow. The triple-pressed designation refers to the pressing operations used to separate the solid mixture of stearic and palmitic acids from the liquid red oil, which is primarily oleic acid.[15] Coconut fatty acids, derived from coconut oil, are also commonly employed in aerosol formulations. The composition of a typical commercial coconut oil is as follows:[16] 4 wt.% capric acid; 55 wt.% lauric acid; 19 wt.% myristic acid; 7 wt.% palmitic acid; 4 wt.% stearic acid; 8 wt.% oleic acid; and 3 wt.% linoleic acid.

Most aerosol formulations contain mixtures of commercial fatty acids. These mixtures are used because experience has shown that they give products with more desirable properties than those obtained with any one fatty acid alone. Most aerosol shaving lathers, for example, are formulated with a mixture of the triethanolamine salts of triple-pressed stearic acid and coconut fatty acid in which the ratio of stearic acid to coconut fatty acid ranges from about 90/10 to 70/30.[17] Triethanolamine stearate by itself gives viscous emulsions that have a sputtery, noisy discharge and produce pasty, unattractive foams. The triethanolamine salts of coconut fatty acids produce attractive foams, but coconut oil can be irritating to a sensitive skin. Therefore, it is used in shaving lathers only to such an extent that skin irritation is not a problem.[18] The combination of triple-pressed stearic acid and coconut fatty acid gives excellent foams with a minimum of skin irritation. A number of formulations for aerosol shaving lathers are given in References 7 and 17.

Other Anionic Agents.   Other anionic surface-active agents that produce aerosol foams are listed in Reference 7. These include aryl-alkyl sulfonates and alkyl sulfates, used from time to time as components of shampoos. In some cases, considerable container corrosion resulted with these compounds.[19]

Self-emulsifying esters of the fatty acids, for example, glycerol monostearate, are often used in aerosol formulations.[8] The esters themselves are nonionic, but the self-emulsifying products contain a low concentration of an anionic agent, that is, potassium stearate, which promotes dispersion of the nonionic ester. The self-emulsifying esters, therefore, are mixtures of nonionic and anionic agents. Some of these esters give excellent emulsions and foams.

Nonionic Agents

Nonionic surfactants are used to a considerable extent in aerosol cosmetic products. Courtney[20] has suggested an aerosol shaving lather formulation based on nonionic surfactants. Some advantages of nonionic surfactants are their generally low toxicity, noncorrosive properties, and effectiveness in the presence of electrolytes. In addition, many nonionic surfactants such as polyoxyethylene ethers are stable in either acidic or alkaline systems.

Nonionic polyoxyethylene (POE) fatty ethers have been investigated in detail in aerosol systems. The advantage of studying this class of compounds is that a series of the ethers, whose members range from those that are predominantly lipophilic (low HLB numbers) to those that are hydrophilic (high HLB numbers), is now commercially available.[21] These products are prepared by condensing varying proportions of ethylene oxide with fatty alcohols. The number of moles of ethylene oxide reacted with 1 mole of the fatty alcohol is indicated in parentheses in the name of the surfactant. For example, polyoxyethylene (4) lauryl ether indicates that the agent was prepared by reacting 4 moles of ethylene oxide with 1 mole of lauryl alcohol. The compounds become more hydrophilic as the number of ethylene oxide units in the molecule increases.

Evaluation of a series of eight polyoxyethylene fatty ethers with HLB numbers ranging from 4.9 to 16.9 showed that only two of the eight compounds were effective as emulsifying and foaming agents. These were POE (2) cetyl ether and POE (2) stearyl ether. These surfactants have the lowest HLB numbers—5.3 and 4.9—and are the only two ethers that are insoluble in water.

Other nonionic surfactants used in aerosol products include polyoxyethylene sorbitan esters, polyoxyethylene fatty ethers, alkyl phenoxy polyethoxy ethanols, fatty acid esters, and the alkanolamides.[1]

Molecular Complexes

The concept of molecular complexes formed between water-soluble surface-active agents and long-chain polar compounds, and the way in which they can alter the properties of conventional, nonaerosol colloidal systems, was discussed in Chapters 16 and 17. One of the most common examples of a molecular complex is that formed from sodium lauryl sulfate and lauryl alcohol.

In recent years, the investigation of molecular complexes has been extended to aerosol systems. Certain combinations of water-soluble or water-dispersible ionic surfactants and long-chain alcohols, known from previous investigations to form molecular complexes in nonaerosol systems, also had a marked influence on the properties of aerosol emulsions and foams: Emulsion stability, foam stability, and viscosity were increased, and the rate of foam drainage was decreased. These effects were analogous to those observed previously in nonaerosol systems. On this basis, it was concluded that complexes between the water-soluble and oil-soluble components were present in the interfacial films of aerosol emulsions and foams.

There is no direct experimental evidence to indicate that stoichiometric molecular complexes are formed between nonionic polyoxyethylene fatty ethers and fatty alcohols or acids in nonaerosol systems. However, in aerosol systems, the effect of combinations of polyoxyethylene fatty ethers and fatty alcohols or acids upon emulsion and foam properties was similar to that produced by molecular complexes in the ionic surfactant systems. Judging by this, it seemed reasonable to assume that complexes between the polyoxyethylene fatty esters and fatty alcohols or acids were present in the interfacial films in aerosol emulsions and foams. Whether the composition of these complexes in the interfacial films is stoichiometric or indefinite remains unknown.

The use of molecular complexes provides a powerful method for controlling and altering the properties of aerosol systems so that they can be tailored to meet the requirements of a variety of applications. Thus, by the proper choice of the surface-active agent and long-chain alcohol or acid, aerosol foams which have many different properties can be obtained. In addition, because complexes often are so much more effective than the surface-active agent by itself, their use may permit a marked reduction in the concentration of surface-active agent without any decrease in the quality of the product.

The results of molecular complex studies in aerosol systems have been reported in a series of papers.[4, 5, 22, 23] The material presented in this chapter is essentially a condensation of the data in these papers. Unless otherwise specified, the aerosols had a composition of 90 wt.% aqueous phase and 10 wt.% fluorocarbon 12/fluorocarbon 114 (40/60) propellant.

Triethanolamine Fatty Acid Salt–Fatty Alcohol Complexes. The investigation of complex formation between triethanolamine salts of the fatty acids and long-chain alcohols is reported in Reference 4. Triethanolamine salts were prepared from lauric, myristic, palmitic, and stearic acid. Each of the salts was evaluated by itself and in combination with lauryl, myristyl, cetyl, stearyl, and oleyl alcohol, and also cholesterol. Cholesterol was included in the series because it had been shown by Schulman and Cockbain to form complexes with the sodium alkyl sulfates.[24] Some general comments are listed below:

1. Molecular complexes usually had an effect on the properties of both the aerosol emulsions and the corresponding foams.
2. Microscopic examination of the foams showed that complexes usually decreased the bubble size of the foam and increased the uniformity of the bubble size.
3. The effect of the complexes in decreasing foam drainage was quite marked. This was consistent with the results reported previously in nonaerosol systems. The decrease in foam drainage results from an increase in viscosity of the aqueous phase.
4. The extent to which complex formation affected the properties of the aerosol systems depended upon such factors as the type and concentration of the long-chain alcohol and of the triethanolamine salt. Specific examples of the effect of these variables are given in the following sections.

VARIATION IN ALCOHOLS   The effectiveness of any specific alcohol was a function of the particular triethanolamine salt with which it was complexed. It was a rather general rule throughout the triethanolamine salt–fatty alcohol series that straight-chain alcohols with about the same chain length as that of the fatty acid salt usually had the most noticeable effect upon emulsion and foam properties. Thus, lauryl and myristyl alcohol were more effective than stearyl alcohol in the triethanolamine laurate system, while the reverse was true in the triethanolamine stearate system. Oleyl alcohol generally had a much slighter effect than the straight-chain alcohols, indicating weak complex formation. This is due to the steric hindrance caused by the double bond in oleyl alcohol.

The effect of various alcohols in the triethanolamine laurate system is illustrated in Table 18-1. Similar data for the triethanolamine myristate, palmitate, and stearate systems are given in Reference 4.

The change in emulsion stability and foam drainage upon addition of the alcohols was very pronounced. Foam drainage results correlated well with wetting properties, and this was generally true in the other systems. Foams from triethanolamine laurate alone or those with stearyl alcohol present had the highest drainage rates and wetted paper almost immediately after discharge. Foams with lauryl, myristyl, or cetyl alcohol did not wet for over an hour.

Foam stability, as judged by foam collapse, was increased by the presence of the alcohols. The foams from triethanolamine laurate alone or those con-

TABLE 18-1   TRIETHANOLAMINE LAURATE[a] SYSTEM—
EFFECT OF ALCOHOLS

| Alcohol[b] | Emulsion Stability[c] | Foam Properties | |
		% Drainage (30 min)	Stiffness (g)
None	< 1 min	82	8
Lauryl	> 5 hr	2	20
Myristyl	> 5 hr	0	33
Cetyl	> 5 hr	0	18
Stearyl	> 5 hr	67	10
Oleyl	15–30 min	2	12

[a] Triethanolamine laurate concentration = 0.10 M.
[b] Soap–alcohol ratio (molar) = 1:1.
[c] Time to observable phase separation after the bottles had been shaken.

taining stearyl alcohol started to collapse within 30 min after discharge. Foams with lauryl, myristyl, or cetyl alcohol retained their structure for over an hour.

The presence of complexes usually resulted in a smaller bubble size in the foam. This is illustrated in Figure 18-1 by photomicrographs of foams from triethanolamine stearate and the triethanolamine stearate–stearyl alcohol complex.

Cholesterol is unique in some respects. There seems to be sufficient evidence that it forms complexes, yet its effect upon some properties is opposite that of the straight-chain alcohols. An increase in emulsion stability and a decrease in the foam drainage occurred upon addition of cholesterol to the triethanolamine laurate and myristate systems. In contrast to this, a decrease in foam stiffness and an increase in foam bubble size accompanied the change in emulsion stability and drainage. However, these apparently anomalous effects may be explained by the nature of the interfacial films formed with the various triethanolamine salt–alcohol complexes. Schulman and Cockbain[24] showed that the interfacial film formed by sodium cetyl sulfate–cholesterol complexes was very fluid in contrast to the solid, condensed films from the sodium cetyl sulfate–alcohol complexes. If the triethanolamine salt–cholesterol interfacial film is also fluid and plastic, this could account for both the increased bubble size and the lower foam stiffness. It seems likely that the interfacial films formed by the triethanolamine salt–straight-chain alcohol complexes are the solid, condensed type.

VARIATION IN ALCOHOL CONCENTRATION   The relationship between the concentration of lauryl alcohol in the triethanolamine laurate system and the change in emulsion and foam properties is shown in Table 18-2.

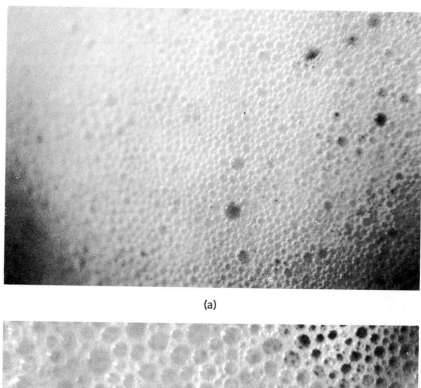

(a)

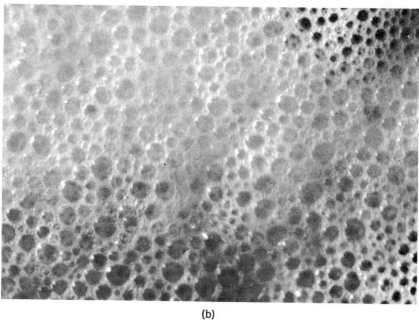

(b)

Figure 18-1 (a) Photomicrograph of a triethanolamine stearate–stearyl alcohol foam. (b) Photomicrograph of a triethanolamine stearate foam. (*Courtesy of the J. Soc. Cosmet. Chem.*)[4]

TABLE 18-2   TRIETHANOLAMINE LAURATE SYSTEM—EFFECT OF
LAURYL ALCOHOL CONCENTRATION

Laurate[a]–Lauryl Alcohol Ratio (Molar)	Emulsion Stability	Foam Properties	
		% Drainage (60 min)	Stiffness (g)
1:0	1–5 min	86.0	8
1:¼	30–60 min	83.0	10
1:½	> 60 min	65.0	16
1:¾	> 16 hr	40.0	18
1:1	> 16 hr	11.0	22

[a] Triethanolamine laurate concentration = 0.10 M.

Although low concentrations of lauryl alcohol affected the properties, maximum efficiency occurred at a 1:1 mole ratio of triethanolamine salt to lauryl alcohol. In other triethanolamine salt systems, the maximum appeared at a 1:½ mole ratio.[4]

Triethanolamine Salt–Fatty Acid Complexes.   An aqueous solution of a triethanolamine salt of a fatty acid contains not only the salt, but also free fatty acid and free base. The latter two components result from hydrolysis of the salt. For example, in an aqueous system containing sodium stearate, free stearic acid and sodium hydroxide are also present. Likewise, in an aqueous system containing triethanolamine myristate, free myristic acid and triethanolamine are present. This effect of hydrolysis is important with regard to molecular complexes because free fatty acids can complex with their salts. One of the best known examples of this type of complex is the sodium stearate–stearic acid complex discussed in Chapter 16.

Complex formation between triethanolamine salts and their corresponding free acids also occurs in aerosol systems. This was indicated from a comparison of the properties of triethanolamine salt systems with two different fatty acid/triethanolamine ratios. One had an excess of triethanolamine, which would minimize hydrolysis. This would decrease complex formation because of the decreased concentration of free fatty acid. The other had an excess of fatty acid with respect to the triethanolamine. This would promote complex formation.

The data obtained with three systems—triethanolamine laurate, triethanolamine myristate, and triethanolamine palmitate—are given in Table 18-3. These data indicate that complex formation between the triethanolamine salt and free fatty acid occurred in the triethanolamine myristate and palmitate systems. The presence of excess lauric acid had little effect in triethanolamine laurate. This is probably a result of the complex being too water-soluble.

TABLE 18-3   EFFECT OF FATTY ACID/TRIETHANOLAMINE RATIO

Acid	Acid–Base Ratio (Molar)	Emulsion Stability	Foam Properties	
			% Drainage (60 min)	Stiffness (g)
Lauric	1:1½[a]	< 1 min	85	9
	1½:1[b]	< 1 min	86	10
Myristic	1:1½[a]	< 1 min	71	21
	1½:1[b]	> 1 hr	5	37
Palmitic	1:1½[a]	15–30 min	0	23
	1½:1[b]	> 1 hr	0	42

[a] 1:1½ ratio = 0.10-M acid/0.15-M base.
[b] 1½:1 ratio = 0.15-M acid/0.10-M base.

The change in foam properties that results with a variation in the acid/base ratio (that is, complex formation) is important in the formulation of foam products such as shaving lathers and shampoos. A system with the strongest complex might not necessarily be the most desirable, since the foam might not have sufficient wetting properties.

Sodium Lauryl Sulfate-Fatty Alcohol Complexes. Sodium lauryl sulfate was included in the investigation of complexes in aerosol systems because the sodium alkyl sulfate-fatty alcohol complexes had been widely studied in nonaerosol systems.

TABLE 18-4   SODIUM LAURYL SULFATE[a] SYSTEM—EFFECT
OF ALCOHOLS

Alcohol[b]	Emulsion Stability	Foam Properties	
		% Drainage (60 min)	Stiffness (g)
None	< 1 min	82	11
Lauryl	> 5 hr	0	40
Myristyl	> 5 hr	0	38
Cetyl	> 5 hr	2	14
Stearyl	< 5 min	71	12
Oleyl	30–60 min	84	10
Cholesterol	1–5 min	86	13

[a] Sodium lauryl sulfate concentration = 0.10 M.
[b] Sodium lauryl sulfate–alcohol ratio (molar) = 1:1.

The change in properties that occurred upon addition of various alcohols to aerosol systems containing sodium lauryl sulfate is shown in Table 18-4. The effect upon emulsion stability, foam drainage, and foam stiffness is particularly noticeable with lauryl and myristyl alcohol, thus indicating fairly strong complex formation.

Microscopic examination of the sodium lauryl sulfate foams showed bubble sizes ranging from about 0.001 to 0.01 in. with lamellae thickness of about 0.001-0.003 in. The foams containing lauryl alcohol had a smaller bubble size and thinner lamellae. Photomicrographs confirmed the difference in bubble size between the two foams (Figure 18-2).

Polyoxyethylene Fatty Ether–Fatty Alcohol Complexes. The investigation of polyoxyethylene fatty ether–fatty alcohol combinations was first reported by Becher and Del Vecchio.[25] They observed that addition of fatty alcohols to polyoxyethylene lauryl ethers resulted in systems which showed film drainage transition phenomena typical of those previously noted in anionic systems. Subsequent investigations in the aerosol field showed that addition of fatty alcohols and fatty acids to systems containing polyoxyethylene fatty ethers often had a considerable effect upon foam properties. Although there was no direct experimental evidence to show that stoichiometric association complexes formed between the relatively complex polyoxyethylene fatty ethers and fatty alcohols, the effect upon foam properties was interpreted as an indication of complex formation between the polyoxyethylene fatty ethers and the long-chain polar compounds in the interfacial film.

The investigation in aerosol systems was carried out using combinations of eight polyoxyethylene fatty ethers with lauryl, myristyl, cetyl, stearyl, and oleyl alcohol.[5]

GENERAL DISCUSSION   Some general comments about the results of the study are given below:

1. The addition of fatty alcohols to the POE fatty ether systems caused a marked change in one or more foam properties. Myristyl and cetyl alcohol generally had the greatest effect.
2. The effect of alcohols was most pronounced with the three polyoxyethylene fatty ethers having intermediate HLB values, that is, POE (4) lauryl ether, POE (10) cetyl ether, and POE (10) stearyl ether. In systems containing POE (2) cetyl ether and POE (2) stearyl ether, the effect of alcohols was less noticeable because these two surface-active agents gave good emulsions and foams by themselves.
3. Emulsion stability was increased in only two out of six systems.
4. Complex formation decreased the bubble size of the foams. The effect was similar to that observed with foams stabilized with anionic complexes.

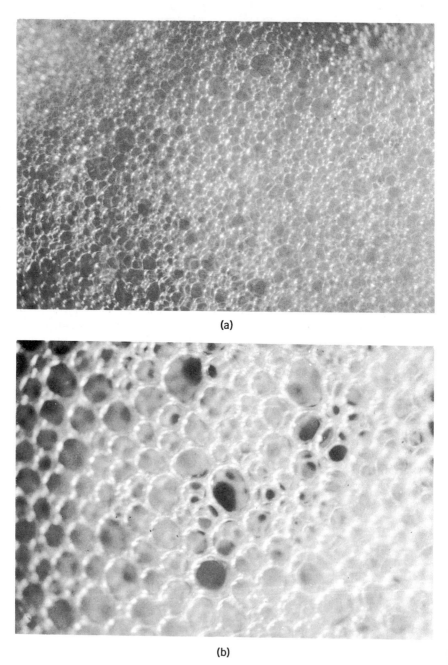

(a)

(b)

Figure 18-2 (a) Photomicrograph of a sodium lauryl sulfate–lauryl alcohol foam. (b) Photomicrograph of a sodium lauryl sulfate foam. (*Courtesy of the J. Soc. Cosmet. Chem.*)[4]

TABLE 18-5   POLYOXYETHYLENE (4) LAURYL ETHER
SYSTEM—EFFECT OF ALCOHOLS

Alcohol	Emulsion Stability	Foam Properties			
		Stability (min)	Stiffness (g)	% Drainage (30 min)	Wetting (min)
None	< 1 min	< 5	5	95	< 1
Lauryl	> 30 min	60–120	12	6	< 1
Myristyl	> 30 min	> 120	23	0	> 120
Cetyl	> 30 min	> 120	20	0	> 120
Stearyl	> 30 min	> 120	8	0	> 120
Oleyl	< 5 min	< 15	5	93	< 1

VARIATION IN ALCOHOLS   The specific properties of the aerosols affected by the alcohols and the extent to which they were changed were determined by both the particular alcohol and the particular POE fatty ether present. The effect of the various alcohols upon the properties of the POE (4) lauryl ether system is shown in Table 18-5. Data for the other seven POE fatty ether systems are given in Reference 5.

Both emulsion and foam stability were increased by the addition of alcohols. Myristyl, cetyl, and stearyl alcohol caused the greatest change in properties in the POE (4) lauryl ether system.   Oleyl alcohol had little effect, and this was generally true with the other POE fatty ethers.

FOAM BUBBLE SIZE   The correlation between bubble size of the foams and other foam properties was very marked in the POE fatty ether series. Whenever strong complex formation occurred and affected other foam properties, there was a corresponding decrease in foam bubble size. This is illustrated in Figure 18-3 by photomicrographs of the POE (4) lauryl ether foams listed in Table 18-5.   The bubble size of the foams containing myristyl, cetyl, and stearyl alcohol was considerably smaller than that of the other foams. This correlates with the effect of these alcohols upon the foam properties listed in Table 18-5.

VARIATION IN FATTY ALCOHOL CONCENTRATION   The relationship between the concentration of cetyl alcohol in POE (4) fatty ether systems and emulsion and foam properties is shown in Table 18-6. The first addition of cetyl alcohol, which gave an ether–alcohol mole ratio of $1:1/4$, caused a significant change in properties. The effect upon foam properties became most pronounced when the ether–alcohol ratio reached $1:1/2$.   Continued addition of cetyl alcohol had a relatively smaller effect, although foam stiffness continued to increase.

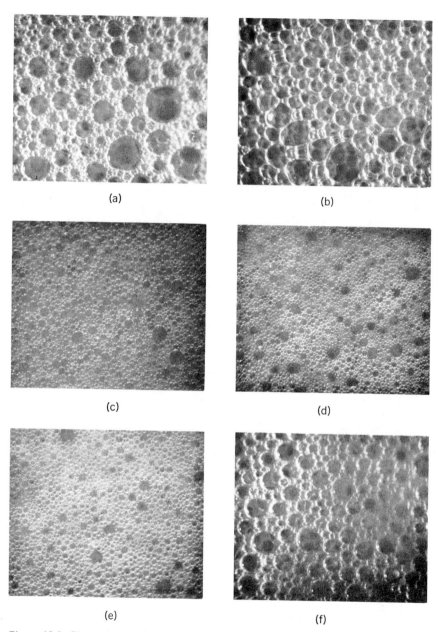

(a)

(b)

(c)

(d)

(e)

(f)

Figure 18-3 Photomicrographs of polyoxyethylene (4) lauryl ether foams—effect of alcohols. (a) No alcohol. (b) Lauryl alcohol. (c) Myristyl alcohol. (d) Cetyl alcohol. (e) Stearyl alcohol. (f) Oleyl alcohol. (*Courtesy of the J. Soc. Cosmet. Chem.*)[5]

TABLE 18-6 POLYOXYETHYLENE (4) LAURYL ETHER
SYSTEMS—EFFECT OF CETYL
ALCOHOL CONCENTRATION

Ether/ Alcohol Ratio (Molar)	Emulsion Stability	Foam Properties			
		Stability (min)	Stiffness (g)	% Drainage (30 min)	Wetting (min)
1:0	< 1 min	< 5	5	89	< 1
1:¼	> 30 min	15–30	7	11	15–30
1:½	> 30 min	> 120	13	0	> 120
1:¾	> 30 min	> 120	21	0	> 120
1:1	> 30 min	> 120	26	0	> 120
1:2	> 30 min	> 120	31	0	> 120

Polyoxyethylene Fatty Ether-Fatty Acid Complexes. Combinations of the eight polyoxyethylene fatty ethers with four fatty acids—lauric, myristic, palmitic, and stearic—were also evaluated. Like fatty alcohols, fatty acids had a considerable effect upon the properties of emulsions and foams in many cases, indicating complex formation between the polyoxyethylene ether and fatty acid.

A survey of the data in Reference 5 shows that the systems most affected by the acids were those containing the POE ethers with intermediate HLB values. These results are similar to those observed with fatty alcohols. The effect upon systems with the two ethers having low HLB numbers was less pronounced, but the fatty acids increased foam stiffness. Systems containing the POE ethers with high HLB numbers were least affected by fatty acids.

VARIATION IN FATTY ACIDS As a general rule, myristic, palmitic, and stearic acid appeared to be somewhat more effective than lauric acid. As with the alcohols, however, the extent to which any particular fatty acid affected the properties of an aerosol system depended upon the particular polyoxyethylene fatty ether present.

FATTY ALCOHOLS VERSUS FATTY ACIDS When compared in the same system, the long-chain alcohols generally had a greater effect upon such properties as foam stiffness, foam drainage, and wetting than the corresponding fatty acids. However, there were some instances where fatty acids were more effective. For example, fatty acids were superior to alcohols in improving the emulsion stability of the POE (10) cetyl ether and POE (10) stearyl ether systems. This suggests that for some products, a combination of all three components—the polyoxyethylene fatty ether, fatty alcohol, and fatty acid—might provide properties superior to those from either the POE ether–fatty alcohol or POE ether–fatty acid complexes alone.

The POE ether–fatty acid complexes had far less effect upon the bubble size of the foams than those containing fatty alcohols. This is additional evidence that the complexes containing the fatty acids are weaker than those containing the fatty alcohols.

Pearlescent and Iridescent Complexes. During the investigation of polyoxyethylene fatty ether-fatty alcohol or fatty acid complexes, certain combinations of the components were observed to give highly pearlescent aqueous concentrates and aerosols. These systems subsequently were studied in some detail because they had considerable aesthetic appeal owing to their optical properties and should be of interest for the formulation of cosmetic and pharmaceutical products. The results of the study of the pearlescent and iridescent complexes are reported in Reference 22.

GENERAL DISCUSSION Some general conclusions drawn from the data are listed below:

1. The degree of pearlescence in the aqueous concentrates was determined both by the particular polyoxyethylene fatty ether and by the specific fatty alcohol or acid present. Only polyoxyethylene fatty ethers that were water-dispersible or water-soluble gave pearlescent concentrates. Myristyl and cetyl alcohol were most effective at producing pearlescence. Fatty acids were relatively ineffective in producing pearlescent structures except for one specific combination involving lauric acid.
2. Pearlescence was affected by the concentrations of the POE fatty ether and fatty alcohol in the aqueous phase. In general, a mole ratio of POE fatty ether to alcohol of 1 : 1 was preferred.
3. Pearlescence of the aqueous concentrates was dependent upon the temperature. In all cases, increasing the temperature ultimately caused pearlescence to disappear. Loss of pearlescence occurred over a temperature range rather than at a specific temperature. This property, in conjunction with the optical characteristics, suggests that these pearlescent complexes belong to the class of substances known as liquid crystals.

THE RELATIONSHIP BETWEEN AEROSOL EMULSIONS AND FOAMS

A considerable number of papers have been published on the properties of aerosol foams, but relatively little information has been available about the emulsions from which the foams were obtained. As a result, little was known about possible relationships between the properties of aerosol emulsions and their foams. Two recent papers,[26, 27] however, have increased the knowledge of this relationship.

In order to study aerosol emulsions, equipment for their microscopic observation was developed. The apparatus was constructed by capping a 4-oz glass bottle with a standard Precision valve having a 0.08-in.-inside-diameter tail piece. A hole with a 0.08 in. diameter (thus corresponding to the diameter of the tail piece) was drilled through the entire valve to the dip tube. This essentially converted the valve into an extension of the dip tube.

A glass pressure cell was prepared by placing an oval brass shim, 1 mil thick, between two pieces of ordinary glass and fusing the edges of the glass together. This left the brass shim sandwiched between the sheets of glass. The brass shim was removed by immersion in nitric acid, leaving a cell with a uniform inside depth of 1 mil. The ends of the oval cell were fused to short lengths of 0.08-in.-inside-diameter glass tubing. The outside dimensions of the cell were as follows: 2 mm thick, 5 mm wide, and 7 mm long.

One end of the cell was connected to the stem of the valve on the bottle and the other end to the tail piece of a standard Precision valve with a 0.018-in. inlet orifice and foam actuator. The apparatus is illustrated in Figure 18-4. The oval cell is located midway between the top valve and the valve on the glass bottle.

The aerosol emulsion sample for microscopic examination was prepared by weighing the concentrate into a 4-oz glass bottle and capping the bottle with a

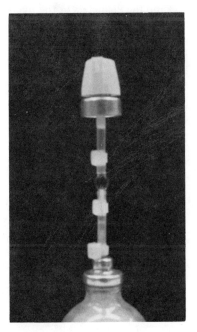

Figure 18-4   Apparatus for microscopic observation of aerosol emulsions. (*Courtesy of the J. Soc. Cosmet. Chem.*)[26]

drilled valve. The glass cell was connected to the valve on the bottle and the op-posite end of the cell to a standard valve. The propellant was pressure-filled through the entire assembly. The sample was shaken thoroughly after filling and allowed to stand overnight. It was then reshaken and 5% of the sample was discharged to eliminate nonemulsified propellant in the cell and dip tubes. For microscopic observations, the entire apparatus was supported horizontally with the cell positioned immediately below the objective of the microscope.

The microscope was a standard Bausch and Lomb monocular dynoptic model equipped with a microipso viewing attachment and a 3¼ X 4¼ Leitz Polaroid camera. Unless otherwise indicated, photographs of the emulsions were normally taken at a magnification of 150X using a 15X eyepiece and a 10X objective.

The cell is most satisfactory with a stable emulsion system. A photomicro-graph of a stable aqueous triethanolamine myristate/fluorocarbon 12/fluorocar-bon 114 (40/60) propellant emulsion with a creaming time greater than 1 h is illustrated in Figure 18-5. Emulsions with a very short creaming time present a problem because the larger dispersed propellant droplets settle to the bottom of the cell before a photomicrograph can be taken.

After the emulsion had been photographed in the cell, the product was dis-charged onto a microscope slide in order to photograph the foam, which was illuminated from the top. The cover glass was placed very lightly on the top of the foam to minimize distortion.[28]

The sizes of the emulsified propellant droplets and the diameters of the foam bubbles were estimated using a scale prepared by photographing a stage microm-

Figure 18-5 Photomicrograph of a stable aqueous triethanolamine myristate/fluorocarbon 12/fluorocarbon 114 (40/60) propellant emulsion. (*Courtesy of the J. Soc. Cosmet. Chem.*)[26]

eter at the same magnification as that used for the emulsions and foams. The scale was calibrated in intervals of 10 μ. Diameters down to about 2 μ could be estimated, while droplets with diameters less than 2 μ could be detected but not measured with any degree of accuracy. Edmundson[29] reported that the smallest particle diameter that could be measured microscopically with any precision lies in the range of 1-2 μ. Augsburger and Shangraw have reported a method for bubble size analysis based upon photomicrographs of aerosol foams.[30]

In order to obtain the range of diameters of emulsion droplets and foam bubbles, the field under the microscope was scanned until the largest droplet or bubble was observed. This portion of the field was then photographed. Diameters of the largest droplets and bubbles were measured, and those of the smallest were estimated. It must be emphasized that the range between the smallest and largest diameters has no direct relationship to the average droplet or bubble size.

Fluorocarbon Propellant/Water Systems

Triethanolamine salts of fatty acids are commonly used as surfactants for aerosol foams. In the first investigation, two different aqueous triethanolamine myristate/ fluorocarbon 12/fluorocarbon 114 (40/60) propellant systems were studied. One had an excess of myristic acid and was known to produce aerosol emulsions and foams of high stability because of the presence of the triethanolamine myristate-myristic acid surfactant complex.[4] The other was formulated with an excess of

TABLE 18-7   PROPERTIES OF AQUEOUS TRIETHANOLAMINE
MYRISTATE/FLUOROCARBON 12/FLUOROCARBON 114
(40/60) PROPELLANT SYSTEMS[a]

	Excess Myristic Acid	Excess Triethanolamine
Emulsion Properties		
Droplet size (μ)	<2–30	<2–100
Stability (phase separation)	>24 hr	<1 min
Viscosity	High	Low
Foam Properties		
Bubble size (μ) 1 min after discharge	20–160	20–210
Stability (% of original height after 60 min)	100	0
Density (1 min after discharge, g/cc)	0.45	0.49
Stiffness (g)	72	20
% Drainage (1 hr)	0	81
Wetting	>1 hr	<30 sec

[a]Aerosol formulation = 90 wt.% aqueous phase and 10 wt.% propellant.
From *J. Soc. Cosmet. Chem.*[26]

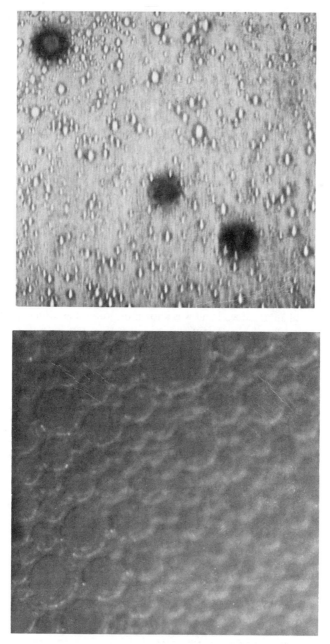

Figure 18-6 Photomicrographs of an emulsion (top) and foam (bottom) from an aqueous triethanolamine myristate/fluorocarbon 12/fluorocarbon 114 (40/60) propellant system with excess myristic acid. (*Courtesy of the J. Soc. Cosmet. Chem.*)[26]

triethanolamine and, therefore, contained a minimum of the complex. This aerosol system was known to produce emulsions and foams having low stability (p. 319).

The properties of the aerosol emulsions and foams from the two systems are given in Table 18-7. Photomicrographs are shown in Figures 18-6 and 18-7.

The excess myristic acid system, with the triethanolamine myristate–myristic acid complex, produced emulsified propellant droplets having smaller diameters than those in the excess triethanolamine system. The increased creaming stability in the excess myristic acid emulsion is probably the result of both the smaller emulsion droplet diameter and the higher viscosity.

The data show that a definite relationship exists between the properties of aerosol emulsions and those of the foams from the two triethanolamine myristate/fluorocarbon 12/fluorocarbon 114 (40/60) propellant systems. Aerosol emulsions with the smaller dispersed propellant droplets had better emulsion stability and, upon discharge, produced foams with smaller bubble size, superior stability, higher stiffness, and decreased rate of drainage and wetting.

Foam Bubble Size Increase after Discharge

After an aerosol emulsion has been discharged, the bubble size of the resulting foam continues to increase as the foam ages. This was reported previously by Augsburger and Shangraw.[30] This effect is shown in Table 18-8 and the photo-

TABLE 18-8   EMULSION DROPLET SIZE AND RATE OF INCREASE IN FOAM BUBBLE SIZE WITH TIME (AFTER 5% DISCHARGE)

	Propellant Emulsion Droplet Size (μ)	
	Excess Myristic Acid <2.0–30	*Excess Triethanolamine* <2–100
	Foam Bubble Size (μ)	
Time after Foam Discharge	*Excess Myristic Acid*	*Excess Triethanolamine*
10 sec	<2–30	–
20 sec	<2–60	–
40 sec	<2–120	–
60 sec	10–170	20–210
2 min	30–200	40–200
5 min	50–350	60–450
10 min	50–440	120–700
30 min	90–450	Foam collapsed
60 min	240–700	

From *J. Soc. Cosmet. Chem.*[26]

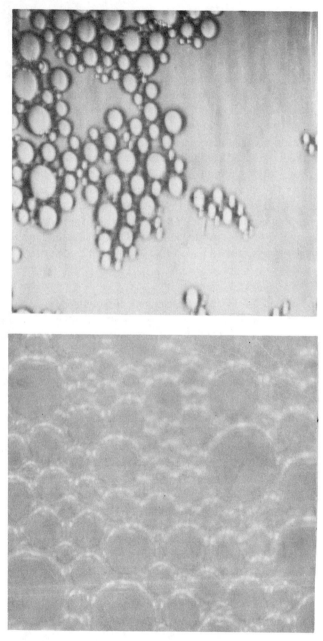

Figure 18-7 Photomicrographs of an emulsion (top) and foam (bottom) from an aqueous triethanolamine myristate/fluorocarbon 12/fluorocarbon 114 (40/60) propellant system with excess triethanolamine. (*Courtesy of the J. Soc. Cosmet. Chem.*)[26]

micrographs of Figure 18-8 for the two aqueous triethanolamine myristate/ fluorocarbon 12/fluorocarbon 114 (40/60) propellant systems.

The excess myristic acid system with smaller propellant droplets produced foams that increased less rapidly in bubble size with age than those from the excess triethanolamine system. This may be due to the higher resistance to expansion of the strong interfacial film formed by the triethanolamine myristate-myristic acid complex.

Augsburger and Shangraw[30] have listed several possible reasons why the bubble size of foams increases with age. One is that the initial discharge contains a mixture of emulsified propellant droplets as well as foam bubbles formed by vaporized propellant. If some of the bubbles contained residual liquefied propellant, continued vaporization of it would increase foam bubble size.

The presence of emulsified droplets in the initial foam discharge is suggested by the work of York and Weiner.[31,32] These investigators showed that when propellant alone is sprayed, only a relatively small proportion of it flashes immediately into vapor after leaving the actuator. The remainder of the propellant droplets in the spray cool down to a temperature at which the vapor pressure of the propellant approximates atmospheric pressure. They continue to evaporate, but at a much slower rate. Considering this, it is very unlikely that all of the emulsified droplets immediately flash into vapor and form bubbles when a foam is discharged.

Microscopic observations of the foam showed that emulsified propellant droplets continued to burst into bubbles more than 2 min after discharge. Quite often, the emulsified propellant droplet was in contact with a bubble. When this propellant droplet vaporized, the resulting bubble would coalesce with the other bubble, forming one single, large bubble. Sometimes individual emulsified propellant droplets were observed vaporizing into single bubbles.

Other processes causing an increase in bubble size are (1) continual coalescence of the bubbles and (2) the differential in pressure between a small bubble and a large bubble, which causes the larger bubble to grow at the expense of the smaller one.

Aqueous Triethanolamine Myristate/Mineral Oil/ Fluorocarbon Propellant Systems

This type of aerosol involves concentrates which themselves are oil-in-water emulsions, for example, emulsions of mineral oil in water. When propellant is added to the emulsion concentrate, the liquefied propellant droplets combine with the already dispersed oil droplets. This type of system can be varied to a much greater extent than one in which the propellant is the only dispersed phase. Since emulsion concentrates are not under pressure, a variety of procedures can be used for their preparation.

In the study of mineral oil emulsions reported in Reference 27, two types

10 sec

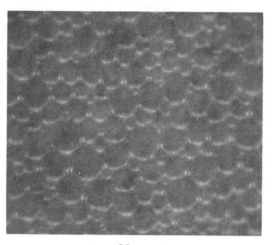

20 sec

Figure 18-8 Increase in bubble size with age of triethanolamine myristate/fluorocarbon 12/ fluorocarbon 114 (40/60) foam (excess myristic acid system). (*Courtesy of the J. Soc. Cosmet. Chem.*)[26]

of aqueous triethanolamine myristate emulsions were used. One had a 50 wt.% excess of myristic acid and the other a 50 wt.% excess of triethanolamine. Both were formulated with fluorocarbon 12/fluorocarbon 114 (40/60) propellant. Previous work had shown that aerosol emulsions prepared with an excess of myristic acid were far more stable due to the triethanolamine myristate–myristic

40 sec

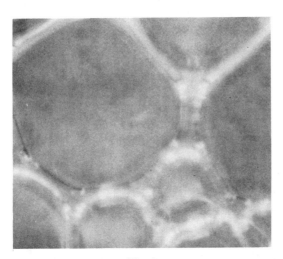

30 min

Figure 18-8 (*Continued*).

acid complex. The final emulsion concentrates had a composition of 90 wt.% aqueous phase and 10 wt.% mineral oil.

Each of the two mineral oil concentrates (one with excess myristic acid and the other with excess triethanolamine) were prepared by 14 different procedures. The different methods of preparation involved such variations as the distribution of triethanolamine and myristic acid between the oil and aqueous

TABLE 18-9   PROPERTIES OF AQUEOUS TRIETHANOLAMINE
MYRISTATE/MINERAL OIL EMULSIONS

Type of System	Method of Preparation	Mineral Oil Droplet Size	Creaming Time	Agglomeration
Excess myristic acid	Best	<2–5	8–24 hr	Low
Excess myristic acid	Poorest	<2–35	3–16 hr	Low
Excess triethanolamine	Best	<2–40	30–60 min	High
Excess triethanolamine	Poorest	15–190	5 min	High

From *J. Soc. Cosmet. Chem.*[27]

phases, temperature of the aqueous phase, and order of addition of the two phases, that is, oil to aqueous or aqueous to oil. Details of the methods of preparation are given later in the section "Preparation of Aerosol Emulsions."

Photomicrographs of the mineral oil concentrates were obtained with the equipment described in the previous section. Aerosol emulsions were photographed in the pressure cell. Photomicrographs were taken at 300X with a 15X eyepiece and a 20X long-range objective.

The properties of the aqueous triethanolamine myristate/mineral oil emulsions prepared by the best and poorest procedures are listed in Table 18-9.

The data show that the method of preparation significantly affects mineral oil droplet size and creaming time. Photomicrographs of the best and poorest emulsions from the two systems are shown in Figure 18-9. These illustrate very clearly the difference in droplet size range resulting from a variation in the method of preparation. Only the emulsions from a given type of system are compared with each other. Thus, the two different emulsions prepared with an excess of myristic acid are compared only with each other and not with emulsions prepared with an excess of triethanolamine. The properties of the mineral oil emulsions, aerosol emulsions, and the corresponding foams are given in Tables 18-10 and 18-11. Aerosol emulsions and foams from the best mineral oil concentrates in each system are superior to those obtained with the poorest concentrates. In aerosol emulsions with excess myristic acid, the superiority is shown by a slightly better stability to phase separation. There was no significant difference in the foam properties.

In the excess triethanolamine system, the superiority of the emulsions and foams from the best concentrate is more evident. The aerosol emulsions have smaller droplet size ranges and better stability. The foams have better stability and lower drainage. Photomicrographs (300X) of the emulsions are presented in Figure 18-10. The superiority of the excess triethanolamine aerosol emulsions from the best concentrate are readily discernible by the smaller droplet diam-

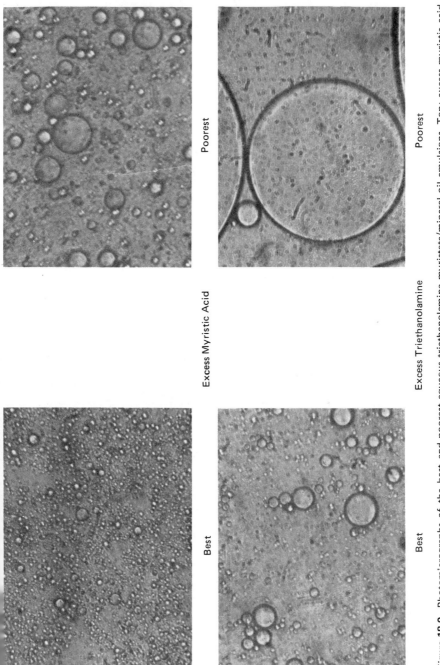

Poorest

Poorest

Excess Myristic Acid

Excess Triethanolamine

Best

Best

Figure 18-9 Photomicrographs of the best and poorest aqueous triethanolamine myristate/mineral oil emulsions. Top: excess myristic acid. Left, best (procedure No. 1); right, poorest (procedure No. 14). Bottom: excess triethanolamine. Left, best (procedure No. 1; right, poorest (procedure No. 14). (*Courtesy of the J. Soc. Cosmet. Chem.*)[27]

TABLE 18-10   PROPERTIES OF AQUEOUS TRIETHANOLAMINE
MYRISTATE/MINERAL OIL/PROPELLANT[a] SYSTEMS

	Excess Myristic Acid	
	Method of Preparation	
	Best	Poorest
Concentrate		
Oil droplet size (μ)	<2–5	<2–40
Aerosol Emulsion		
Emulsion droplet size (μ)	<2–40	<2–40
Emulsion stability	1.5% in 24 hr	15% in 24 hr
(% phase separation)		
Foam		
Stability	8 hr	8 hr
Wetting	8 hr	8 hr
Stiffness (g)	50	46
% Drainage (30 min)	0	0

[a] 10 wt.% fluorocarbon 12/fluorocarbon 114 (40/60).
From *J. Soc. Cosmet. Chem.*[27]

TABLE 18-11   PROPERTIES OF AQUEOUS TRIETHANOLAMINE
MYRISTATE/MINERAL OIL/PROPELLANT[a] SYSTEMS

	Excess Triethanolamine	
	Method of Preparation	
	Best	Poorest
Concentrate		
Oil droplet size (μ)	<2–40	15–190
Aerosol Emulsion		
Emulsion droplet size (μ)	<2–80	<2–120
Emulsion stability	100% in 2.5 hr	100% in 15–30 min
(% phase separation)		
Foams		
Stability	3.5–4 hr	1 hr
Wetting	1 min	1 min
Stiffness (g)	20	18
% Drainage (15 min)	6	83

[a] 10 wt.% fluorocarbon 12/fluorocarbon 114 (40/60).
From *J. Soc. Cosmet. Chem.*[27]

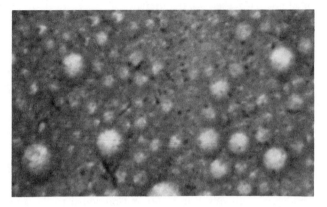

Best

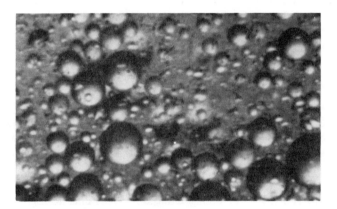

Poorest

Figure 18-10 Photomicrograph (300X) of an aqueous triethanolamine myristate/mineral oil/fluorocarbon 12/fluorocarbon 114 (40/60) aerosol emulsion with excess triethanolamine. Top, aerosol emulsion from best mineral oil concentrate. Bottom, aerosol emulsion from poorest mineral oil concentrate. (*Courtesy of the J. Soc. Cosmet. Chem.*)[27]

eters. The data also show that concentrates and aerosol emulsions with the smallest droplet size range produce foams with the smallest bubble size range.

The increase in bubble diameter with age was also observed with foams from the mineral oil systems. Foams from the best concentrates initially had the smaller bubble diameter ranges, and these remained smaller as the foams aged.

PREPARATION OF AEROSOL EMULSIONS

General Discussion

An emulsion can be considered to consist simply of water, oil, and an emulsifying agent. The way in which they are combined can have a considerable effect upon the properties of the resulting emulsion.

Becher, in a discussion of the techniques of emulsification for nonaerosol systems, lists the following four standard procedures for combining the oil phase, aqueous phase, and emulsifying agent.[13]

Agent-in-Water Method

The emulsifying agent is dissolved in water and oil added to the aqueous phase with vigorous agitation. This forms an oil-in-water emulsion. At very high concentrations of the oil phase, inversion of the oil-in-water emulsion to a water-in-oil emulsion may occur. This is one of the simplest and easiest procedures, but the least preferred of the four methods because it tends to give coarse emulsions with a wide range of droplet sizes. The droplet size can be decreased to produce a more stable emulsion by subsequent passage of the coarse emulsion through a colloid mill or homogenizer.

Agent-in-Oil Method

The emulsifying agent is dissolved or dispersed in the oil phase, which is then added to the water with agitation, thus forming the oil-in-water emulsion. If the order of mixing is reversed, with the water being added to the oil phase, a water-in-oil emulsion may form initially. Continued addition of water will cause inversion to an oil-in-water emulsion.

The agent-in-oil method, also called the Continental Method, is one of the preferred procedures for preparing emulsions. According to DeNavarre,[33] it usually gives good emulsions with a uniform distribution of droplet sizes. This is particularly true if the inversion procedure is used.

Nascent Soap (in situ) Method

This method is used in the preparation of emulsions stabilized with salts of fatty acids. The fatty acid is dissolved in the oil phase and the base in the aqueous phase. When the two phases are mixed, the salt of the fatty acid is formed at the oil–water interface. This is the preferred method when salts of fatty acids are used as the emulsifying agent. It is not applicable to aerosol emulsions because the higher fatty acids are not soluble in the propellants. It can be used to prepare concentrates that are oil-in-water emulsions.

Alternate Addition Method

This is also called the English Method and involves adding portions of the aqueous phase and oil phase alternately to the emulsifying agent. This procedure is reported to be particularly advantageous for the preparation of food emulsions containing vegetable oil.

The droplet size of an emulsion usually can be reduced by passage through a colloid mill or homogenizer. Reference 13 is suggested for those interested in a detailed treatment of this subject.

One of the most effective procedures for emulsification in the laboratory is the Briggs intermittent method of shaking. Briggs found that intermittent shaking, with periods between shakes, was much more effective than continuous shaking.[73] Gopal[74] reported, for example, that about 3,000 shakes in a machine were necessary to emulsify 60 vol.% of benzene in 1 wt.% sodium oleate. This required about 7 min. The same mixture could be completely emulsified with only 5 shakes by hand if a rest period of 20–30 sec was allowed after each shake.

Aerosol Emulsions

The properties of liquefied gas propellants cause some unique problems in the preparation of aerosol emulsions and limit the methods that can be used. Liquefied gas propellants must be filled into the containers either under pressure or at temperatures below the boiling points of the propellants. Since the boiling points are considerably below the freezing point of water, the cold-filling procedure is not satisfactory. Practically all aerosol emulsion products are packaged at room temperature and the propellants pressure-filled.

When the propellant is the only organic phase in an aerosol emulsion, the method for preparing the emulsion is essentially restricted to adding the propellant to the aqueous phase at room temperature last. The agitation necessary for achieving emulsification of the propellant is obtained by turbulence created by pressure-filling or by hand- or machine-shaking during production, during transportation, or when the consumer shakes the product immediately before use. In view of the limited methods of achieving emulsification of the propellant, it is essential to select a surfactant system that promotes emulsification of the propellant with the minimum amount of agitation. The surfactant should produce dispersed droplets of the smallest possible size (to reduce creaming) and provide strong interfacial films (to minimize coalescence).

When the aerosol concentrate itself is an emulsion, its properties can be modified by varying the method of preparation. Since the concentrate is not under pressure, a variety of procedures can be used to prepare it. The procedure that produces the best emulsion can be determined by conventional methods.

In order to determine the most effective procedure for obtaining aerosol emulsion concentrates, two aqueous mineral oil emulsions were prepared under a variety of conditions.[27] The emulsions were prepared from two aqueous tri-ethanolamine myristate systems known to produce emulsions with considerably different stabilities (these have been described previously). One had a 50 wt.% excess of myristic acid and the other a 50 wt.% excess of triethanolamine. The final emulsion had a composition of 90 wt.% aqueous phase and 10 wt.% mineral oil.

Each of the two concentrates was prepared by 14 different procedures. The different methods of preparation involved variations in the distribution of tri-ethanolamine and myristic acid between the oil and aqueous phases, temperature of the aqueous phase, and order of addition of the two phases, that is, oil to aqueous or aqueous to oil. When the mineral oil and myristic acid were com-bined, the mixture was maintained at 54.4°C in order to keep the myristic acid in solution.

Details of the 14 methods used to prepare the mineral oil concentrates with excess myristic acid and the properties of the resulting emulsions are given in Table 18-12. The emulsions are listed in order of increasing droplet size range. Method No. 1 gave the best emulsion in each case, judging by both droplet size range and creaming time. Method No. 14 gave the poorest emulsion on the basis

TABLE 18-12 EFFECT OF METHOD OF PREPARATION UPON THE
PROPERTIES OF AQUEOUS TRIETHANOLAMINE
MYRISTATE/MINERAL OIL EMULSIONS
(EXCESS MYRISTIC ACID SYSTEM)

Method of Preparation (No.)	Phase Composition[a]		Order of Mixing	Temper-ature of Mixing	Range of Droplet Diameters (μ)	Creaming Time (h)	Agglom-eration[b]
	Aqueous	Oil					
1	H_2O,TEA	Oil,MA	Aq. to oil	RT	<2–5	8–24	L
2	H_2O,TEA	Oil,MA	Aq. to oil	54.4°C	<2–5	2–2.5	L
3	H_2O,TEA,MA	Oil	Aq. to oil	RT	<2–5	2–2.5	H
4	H_2O,TEA,MA	Oil	Oil to aq.	RT	<2–10	1–1.5	H
5[c]	–	–	–	54.4°C	/ <2–10	–	–
6	H_2O,TEA,MA	Oil	Aq. to oil	54.4°C	<2–15	0.5–1.0	H
7	All components mixed together			54.4°C	<2–20	6–16	M
8	H_2O,TEA	Oil,MA	Oil to aq.	RT	<2–20	1.5–2.0	M
9	H_2O,TEA	Oil,MA	Oil to aq.	54.4°C	<2–20	1.5–2.0	L
10	H_2O,TEA,MA	Oil	Oil to aq.	54.4°C	<2–20	0.5–1.0	H
11	H_2O	Oil,TEA,MA	Oil to aq.	RT	<2–25	5–16	L
12	H_2O	Oil,TEA,MA	Aq. to oil	54.4°C	<2–25	3–16	L
13	H_2O	Oil,TEA,MA	Aq. to oil	RT	<2–25	5–16	L
14	H_2O	Oil,TEA,MA	Oil to aq.	54.4°C	<2–35	3–16	L

[a]TEA = triethanolamine; MA = myristic acid.
[b]L = low; M = medium; H = high.
[c]Preparation No. 5: MA heated to 54.4°C; TEA,H_2O heated to 54.4°C; aqueous phase added to MA at 54.4°C; mixture cooled to RT; oil added at RT. This simulates the preparation of an aerosol.
From *J. Soc. Cosmet. Chem.*[27]

of droplet size range. Similar details for the excess triethanolamine emulsions are given in Reference 27. In some instances, droplet size range and creaming times of some of the intermediate emulsions did not correlate. This may be due to differences in agglomeration (which would affect creaming time), average droplet size, or emulsion viscosity. Photomicrographs of the best and poorest emulsions have previously been illustrated in Figure 18-9. The procedures giving the best and poorest emulsions and the properties of the emulsions are summarized in Table 18-13.

The preferred method for preparing the mineral oil emulsions is to add the aqueous triethanolamine solution at room temperature to the mineral oil/myristic acid solution at 54.4°C. A possible explanation for the efficiency of this method is that triethanolamine myristate–myristic acid complex formation is promoted during the initial stages of emulsification. Previous work had shown that the presence of the complex in an emulsion resulted in increased emulsion stability and a smaller droplet size.[4,5,26,27]

When the aqueous phase is added to the oil phase, initially the concentration of myristic acid in the emulsion exceeds that of triethanolamine. This favors complex formation. In the system with excess myristic acid, the myristic acid remains at a higher molar concentration throughout addition of triethanolamine. Complex formation is thus promoted at all times. In the system with excess triethanolamine, myristic acid is at a higher concentration than triethanolamine during the initial stages of addition of the aqueous phase. Ultimately, the concentration of triethanolamine exceeds that of the myristic acid as addition continues. When this point is reached, complex formation is no longer favored.

When the oil phase containing myristic acid is added to aqueous triethanolamine solution, complex formation is not favored initially because triethanolamine is at a higher concentration than myristic acid. In the system with excess triethanolamine, the latter remains at a higher concentration than myristic acid throughout addition of the oil phase, and complex formation is never favored. In the system with excess myristic acid, complex formation is ultimately favored only when the concentration of myristic acid exceeds that of triethanolamine.

In order to test this theory, two experiments were carried out. In the first, involving the excess triethanolamine system, aqueous triethanolamine was added stepwise in three equal portions to the mineral oil/myristic acid solution. Photomicrographs were obtained after each addition. The data in Table 18-14 and the photomicrographs in Figure 18-11 show that the emulsion droplet size range increases as the concentration of triethanolamine in the emulsion increases. This is because complex formation is less favored as addition continues.

In the second experiment, involving the excess myristic acid system, mineral oil/myristic acid solution was added in three equal portions to aqueous trietha-

TABLE 18-13 BEST AND POOREST METHODS FOR PREPARING AQUEOUS TRIETHANOLAMINE MYRISTATE/MINERAL OIL EMULSIONS AND THE EMULSION PROPERTIES

Method of Preparation (No.)	Phase Composition[a]		Type of System	Order of Mixing	Temperature of Mixing	Range of Droplet Diameters (μ)	Creaming Time	Agglomeration[b]
	Aqueous	Oil						
1–Best	H_2O,TEA	Oil,MA	Excess myristic acid	Aq. to oil	RT	<2–5	8–24 hr	L
14–Poorest	H_2O	Oil,TEA,MA	Excess myristic acid	Oil to aq.	54.4°C	<2–35	3–16 hr	L
1–Best	H_2O,TEA	Oil,MA	Excess triethanolamine	Aq. to oil	RT	<2–40	30–60 min	H
14–Poorest	H_2O	Oil,TEA,MA	Excess triethanolamine	Oil to aq.	54.4°C	15–190	5 min	H

[a]TEA = triethanolamine; MA = myristic acid.
[b]L = low; H = high.
From J. Soc. Cosmet. Chem.[27]

TABLE 18-14   STEPWISE ADDITION OF AQUEOUS PHASE
TO OIL PHASE AND VICE VERSA

Excess Triethanolamine System—Stepwise Addition of
Triethanolamine to Mineral Oil/Myristic Acid Solution

Proportion of Triethanolamine Added (wt.%)	Emulsion Droplet Size Range (μ)
1/3	<2–5
2/3	<2–10
3/3	<2–30

Excess Myristic Acid System—Stepwise Addition of
Myristic Acid/Mineral Oil Solution to Aqueous Triethanolamine

Proportion of Myristic Acid Added (wt.%)	Emulsion Droplet Size Range (μ)
1/3	<2–30
2/3	<2–10
3/3	<2–10

From *J. Soc. Cosmet. Chem.*[27]

nolamine solution. The data in Table 18-14 and photomicrographs in Figure 18-12 show that the range of droplet size in the emulsion decreases as addition of the oil phase continues. This is because the concentration of the complex increases with continued addition.

The results also indicated that better emulsification was obtained when the aqueous phase was at room temperature during addition rather than at 54.4°C. A possible explanation is that the triethanolamine myristate–myristic acid complex is decomposed at the higher temperature and does not form until the emulsion is cooled below the decomposition-point temperature of the complex. The fact that complexes decompose at higher temperatures has been reported by Epstein and his co-workers.[34] The decomposition point is referred to as the film drainage transition temperature.

The worst emulsions were obtained when hot water was added to the hot mixture of triethanolamine, myristic acid, and mineral oil. Apparently, the complex does not form in mineral oil alone. The temperature probably is too high and, also, the ion–dipole and dipole–dipole interactions involved in complex formation are not favored in the oil phase.

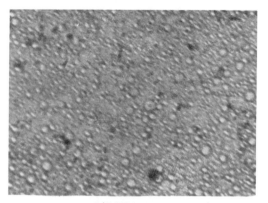

1/3 TEA added

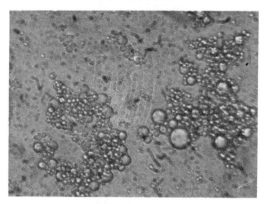

2/3 TEA added

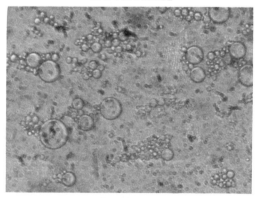

3/3 TEA added

Figure 18-11 Stepwise addition of aqueous triethanolamine (TEA) to mineral oil/myristic acid solution (excess TEA system). (*Courtesy of the J. Soc. Cosmet. Chem.*)[27]

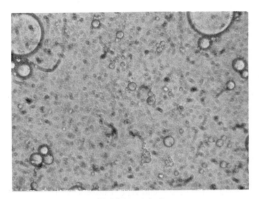

1/3 MA added

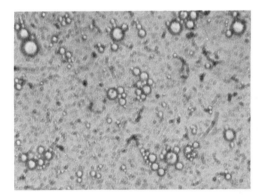

2/3 MA added

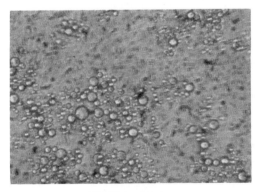

3/3 MA added

Figure 18-12 Stepwise addition of mineral oil/myristic acid (MA) solution to aqueous triethanolamine (excess myristic acid system). (*Courtesy of the J. Soc. Cosmet. Chem.*)[27]

PROPELLANTS

Hydrocarbons

Hydrocarbon propellants, such as isobutane and propane, are used extensively in foam products and are the predominant propellants for aerosol shaving lathers. They are used at concentrations of about 3.5–5.0 wt.%. This low concentration is adequate because of the low molecular weight of the hydrocarbons. The propellant used for shaving lathers is mostly isobutane, but it often contains small percentages of propane.[6] Because of its high vapor pressure, propane is not used alone.

Except for their flammability, the hydrocarbons are satisfactory propellants.

Fluorocarbons

A number of the fluorinated hydrocarbon propellants have properties that make them desirable for aerosol foams. Fluorocarbons 12 and 114, for example, not only give excellent foams, but are nonflammable, low in toxicity, essentially odorless, and stable in aqueous systems under most conditions. Therefore, they are particularly useful for cosmetic and pharmaceutical products. Individual fluorinated hydrocarbon propellants are discussed below.

Fluorocarbon 12

Fluorocarbon 12 has too high a vapor pressure (70.2 psig at 70°F) to be used alone as the propellant in aqueous systems when no other organic solvents are present. Thus, it is usually combined with a vapor pressure depressant. The vapor pressure depressant can be a high-boiling organic compound (such as mineral oil) or a higher-boiling propellant (such as fluorocarbon 114). Combinations of fluorocarbon 12 with a solvent like mineral oil have a number of advantages. For example, the composition of the mixture can be adjusted to have a density about the same as that of the aqueous phase, which will reduce creaming. In addition, mineral oil is more easily emulsified than fluorocarbon 12 and gives more stable emulsions. As a result, combinations of fluorocarbon 12 and mineral oil generally have superior stability. About 5–7 wt.% fluorocarbon 12 is sufficient to give a good foam.

Fluorocarbon 12/Fluorocarbon 114 Blends

Blends of fluorocarbon 12 and fluorocarbon 114 are the most commonly used propellants of the fluorinated hydrocarbon group. Mixtures with vapor pressures varying from 12.9–70.2 psig at 70°F can be obtained by varying the fluorocarbon 12/fluorocarbon 114 ratio.[20] The fluorocarbon 12/fluorocarbon 114 (15/85) blend is often used for glass bottle products, while the fluorocarbon 12/fluorocarbon 114 (40/60) blend is often used for products in metal containers.

Fluorocarbon 114

Fluorocarbon 114 has a relatively high boiling point and a low vapor pressure—properties which make it useful for products that discharge as a liquid and subsequently expand into a foam. It is also utilized in foam products containing high concentrations of propellant (50–90 wt.%). Fluorocarbon 114 can give very stable foams, and because of its stability and low order of toxicity, is useful for cosmetic and pharmaceutical foams.

Fluorocarbon 11

Fluorocarbon 11 is not suitable for oil-in-water emulsions in metal containers because metal catalyzes its hydrolysis and the hydrolysis products can cause corrosion.[2]

Aqueous foams prepared with fluorocarbon 12/fluorocarbon 11 propellant blends usually are much less stiff and stable and have a lower density than corresponding foams prepared with fluorocarbon 12/fluorocarbon 114 blends with about the same vapor pressure.[35]

Fluorocarbon 142b

This propellant generally produces foams that are less stiff and stable and have a lower density than those produced with fluorocarbon 12/fluorocarbon 114 (40/60) blends. In some products, this is an advantage. Fluorocarbon 142b has a lower molecular weight than fluorocarbon 12/fluorocarbon 114 blends and, therefore, less fluorocarbon 142b is required. It is stable to hydrolysis, but has the disadvantage of being flammable.

Fluorocarbon 152a

Foams with fluorocarbon 152a as the propellant have properties similar to those formulated with fluorocarbon 142b. Fluorocarbon 152a has too high a vapor pressure to be used alone in aqueous systems, and a vapor pressure depressant is necessary. Fluorocarbon 152a is stable to hydrolysis, but is flammable.

Fluorocarbon 123

Fluorocarbon 123 has too low a vapor pressure to be useful by itself as a propellant. It also is unstable in aqueous alkaline systems. Whether or not it could be used in blends in aqueous systems formulated with nonionic surfactants has not been determined.

Fluorocarbon 124

Fluorocarbon 124 is stable in aqueous alkaline systems and produces excellent foams. No other propellant is needed with fluorocarbon 124.

Fluorocarbon 133a

Fluorocarbon 133a is somewhat more stable in aqueous alkaline systems than fluorocarbon 123, but not stable enough to be used in such products. Whether it is sufficiently stable in neutral or acidic systems has not been determined. Fluorocarbon 133a has too low a vapor pressure to be used by itself in most products.

Propellant Concentration

Increasing propellant concentration in an aqueous foam product increases foam stiffness. This was first reported by Reed,[36] who observed that high propellant concentrations with a shampoo concentrate resulted in stiff, dry, elastic foams, while low propellant concentrations gave soft and less resilient foams.

In a subsequent experiment, the stiffness and density of a series of foams with propellant concentrations varying from 5 to 50 wt.% were determined. The aerosols were prepared with a typical aqueous shaving lather concentrate and a fluorocarbon 12/fluorocarbon 114 (40/60) blend as the propellant.[35] The results are presented in Table 18-15.

The data in Table 18-15 show how increasing the propellant concentration increases foam stiffness and decreases foam density. The same type of effect can be observed with any other liquefied gas.

Effect of Discharge

Foresman,[19] in a discussion of foam properties, pointed out that there is a drop in foam strength as increasing amounts of material are dispensed. As an aerosol product is discharged from the container, the volume of the liquid phase decreases, and the volume of the vapor phase increases correspondingly. In order to maintain a constant pressure in the aerosol container during discharge, propel-

TABLE 18-15 EFFECT OF VARIATION IN PROPELLANT CONCENTRATION UPON FOAM STIFFNESS AND DENSITY

Propellant Concentration (wt %)	Stiffness (g)	Density (g/cc)
5	43	0.138
10	65	0.078
15	94	0.053
25	114	0.045
50	126	0.036

lant in the liquid phase vaporizes and migrates to the vapor phase. This decreases the concentration in the liquid phase.

In spray products containing a relatively high proportion of propellant, loss of propellant from the liquid phase to the vapor phase has little noticeable effect upon spray characteristics, since the percentage loss of propellant is small. Foam products, on the other hand, are usually formulated with a low proportion of propellant. Thus, the change in concentration of propellant in the liquid phase as the product is discharged is significant and causes a noticeable change in foam properties.

It can be predicted from Table 18-15 that as the concentration of propellant in the liquid phase decreases during product use, foam stiffness will decrease and density will increase. This was verified experimentally by repeatedly discharging a 12-oz container of an aqueous shaving lather containing 10 wt.% of a fluorocarbon 12/fluorocarbon 114 (40/60) propellant blend. Density and stiffness determinations were made after each discharge.

The variation of foam stiffness and density with percent discharge for this particular system is illustrated in Figure 18-13.

The extent to which discharge affects these properties in other systems will depend upon the quantity of propellant initially present and the type of propellant used.

Richman and Shangraw[37] calculated the loss of foam consistency with discharge of a typical shaving lather formulation. Two propellant systems were evaluated: a single propellant—fluorocarbon 12—and a propellant blend of 57 wt.% fluorocarbon 12 and 43 wt.% fluorocarbon 114. They reported that the

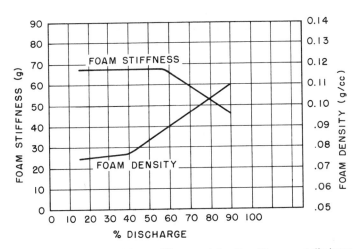

Figure 18-13 Variation of foam stiffness and density with percent discharge.

change in foam consistency during product use could be predicted by correlating the change in propellant concentration with known rheological and density data.

When mixtures of propellants are employed, fractionation takes place as a result of the difference in propellant vapor pressures. In one experiment reported by Richman and Shangraw, the composition of the fluorocarbon 12/fluorocarbon 114 blend changed from an initial weight percent ratio of 57/43 to 43/57 during discharge of 90% of the product. This caused a pressure drop from 64.5 to 59 psia. Therefore, there was a loss in pressure in the container as well as a decrease in propellant concentration during discharge.

If the aerosol foam product contains appreciable concentrations of an organic solvent such as VM&P naphtha, mineral oil, isopropyl myristate, or odorless mineral spirits, the decrease in pressure with product discharge will be much more marked. The solvents have a relatively low vapor pressure, and their concentration in the aqueous phase remains essentially unchanged as the product is discharged. Therefore, only the propellant decreases in concentration, and the ratio of solvent to propellant in the liquid phase can change drastically.

Propellant concentration in the liquid phase decreases with discharge as a result of loss of propellant to the vapor phase. Emulsion droplet and foam bubble size also decrease. This was verified experimentally. Photomicrographs of emulsions and foams from two aqueous triethanolamine myristate/fluorocarbon systems—one with a 50 wt.% excess of myristic acid and the other with a 50 wt.% excess of triethanolamine—were taken after product discharges of 5, 50, and 75%. The same container was used for all three discharges. Ranges of emulsion droplet diameters and average bubble size are listed in Table 18-16. The

TABLE 18-16 EFFECT OF DISCHARGE ON EMULSION DROPLET AND FOAM BUBBLE SIZE

	Emulsion Droplet Size Range (μ)[a]	
Percent Discharge	Excess Myristic Acid	Excess Triethanolamine
5	<2.0–30	<2.0–100
50	<2.0–10	<2.0–70
75	<2.0–10	<2.0–40
	Average Bubble Size (μ)[b]	
5	52	63
50	41	61
75	41	55

[a] Average of five determinations.
[b] Average of diameters of 50 bubbles on a photomicrograph; one measurement.
From *J. Soc. Cosmet. Chem.*[26]

TABLE 18-17 EFFECT OF DISCHARGE UPON DROPLET SIZE RANGES
FROM AQUEOUS TRIETHANOLAMINE MYRISTATE/
MINERAL OIL/PROPELLANT SYSTEMS

Aqueous Concentrate Method of preparation[a] TEA/MA ratio	1 Excess MA	14 Excess MA	1 Excess TEA	14 Excess TEA
Aerosol Emulsions Diameter range of emulsion droplets				
5% Discharge	<2–40	<2–40	<2–80	<2–120
75% Discharge	<2–30	<2–30	<2–60	<2–110

[a]See Table 18-13.
From *J. Soc. Cosmet. Chem.*[27]

emulsified propellant droplets decreased both in size and concentration. The smaller droplets probably disappear completely.

Augsburger and Shangraw reported that as an aerosol foam is discharged, the average bubble size decreases.[30] This would follow as a consequence of the decrease in the size of the emulsified droplets. This was also observed with the present triethanolamine myristate systems. The relationship is more noticeable in the system with excess triethanolamine. The change in emulsion droplet size is larger than that with excess myristic acid.

The decrease in the range of emulsion droplet size with discharge also occurs when mineral oil is a component of the emulsion system. This is shown by the data in Table 18-17 and the photomicrographs of Figure 18-14. The change is not as marked as with emulsions containing only propellant as the dispersed phase. This is probably because the concentration of mineral oil in the emulsions does not change with discharge as does that of the liquefied gas propellant.

STABILIZATION OF AEROSOL OIL-IN-WATER EMULSIONS AND FOAMS

The work of Goddard, Kung, and others established the compositions of complexes that were prepared and isolated in various ways, but the actual structure of a complexed interfacial film around an emulsified oil droplet or at the interfacial regions in foams is considerably less certain. It could have a definite composition with the surfactant and fatty alcohol (or acid) present in the same stoichiometric ratios as those found in the complexes that were isolated and analyzed, or it could have an indefinite and varying composition. According to Alexander,[38] Lawrence has suggested a liquid crystal structure at the interface for the ionized surfactant and fatty alcohol complex. Alexander also pointed out that in this event, the interface, instead of consisting of a mixed monolayer,

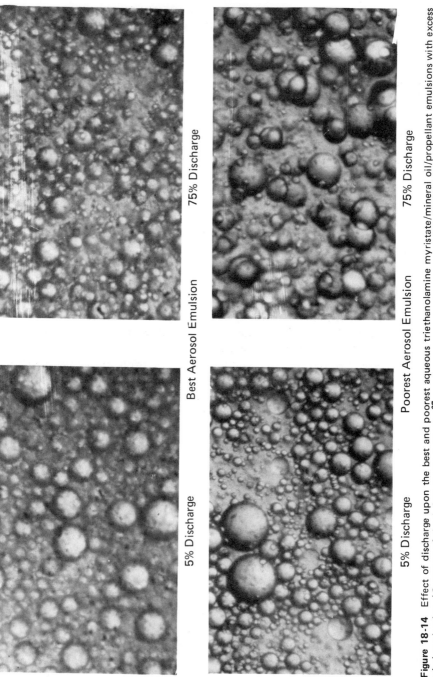

75% Discharge

75% Discharge

Best Aerosol Emulsion

Poorest Aerosol Emulsion

5% Discharge

5% Discharge

Figure 18-14 Effect of discharge upon the best and poorest aqueous triethanolamine myristate/mineral oil/propellant emulsions with excess triethanolamine. (*Courtesy of the J. Soc. Cosmet. Chem.*)[27]

probably would have a layered structure closer to that of solid stabilizers such as zinc and aluminum stearate.

Little basic information about aerosol emulsions and foams is available, but some of their general properties suggest that in the interfacial regions in aerosol systems, molecular complexes could have the type of liquid crystal structure proposed by Lawrence.

Many molecular complexes from surface-active agents and long-chain alcohols possess liquid crystal structures in aqueous systems, even at very low concentrations. It seems reasonable to assume, therefore, that molecular complexes also have a liquid crystal structure around emulsified propellant droplets or in aerosol foams.

The most effective surface-active agents for stabilizing aerosol emulsions and foams are practically insoluble in both the aqueous and propellant phases. From this, it is concluded that the surface-active agents are present in the interfacial regions essentially in solid form, and stabilize emulsions and foams in the same way as finely divided inorganic stabilizers. Since molecular complexes are effective stabilizers for aerosol systems, this suggests that the molecular complexes are also acting as solid stabilizers. The Lawrence–Alexander hypothesis that molecular complexes have a liquid crystal structure in the interfacial regions and act as finely divided solids appears to be compatible with the experimental data for aerosol emulsions and foams.

One reason that certain molecular complexes may be effective as stabilizers for aerosol systems is that addition of the fatty alcohol or acid converts the normally water-soluble or water-dispersible surface-active agent into a structure, considerably less hydrophilic than the surface-active agent itself, which is able to function at the interface as a solid stabilizer.

Evidence for the polymolecular structure in the interfacial regions is based primarily upon the known polymolecular orientation of the molecules in liquid crystals and experimental data which indicate that many common emulsion systems have polymolecular interfacial films.

Liquid Crystal Structures of Surface-Active Agents

There is considerable evidence that molecular complexes from surface-active agents and long-chain alcohols or acids form liquid crystal structures in aqueous systems. Most of the data were obtained from aqueous systems without a dispersed oil phase. In order for the results to apply to the Lawrence–Alexander hypothesis, it is necessary to assume that the same forces that cause molecular complexes to form liquid crystal structures in a bulk aqueous phase also cause the formation of liquid crystal structures in the interfacial regions around dispersed oil droplets. The structures in the bulk phase and at the oil–water interface probably would be different, but possibly still liquid crystalline in both cases.

All three types of surface-active agents (anionic, cationic, and nonionic) form liquid crystals in aqueous systems regardless of whether they are complexed with a long-chain polar compound or not.[39,40,41] Surface-active agents generally form the "smectic" type of liquid crystal structure, which consists of regular layers of surface-active agent molecules separated by water molecules.[42] The polar heads of the surface-active agents are oriented toward the aqueous phase. The "smectic" type of structure is regarded by Mulley[43] as a fully developed McBain lamellar micelle. The same type of structure in a modified form could exist around a dispersed oil droplet and function essentially as a solid stabilizer.

Ionic surface-active agents with short alkyl chains may not separate as a liquid crystal phase until the concentration reaches 20–30 wt.%. As the chain length increases, the concentrations at which liquid crystal structures form decrease rapidly. Liquid crystals may occur at concentrations of 1–2 wt.% when the chain length reaches C_{18}.[43]

In aqueous systems containing a surface-active agent and long-chain fatty alcohol or acid complex, liquid crystal structures form at low concentrations, possibly because the repulsion between the ionized heads of the surface-active agent is reduced by the presence of the alcohol molecules. This allows closer packing of the molecules.[43] Ekwall, Danielson, and Mandell[44] have indicated that mesomorphic phases may be formed below the critical micelle concentration (CMC) in systems containing surface-active agents and polar compounds. Lawrence and Hyde[45] have also suggested that the breaks occurring in the conductivity curves of cationic detergents in the presence of organic additives near the CMC may be due to the formation of a mesomorphic phase rather than changes in the CMC. When the third component is nonpolar, liquid crystals form only at high concentrations because the nonpolar component does not aid in the orientation.

Mulley[43] visualizes that addition of water-insoluble compounds to aqueous systems of surface-active agents containing Hartley micelles gradually changes the micelle structure until precipitation of the liquid crystal occurs. This concept has been discussed by Winsor.[46]

Liquid Crystal Structures and Pearlescence

The pearlescence in aqueous systems containing certain molecular complexes is often associated with liquid crystal structures. For example, the pearliness in creams formulated with an excess of stearic acid in sodium or potassium stearate systems is considered to be due to the ordered structure of the acid soap complex of free stearic acid and sodium stearate.[47] The 1:1 complex of sodium palmitate and palmitic acid has a crystal pattern and a liquid crystal structure in aqueous systems.[48] Fatty acid salts containing free fatty acids also show a slight pearlescence by themselves in the aqueous phase and are much more effective stabilizers for aerosol emulsions and foams than the triethanolamine salts by themselves.

TABLE 18-18   TEMPERATURE DEPENDENCE OF PEARLESCENT
STRUCTURES–POLYOXYETHYLENE (10) CETYL
ETHER SYSTEMS

Fatty Alcohol	Transition Temperature Range (°F)
Lauryl	77–82
Myristyl	93–101
Cetyl	95–115
Stearyl	100–125

Pearlescent aqueous systems are produced by certain nonionic polyoxy-ethylene fatty ether–fatty alcohol combinations. The pearlescence is believed to result from the liquid crystal structure of complexes in the aqueous phase. The complex structures were judged to be liquid crystalline in nature, not only because of their pearlescent properties, but also because the pearlescence disappeared over a specific transition temperature range (Table 18-18), which apparently is due to the gradual "melting" of the liquid crystalline structures responsible for the pearlescence.[22] This change is reversible and can be considered analogous to the two melting point transitions exhibited by typical liquid crystals.

Becher and Del Vecchio[25] determined film drainage transition temperatures for several polyoxyethylene lauryl ethers in the presence of lauryl and cetyl alcohols by means of surface viscosity measurements. The curves illustrating the change in surface viscosity with temperature were regarded as being in the nature of melting point depression curves, possibly attributable to complex formation. Conceivably, the two-dimensional melting points could also be interpreted on the basis of the liquid crystalline nature of the structures.

Stabilization by Solid Emulsifiers

Both emulsions and foams can be stabilized by finely divided inorganic solids. In order to function effectively in emulsion systems, the solid must be wetted properly by both the aqueous and oil phases.[49] If the solid remains suspended in either the aqueous or oil phase, it will not concentrate at the oil–water interface and act as a stabilizer. Becher[13] has suggested that the stabilizing action of solids may result from the formation of a strong interfacial film around the droplets. Verwey[50] has noted that emulsions increase in stability as the interfacial regions around the dispersed emulsion droplets become more solid.

Foams also can be stabilized by solids. Bikerman[51] refers to these as three-phase foams. In such foams, coalescence of the bubbles is prevented or retarded by solid substances partially immersed in the liquid phase. The solid must possess the correct degree of wettability by the liquid phase so that it will remain at the gas–water interface rather than in the bulk of the liquid phase.

Aqueous Aerosol Systems

There is considerable evidence that stabilization of essentially all aqueous aerosol emulsions and foams occurs through the mechanism of surface-active agents acting as solid stabilizers. The most useful surfactants for aerosol emulsions are not soluble in either the liquefied gas propellants or the aqueous phase at concentrations where they are effective. If the concentration of surfactant is sufficiently low so that it is soluble in either the propellant or the aqueous phase, then adequate stabilization does not occur. This indicates that it is necessary for the surfactants to be present in the solid or undissolved state in order to provide sufficient stabilization. The surfactants, therefore, are functioning as solid stabilizers in a manner similar to other solids known to stabilize emulsions. The best stabilizers for foams likewise are insoluble in either the propellant or the aqueous phase at concentrations where they are effective.

This is illustrated in Table 18-19, which shows the relationship between the solubility of a series of polyoxyethylene (POE) fatty ethers in the aqueous phase and their effectiveness as stabilizers for aerosol emulsions and foams.[23] None of the ethers is soluble in the liquefied gas propellant at the concentrations used. The concentrations (wt.%) were approximately 0.05 M based on the aqueous phase.

Only two polyoxyethylene fatty ethers—POE (2) cetyl ether and POE (2) stearyl ether—were effective emulsifying and foaming agents for the aerosol systems. These two agents also are the only surfactants insoluble in water. They could be dispersed in the propellant, which indicated that they were wetted to

TABLE 18-19   STABILIZATION OF AQUEOUS AEROSOL SYSTEMS[a] BY POLYOXYETHYLENE FATTY ETHERS

| Composition | | Wt.% (Aqueous Phase) | HLB Value | Stability (min) | |
Polyoxyethylene Fatty Ether	Water Solubility[b]			Emulsion[c]	Foam[d]
POE (4) lauryl ether	D	1.8	9.7	< 1	< 5
POE (23) lauryl ether	S	6.0	16.9	< 1	< 5
POE (2) cetyl ether	I	1.6	5.3	>30	>120
POE (10) cetyl ether	D	3.3	12.9	< 1	< 15
POE (20) cetyl ether	S	5.4	15.7	< 1	< 5
POE (2) stearyl ether	I	1.8	4.9	>30	>120
POE (10) stearyl ether	D	3.4	12.4	< 1	< 15
POE (20) stearyl ether	D	5.5	15.3	< 5	< 5

[a] Aerosol formulation: 90 wt.% aqueous phase and 10 wt.% fluorocarbon 12/fluorocarbon 114 (40/60) propellant.
[b] S, soluble; D, dispersible; I, insoluble.
[c] Time to initial phase separation after shaking.
[d] Time to first indication of foam collapse.

TABLE 18-20   EFFECT OF ALCOHOLS IN POLYOXYETHYLENE (4) LAURYL ETHER SYSTEMS[a]

| Alcohol | Stability (min) | |
	Emulsion	Foam
None	< 1	< 5
Myristyl	>30	>120
Cetyl	>30	>120
Stearyl	>30	>120

[a]Aerosol formulation: 90 wt.% aqueous phase and 10 wt.% fluorocarbon 12/fluorocarbon 114 (40/60) propellant. The mole ratio of surfactant and alcohol is 1 : 1.

some degree. According to their HLB values,[21] these two surfactants should be useful for the preparation of water-in-oil emulsions, but in the present case they stabilize oil-in-water emulsions. These data suggest strongly that these surfactants function as solid stabilizers in aqueous aerosol emulsions and foams. The other surfactants, all of which are either water-soluble or water-dispersible, are probably ineffective because they do not have the proper wettability characteristics by the two phases. They are wetted too much by the aqueous phase and not enough by the propellant phase, so that they remain dispersed or dissolved in the aqueous phase instead of concentrating at the interface.

The addition of certain fatty alcohols to some of the unstable polyoxyethylene fatty ether systems increases emulsion and foam stability, as shown in Table 18-20.[5]

The stabilizing effect of the alcohols is attributed to complex formation with the polyoxyethylene fatty ethers and subsequent formation of a strong interfacial film. It seems likely that part of the reason the complexes are effective is that they are more hydrophobic than the surfactants alone and possess solubility and wettability characteristics which cause them to collect at the propellant–water interface where they can function as solid stabilizers.

Interfacial Film Thickness

The Lawrence–Alexander hypothesis of a liquid crystal structure for interfacial films from molecular complexes implies a polymolecular rather than monomolecular film thickness because of the orientation and structure in liquid crystals. In this connection, Alexander cites the similarity between liquid crystal interfacial films and the layered structures of zinc and aluminum stearate.

Stabilizers such as gums and proteins are known to form polymolecular interfacial films.[49] It is the question of the thickness of interfacial films from other types of stabilizers that remains unsettled. However, there is also experimental evidence that the interfacial films in some of the common emulsion systems may be more than monomolecular in thickness. Martynov,[52] for example, estimated the thickness of the interfacial film in an aqueous sodium oleate-benzene emulsion by density measurements. He assumed that if the density of the emulsion differed from that of the benzene and aqueous phases, the difference was due to the interfacial film, which formed a third phase. He concluded from his data that the interfacial film in the emulsion was polymolecular rather than monomolecular.

Cockbain[53] reported a study of the aggregation of dispersed benzene and paraffin hydrocarbon droplets in soap-stabilized emulsions using the rate of creaming as the criterion of aggregation. The aggregation and disaggregation phenomenon became very complex as the soap concentration increased beyond the CMC. The most reasonable explanation Cockbain found for this behavior was that polymolecular adsorbed films formed when the soap concentration exceeded the critical micelle concentration.

Additional evidence for polymolecular interfacial films was obtained by Dixon and his co-workers,[54] who showed by a radiotracer method that the adsorption of the anionic agent—di-n-octyl sodium sulfosuccinate—and the cationic stearamido-propyldimethyl-2-hydroxyethyl ammonium sulfate was greater than that which could be accounted for by a monolayer. The adsorption of the latter was about ten times that calculated for a monolayer. Other workers, such as Defay,[55] have also postulated polymolecular interfacial films in order to interpret and explain interfacial phenomena.

Structure of Water in Interfacial Regions

Thus far, very little has been said about the contribution of water to the structure of the interfacial films. There is little doubt that the structure of water in the immediate vicinity of interfaces is different from that in the bulk phase. Derjaguin and Landau[56] believe that thick layers of water are immobilized at interfaces, while Davies and Rideal[57] report that layers of water may be oriented at liquid surfaces to form what they term "soft ice" with a viscosity similar to that of butter and a density greater than one. Drost-Hansen[58] states that the thickness of surface layers at aqueous interfaces is probably larger than is generally realized, and may be greater than 20 molecular layers. Henniker[59] reports that in many liquids other than liquid crystals, the orientation may extend 1000 Å or more below the surface, the precise depth being determined by the specific material with which the liquid is in contact.

Ions have a considerable effect upon the orientation of water molecules. McBain[60] suggests that the dipoles of a surface-active agent in the interface can orient dipoles of water around the polar head groups of the surfactants. Low[61] has also discussed a model for ion–water interactions. In this structure, an ion is surrounded by three regions. In the region immediately adjacent to the ion, the water is strongly oriented and immobilized by the electric field of the ion. The second region contains water in which the structure is broken down and more random than that of normal water. Finally, the third region consists of normal water polarized by the ionic field, which is relatively weak at this distance.

Thus, considering the orientation of water in the vicinity of interfaces and ions, it is not surprising that hydration layers separate the alternating layers of surface-active agents in the liquid crystal structures. Boffey, Collison, and Lawrence[42] report that the thickness of the water layer in the "smectic" liquid crystal may reach 110 Å, and they consider that it is held by solution forces to the ionic heads of the surface-active agents and the hydroxyl groups of the polar additives. Similar regions are shown by systems with nonionic surface-active agents.

Discussion and Summary

The most effective emulsion and foam stabilizers for aerosol systems are surface-active materials that form an oriented, polymolecular structure at the interface with essentially solid properties. The surface-active materials must have a low solubility in both the aqueous and propellant phases and, also, the proper wettability characteristics so that they remain in the interfacial regions instead of dispersing or dissolving in either of the two phases. It is possible that the poor solvent properties of the fluorocarbon propellants are a factor in the necessity of having a stabilizer to function as a finely divided solid.

Many molecular complexes derived from combinations of surface-active agents and long-chain alcohols or acids are excellent stabilizers for aerosol systems. Since it has been established that molecular complexes form liquid crystals in aqueous systems at the concentrations used for stabilizing aerosol systems, it seems reasonable to assume that the complexes also form liquid crystal structures at the propellant–water interface. The liquid crystal molecular complexes, with their oriented polymolecular structure, would be expected to stabilize aerosol emulsions and foams by somewhat the same mechanism as finely divided inorganic solids.

Many water-soluble or water-dispersible surface-active agents are relatively poor stabilizers for aerosol systems. The addition of a long-chain polar compound to such systems often forms a complex with the surface-active agent, which produces a much more stable emulsion or foam. The evidence suggests that the resulting complexes form liquid crystal structures that are much less hydrophilic and more solid than the surface-active agents themselves and, therefore, have a greater tendency to remain at the propellant–water interface.

The oriented liquid crystal nature of molecular complexes with their attendant layers of oriented water molecules suggests that the interfacial region around an emulsified propellant droplet can be viewed as consisting of alternating shells of oriented water and molecular complex molecules. The propellant interface would consist of a monolayer of adsorbed molecular complex molecules with the polar heads oriented toward an adjacent hydration layer. The hydration layer of water molecules, in turn, would be surrounded with a bimolecular shell of complex molecules with the polar heads on one side of the shell oriented toward the inner hydration layer and the polar heads on the other side oriented toward an outer hydration shell. This configuration of alternating layers of oriented water and bimolecular complex molecules would extend into the bulk phase with diminishing orientation until it disappeared.

This type of structure is patterned after the interfacial structures proposed by Cockbain[53] for soap-stabilized emulsions. It could almost be considered a large spherical micelle except that the propellant is not solubilized and constitutes a major portion of the structure. The propellant droplet is the core of the spherical structure with alternating layers of oriented water and surface-active agent complex radiating outward into the bulk phase.

WATER-IN-OIL EMULSIONS

In water-in-oil emulsions, the water is dispersed in the oil phase. The oil phase may consist only of propellant, or it may be composed of a mixture of propellants, solvents, and various active ingredients. One of the earliest references (1950) to aerosol water-in-oil emulsions is a patent by Boe.[62] Two later publications that describe basic studies on water-in-oil emulsions with the Freon® propellants appeared in 1956 and 1958.[63,64] Although the new system was suggested at this time for a variety of products (room deodorants, mothproofers, etc.), there was little activity in this field until about 1962. Some of the problems in formulating room deodorants, hair sprays, snows, and insecticides were mentioned by Fowks,[65] and a number of publications dealing with insecticides formulated as water-in-oil emulsions appeared shortly thereafter.[66-68] Aerosol insecticides have since become one of the major aqueous-based products. Other water-in-oil emulsion products that appeared on the market included room deodorants and furniture polishes.

There are still many areas in which the use of aerosol water-in-oil emulsions has been limited. This is particularly true in the cosmetic and pharmaceutical fields. Water-in-oil emulsion systems for products in these fields may be utilized to a greater extent in the future. Some advantages of water-in-oil emulsions for aerosol products are as follows:

1. Nonfoaming sprays which range from soft space sprays to wet residual sprays can be obtained by proper formulating techniques. Water-in-oil systems, therefore, can be used for a variety of products.
2. Water-in-oil emulsions utilize substantial proportions of water, which is inexpensive, low in toxicity, and an excellent solvent. There are many materials which are soluble in water but not in organic solvents, and the water-in-oil system provides a way of spraying these materials.

3. The sprays are warm and do not have an undesirable cooling effect when applied to the skin.
4. Corrosive characteristics are low. Even fluorocarbon 11 can be used as one of the propellants in the water-in-oil system.

Surface-Active Agents

Probably the most widely used agent is polyglycerol oleate.*[63,64] The suppliers of surface-active agents may be able to suggest additional agents.

Polyglycerol oleate is used at concentrations from about 0.5 to 4.0 wt.%, based upon the entire aerosol. Generally, as little as possible is used and the most effective concentration for any product can only be determined by experimentation. The normal concentration is 1.0-2.0 wt.%.

Propellants

Hydrocarbons and Hydrocarbon/Fluorocarbon Blends

Hydrocarbons, or blends of hydrocarbons with fluorocarbon 12, are used extensively as the propellants in commercial water-in-oil emulsion products. Baker and Moore[67] used an isobutane/propane (87/13) mixture for insecticides at concentrations of 30-35 wt.%. Glessner[68] mentions that either hydrocarbons or fluorocarbon 12/hydrocarbon blends are satisfactory for use with aqueous-based insecticides. Jones and Weaning[66] used fluorocarbon 12/butane (70/30 or 65/35) mixtures for aqueous-based insecticides with a propellant concentration of about 39-44 wt.%. Clayton and Scott[69] suggest fluorocarbon 12/isobutane (30/70 or 25/75) mixtures at concentrations of 45-85 wt.% as a starting point in formulating aqueous-based products.

Fluorocarbons

Although hydrocarbon propellants are used to a considerable extent in aqueous-based products, little basic information on water-in-oil systems containing these propellants has been published. Since much more fundamental information on the use of fluorocarbon propellants in water-in-oil systems is available, the discussion of these systems will, of necessity, be centered around these propellants.

Emulsion Stability. The stability of water-in-oil emulsions prepared with fluorocarbons 11, 12, 113, and 114 is given in Table 18-21[64] (fluorocarbons 11 and 113 are considered as propellants to simplify the discussion in the present case). The emulsions were prepared at four different propellant/water ratios using 4 wt.% polyglycerol oleate as the emulsifying agent. Emulsion stability was determined by noting the time after the bottles had been shaken before phase separation was observed.

_____

*"Witconol" 14, formerly "Emcol" 14.

TABLE 18-21   STABILITY OF WATER-IN-OIL EMULSIONS WITH
FLUOROCARBONS

Fluorocarbon	Kauri-Butanol Value of Propellants	Emulsion Stability			
		Propellant–Water Ratio (wt %)			
		90/10	80/20	60/40	40/60
12	18	< 1 min	< 1 min	< 1 min	1–5 min
114	12	< 1 min	< 1 min	< 1 min	15–30 min
11	60	15–30 min	15–30 min	> 1 hr	> 1 hr
113	32	1–5 min	1–5 min	> 1 hr	> 1 hr

The relative stability of the emulsions prepared with the various fluorocarbon propellants corresponds roughly to the solvent properties of the propellants, as indicated by their Kauri-Butanol values (see Chapter 10). Fluorocarbons 12 and 114 are comparatively poor solvents. The solubility of polyglycerol oleate in these propellants may be too low to form a good emulsion. Water-in-oil emulsions prepared with fluorocarbon 12, fluorocarbon 114, or blends of the two have poor stability when no other solvents are present. However, by using another compound with better solvent properties, such as fluorocarbon 11, fluorocarbon 113, or an organic solvent, in conjunction with fluorocarbon 12 or fluorocarbon 114, emulsions with satisfactory stability can be obtained.

FLUOROCARBON 12/FLUOROCARBON 11/WATER EMULSIONS   Stable emulsions can be obtained by combining fluorocarbon 11, which has relatively good solvent powers, with fluorocarbon 12. The spray characteristics of the emulsions may be varied from fine to very wet by changing either the ratio of fluorocarbon 12 to fluorocarbon 11 or the ratio of propellant to water.

Variation in the Fluorocarbon 12/Fluorocarbon 11 Ratio.   The effect of varying the fluorocarbon 12/fluorocarbon 11 ratio upon emulsion stability and spray characteristics (standard valves and actuators) at a propellant/water weight percent ratio of 60/40 is shown in Table 18-22.[64] Emulsion stability increased as the proportion of fluorocarbon 11 in the propellant increased.

Spray characteristics and emulsion stability may be affected by the addition of other components of the aerosol.

Variation in Propellant/Water Ratio.   The effect of varying the propellant/water ratio is shown in Table 18-23.[64] The propellant in this case was fluorocarbon 12/fluorocarbon 11 (30/70).

The increase in emulsion stability that occurs with an increase in the concentration of water probably is the result of an increase in viscosity.

TABLE 18-22 EFFECT OF FLUOROCARBON 12/FLUOROCARBON 11
RATIO UPON EMULSION STABILITY AND SPRAY
CHARACTERISTICS

	(60/40 Propellant/Water Ratio)	
FC-12/FC-11 Ratio *(wt.%)*	*Emulsion* *Stability*	*Spray* *Characteristics*
100/0	< 1 min	- - - - - - - -
70/30	30–60 min	Fine
50/50	30–60 min	Medium–fine
30/70	30–60 min	Medium–coarse
0/100	> 1 hr	No spray

TABLE 18-23 EFFECT OF PROPELLANT[a]/WATER RATIO UPON
EMULSION STABILITY AND SPRAY CHARACTERISTICS

Propellant–Water Ratio *(wt %)*	*Emulsion* *Stability*	*Spray* *Characteristics*
90/10	1–5 min	Very fine
80/20	5–15 min	Fine
60/40	30–60 min	Medium
40/60	> 1 hr	Very coarse— almost foams
20/80	< 1 min	Stream-foams

[a]Fluorocarbon 12/fluorocarbon 11 (30/70).

Water-in-Oil Emulsions with Auxiliary Solvents

Hydrocarbon and Hydrocarbon/Fluorocarbon Blends

Many publications on aqueous-based products are concerned primarily with aerosol insecticides. In order to obtain the requisite particle size in the spray, the presence of an organic solvent is necessary.[67] Solvents used include methyl chloroform,[67,68] methylene chloride,[67] petroleum distillate,[67,68] and kerosene and isoparaffinic oils.[66] Clayton and Scott[69] evaluated water-in-oil emulsions prepared with a variety of organic solvents, including dipropylene glycol, hexylene glycol, isopropyl alcohol, mineral oil, and triethylene glycol. Mineral oil was the preferred solvent on the basis of the type of spray obtained.

Fluorocarbons

Some auxiliary solvents used in water-in-oil emulsions with fluorocarbon propellants are mineral oil, isopropyl myristate, cottonseed oil, odorless mineral spirits, and VM&P naphtha. In products formulated to have a fine spray, the use of high-boiling organic solvents should be avoided if there is any chance of inhalation of the small droplets in the spray. High-boiling, water-insoluble compounds such as mineral oil can cause lipoid pneumonia if a sufficient quantity in the form of small droplets is inhaled.

Some advantages of using auxiliary solvents are given below:

1. Added solvents generally increase emulsion stability. This is particularly noticeable in emulsions with fluorocarbon 12 or fluorocarbon 12/fluorocarbon 114 blends as the propellant. In fact, in these cases, the presence of an auxiliary solvent is almost a necessity in order to have sufficient emulsion stability. In addition, the composition of the propellant/solvent combinations can be adjusted to have a density about the same as that of the aqueous phase. This may increase emulsion stability. Often, fluorocarbon 12 is used alone as the propellant because of the vapor pressure depressant effect of the solvents. The density adjustment is one of the reasons why solvents with a density less than that of water are preferred.
2. Some fluorocarbon 12/fluorocarbon 11 water-in-oil emulsions tend to bubble when sprayed on a surface. The addition of a small quantity of an auxiliary solvent may minimize or eliminate the bubbling.
3. Efficiency of emulsification may be increased by the use of solvents. For example, the solvent may be used to dissolve the water-insoluble surface-active agent. A water-in-oil emulsion is then prepared by adding water to the solution of surface-active agent. This emulsion is loaded into the container. Addition of propellant merely extends the continuous oil phase.

Data for water-in-oil emulsions containing combinations of odorless mineral spirits with fluorocarbon 12/fluorocarbon 114 or fluorocarbon 12/fluorocarbon 11 blends are given in Reference 64.

Effect of Ethyl Alcohol. Some water-in-oil emulsions can tolerate a fairly high concentration of alcohol without an appreciable effect upon stability. For example, concentrations of ethyl alcohol in the aqueous phase up to 30 wt.% had no effect upon the emulsion stability of fluorocarbon 12/fluorocarbon 11/odorless mineral spirits/water systems emulsified with polyglycerol oleate. Higher concentrations caused a definite decrease in stability. The effect of ethyl alcohol upon fluorocarbon 12/fluorocarbon 11/water systems was not determined.[64]

Effect of Sodium Chloride. Concentrations of sodium chloride of 10 and 25 wt.% in the aqueous phase decreased emulsion stability in systems formulated with odorless mineral spirits. In most cases, however, the emulsion stability was sufficient for practical purposes.[64]

Viscosity of Water-in-Oil Emulsions

One of the most important factors that affects the viscosity of water-in-oil emulsions is the propellant/water ratio. As the concentration of the aqueous phase is increased, there is an accompanying increase in the viscosity of the emulsion.[64] This improves the stability of emulsions because of the increased difficulty of movement of the dispersed water droplets.

When the concentration of the aqueous phase is increased to 60–80 wt.%, inversion of the emulsion from water-in-oil to oil-in-water generally occurs. Inversion is accompanied by a sharp decrease in emulsion viscosity. Electrical conductivity measurements now indicate the emulsion to be oil-in-water rather than water-in-oil, and creaming occurs from the top rather than from the bottom. Other indications of the phase change are the foaming that occurs in the bottles when they are shaken in contrast to the lack of foaming when water-in-oil emulsions are shaken. In addition, the emulsion now produces a stream that foams on contact, whereas the previous water-in-oil emulsion gave a non-foaming spray. Thus, the change from water-in-oil emulsion to an oil-in-water emulsion is unmistakable.

These results are similar to those obtained by Sherman,[70] who noted that the viscosities of water-in-oil emulsions prepared with distilled water in mineral oil reached a maximum value between 77 and 82 wt.% water. At higher concentrations of water, inversion of the emulsion occurred. The specific concentration at which inversion occurred was a function of the amount of emulsifying agent present.

Occasionally, it was observed that of two supposedly duplicate samples prepared at the Freon® Products Laboratory with 20 wt.% fluorocarbon propellant and 80 wt.% water, one was a very viscous water-in-oil emulsion, while the other was a less viscous oil-in-water type. The emulsion stability of the former was greater than 1 h with creaming starting from the bottom, while the emulsion stability of the latter was less than 1 min with creaming starting from the top. This was probably a case of the formation of dual emulsions (emulsions with the same compositions but having opposite phase types), where the system is such that the type of emulsion obtained depends upon how the bottle is shaken. This phenomenon was first observed by Ostwald[71] and later confirmed by Cheesman and King.[72]

Storage Stability

EMULSIONS WITH HYDROCARBONS OR HYDROCARBON/FLUOROCARBON BLENDS Aerosol insecticides had adequate storage stability when properly formulated according to Glessner[68] and Jones and Weaning.[66] The insecticide had to be chosen with care, since some are affected by water and will produce chlorides or acids, which in turn cause corrosion.

Millions of cans of room deodorants and space insecticides have been packaged as water-in-oil emulsions and successfully marketed. This shows that basically water-in-oil emulsions formulated with hydrocarbons or hydrocarbon/fluorocarbon blends have sufficient storage stability.

EMULSIONS WITH FLUOROCARBON PROPELLANTS   The storage stability of fluorocarbon 12/fluorocarbon 114/solvent/water systems appears to be good.[64] Fluorocarbon 12/fluorocarbon 11/water emulsions, however, are fairly corrosive. The presence of nitromethane stabilizes the system sufficiently so that the storage stability is excellent.

The effectiveness of nitromethane as a stabilizer and corrosion inhibitor for water-in-oil emulsions containing fluorocarbon 11 compared to its relatively poor activity with oil-in-water emulsions may be due to several factors. Boe[62] has suggested that water-in-oil systems might be expected to show less corrosion than oil-in-water emulsions because in the former the dispersed water has less tendency to come into contact with the container and valve. It certainly is true that in a water-in-oil emulsion, the dispersed water droplets are surrounded by an interfacial film of surface-active agent that would minimize contact of the water droplets with metal. This is important because it has been shown that metal catalyzes the decomposition of fluorocarbon 11 by water. The interfacial film surrounding the droplets must have sufficient strength so that coalescence of the droplets does not occur, even if creaming does. Otherwise, a separate water layer will form after coalescence and contact with metal will then take place. Nitromethane probably functions as a stabilizer by forming a monomolecular film on the metal. This further minimizes contact of water with metal.

REFERENCES

1. M. J. Root, *Am. Perfum. Aromat.* **71**, 63 (1958).
2. P. A. Sanders, *Soap Chem. Spec.* **41**, 117 (December 1965).
3. P. A. Sanders, *Soap Chem. Spec.* **39**, 63 (September 1963).
4. P. A. Sanders, *J. Soc. Cosmet. Chem.* **17**, 801 (1966).
5. P. A. Sanders, *Soap Chem. Spec.* **43**, 68 (July 1967).
6. J. W. Hart and A. C. Cook, *Am. Perfum. Aromat.* **77**, 49 (1962).
7. J. G. Spitzer, I. Reich, and N. Fine, U.S. Patent 2,655,480 (October 1953).
8. H. R. Shepherd, "Aerosol Cosmetics," in E. Sagarin (Ed.), "Cosmetics: Science and Technology," Interscience Publishers, Inc., New York, 1957.
9. A. Herzka and J. Pickthall, "Pressurized Packaging," Academic Press, Inc., New York, 1958.
10. H. R. Shepherd (Ed.), "Aerosols: Science and Technology," Interscience Publishers, Inc., New York, 1961.
11. A. Herzka (Ed.), "International Encyclopaedia of Pressurized Packaging" (Aerosols), Permagon Press, Inc., New York, 1966.

12. A. M. Schwartz, J. W. Perry, and J. Berch, "Surface Active Agents and Detergents," Vol. 2, Interscience Publishers, Inc., New York, 1958.
13. P. Becher, "Emulsions: Theory and Practice," Reinhold Publishing Corp., New York, 1958.
14. McCutcheon's Detergents and Emulsifiers, 1977 North American Edition, MC Publishing Company, Ridgewood, N.J.
15. S. J. Strianse, "Hand Creams and Lotions," in E. Sagarin (Ed.), "Cosmetics: Science and Technology," Interscience Publishers, Inc., New York, 1957.
16. Drew Chemical Corp., New York.
17. P. Carter and H. M. Truax, *Proc. Sci. Sec. Toilet Goods Assoc.* **35**, 37 (1961).
18. M. H. Guest, "Shaving Soaps and Creams," in E. Sagarin (Ed.), "Cosmetics: Science and Technology," Interscience Publishers, Inc., New York, 1957.
19. R. A. Foresman, Jr., *Proc. 39th Ann. Meet. Chem. Spec. Manuf. Assoc.*, 32 (December 1952).
20. D. L. Courtney, Paper presented at Society of Cosmetic Chemists Meeting, New York City, December 1975.
21. Cosmetic Bulletin, "Brij" Surfactants (Polyoxyethylene Fatty Ethers), Atlas Chemical Industries, Inc., November 1964.
22. P. A. Sanders, *J. Soc. Cosmet. Chem.* **20**, 577 (1969).
23. P. A. Sanders, *J. Soc. Cosmet. Chem.* **21**, 377 (1970).
24. J. H. Schulman and E. G. Cockbain, *Trans. Faraday Soc.* **36**, 651 (1940).
25. P. Becher and A. J. Del Vecchio, *J. Phys. Chem.* **68**, 3511 (1964).
26. P. A. Sanders, *J. Soc. Cosmet. Chem.* **24**, 87 (1973).
27. P. A. Sanders, *J. Soc. Cosmet. Chem.* **24**, 623 (1973).
28. E. Levy, Personal Communication, The Gillette Company, Boston, Mass., 1972.
29. I. C. Edmundson, "Particle Size Analysis," in H. S. Bean, A. H. Bechett, and J. E. Carless (Eds.), "Advances in Pharmaceutical Science," Vol. 2, Academic Press, Inc., London and New York, 1967.
30. L. J. Augsburger and R. F. Shangraw, *J. Pharm. Sci.* **57**, No. 4 (April 1968).
31. J. L. York, *J. Soc. Cosmet. Chem.* **7**, 204 (1956).
32. M. V. Wiener, *J. Soc. Cosmet. Chem.* **9**, 289 (1958).
33. M. G. DeNavarre, "Chemistry and Manufacture of Cosmetics," D. Van Nostrand and Company, Inc., New York, 1941, p. 191.
34. M. B. Epstein, A. Wilson, C. W. Jakob, L. E. Conroy, and J. Ross, *J. Phys. Chem.* **58**, 60 (1954).
35. P. A. Sanders, *Aerosol Age,* **8**, 33 (July 1963).
36. F. T. Reed, *Proc. 39th Ann. Meet. Chem. Spec. Manuf. Assoc.*, 30 (December 1952).
37. M. D. Richman and R. F. Shangraw, *Aerosol Age* **11**, 36, 28 (July/November 1966).
38. A. E. Alexander, "Emulsions in Theory and Practice," *Amer. Perfum. Cosmet.* **84**, 27 (January 1969).
39. F. B. Rosevear, *Amer. Chem. Soc.* **31**, 628 (1954).
40. F. B. Rosevear, *J. Soc. Cosmet. Chem.* **19**, 58 (1968).
41. R. S. Porter and J. F. Johnson, "Ordered Fluids and Liquid Crystals," Advan. Chem. Ser., No. 63, American Chem. Soc. (1967).
42. J. Boffey, R. Collison, and A. S. C. Lawrence, *Trans. Faraday Soc.* **55**, 654 (1959).
43. B. A. Mulley, "Solubility in Systems Containing Surface Active Agents," in H. S. Bean, A. H. Bechett, and J. E. Carless (Eds.), "Advances in Pharmaceutical Science," Vol. 1, Academic Press, Inc., London and New York, 1964.
44. P. Ekwall, I. Danielson, and I. Mandell, *Kolloid Z.* **169**, 113 (1960).
45. A. S. C. Lawrence and A. J. Hyde, Proceedings of the Third International Congress on Surface Activity, Cologne, 1960, Sect. A, Vol. I, 21 (1960).

46. P. A. Winsor, "Solvent Properties of Amphiphilic Compounds," Butterworths, London, 1954.
47. F. Atkins, *Perfum. Essent. Oil Rec. 332* (March 1934).
48. R. Kohlhaas, *Chem. Rev.* **82**, 487 (1949).
49. J. A. Kitchener and P. R. Mussellwhite, "The Theory of Stability of Emulsions," in P. Sherman (Ed.), "Emulsion Science," Chap. 2, Academic Press, Inc., New York, 1968.
50. E. J. W. Verwey, "Electrical Double Layer and Stability of Emulsions," *Trans. Faraday Soc.* **36**, 192 (1940).
51. J. J. Bikerman, "Foams," Springer-Verlag, New York, Heidelberg, and Berlin 1973.
52. V. M. Martynov, *Kolloidn. Zh.* **11**, 225 (1949).
53. E. G. Cockbain, *Trans. Faraday Soc.* **48**, 185 (1952).
54. J. K. Dixon, D. J. Salley, N. A. Argyle, and A. J. Weith, *Proc. Roy. Soc.* (London) **203A**, 42 (1950); J. K. Dixon, D. J. Salley, N. A. Argyle, and A. J. Weith, *J. Chem. Phys.* **19**, 378 (1951).
55. V. J. Defay, *J. Chem. Phys.,* **46**, 375 (1948).
56. B. V. Derjaguin and L. Landau, *Acta. Phys. Chim. USSR* **14**, 633 (1941).
57. J. T. Davies and E. K. Rideal, "Interfacial Phenomena," 2nd ed., Chap. 5, Academic Press, Inc., New York and London, 1963.
58. W. Drost-Hansen, "Aqueous Interfaces," in S. Ross (Ed.), "Chemistry and Physics of Interfaces," Chap. 3, American Chemical Society Publications, Washington, D.C., 1965.
59. J. C. Henniker, *Rev. Mod. Phys.* **21**, 322 (1949).
60. J. W. McBain, "Colloid Science," D. C. Heath and Company, Boston, Mass., 1950.
61. P. F. Low, "Physical Chemistry of Clay-Water Interactions," in A. G. Norman (Ed.), "Advances in Agronomy," Vol. 13, Academic Press, Inc., New York and London, 1961.
62. C. F. Boe, U.S. Patent 2,524,590.
63. "Aerosol Emulsions with Freon® Propellants," Freon® Aerosol Report FA-21.
64. P. A. Sanders, *J. Soc. Cosmet. Chem.* **9**, 274 (1958).
65. M. E. Fowks, *Aerosol Age* **8**, 25 (March 1963).
66. C. D. G. Jones and A. J. S. Weaning, *Aerosol Age* **11**, 21 (February 1966).
67. C. J. Baker and J. B. Moore, *Proc. 50th Mid-Year Meet. Chem. Spec. Manuf. Assoc.,* 63 (May 1964); *Aerosol Age* **9**, 28 (November 1964).
68. A. S. Glessner, *Aerosol Age* **9**, 98 (October 1964).
69. M. E. Clayton and R. J. Scott, *Proc. 50th Mid-Year Meet. Chem. Spec. Manuf. Assoc.,* 68 (May 1964).
70. P. Sherman, *J. Soc. Chem. Ind.* **69**, 570 (1950).
71. W. O. Ostwald, *Kolloid Z.* **6**, 103 (1910).
72. D. F. Cheesman and A. King, *Trans. Faraday Soc.* **34**, 594 (1938).
73. R. T. Briggs, *J. Phys. Chem.* **24**, 120 (1920).
74. E. S. R. Gopal, "Principles of Emulsion Formation," in P. Sherman (Ed.), "Emulsion Science," Chap. 1, Academic Press, Inc., New York, 1968.

19

AQUEOUS ALCOHOL FOAMS

Aqueous alcohol foams consist basically of a mixture of water, alcohol, surface-active agent, and propellant. The aqueous alcohol foam system is unique in several respects. Generally, the combinations of water, alcohol, and propellant are mutually soluble, so that the aerosol is not an emulsion system. The system can produce foams with over 50 wt. % ethyl alcohol. This has made possible the development of products which previously could not be formulated as foams using conventional aqueous systems.

Some typical aerosols based upon the aqueous ethyl alcohol foam system are listed in Table 19-1.

Aqueous ethyl alcohol foams are discussed in detail in References 1 and 2. Unless otherwise indicated, the data in the subsequent sections were obtained from these two sources.

COMPOSITIONS OF AQUEOUS ETHYL ALCOHOL FOAMS

Water/Ethyl Alcohol/Propellant Ratios

Most combinations of water, ethyl alcohol, and propellant are not miscible and form two liquid layers when mixed. Aqueous ethyl alcohol foams can be prepared from these immiscible combinations, but the products are emulsions and do not have the advantages of a homogeneous system. However, there is a much smaller region in the water/ethyl alcohol/propellant system where the three components are mutually soluble and form a clear solution. Compositions in the miscible region are of particular interest for the formulation of aqueous ethyl alcohol foams, since these products are not emulsion systems. The solubility characteristics of water/ethyl alcohol/propellant combinations are, therefore, of considerable importance because they determine whether a given composition will be in the miscible or immiscible region.

TABLE 19-1   SOME AQUEOUS ETHYL ALCOHOL FOAM PRODUCTS

1. After shave	5. Hair setting foam
2. Athlete's Foot Relief	6. Insect repellent
3. Cologne	7. Skin moisturizer
4. Hair dressing	8. Sunscreen

Miscibility Characteristics with Different Propellants

A wide variety of propellants, including fluorocarbons, hydrocarbons, and mixtures of the two, have been proposed for use with aqueous ethyl alcohol foams.[1,3,4] The miscibility characteristics of water/ethyl alcohol/propellant systems vary considerably, depending upon the particular propellant present. The location of the solubility limit boundary that divides the immiscible and the miscible regions depends upon the propellant.

The miscibility characteristics of a number of water/ethyl alcohol/fluorocarbon propellant mixtures are illustrated on the triangular coordinate chart of Figure 19-1. Compositions to the left of the solubility limit boundaries are not miscible and form two liquid phases. Products with compositions in these regions have to be formulated as emulsion systems. Compositions to the right of the

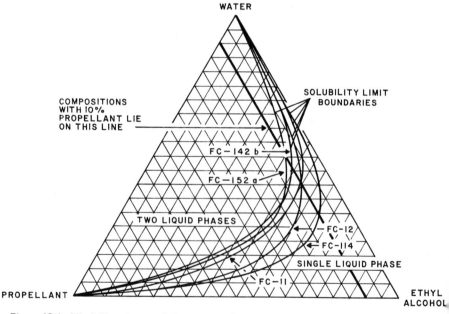

Figure 19-1 Miscibility characteristics of water/ethyl alcohol/fluorocarbon propellant mixtures at 70° F.

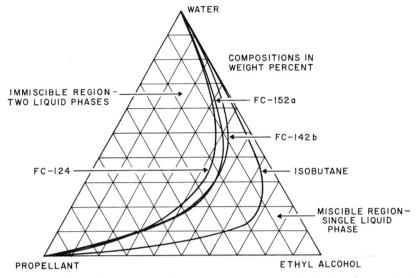

Figure 19-2 Miscibility characteristics of water/ethyl alcohol/propellant (isobutane and fluorocarbons 124, 142b, and 152a) mixtures at 70°F.

curves are mutually soluble and form a single liquid phase. The miscibility diagrams indicate only whether a given mixture of water, ethyl alcohol, and propellant will be immiscible and form two liquid layers or will be miscible and form a single liquid phase. They do not indicate the areas in either region where foams can be obtained.

The miscibility characteristics of fluorocarbon 124 and isobutane in water/ ethyl alcohol solutions in comparison with those of fluorocarbons 142b and 152a are illustrated in Figure 19-2. Fluorocarbon 124 is more soluble than fluorocarbons 142b and 152a, while isobutane is considerably less soluble.

Miscibility Characteristics with Different Alcohols and Acetone

Although ethyl alcohol is the only alcohol used to any extent in aqueous alcohol foam products, foams can also be obtained with other alcohols and acetone. The particular alcohol present in a water/alcohol/propellant system has a considerable effect upon the miscibility properties of the system, and thus affects the location of the miscible and immiscible regions.

The miscibility characteristics of three-component mixtures of water and fluorocarbon 12/fluorocarbon 114 (40/60) propellant with methyl alcohol, ethyl alcohol, isopropyl alcohol, and acetone are illustrated on the triangular coordinate chart of Figure 19-3. The largest miscible region is obtained with isopropyl alcohol, but it gives the poorest foams. Methyl alcohol gives the small-

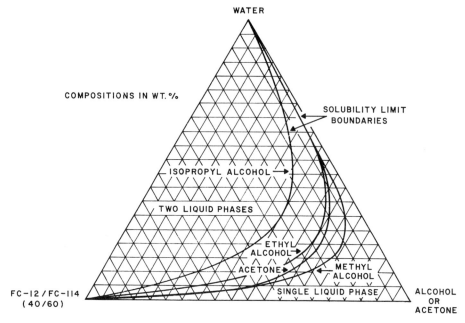

Figure 19-3 Miscibility characteristics of fluorocarbon 12/fluorocarbon 114 (40/60) propellant in water/methyl alcohol, water/ethyl alcohol, water/isopropyl alcohol, and water/acetone solutions at 70°F.

est miscible area, while ethyl alcohol and acetone are in between with fairly similar miscibility characteristics.

When an aqueous alcohol foam system is limited to a specific alcohol, surface-active agent, and propellant, the proportions of the components that give both a homogeneous system and a satisfactory foam become even more limited. This is illustrated by the small area of acceptable foam compositions on the triangular coordinate chart of Figure 19-4. Only water/ethyl alcohol weight percent ratios in the range from about 38/62 to 30/70 give acceptable foams. If, with this particular system, the water/ethyl alcohol/propellant ratios deviate from the limited area of satisfactory compositions, the product will end up either as an emulsion system or a nonfoaming aerosol.

Foam Compositions in the Miscible Regions

Foams with adequate stability can be obtained only in a very limited part of the miscible region of each water/alcohol/propellant system. The boundaries of this limited area depend upon the particular surface-active agent, alcohol, and propellant present. They also depend upon the criteria used to define adequate stability of a foam. Lanzet[3] believes that foams should be of the self-collapsing type with durations of from 10–60 sec and reports that consumers prefer foams

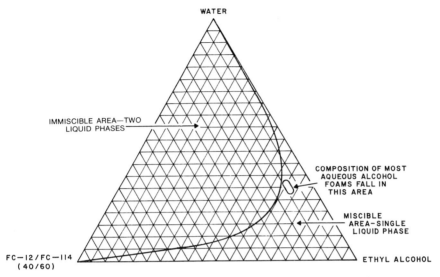

Figure 19-4 Miscibility characteristics of the water/ethyl alcohol/"Polawax"/fluorocarbon 12/fluorocarbon 114 (40/60) propellant system at 70° F ("Polawax," a surface-active agent, will be discussed later in the chapter).

of this type to the more stable ones. He has obtained foams that fall within his classification from homogeneous aqueous ethyl alcohol systems containing as little as 30.4 vol. % ethyl alcohol.

Klausner[4] refers to aqueous alcohol foams as foams of limited stability and defines them as foams that do not become completely liquefied when exposed to the atmosphere for 15 min or longer, but which revert to a liquid within about 2 sec when rubbed. Klausner gives the following critical ratios of components for aqueous foams that meet his specifications (Table 19-2).

The composition of a typical basic aqueous ethyl alcohol foam that falls within the specified limits in Figure 19-4 is given in Table 19-3.

The water/ethyl alcohol ratio in the above formulation is 35/65. Excellent foams can be obtained with compositions falling in the miscible regions of

TABLE 19-2 CRITICAL RATIOS FOR AQUEOUS ALCOHOL FOAMS

Component	Composition (wt. %)
Alcohol	46–66
Water	28–42
Surfactant	0.5–5
Propellant	2–15

From Klausner.[4]

TABLE 19-3   COMPOSITION OF A TYPICAL BASIC
AQUEOUS ETHYL ALCOHOL FOAM

Component	Concentration (wt. %)
Water	30.5
Ethyl alcohol	56.5
"Polawax"	3.0
FC-12/FC-114 (40/60) propellant	10.0

aqueous systems based upon methyl alcohol, ethyl alcohol, or acetone, but not from isopropyl alcohol. The stability of the aqueous alcohol or acetone foam appears to be related to the relative areas of the miscible regions of the water/alcohol/propellant systems. The miscibility data in Figure 19-3 show that isopropyl alcohol produces the largest miscible region and methyl alcohol the smallest.

Foam Compositions in the Immiscible Regions

Stable foams can be obtained over a wide range of water/alcohol ratios throughout the immiscible regions. Products formulated in the immiscible regions are emulsions and have to be shaken before use. Quite often these products have an initial wet, sloppy discharge. In these cases, shaking emulsifies the propellant in the bulk phase, but not in the dip tube. This problem can be minimized by using a capillary dip tube or a valve without a dip tube.

Surface-Active Agents

Surface-active agents for producing foams from aqueous alcohol systems can be divided into two groups: those that are effective for the homogeneous compositions in the miscible region and those that are satisfactory for the emulsion systems in the immiscible region. The group of surfactants for the miscible region is much smaller than that for the emulsion systems.

Surfactants for the Miscible Region

Although a number of surface-active agents have been reported to be satisfactory for homogeneous aqueous ethyl alcohol foams,[1,3,4] the most effective agents are fatty alcohol combinations, ethoxylated fatty alcohols, and mixtures of the fatty alcohols with their ethoxylated derivatives.

"Polawax." "Polawax" was one of the first surface-active agents reported for aqueous ethyl alcohol foams,[5,6] and it is still one of the most commonly used. As a result, much of the basic information on aqueous ethyl alcohol foams has been obtained with "Polawax" as the surface-active agent. "Polawax" is described

as an ethoxylated stearyl alcohol with auxiliary emulsifying agents, but its composition has not been disclosed by Croda, Inc. It is probably the best single product for aqueous ethyl alcohol foams.

Fatty Alcohol Combinations. Cetyl alcohol/stearyl alcohol combinations give excellent pressure-sensitive foams, but they generally are not soluble in the water/ethyl alcohol/propellant mixtures. This produces a hazy product or one with suspended solids. If the aerosol product is packaged in an opaque bottle, as most aqueous ethyl alcohol foam products are, the haziness is of little consequence. However, the formulation must be checked to be certain that the suspended particles do not cause valve clogging.

One of the factors that controls the stiffness of foams with these combinations is the ratio of cetyl alcohol to stearyl alcohol, as shown in Table 19-4. The data were obtained with an aerosol formulation of 88 wt. % water/ethyl alcohol (35/65) solution, 2 wt. % total surface-active agent, and 10 wt. % fluorocarbon 12/fluorocarbon 114 (40/60) propellant.

One of the most interesting features of the cetyl alcohol/stearyl alcohol surfactant combination is that neither fatty alcohol by itself is effective. The purity and composition of commercial cetyl and stearyl alcohol varies, depending upon the source, and fatty alcohols from different suppliers may give different results.

Ethoxylated Fatty Alcohols. The only ethoxylated fatty alcohol which by itself gave a satisfactory foam was "Siponic" E-O (American Alcolac Corp.). This surfactant is the condensation product of 1 mole of ethylene oxide with 1 mole of stearyl alcohol. Other ethoxylated fatty alcohols containing higher proportions of ethylene oxide were less effective, possibly because of their increased water solubility.

Fatty Alcohol/Ethoxylated Fatty Alcohol Combinations. Very stiff, stable foams can be obtained with cetyl alcohol/stearyl alcohol/ethoxylated fatty

TABLE 19-4 EFFECT OF CETYL ALCOHOL/STEARYL
ALCOHOL RATIO UPON FOAM STIFFNESS

Fatty Alcohol Concentration in Aerosol (wt %)		Foam Stiffness (g)
Cetyl Alcohol	Stearyl Alcohol	
2.0	0.0	0
1.5	0.5	14
1.0	1.0	52
0.5	1.5	36
0.0	2.0	0

TABLE 19-5   SURFACE-ACTIVE AGENTS FOR AQUEOUS ALCOHOL (50/50) EMULSION SYSTEMS

1. Hydroxylated lecithin	6. Polyoxyethylene (2) stearyl ether
2. Emulsifying wax #215	7. Propylene glycol monostearate (SE)
3. Glycerol monostearate (SE)	8. Sorbitan monopalmitate
4. Kesscowax A-33	9. Sorbitan monostearate
5. "Polawax"	

alcohol combinations. The main disadvantage of these combinations is their general lack of solubility in the water/ethyl alcohol/propellant system, so that suspended solid material is present.

The stiffness of the foams depends upon the ratios of the three components in the surfactant blend.[1]

Surfactants for the Immiscible Region (Emulsion Systems)

In the immiscible region of emulsions, the water/alcohol weight percent ratio varies from 100/0 to about 38/62 with a fluorocarbon 12/fluorocarbon 114 (40/60) propellant concentration of 8-10 wt.%. The ethyl alcohol concentration in the aerosols varies from 0 wt.% to slightly over 53 wt.%. At higher alcohol concentrations the water/ethyl alcohol/propellant system becomes miscible in all proportions.

At low alcohol concentrations, the emulsion system is essentially an aqueous system and most of the surface-active agents that are effective in aqueous systems can be used. As the concentration of ethyl alcohol in the emulsion system increases, the number of surface-active agents that give adequate foams decreases. Surface-active agents that gave acceptable foams from an aerosol system consisting of 86 wt.% water/ethyl alcohol (50/50) solution, 4 wt.% surfactant, and 10 wt.% fluorocarbon 12/fluorocarbon 114 (40/60) are listed in Table 19-5.

Propellants

The most commonly used propellants for the aqueous ethyl alcohol foam products are blends of fluorocarbon 12 and fluorocarbon 114 with weight percent ratios varying from about 15/85 to 40/60. Pressures in the bottles generally do not exceed 25 psig at 70°F with these propellants as long as the water/ethyl alcohol/propellant mixtures are miscible. Fluorocarbon 142b is also used to a limited extent to provide quick-breaking foams. Fluorocarbon 124 has not been investigated in this system, but fluorocarbon 152a has been used to a slight extent. Fluorocarbons 123 and 133a have too low a vapor pressure to be effective.

FOAM FORMATION AND STABILIZATION

The formation of a foam from a clear aqueous ethyl alcohol aerosol can be explained by two concepts: the stabilization of foams by solid materials and solubility parameters. Data obtained for aqueous alcohol foam systems indicate that these foams belong to the class known as three-phase foams. In this type of foam, stability is achieved by the presence of solid materials that are attracted or deposited at the gas–liquid interface. According to Bikerman,[7] the solid materials are partially immersed in the liquid phase and thus prevent the gas bubbles from coming into contact with each other and coalescing. The solid material that performs this function in aqueous ethyl alcohol foams is "Polawax."

The application of solubility parameters to the aqueous ethyl alcohol foam system is useful in gaining an understanding of how solid "Polawax" is deposited at the gas–liquid interface during discharge of the aerosols. The theory of solubility parameters is covered in Chapter 10 and in References 2, 8, and 9; however, it might be helpful to mention a few points about solubility parameters at this time. A solubility parameter is a fundamental physical property of a compound, just like density, boiling point, or melting point. Solubility parameters for a large number of liquids and solids have been calculated or determined experimentally and are available in tables.[8, 9] Solubility parameters can be used to predict if an amorphous solid will dissolve in a liquid: If the solubility parameters of the solid and liquid are close to each other, the solid will usually dissolve in the liquid.

One important feature of the solubility parameter concept that applies to an aqueous ethyl alcohol foam system containing "Polawax" is that it explains why two liquids, neither of which by itself is a solvent for a given compound, may dissolve the compound when mixed together. For example, although neither ether nor ethyl alcohol alone is a solvent for nitrocellulose, the mixture of the two liquids will dissolve nitrocellulose. The explanation for this is that ethyl alcohol has too high a solubility parameter to dissolve the nitrocellulose, while ether has too low a solubility parameter. However, the average solubility parameter of the mixture of ether and ethyl alcohol is sufficiently close to that of nitrocellulose so that the mixture becomes a solvent for nitrocellulose.

Solubility Properties of "Polawax"

The solubility properties of "Polawax" in a basic aqueous ethyl alcohol foam system explain why it dissolves in the water/ethyl alcohol/propellant mixture but precipitates from solution during discharge of the aerosol. The solubility of "Polawax" in the components of a typical aqueous ethyl alcohol foam aerosol with a composition of 87 wt.% water/ethyl alcohol (35/65) solution, 3 wt.% "Polawax," and 10 wt.% fluorocarbon 12/fluorocarbon 114 (40/60) propellant is shown in Table 19-6.

"Polawax" is insoluble in either water/ethyl alcohol (35/65) solution or fluorocarbon 12/fluorocarbon 114 (40/60) propellant. However, "Polawax" is soluble in the mixture of water/ethyl alcohol (35/65) solution and propellant. This is analogous to the ether/ethyl alcohol/nitrocellulose system described previously. In the present case, the solubility parameter of water/ethyl alcohol

TABLE 19-6 SOLUBILITY OF "POLAWAX"

Insoluble	Soluble
Water/ethyl alcohol (35/65) solution FC-12/FC-114 (40/60) propellant	Mixture of water/ethyl alcohol solution (35/65) plus FC-12/FC-114 (40/60) propellant

(35/65) solution is too high to dissolve the "Polawax" and that of the propellant is too low. Apparently, the average solubility parameter of the mixture of water/ethyl alcohol solution and fluorocarbon 12/fluorocarbon 114 (40/60) propellant is close enough to that of "Polawax" so that the combination of the three liquids dissolves "Polawax."

To verify this theory, the solubility of "Polawax" was determined in a variety of solvents with known solubility parameters. Burell[8] classified solvents into three groups according to their hydrogen-bonding capacity. The solubility of "Polawax" in these three groups of solvents is shown in Table 19-7.

Thus, "Polawax" was soluble in solvents having solubility parameters in the range of 7.5–14.5, depending upon the hydrogen-bonding capacity of the solvent. It is insoluble in fluorocarbon 12/fluorocarbon 114 (40/60) because the propellant has a solubility parameter of about 6.2 (Chapter 10), which is too low. It is also insoluble in the water/ethyl alcohol (35/65) solution, which has a solubility parameter greater than 14.5 (the solubility parameter of ethyl alcohol is 12.7 and that of water is 23). When propellant is added to the water/ethyl alcohol (35/65) solution, the solubility parameter of the resulting mixture is shifted downward to where the mixture becomes a solvent for "Polawax." The solubility parameter of water/ethyl alcohol (35/65) solution apparently is very close to the upper limit for that of "Polawax," so that only a comparatively minor change in solubility parameter is enough to cause the mixture to become a solvent for it.

TABLE 19-7 SOLUBILITY OF "POLAWAX" IN VARIOUS SOLVENTS

Hydrogen-Bonding Capacity of Solvents	Insoluble in Solvents with these Solubility Parameters	Soluble in Solvents with these Solubility Parameters
Low hydrogen bonding	7.4, 10.7	7.5–10.0
Medium hydrogen bonding	14.8	8.4–10.8
High hydrogen bonding	9.1, 16.5	9.5–14.5

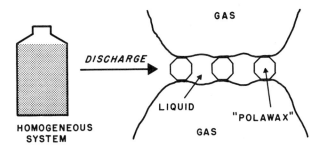

Figure 19-5   Foam formation from aqueous ethyl alcohol systems.

Mechanism of Foam Formation

The mechanism of foam formation can now be understood on the basis of the solubility properties of "Polawax" in the water/ethyl alcohol/propellant system, assuming that solid "Polawax" is necessary to stabilize the foam structure. As shown in the previous section, the presence of both the aqueous ethyl alcohol solution and the propellant is necessary in order to dissolve "Polawax." However, when the aerosol is discharged, the propellant changes from a liquid to a gas and is no longer available as a solvent. Since "Polawax" is not soluble in the remaining water/ethyl alcohol (35/65) solution, it precipitates, thus providing the solid form of the surfactant required for stabilization of the foam (Figure 19-5).

The mechanism of foam formation and stabilization described above applies to a clear homogeneous aerosol system. It is not necessary for "Polawax" to be in solution initially in order to obtain a stable foam. All that is required is for "Polawax" to be available in solid form after the propellant has been vaporized. If "Polawax" is insoluble in the aerosol initially, foam stabilization will still occur because insoluble "Polawax" is present for foam stabilization.

FACTORS AFFECTING FOAM STABILITY

One requirement for a stable foam is the presence of solid surfactant after the product had been discharged. If the surfactant is soluble in the aqueous alcohol concentrate and remains in solution after the product has been discharged, no foam will be obtained. This explains why changes in the water/alcohol ratio and temperature affect foam stability. It also explains why the addition of certain solvents to the aqueous ethyl alcohol foam system destroys the foam.

Variation in Alcohols

Ethyl Alcohol

Stable foams were obtained with water/ethyl alcohol weight percent ratios varying from 100/0 to about 40/60. At lower water/alcohol weight percent ratios,

TABLE 19-8 FOAM STABILITY AND STIFFNESS AS A FUNCTION OF THE WATER/ETHYL ALCOHOL RATIO

Water–Ethyl Alcohol Ratio (wt. %)	Foam Stiffness (g)	Foam Stability	Type of System	Solubility of "Polawax"	
				Aerosol	Aqueous Alcohol Concentrate
100/0	50	Stable	Emulsion	Insoluble	Insoluble
75/25	47	"	"	"	"
50/50	61	"	"	"	"
40/60	72	"	"	"	"
35/65	61	Medium stability	Clear, homogeneous	Soluble	"
30/70	25	Low stability	"	"	Almost soluble
25/75	0	No foam	"	"	Soluble

* The dotted line separates the immiscible and miscible regions.

the alcohol concentration becomes very important, and no foam is obtained at a ratio lower than 30/70 (Table 19-8).

The solubility of "Polawax" in both the aerosols and the aqueous ethyl alcohol concentrates is also listed in Table 19-8. The data indicate that the solubility of "Polawax" in the aqueous ethyl alcohol concentrate is the factor that controls foam stability. Throughout the immiscible region, "Polawax" is insoluble in both the aqueous ethyl alcohol concentrates and the aerosols; therefore, solid surfactant is always available to stabilize the foam after the product is discharged.

In the miscible region, "Polawax" is soluble in the aerosol with the water/ ethyl alcohol weight percent ratio of 35/65, but insoluble in the concentrate. It precipitates out when the product is discharged and stabilizes the foam. However, it is almost completely soluble in the concentrate with the 30/70 weight percent ratio and, therefore, most of the "Polawax" remains in solution when the product is discharged. The resulting foam is thin and unstable. No foam is obtained with the water/alcohol weight percent ratio of 25/75 because "Polawax" is soluble in this concentrate.

Other Alcohols and Acetone

Stable foams may be obtained from aqueous solutions of methyl alcohol and acetone at the same (or even lower) water/alcohol or water/acetone weight percent ratios as those with ethyl alcohol. Isopropyl alcohol gives only thin foams at very high water/alcohol weight percent ratios.

The stability and stiffness of foams from the three alcohols and acetone at four water/solvent weight percent ratios are given in Table 19-9. Although foam stiffness values do not correlate directly with foam stability, they are an indication of foam stability to the extent that a foam has to have a certain degree of stability in order to have a stiffness value.

TABLE 19-9   FOAM STIFFNESS WITH VARIOUS ALCOHOLS
AND ACETONE[a]

Water–Alcohol or Water–Acetone Ratio (wt %)	Foam Stiffness (g)			
	Methyl Alcohol	Acetone	Ethyl Alcohol	Isopropyl Alcohol
35/65	46	96	61	No foam
30/70	52	51	25	No foam
20/80	44	No foam	No foam	No foam
15/85	No foam	No foam	No foam	No foam

[a] Aerosol formulation: 86 wt.% water/alcohol or acetone solution, 4 wt.% "Polawax," and 10 wt.% fluorocarbon 12/fluorocarbon 114 (40/60) propellant.

Although no data are available regarding the solubility of "Polawax" in aqueous methyl alcohol, acetone, or isopropyl alcohol concentrates, it is probably this factor that determines the stability of foams from systems containing these solvents.

Temperature

In the miscible region, increasing the temperature at which the aerosol is discharged decreases foam stability and, conversely, lowering the temperature increases foam stability. This is illustrated by the data in Table 19-10, obtained with an aerosol formulation of 87 wt.% water/ethyl alcohol solution, 3 wt.% "Polawax," and 10 wt.% fluorocarbon 12/fluorocarbon 114 (40/60) propellant.

The effect of temperature upon foam stability is related to its effect upon the solubility of "Polawax" in the water/ethyl alcohol concentrate. The formulation with the water/alcohol weight percent ratio of 35/65 does not foam at 110°F because "Polawax" is soluble in the concentrate at this temperature and, consequently, does not precipitate when the product is discharged. This is also why aqueous ethyl alcohol foams break so rapidly when discharged on the skin. The warmth of the skin causes the solid "Polawax" to dissolve in the aqueous alcohol phase, and this leads to collapse of the foam.

When the temperature of the products with the water/alcohol weight percent ratios of 30/70 or 25/75 is decreased, foam stability is increased because "Polawax" is less soluble at the lower temperatures. Therefore, more "Polawax" precipitates from the aerosol when it is discharged at the lower temperatures and increased foam stabilization results.

Concentration of Surface-Active Agent

Since the presence of a surface-active agent is necessary in order to obtain a stable foam, the concentration of the agent would be expected to be one of the factors controlling foam stability. The variation of foam stiffness with a change

TABLE 19-10   EFFECT OF TEMPERATURE UPON FOAM STABILITY

Water–Ethyl Alcohol Ratio (wt %)	Foam Stability	
	Room Temperature	Other Temperatures
35/65	Good	No foam at 110°F
30/70	Poor	Fair at 60°F
25/75	No foam	Unstable, large bubble size foam at 32°F

TABLE 19-11   EFFECT OF "POLAWAX" CONCENTRATION UPON FOAM STIFFNESS

Components	Composition (wt %)			
	#1	#2	#3	#4
Water–ethyl alcohol (35/65) solution	86.0	88.0	88.5	89.0
"Polawax"	4.0	2.0	1.5	1.0
FC-12/FC-114 (40/60) propellant	10.0	10.0	10.0	10.0
Foam stiffness (g)	73	50	42	Foam too thin and unstable to measure

in concentration of "Polawax" in aqueous ethyl alcohol systems is shown in Table 19-11.

Most formulators try to use as low a concentration of surface-active agent as possible.

Variation in Propellants

The type of propellant can have a considerable effect upon foam properties. A number of propellants were evaluated in an aerosol system consisting of 86 wt.% water/ethyl alcohol (35/65) solution, 4 wt.% "Polawax," and 10 wt.% propellant. The results are presented below.

Fluorocarbon 12

This gave a sputtery discharge and a very dense, cake-icing type of foam. The foam wet paper rapidly, but retained its structure.

Fluorocarbon 114

This propellant gave a product with an excellent discharge and a stable foam with a stiffness of 78 g.

Fluorocarbon 12/Fluorocarbon 114 (40/60)

The discharge characteristics were excellent and the stable foam had a stiffness of 68 g.

Fluorocarbon 12/Fluorocarbon 11 (50/50)

This propellant produced a thin, unstable foam that collapsed in a few minutes on paper.

Fluorocarbon 142b

The aerosol with this propellant produced a thin foam with a stiffness of approximately 19 g. The foam collapsed in about 20 min.

Fluorocarbon 152a

The effect of fluorocarbon 152a was similar to that of fluorocarbon 12. It produced a very dense, thick foam that caused rapid wetting of paper, but exhibited good stability. "Polawax" was not soluble in the aerosol formulation and crystallized out, resulting in an unattractive-appearing product.

Fluorocarbon 124

Fluorocarbon 124 has not been tested in aqueous ethyl alcohol systems, but should give excellent foams.

Effect of Other Solvents

Solvents can either enhance foam properties or destroy them. The effect of various solvents upon the foam properties of an aqueous ethyl alcohol system with a composition of 76 wt.% water/ethyl alcohol (35/65) solution, 10 wt.% solvent, 4 wt.% "Polawax," and 10 wt.% fluorocarbon 12/fluorocarbon 114 (40/60) propellant is shown in Table 19-12.

All solvents that gave good foams had two characteristics in common: They were soluble in the aqueous ethyl alcohol concentrate and their presence in the concentrates did not give water/ethyl alcohol/solvent mixtures in which "Polawax" was soluble. Solubility in the propellant was not a determining factor, since some of the solvents were soluble in the propellants while others were not. Some of the solvents produced exceptionally stiff foams.

The adverse effect of other solvents was a result of several factors. In some cases, "Polawax" became soluble in the concentrate after the addition of the

TABLE 19-12   EFFECT OF SOLVENTS UPON FOAM PROPERTIES

Solvents Giving Good Foams	Solvents Having an Adverse Effect upon Foam Formation and Stability
Acetone	Chloroform
Diethylene glycol	Hexane
Ethylene glycol	Isopropyl myristate
Ethyl acetate	Methyl isobutyl ketone
Glycerine	Odorless mineral spirits
Methyl ethyl ketone	Silicone fluid DC 200 (500 cps)
Propylene glycol	Toluene
	Trichloroethylene

solvent, so that the "Polawax" remained in solution after the product was discharged. In other cases, the solvents were not miscible with the water/alcohol concentrate, but "Polawax" dissolved in the added solvent and therefore did not precipitate when the product was discharged. The addition of some solvents resulted in conversion of the homogeneous system to an emulsion system.

PREPARATION OF AQUEOUS ETHYL ALCOHOL FOAMS

The insolubility of surface-active agents such as "Polawax" in the aqueous ethyl alcohol concentrate at room temperature necessitates special procedures for the commercial production of these aerosols. There are essentially three methods by which the concentrate can be loaded into the containers.

1. "Polawax" can be dissolved in the aqueous ethyl alcohol solution at about 110°F, and the warm solution of "Polawax" can be added to the containers in one step. This would require equipment to keep the solution hot unitl it was added to the containers.

2. "Polawax" can be dissolved in ethyl alcohol at room temperature and the concentrate loaded into the containers in two steps. The aqueous phase is loaded in the first step and the alcoholic solution of "Polawax" is loaded in the second step. "Polawax" will precipitate in the container when the aqueous and ethyl alcohol solutions are mixed, but the surfactant will redissolve after the propellant has been added.

3. "Polawax" can be dispersed in the aqueous ethyl alcohol concentrate at room temperature and the dispersion added to the containers. The dispersion would have to be sufficiently stable so that it remained homogeneous during the addition. The use of other surfactants as an aid in obtaining a stable dispersion of "Polawax" might be helpful.

STORAGE STABILITY OF AQUEOUS ETHYL ALCOHOL FOAMS

High-spot storage stability tests with a typical aqueous ethyl alcohol after-shave foam indicated that tinplate containers should not be used with this product as a result of corrosion and discoloration. Extensive storage stability tests should be carried out with any aqueous ethyl alcohol product packaged in tinplate containers. Most aqueous ethyl alcohol foam products on the market are packaged in glass and a few in aluminum containers.

SOME PROBLEMS WITH AQUEOUS ETHYL ALCOHOL FOAMS

If aqueous ethyl alcohol foam products are cooled slightly below room temperature, "Polawax" may precipitate from solution. This causes an unsightly appearance and is one reason why many aqueous ethyl alcohol foam products are packaged in opaque glass bottles. Precipitated "Polawax" does not redissolve easily unless the samples are warmed up to room temperature or above and shaken. If the aerosol is used before the surfactant is redissolved, valve clogging may occur.

Attempts to solve this problem by adding additional solvents to prevent "Polawax" from precipitating at the lower temperatures generally have been unsuccessful. Any solvent that achieved this objective also increased the solubility of "Polawax" in the aqueous ethyl alcohol concentrate at room temperature. This destroyed foam stability. One partial solution to this problem may be the use of additional dispersing agents to keep precipitated "Polawax" in a finely divided form so that valve clogging does not occur.

Another problem that arises occasionally is precipitation of "Polawax" after most of the product has been used. The reason for this is as follows. As the vapor phase in the container increases (a result of product use), propellant vaporizes from the residual liquid phase and enters the vapor phase. This reduces the concentration of propellant in the liquid phase, and ultimately it becomes too low to keep "Polawax" in solution. This problem can be minimized by keeping the surfactant concentration as low as possible or by increasing the propellant concentration in the aerosol.

AEROSOL FORMULATIONS

A number of formulations for various aqueous ethyl alcohol foam products have been published and are available in References 1, 3, 4, 5, and 6.

REFERENCES

1. P. A. Sanders, *Drug Cosmet. Ind.* **99**, 56, 48 (August/September 1966).
2. P. A. Sanders, *Aerosol Report/AER* **8**, 202 (May 1969) (Freon® Aerosol Report A-75).
3. M. Lanzet, U.S. Patent 1,121,563.
4. K. Klausner, U.S. Patent 3,131,152.
5. "Quick-Breaking Foam Aerosols," *Aerosol Age* **5** (May 1960).
6. T. A. Wallace, *Am. Perfum. Cosmet.* **77**, 85 (1962).
7. J. J. Bikerman, "Foams: Theory and Industrial Applications," Reinhold Publishing Corp., New York, 1953, p. 184.
8. H. Burrell, *Official Digest* **27**, 726 (1955).
9. J. D. Crowley, G. S. Teague, and J. W. Lowe, *J. Paint Technol.* **38**, 269 (May 1966).

20

NONAQUEOUS FOAMS

In contrast to the large number of investigations of conventional aqueous emulsions and foams, relatively little information has been published on nonaqueous systems. Most of the work in the nonaqueous field has involved studies of the formation and behavior of micelles and evaluations of mixtures of surfactants in organic solvents for applications such as dry cleaning and lubrication. The surfactants used for these systems are discussed by Schwartz, Perry, and Berch.[1]

King[2] studied the foaming properties of a considerable number of organic liquids containing commercial surface-active agents. Methyl alcohol, ethylene glycol, glycerol, benzene, nitrobenzene, and chlorinated naphthalene were typical of the liquids investigated. Several combinations of surface-active agents and organic liquids gave foams that compared favorably in volume and stability with those from aqueous systems.

Recently, Petersen and Hamill[3] investigated nonaqueous emulsions of olive oil with glycerine, propylene glycol, and polyethylene glycol 400. Typical anionic, nonionic, and cationic surfactants were used as emulsifying agents. Both olive oil-in-polyol and polyol-in-olive oil emulsions were obtained, and some emulsions were quite stable at very low surfactant concentrations.

In the aerosol field, studies of nonaqueous systems have also been limited. Klausner[4] obtained nonaqueous foams from various combinations of alcohols, dialkyl ketones, glycerol, alkylene glycols, surfactants, and propellants. Nonaqueous foams have also been produced from combinations of surface-active agents and fluorocarbon propellants with glycols, glycol derivatives, or mineral oils. These investigations have been reported in References 5 and 6. Unless otherwise noted, the data and results on nonaqueous foams presented in the following sections of this chapter were obtained from the latter two publications.

There are relatively few aerosol products on the market based on the nonaqueous foam system, possibly because nonaqueous foams are still relatively new. However, they have a number of advantages and should be useful for many applications in the cosmetic, pharmaceutical, household, and industrial fields. Foam properties can be varied over a wide range, and extremely stable (as well as quick-breaking) foams can be obtained. The nonaqueous foam systems are particularly useful for active ingredients that are sensitive to moisture or are more soluble in glycols or mineral oils than in water. Mineral oil foams provide hydrophobic films on the skin that resist removal by water and, therefore, are desirable for products such as suntans etc. Compositions for a variety of products formulated as nonaqueous foams are given in References 4 and 5.

GLYCOL FOAMS

Glycol foams are usually formulated as emulsion systems with the propellant dispersed in the glycol or glycol derivative. The surface-active agent is normally present as a suspended solid. In a few instances, the propellant may be soluble in the glycol concentrate. In some formulations the surfactant may be soluble in the glycol/propellant mixture, but in all cases it is insoluble in the glycol concentrate.

Glycol foams belong to the class of three-phase, solid-stabilized foams, thus being similar in this respect to the aqueous ethyl alcohol foams. Some glycol foams are much stiffer than either aqueous or aqueous ethyl alcohol foams. Stiffness, as well as other properties of the glycol foams, depends upon the type of glycol, surface-active agent, and propellant.

Type of Glycol (or Glycol Derivative)

The type of glycol is one of the major factors that determines whether a foam is produced. The stability of foams from 25 different glycols and glycol derivatives varied from over 48 h to less than 1 min. Seven glycols gave no foam. The glycols were evaluated in a basic system with a composition of 86 wt.% glycol, 4 wt.% propylene glycol monostearate, and 10 wt.% fluorocarbon 12/flurocarbon 114 (40/60) propellant.

The specific properties of a glycol that determine whether it will foam with a given surfactant and propellant have not been completely clarified. A considerable amount of evidence indicates that the surfactant must be insoluble in the glycol concentrate for foam formation. Therefore, the solubility of the surfactant in the glycol is one the controlling factors. However, this is not the only factor because insolubility of the surfactant in the glycol is not sufficient itself to ensure foam formation. Some surfactants do not produce foams even though they are insoluble in the glycol. The extent to which the surfactant is wetted by the glycol and propellant is probably one of the major factors that determines the effectiveness of the surfactant.

The solubility of the propellant in the glycol may play a part in foam forma-

TABLE 20-1   FOAMING CHARACTERISTICS OF VARIOUS GLYCOLS[a]

Glycol	Solubility of "Polawax" in Glycol	Solubility of Propellant in Glycol	Foam[b] Stability
Propylene Glycol	Insoluble	Insoluble	> 16 hr
Diethylene Glycol	Insoluble	Insoluble	> 16 hr
Triethylene Glycol	Insoluble	Insoluble	> 16 hr
Tetraethylene Glycol	Insoluble	Insoluble	> 16 hr
Polyethylene Glycol 400	Insoluble	Insoluble	> 16 hr
Dipropylene Glycol	Soluble	Soluble	No foam
Tripropylene Glycol	Soluble	Soluble	No foam
2-Ethyl Hexanediol-1, 3	Soluble	Soluble	No foam
Hexylene Glycol	Soluble	Soluble	No foam

[a] Aerosol formulation: 86 wt.% glycol, 4 wt.% "Polawax," and 10 wt.% fluorocarbon 12/ fluorocarbon 114 (40/60) propellant.
[b] Time before observable collapse occurred.

tion, since the glycols in which the propellants are least soluble give the most stable foams.[1,7] Propellant and surfactant solubility in the glycols appear to go hand-in-hand. This is shown by the data in Table 20-1, which lists the foaming characteristics of nine glycols in an aerosol formulation consisting of 86 wt.% glycol, 4 wt.% "Polawax" (an ethoxylated stearyl alcohol), and 10 wt.% fluorocarbon 12/fluorocarbon 114 (40/60) propellant. These data do not indicate whether the important factor in foam stability is the solubility of the surfactant in the glycol or the solubility of the propellant. However, other evidence indicates that it is the solubility of the surfactant.

The choice of a particular glycol for a product depends not only upon the foam stability and stiffness desired, but also upon other factors such as the toxicity of the glycol and its solubility for the active ingredients. Toxicity considerations are extremely important, particularly for topical applications. Propylene glycol, 1,3-butylene glycol, and the polyethylene glycols are relatively low in toxicity, judging by the data which have been reported for these compounds.[8-12] These glycols also give stable foams.

Glycerine, although low in toxicity, is not satisfactory by itself for nonaqueous foams. Glycerine forms extremely viscous systems or gels with many surfactants. Glycerine has many excellent properties, however, and is used widely in cosmetics and pharmaceuticals. It certainly could be used in combination with the glycols in the nonaqueous foam systems.

Emulsions prepared with propylene glycol, 1,3-butylene glycol, or the low-molecular-weight polyethylene glycols have relatively low viscosities. This permits easy redispersion of the surfactant and propellant if phase separation occurs during storage. The discharge characteristics of the emulsions are satisfactory.

Surface-Active Agents

Many surface-active agents effective in aqueous ethyl alcohol foams are also good foaming agents for nonaqueous glycol foams. Ethoxylated alcohols and the fatty acid esters are particularly useful. "Polawax," an ethoxylated stearyl alcohol, is one of the preferred surfactants. Combinations of cetyl and stearyl alcohols give excellent foams.

Ethoxylated Fatty Alcohols

The efficiency of ethoxylated fatty alcohols as foaming agents in nonaqueous systems is related to the number of ethylene oxide groups in the surfactant molecule. Foam stability decreases as the ethylene oxide content increases; therefore, the most hydrophobic surfactants are the most effective. This is illustrated by the data in Table 20-2, which gives the properties of foams obtained with a series of ethoxylated fatty alcohols in an aerosol system consisting of 86 wt.% propylene glycol, 4 wt.% ethoxylated fatty alcohol, and 10 wt.% fluorocarbon 12/fluorocarbon 114 (40/60) propellant. The number in parentheses in the designation of the ethoxylated fatty alcohol surfactant indicates the average number of moles of ethylene oxide per mole of the fatty alcohol in the surfactant.

As a general rule, the least stable foams had the lowest stiffness values. Fairly stiff, stable foams can be obtained at surfactant concentrations as low as 1 wt.%, as shown by the data in Table 20-3. A typical aqueous shaving lather foam will have a stiffness value in the range of about 60-100 g.

TABLE 20-2 PROPERTIES OF ETHOXYLATED FATTY ALCOHOL/ PROPYLENE GLYCOL/FLUOROCARBON 12/FLUOROCARBON 114 (40/60) FOAMS

Ethoxylated Fatty Alcohol	Trade Name**	Foam Stiffness (g)	Foam Stability
Ethoxylated stearyl alcohol	"Polawax"	222	> 24 hr
POE* (1) stearyl–cetyl alcohol	"Siponic" E-0	133	> 24 hr
POE (2) stearyl–cetyl alcohol	"Siponic" E-1	134	> 24 hr
POE (4) stearyl–cetyl alcohol	"Siponic" E-2	112	> 24 hr
POE (6) stearyl–cetyl alcohol	"Siponic" E-3	27	< 10 min
POE (8) stearyl–cetyl alcohol	"Siponic" E-4	17	< 10 min
POE (2) cetyl ether	"Brij" 52	58	> 24 hr
POE (2) stearyl ether	"Brij" 72	126	> 24 hr
POE (10) cetyl ether	"Brij" 76	9	< 10 min

*POE = Polyoxyethylene.
**The "Siponic" and "Brij" series of surfactants are manufactured by the Alcolac Chemical Corp. and ICI United States, respectively. "Polawax" is supplied by Croda, Inc.

TABLE 20-3   EFFECT OF CONCENTRATION OF "POLAWAX"
UPON FOAM PROPERTIES

| | Composition (wt.%) | | Foam | |
"Polawax"	Propylene Glycol	FC-12/FC-114 (40/60)	Stiffness (g)	Foam Stability
1	89	10	88	< 8 hr
2	88	10	154	> 16 hr
3	87	10	240	> 16 hr
4	86	10	222	> 16 hr

The solubility of the ethoxylated fatty alcohol in the glycol appears to be one of the major factors that controls the stiffness and stability of nonaqueous foams. This is indicated by the data in Table 20-4, which gives the properties of foams obtained from a nonaqueous system of 86 wt.% polyethylene glycol 400, 4 wt.% ethoxylated fatty alcohol, and 10 wt.% fluorocarbon 12/fluorocarbon 114 (40/60) propellant. Only surfactants that were insoluble in the polyethylene glycol 400 gave stable foams.

Other Surfactants

Propylene glycol monostearate (self-emulsifying) is an excellent surfactant for glycol foams, but a little more difficult to disperse in the glycols than compounds such as "Polawax." Other surfactants that give stable foams are listed in Reference 5 and include products such as sorbitan monostearate and cetyl alcohol/stearyl alcohol mixtures.

Propellants

Type of Propellant

Almost any fluorocarbon 12/fluorocarbon 114 blend will give excellent foams from the glycols, and the choice of a particular blend is usually determined by pressure and economic considerations. Generally, it is advisable to use a blend with at least 15 wt.% fluorocarbon 12 so that the vapor pressure will be high enough to obtain a satisfactory discharge. There appears to be a tendency for foam stability to increase with increasing concentrations of fluorocarbon 12 in the blend.

Fluorocarbon 114 gave a slow discharge as a result of its low vapor pressure and produced a shiny, pasty foam that continued to expand after discharge. Fluorocarbon 12/fluorocarbon 11 combinations gave less attractive and less stable foams than the fluorocarbon 12/fluorocarbon 114 blends and were generally of little interest.

TABLE 20-4 EFFECT OF SURFACTANT SOLUBILITY UPON STIFFNESSS
AND STABILITY OF POLYETHYLENE GLYCOL 400 FOAMS

Surfactant	Trade Name[a]	Solubility of Surfactant in Glycol	Foam Stiffness[b] (g)	Foam Stability
Ethoxylated stearyl alcohol	"Polawax"	Insoluble	192	4-16 hr
POE (4) lauryl ether	"Brij" 30	Soluble	—	< 3 min
POE (23) lauryl ether	"Brij" 35	Insoluble	26	4-16 hr
POE (2) cetyl ether	"Brij" 52	Insoluble	162	> 16 hr
POE (10) cetyl ether	"Brij" 56	Soluble	—	< 3 min
POE (20) cetyl ether	"Brij" 58	Insoluble	84	4-16 hr
POE (2) stearyl ether	"Brij" 72	Insoluble	118	> 16 hr
POE (10) stearyl ether	"Brij" 76	Insoluble	38	> 16 hr
POE (2) oleyl ether	"Brij" 92	Soluble	—	< 3 min
POE (10) oleyl ether	"Brij" 96	Soluble	—	< 3 min

[a]"Polawax," Croda, Inc.; "Brij" compounds, ICI United States.
[b]Stiffness measurements could not be carried out on foams with a stability less than 3 min.

The vapor pressures of the glycol emulsion systems were slightly less than those of the pure propellants.

Concentration

Propellant concentrations of 8-10 wt.% are generally used for nonaqueous foams. These concentrations give a satisfactory discharge and excellent foams. At lower propellant concentrations, for example, 5 wt.%, the product may discharge as a liquid, but the liquid will subsequently expand into a foam. At propellant concentrations higher than 10 wt.%, the coherence of the foam increases markedly, and at concentrations of 60-90 wt.%, the nonaqueous systems can be sprayed to give a product similar to cold cream. This product subsequently expands into a stable foam.

Effect of Ethyl Alcohol

Ethyl alcohol increases the pressure sensitivity of the foams. A nonaqueous system consisting of 76 wt.% polyethylene glycol 400, 10 wt.% ethyl alcohol, 4 wt.% propylene glycol monostearate or "Polawax," and 10 wt.% fluorocarbon 12/fluorocarbon 114 (40/60) propellant gave an excellent foam and one in which the pressure of the fingers caused an immediate breakdown of the foam structure with liquefaction. The extent to which the foam is pressure-sensitive can be regulated by the concentration of ethyl alcohol, and this depends upon the glycol and surfactant present. In the foam system described above, increasing the concentration of ethyl alcohol beyond 15 wt.% destroyed the foam.

Type of Discharge

Some glycol systems foam as soon as the product leaves the container, while others discharge a liquid stream that subsequently expands into a foam. The way in which the products are discharged is determined by the propellant and surfactant present.

Surfactants that are insoluble in both the glycol concentrate and the glycol/propellant mixture will produce foams as soon as the product is discharged. When the surfactant is insoluble in the glycol concentrate but soluble in the glycol/propellant mixture, the product discharges as a liquid which expands into a foam. The surfactant "Siponic" E-O is insoluble in propylene glycol but essentially soluble in the glycol/propellant mixture. The glycol/propellant emulsion is hazy at room temperature, but there is no other evidence of solid surfactant. This product discharges as a liquid which slowly expands into an extremely dense and stable foam. Expansion is essentially complete in a few minutes.

The type of discharge for combinations of "Siponic" E-O with various glycols is shown in Table 20-5. In both cases where "Siponic" E-O was soluble in the glycol/propellant mixture, the product discharged as a liquid which foamed later.

The "Siponic" E-O/propylene glycol/propellant system will exhibit either type of discharge, depending upon the temperature. If this system, which discharges a liquid at room temperature, is cooled a few degrees below room temperature, some of the surfactant separates from solution. The cooled product with solid surfactant now foams immediately during discharge. If the product is warmed, so that most of the surfactant dissolves, the product again discharges a liquid stream that slowly expands into a foam. This is further evidence that solid surfactant must be present to stabilize the foam.

TABLE 20-5   EFFECT OF "SIPONIC" E-O UPON DISCHARGE OF GLYCOL FOAMS[a]

| | | Solubility of "Siponic" E-0 | |
Glycol	In Glycols	In Glycol–Propellant Mixture	Type of Discharge
Propylene Glycol	Insoluble	Essentially sol.	Liquid
Diethylene Glycol	Insoluble	Insoluble	Foam
Triethylene Glycol	Insoluble	Insoluble	Foam
Tetraethylene Glycol	Insoluble	Essentially sol.	Liquid
Polyethylene Glycol 400	Insoluble	Some solid pres.	Liquid-foam

[a]Aerosol formulation: 86 wt.% glycol, 4 wt.% "Siponic" E-O, and 10 wt.% fluorocarbon 12/fluorocarbon 114 (40/60) propellant.

The substitution of fluorocarbon 114 for a fluorocarbon 12/fluorocarbon 114 blend in a system that normally produces a foam discharge usually results in a product which gives a slow foamy liquid type of discharge. The slow discharge is due to the low vapor pressure of the fluorocarbon 114.

Formation and Stabilization

All evidence indicates that the aerosol glycol foams are three-phase, solid-stabilized foams, which puts them in the same category as the aqueous and aqueous ethyl alcohol foams. The results listed in Table 20-4, which show the effect of various surfactants upon the stability of foams from polyethylene glycol 400, indicate that foams of significant stability are obtained only when the surfactant is insoluble in the glycol. The effect of the solubility of the surfactant in the glycol/propellant mixture upon the type of discharge is a further indication that solid surfactant must be present for foam formation.

If the surfactant is insoluble in both the glycol and the glycol/propellant mixture, foam formation occurs immediately during discharge because solid surfactant is present to stabilize the foam. If the surfactant is insoluble in the glycol but soluble in the glycol/propellant mixture, the vaporization of the propellant during discharge causes the surfactant to separate from solution. The solid surfactant resulting stabilizes the foam. This is similar to the way in which foams are formed and stabilized in the aqueous ethyl alcohol system. The fact that glycol systems of this type discharge as liquids that expand later into foams may be due to the higher viscosity of the glycols and the slower separation of the solid surfactant from solution.

Many nonaqueous foams that are stable at room temperature lose their stability at higher temperatures. This is because the surfactant becomes soluble in the concentrate.

Preparation

The surfactants are insoluble in the glycols and, as a result, the glycol concentrate consists of a dispersion of the surfactant in the glycol. The surfactant should be as finely divided as possible in order to minimize the possibility of the valve clogging when the aerosol is discharged.

The standard laboratory procedure for preparing concentrates with ethoxylated fatty alcohols such as "Polawax" is to heat the mixture of surfactant and glycol until the surfactant dissolves. The solution is allowed to cool to room temperature with gentle stirring, during which time the surfactant separates from solution. In many cases, the dispersion of the surfactant is sufficiently finely divided so that the resulting concentrate can be loaded directly into the aerosol containers. Some surfactants, however, gave a coarse dispersion in the glycol. These concentrates were passed through a hand homogenizer that reduced the particle size of the surfactant sufficiently to give a fine dispersion.

In some systems, the concentrates gelled at room temperature or were too viscous to pour into glass bottles. Concentrates of this type had to be loaded into bottles while still warm enough to be fluid. The propellant was added to the warm concentrate and the samples were shaken while they were cooling to room temperature in order to emulsify the propellant.

MINERAL OIL FOAMS

Aerosol foams may also be obtained from mineral oils. Potential cosmetic and pharmaceutical products that might be formulated with mineral oil foam systems include suntans, hair products, baby oils, hormone creams, vitamin creams, and various other ointments.

In most cases, the propellants are soluble in the mineral oils, but the surfactants must be insoluble in the mineral oil in order to obtain a foam. The surfactants can either be soluble or insoluble in the propellant/mineral oil solution. The properties of mineral oil foams vary widely, depending upon such factors as viscosity of the mineral oil, type of surfactant, and type and concentration of the propellant. As a general rule, the foams from the mineral oils have a fairly large bubble size and low stiffness values. High-viscosity mineral oil gives stiffer foams than low-viscosity oil.

Surfactants

Ethoxylated fatty alcohols or mixtures of cetyl and stearyl alcohol are the most effective surfactants for mineral oil foams. Cetyl alcohol or stearyl alcohol alone gave unstable foams. The 50/50 mixture of the two alcohols gave excellent foams, however. The 50/50 mixture is one of the preferred surfactants for mineral oil foams.

Propellants

Mineral oils are miscible with the propellants and, therefore, depress their vapor pressures. Generally, about 10-15 wt.% fluorocarbon 12 is required to give an adequate foam with good discharge. Fluorocarbon 114 gave essentially liquid discharges as a result of its low vapor pressure.

Fluorocarbon 12/fluorocarbon 114 (40/60) blends gave good foams, and the discharge was somewhat quieter than that with fluorocarbon 12 alone. Fluorocarbon 142b gave a noisy discharge, while fluorocarbon 152a gelled the mineral oil.

Foam Formation and Stabilization

Although the data are somewhat limited, the evidence indicates that mineral oil foams, like aqueous ethyl alcohol and glycol foams, are three-phase, solid-stabilized foams. The mineral oil systems are quite similar to the aqueous ethyl alcohol systems, since in both cases the propellant is miscible with the concentrate, resulting in a homogeneous system.

COMBINATIONS OF ALCOHOLS OR KETONES WITH GLYCERINE OR GLYCOLS

Klausner[4] has reported that foams of limited stability can be obtained from homogeneous liquid compositions consisting of an alcohol or ketone, glycerine or glycol, surfactant, and propellant. A foam of limited stability is defined as a foam that does not become completely liquefied when exposed to the atmosphere for about 15 min or longer, but which will revert to a liquid in about 2 sec when heated or rubbed.

The following proportions of the compounds are considered to be critical for achieving the homogeneous systems that produce foams of limited stability:

Alcohol or dialkyl ketone	26–64 wt.%
Glycerol or alkylene glycol	28–64 wt.%
Surface-active agents	0.5–5 wt.%
Propellant	2–30 wt.%

The surface-active agent may be anionic, nonionic, or cationic. It can be soluble in either the alcohol or dialkyl ketone, or in the glycerol or alkylene glycol, but not in both. Nonionic agents such as the polyoxyethylated fatty alcohols and the fatty acid esters are the preferred surfactants. Klausner also gives formulations for nonaqueous foams for suntans, pre-electric shaves, body colognes, and paint removers.

MISCELLANEOUS NONAQUEOUS FOAMS

Aerosol foams may be obtained from a wide variety of nonaqueous materials. Formulations for a linseed oil foam, Neatsfoot Compound foam, and a lubricant for bicycle and motorcycle chains are listed below:

Linseed Oil Foam		*Chain Lubricant Foam*	
Concentrate		*Concentrate*	
Linseed oil (boiled)	96.0 wt.%	Motor oil (10-40W)	84.0 wt.%
Cetyl alcohol	2.0 wt.%	DAG 152 graphite	
Stearyl alcohol	2.0 wt.%	suspension	10.0 wt.%
		Cetyl alcohol	3.0 wt.%
Aerosol		Stearyl alcohol	3.0 wt.%
Concentrate	70.0 wt.%	*Aerosol*	
Fluorocarbon 12	30.0 wt.%		
		Concentrate	70.0 wt.%
		Fluorocarbon 12	30.0 wt.%

Neatsfoot Compound Foam

Concentrate

Neatsfoot compound	96.0 wt.%
Cetyl alcohol	2.0 wt.%
Stearyl alcohol	2.0 wt.%

Aerosol

Concentrate	70.0 wt.%
Fluorocarbon 12	30.0 wt.%

REFERENCES

1. A. M. Schwartz, J. W. Perry, and J. Berch, "Surface Active Agents and Detergents," Vol. 2, Interscience Publishers, Inc., New York, 1958.
2. E. G. King, *J. Phys. Chem.* **48**, 141 (1944).
3. R. V. Petersen and R. D. Hamill, *J. Soc. Cosmet. Chem.* **19**, 627 (1968).
4. K. Klausner, U.S. Patent 3,131,153 (April 1964).
5. P. A. Sanders, *Aerosol Age* **5**, 33 (November 1960).
6. P. A. Sanders, *Amer. Perfum. Cosmet.* **81**, 31 (1966).
7. P. A. Sanders, *Aerosol Age* **5**, 26 (November 1960).
8. G. O. Curme and F. Johnston, American Chemical Society Monograph Series, No. 114, "Glycols," Reinhold Publishing Corp., New York, 1952.
9. Dow Chemical Company Technical Bulletin, "Dow Propylene Glycol, U.S.P."
10. Union Carbide Chemicals Company Technical Bulletin, "'Carbowax' Polyethylene Glycols."
11. Celanese Chemical Company Specification Bulletin, N-26-4, "1,3-Butylene Glycol."
12. Dow Chemical Company Technical Bulletin, "Toxicological Information on Dow Glycols."

21

AEROSOL
SUSPENSIONS
(POWDERS)

An aerosol powder consists primarily of a suspension of a finely divided solid in a liquefied gas propellant. Other materials are usually present (for example, auxiliary solids, suspending agents, and perfumes and oils), but the active ingredient is a dispersed solid. When the product is sprayed, the propellant evaporates rapidly, leaving the dry powder. The sprays may be cool, but generally are not as cold as alcohol-based sprays.

Powder aerosols were suggested as early as 1954.[1-3] There was considerable activity in this field at that time because of the obvious advantages of the powder system for many products. Suspensions of pharmaceutical agents in propellants, designed for inhalation therapy, were among the first aerosol powders to be marketed.[4] Aerosol bath powders also appeared very early.[5] Other powder aerosols that followed included dry shampoos, foot powders, fungicides, spot removers, poison ivy sprays, lubricants, and dry-type antiperspirants. Dry-type antiperspirants constitute the largest single group of powder aerosols on the market today. They are essentially suspensions of antiperspirant salts in a liquefied gas propellant.

PROBLEMS WITH POWDER AEROSOLS

Some problems that may be encountered with powder aerosols are listed below.

Valve Clogging

Clogging of the valve occurs when the concentration of powder is too high and is due to packing of the powder around and in the valve orifices. This can take place even if the initial particle size of the powder is quite small.

According to Beard,[1] clogging results from one or more of the following causes.

400

Large and/or Needle-Shaped Particles

Needle-shaped, fibrous, or large particles can mat or form filter cakes and clog a valve, even at low powder concentrations.

Partially Soluble, Resinous, or Crystalline Materials

Particles that are partially soluble in the propellant tend to increase in size as a result of solution and recrystallization during storage. This can lead to the formation of aggregates that clog the valve. In addition, partially soluble compounds tend to be deposited at the expansion orifices as a result of the rapid evaporation of propellant at these locations.

Agglomerative Sedimentation

Agglomerative sedimentation is defined by Beard as the formation of hard-packed cakes of powder in the bottom of the container and hard plugs in the lower end of the dip tube. Powders generally have a density different from that of the liquefied propellant in which they are suspended. If the powder particles are heavier than the liquid phase, they will settle to the bottom under the influence of gravitational forces. The close packing of the particles at the bottom of the container may lead to formation of agglomerates and hard cakes that are much larger in size than the original particles.

Agglomeration of particles can take place before the particles have settled to the bottom, and the increase in particle size of the agglomerates will increase the rate of settling. Kanig and Cohn[6] have shown that agglomeration occurs relatively soon after preparation of the aerosols. They reported that the rate of agglomeration of the suspensions with which they worked was practically nil after the first week.

Leakage

Failure of the valve to close completely after operation will result in leakage. This is caused by deposits of powder on the valve seats and is more likely to occur with a hard crystalline material than with a soft product. This problem can be deceptive because the leak can be very slight and almost unnoticeable, but enough so that complete loss of propellant will occur during extended storage.

Nonuniform Delivery

In products designed for inhalation therapy, a specified dosage is delivered by means of a metering valve. Agglomeration of the particles can lead to nonuniform delivery because of the change in particle size. This can occur even if the agglomeration is insufficient to cause valve clogging. In addition, since the therapeutic value of the product depends upon the particle size, agglomeration can reduce the efficacy of the product.

Caking and Wall Deposits

Powders with a specific gravity lower than that of the liquid phase will rise to the top of the liquid during storage. In some cases, the powder will form deposits or cakes on the container wall above the liquid level. According to Thiel and Young,[7] this is due to a mutual repulsion between the individual particles, which causes them to creep up the wall above the liquid phase, where they form deposits. Mutual repulsion can arise if all of the particles have the same electric charge. This is particularly serious with a pharmaceutical product formulated to produce a specific dose with a metering valve.

Considering the problems that can occur with powder aerosols, any potential product must be thoroughly tested. Extensive spraying tests are particularly important to make certain that valve clogging or leakage will not occur during use. Various methods, which have been reported to minimize some of these problems, are discussed in a later section.

BASIC COMPONENTS OF AEROSOL POWDER SPRAYS

Powders

Type of Powder

The type of powder is determined by the use for which the product has been developed. Typical powders include talcs, insecticides, pharmaceuticals, antiperspirants, clays, iron oxides, and various lubricants. A number of aerosol formulations are listed in References 1 and 8–12. These give an indication of the variety of powders used as active ingredients.

One of the major constituents of many powder formulations is talc. Before World War II, the best available talcs came from France and Italy, and these talcs are still used extensively in face and bath powders. Some commercial talcs have an average particle diameter as small as 3 μ. Prussin[13] has pointed out that talcs with a platelet structure are preferred to those with needle-like particles, since the platelet structure allows the particles to slide or cascade over each other and have less tendency to clog valves.

The active ingredient for most dry-type antiperspirants is aluminum chlorhydrate, marketed by the Reheis Chemical Company. Two grades are available: the "Micro-Dry" and the "Macro-Dry." The average particle size of the former is 15 μ, and 100% of this product will pass through a 140-mesh screen, while 97% will pass through a 325-mesh screen (about 43 μ). The "Macro-Dry" grade has a larger particle size: a minimum of 95% of the particles have a diameter greater than 10 μ; 1.0% have a diameter larger than 74 μ; and less than 0.1% have a diameter larger than 150 μ.[16]

Concentration

Practically all powders will cause valve clogging or leakage if the concentration is sufficiently high. The specific concentration at which this occurs depends upon the type of powder, particle size of the powder, type of valve, and other additives that are present. Therefore, the maximum concentration of powder that can be sprayed without valve clogging has to be determined for each product.

Traditionally, for powders such as talc, 5-10 wt.% was considered to be about the highest practical concentration because of the possibility of valve clogging and leakage.[3] However, as a result of the improvements that have been made in powder systems, concentrations of 20 wt.% are commonly mentioned in the literature, and some of the published formulations contain as much as 33 wt.% talc. One system uses 90 wt.% powder, but this is a different type of formulation and is discussed later. The concentrations of active ingredients in aerosols designed for inhalation therapy are generally quite low and seldom exceed 3 wt.%. The concentration of aluminum chlorhydrate in dry-type antiperspirants is about 3.5 wt.%.

Particle Size

There is some variation in the literature with respect to the maximum possible particle size for powders: 100 μ has been mentioned several times as about the upper limit, but it is rather generally agreed that for most products the particle size should be below 50 μ and preferably below 30 μ.[7-10,12] A 325-mesh screen is commonly used for preparing aerosol powders because the maximum diameter of a spherical particle that will pass through the screen is 43 μ. Reed[3] has pointed out that since typical aerosol valves do not have orifices much smaller than 400 μ, it is evident that clogging of the valve usually is not the result of the particles being too large, but more the result of an accumulation of the powder around and in the valve orifices.

With pharmaceutical products for inhalation therapy, the particles should be in the 1-10 μ range.[6,10] Particles with a diameter greater than 10 μ may not reach the area to be treated, and particles with a diameter smaller than 1 μ are exhaled.

Solubility

The aerosol powder should be as nearly insoluble in the liquid phase as possible in order to avoid intermittent solution and recrystallization during storage.[3] This leads to an increase in particle size.

The solubility of the powder in water also determines the extent to which the powder is affected by moisture in the formulation. Kanig and Cohn[6] determined the effect of varying quantities of moisture upon the agglomeration of three powders with different water solubilities and found that the powder with the highest solubility in water was the least stable in the presence of moisture. For example, in suspensions stabilized with "Arlacel" 83 (sorbitan sesquioleate), talc, which is insoluble in water, exhibited signs of agglomeration only at moisture levels of 1.0 wt.% or above. Prednisolone, which is slightly soluble in water, was affected by moisture concentrations above the 0.1 wt.% level, while the water-soluble isoproterenol hydrochloride was affected at the 0.005 wt.% moisture level.

Thiel and Young[7] consider 300 ppm moisture to be the limit for aerosol powder formulations where the powder is water-soluble.

Propellants

Type of Propellant

A large variety of propellant blends have been suggested for use with aerosol powders. Fluorocarbon 12/fluorocarbon 11 mixtures with weight percent ratios of fluorocarbon 12 to fluorocarbon 11 varying from 20/80 to 50/50 are used. The fluorocarbon 12/fluorocarbon 11 (30/70) blend is a common propellant for dry-type antiperspirants.[1, 3, 7-9, 14] Fluorocarbon 12/fluorocarbon 114 blends are also employed and tend to be warmer than fluorocarbon 12/fluorocarbon 11 mixtures. Thiel and his co-workers have suggested three-component blends of fluorocarbons 12, 114, and 11, as well as blends with isobutane.[7, 10] The advantage of incorporating the fluorocarbon 11 is that it serves as a vehicle for preparing the slurry of active ingredients etc. According to Reed,[3] a fluorocarbon 12/fluorocarbon 114 (20/80) blend gives soft, dry sprays suitable for cosmetic products, while the fluorocarbon 12/fluorocarbon 11 (30/70) blend gives heavier sprays useful for fungicidal and insecticidal products.

Many other fluorocarbons are excellent propellants for dry-type antiperspirants. Some of these are discussed below.

Fluorocarbon 142b

Fluorocarbon 142b is an excellent propellant for powders and dry-type antiperspirants. Although it is flammable, its flammability is relatively low. Fluorocarbon 142b does not extend a flame and gives a negative closed-drum test. It does not have a flash point. Since it is a single compound, it does not change composition during use, as do propellant blends. It is slightly less dense than a fluorocarbon 12/fluorocarbon 11 (30/70) blend, and the settling rate of antiperspirant salts is somewhat greater than that with the latter propellant.

Fluorocarbon 152a/Fluorocarbon 133a (10/90)

This blend has a vapor pressure of about 17–18 psig at 70°F and a density of 1.27 g/cc. The blend is nonflammable. The low-pressure, nonflammable fluorocarbon 133a provides an excellent vehicle for preparing the slurry of active ingredients.

Fluorocarbon 124/Fluorocarbon 123 (50/50)

This nonflammable blend gives very good dry-type antiperspirants. The blend has a vapor pressure of about 20 psig at 70°F. Fluorocarbon 123 is nonflammable and is a good propellant for preparing the slurry of active ingredients because of its low pressure.

Fluorocarbon 22/Isopentane (35/65)

This particular blend was of considerable interest for dry-type antiperspirants because of its low cost. It has the disadvantages of being very flammable and having a low density, which promotes rapid settling of the active ingredients. In addition, the properties of the blend, such as vapor pressure, flammability, etc., change significantly during discharge because of the difference in vapor pressures of the two components.

Hydrocarbons

Hydrocarbons have two serious disadvantages as propellants for dry-type antiperspirants: They are extremely flammable and they have a low density. The latter results in a rapid settling of the antiperspirant powder. Nevertheless, antiperspirants formulated with hydrocarbons are under serious consideration. The question that remains with these products is to what extent they present a consumer hazard.

Propellant Density and Viscosity

Stokes' Law (Equation 21-1), which gives the sedimentation rate (u) of a spherical particle in a liquid, shows that three factors influence the sedimentation rate of a powder aerosol. These are the relative densities of the dispersed solid particles (d_1) and the propellant phase (d_2), the particle radius (r), and the viscosity of the propellant phase (η) (g is the acceleration of gravity):

$$u = \frac{2gr^2(d_1 - d_2)}{9\eta} \tag{21-1}$$

If d_1 is larger than d_2, the dispersed phase will settle to the bottom. If d_1 is less than d_2, the rate of sedimentation will have a minus sign, indicating that the dispersed solid will rise to the top. Equation 21-1 shows that the sedimentation rate can be reduced by decreasing the particle radius, decreasing the density difference between the dispersed and propellant phases, and increasing the viscosity of the propellant phase.

The effect of variations in propellant density and viscosity upon the settling rate (sedimentation rate) of the solid ingredients (aluminum chlorhydrol; "Cab-O-Sil"—see the section "Miscellaneous Additives") in a dry-type antiperspirant is shown by the data in Table 21-1. Sedimentation rates are indicated by the time required for the suspended solids to settle 50% of the distance to the bottom of the container. The settling rates are compared for propellants having the same density but different viscosities (methylene chloride, fluorocarbon 133a, and fluorocarbon 12) and for propellants having essentially the same viscosity but different densities (methylene chloride versus fluorocarbon 123 and isopentane versus fluorocarbons 152a and 12). Since aluminum chlorhydrol settles to the bottom in all cases, it has a higher density (d_1) than that of any of the propellants (d_2). The particle density (d_1) is constant, since the same concentrate was used in all experiments.

The data indicate that sedimentation rates can be decreased by using propellants with a high density and viscosity. The extent to which the density of the propellant phase can be increased is limited (see Table 11-3 in Chapter 11). Probably the most effective method for decreasing the sedimentation rate is to increase the viscosity of the propellant phase by the addition of thickening or gelling agents.

The viscosity of fluorocarbon 133a is not available. However, judging by the data in Table 21-1, in which the sedimentation rate of fluorocarbon 133a is compared with those of two propellants having known viscosities (methylene chloride and fluorocarbon 12), the viscosity of fluorocarbon 133a is approximately 0.40 centipoise.

TABLE 21-1   SEDIMENTATION RATES OF DRY-TYPE
ANTIPERSPIRANTS IN VARIOUS PROPELLANTS

Compound	Density (g/cc) at 70°F	Viscosity (centipoise)	Time to Settle 50% (sec)
Same Density–Different Viscosity			
Methylene chloride	1.32	0.441	50
Fluorocarbon 133a	1.33	–	46
Fluorocarbon 12	1.32	0.214	23
Essentially Same Viscosity– Different Density			
Methylene chloride	1.32	0.441	50
Fluorocarbon 123	1.47	0.450	62
Isopentane	0.62	0.224	13
Fluorocarbon 152a	0.91	0.227	16
Fluorocarbon 12	1.32	0.214	23

Surface-Active Agents

Many formulators believe that a surface-active agent is a necessity for a stable powder suspension. One function of the surfactant is to retard agglomeration, which leads to rapid sedimentation and consequent valve clogging and leakage.

Thiel and his co-workers[7,10] have been very active in the field of surfactants for powders. They report that nonionic surfactants with an HLB value less than 10 and preferably in the range of 1–5 are the most effective. The surfactant should be liquid and soluble or dispersible in the liquid phase. The preferred surfactants were "Arlacel" C (sorbitan sesquioleate), "Span" 80 (sorbitan monooleate), and "Span" 85 (sorbitan trioleate). These are products of ICI United States, Wilmington, Del. A survey of the literature indicated that "Span" 85 was the most commonly used surfactant. Concentrations varied from 0.25 to 1.0 wt.%.[7,9,10]

"Tween" 80 (polyoxyethylene sorbitan monooleate) and "Witconol" 14 (polyglycerol oleate) have also been reported to be helpful in retarding agglomeration.

Miscellaneous Additives

High-boiling organic liquids such as mineral oils and fatty acid esters are often used in powder formulations. These additives must be soluble in the liquid propellant phase to be effective. Geary and West[8] have listed three ways in which additives of this type are useful in powder formulations: (1) They retard agglomeration; (2) they serve as lubricants for the particles and the valve components; and (3) they reduce dusting.

The oils also increase the adherence of the powder to the skin.[3]

The concentrations of mineral oil are normally less than 1 wt.%, although some formulations with higher concentrations have been published. The concentration of the mineral oil should be kept as low as possible because the inhalation of droplets of it can result in lipoid pneumonia.

One of the most commonly used esters is isopropyl myristate. According to DiGiacomo,[14] purified isopropyl myristate is an ideal suspending agent for powdered aerosols. It is reported to prevent the crystallization of certain bacteriological agents, perfume constituents, and other ingredients that may crystallize out on long standing. Adhesion of the powder to the skin is also increased. Concentrations of 0.5–1.0 wt.% are usually adequate.

"Solulan" 97,* an acetylated lanolin derivative, was found to be particularly useful in retarding agglomeration of talcum powder (unpublished Freon® Products Laboratory data). "Solulan" 97 appears to wet the powder particles and provide a physical barrier to agglomeration. It also adds lubricity to the system. A concentration of 1.0 wt.% is usually sufficient.

"Cab-O-Sil,"† colloidal silica, is negatively charged, and when added to a

* "Solulan" 97 is manufactured by the American Cholesterol Company, Edison, N.J.
† "Cab-O-Sil" is available from Cabot Corp., Boston, Mass.

powder aerosol it gives the system a negative charge. This retards agglomeration. "Cab-O-Sil" also increases the viscosity of the system and thus decreases the rate of settling. The normal concentration in the aerosol is about 0.5 wt.%. Whether the "Cab-O-Sil" increases creeping and caking on the container wall as a result of the repulsion between the particles has not been reported.

Zinc stearate increases the adhesion of the powder to the body and also is reported to increase the slip of the talc. A concentration of 0.5–1.0 wt.% should be adequate.

Valves

Beard[1] has given an excellent summary of the requirements for a powder valve. The path from the valve seat to the spray orifice should be very short. This minimizes the space in which the product can accumulate to collect moisture or fall back to cause clogging. The valve should have a sharp shut-off and a high valve seating pressure in order to minimize leakage. Beard suggests that the valve be operated wide open without any attempt to control the spray rate by varying the pressure on the actuator. Otherwise, the product will build up on the valve seat.

Companies that manufacture valves specifically for powder sprays are listed below:

Company	Valve Designation	Comments
Dispenser Products Division of Ethyl Corp. (formerly VCA/ARC)	PARC-39	The sealing properties of the PARC-39 are accomplished by the wiping action of the valve gasket over the stem orifice and by the penetrating action of a series of concentric rings into the gasket surface
Emson	Valve S-32 VP Valve S-32 V	Powders Antiperspirants
Risdon	Style 510	This valve has a larger-than-normal dip tube to provide better flow for powder products
Summit	S-68 Tilt Action	
Seaquist	NS-40 "Powder-Mate"	The stem is designed with both a primary and secondary seal that function independently. The design of the stem also provides a method that insures that sealing surfaces are wiped clean on each deactuation

The Precision Valve Corporation has a number of valves which it recommends for powder sprays, depending upon the particular product. An antiperspirant valve with a nondegenerative sealing action, a more positive valve stroke, and an optional Pre-Vent filter is claimed to prevent seepage and clogging.

SOME MODIFIED POWDER SYSTEMS

From time to time, various modifications in powder systems have been suggested as a means for minimizing problems such as agglomeration and valve clogging. Some of these modified systems are discussed briefly in the following section. Since not all the systems have a particular designation, they are discussed under the name or names of the inventors.

W. C. Beard[1]

Beard reports that valve clogging due to agglomerative sedimentation can be eliminated or reduced by use of certain selected propellants or propellant blends. He refers to this as the selected propellant process. The system is based upon the discovery that if the propellant has a specific gravity approximately equal to that of the powder, or if the difference between the specific gravities of the powder and the propellant is no greater than 0.4 or 0.5, then the powder will remain in a soft and flocculated condition regardless of whether it rises to the top or settles to the bottom. The powder can readily be dispersed throughout the propellant by a mild shaking or swirling action. In some cases, a difference of as much as 0.8 can be tolerated.

D. C. Geary and R. D. West[8]

Geary and West conceive the volume of a powder as consisting of two parts: the volume of the solid particles and the void volume between the particles. The powder system they suggest is based upon the concept that if the void volume between the particles is either filled or enlarged by substances which prevent them from agglomerating, a stable suspension will result.

The two types of materials used for filling the void volumes are bulking agents and liquids, including propellants. Bulking agents are themselves very finely divided powders. Geary and West mention that the use of bulking agents to impart flow characteristics to powders is well known. The preferred bulking agent is "Santocel" 54, a silica gel with a particle size of $0.5-3.0 \mu$. The amount of bulking agent required is calculated from the bulk and absolute densities of the powders.

The quantities of bulking agents required are generally quite low, being less than 2 wt.% in most cases. However, according to Geary and West, the use of such agents allows the formulation of powder aerosols containing as much as 33 wt.% talc.

P. E. Gunning and D. R. Rink[11]

The powder system proposed by Gunning and Rink is the reverse of the typical powder system in that the powder constitutes the major portion of the aerosol: It is the supporting phase, while the propellant is the minor portion. The system is based primarily upon the concept that some powders can absorb sufficient liquefied gas propellant so that they can be discharged while still remaining as free-flowing powders. If the active powder itself does not have sufficient capacity for the propellant, a carrier which possesses absorbing power is added. Typical carriers include amorphous silica, crystalline silicates, and metals such as aluminum powder.

The system is illustrated with an insecticidal dust with a composition of 90 wt.% insecticide and 10 wt.% fluorocarbon 12. In another example, a medicated foot powder with a composition of 87.55 wt.% talc, 1.75 wt.% zinc stearate, 0.70 wt.% dichlorophene, and 10 wt.% fluorocarbon/12 fluorocarbon 11 (50/50) propellant is disclosed.

Shulton, Inc.[9]

In this system, a stable powder aerosol is obtained by gelling the propellant. This makes it difficult for the powder to move sufficiently to agglomerate. Gel formation is obtained by adding a colloidal silica with a particle size below 0.03 μ to the aerosol system. The gelled system is illustrated with examples of antiperspirant-deodorant aerosols.

C. G. Thiel, J. G. Young, I. N. Porush, and R. D. Law[7,10]

In suspensions where the specific gravity of the powder is less than that of the propellant, cake-out or deposition of the powder on the wall above the liquid may occur. Thiel and his co-workers report that this problem may be minimized by the addition of an auxiliary solid which has a specific gravity greater than that of the liquid. Examples of solids that are suitable for this purpose include compounds such as sodium sulfate, sodium chloride, calcium chloride, and sucrose.

Thiel, Porush, and Law also report that the addition of a quaternary compound such as cetyl pyridinium bromide to a powder formulation not only will minimize agglomeration, but will also retard wall deposits above the liquid surface. The suggested concentration of the quaternaries varies from 0.05 to 1.0 wt.%.

FILLING OF POWDER AEROSOLS

Filling powder aerosols is not a simple operation. Powders can be difficult to handle, and contamination by moisture must be kept to a minimum. In some products, the quantities of active ingredients are very small. The problems in filling powder aerosols and the loading procedures are discussed by Mintzer,[4] and his article should be consulted for details.

The earliest powder aerosols were pharmaceuticals designed for inhalation. The concentration of active ingredients in these products was quite low, and the loading of a dry concentrate was impractical. In addition, it was usually considered advisable to combine any suspending agents with the micronized powders before loading them into the aerosol container, and it was difficult to obtain a homogeneous mixture of the dry ingredients. In order to avoid these difficulties, a slurry or suspension of the active ingredients and additives was prepared in a high-boiling propellant such as fluorocarbon 11. The slurry was loaded into the containers and the remainder of the propellant added. This procedure has been used successfully for many years.[15]

When a sufficient quantity of an emollient such as isopropyl myristate was used, a concentrate was prepared by suspending the powders in the emollient. This concentrate was loaded into the containers and the propellant was either cold- or pressure-filled.

Many of the present aerosols contain 5–15 wt.% powder, a concentration which allows the powder to be loaded in the dry state. This procedure has the advantage that the possibility of contamination by moisture is decreased. Equipment for this operation is available and can be obtained from various manufacturers. Other additives, such as perfumes, emollients, etc., can be combined in a different operation and added at a separate stage using accurate positive-displacement filling devices.

The propellant preferably should be pressure-filled to minimize the possibility of contamination of the product by moisture.

REFERENCES

1. W. C. Beard, *Proc. 41st Ann. Meet. Chem. Spec. Manuf. Assoc.*, 75 (December 1954); *Soap Chem. Spec.* **31**, 139 (January 1955); U.S. Patent 2,959,325 (1960).
2. *Soap Chem. Spec.* **30**, 105 (August 1954).
3. F. T. Reed, Freon® Aerosol Report FA-17, "Aerosol Powders."
4. H. Mintzer, Aerosol Technicomment **10**, No. 3 (October 1967); Aerosol Techniques, Inc., Milford, Conn.
5. *Aerosol Age* **3**, 37 (February 1958).
6. J. J. Kanig and R. Cohn, *Proc. Sci. Sec. Toilet Goods Assoc.* **37**, 35 (1962).
7. C. G. Thiel and J. G. Young, U.S. Patent 3,169,095 (1965).
8. D. C. Geary and R. D. West, *Soap Chem. Spec.* **37**, 79 (June 1961); *Aerosol Age* **6**, 25 (August 1961); U.S. Patent 3,088,874 (1963).
9. Shulton, Inc., British Patent 987,301 (1962).
10. C. G. Thiel, I. N. Porush, and R. D. Law, U.S. Patent 3,014,844 (1961).
11. P. E. Gunning and D. R. Rink, U.S. Patent 3,081,223 (1963).
12. E. Huber, U.S. Patent 3,161,450 (1964).

13. S. Prussin, "Cosmetics: Fragrance and Personal Hygiene Products," in H. R. Shepherd (Ed.), "Aerosols: Science and Technology," Interscience Publishers, Inc., New York, 1961.
14. V. DiGiacomo, *Soap Chem. Spec.* **32,** 164 (June 1956).
15. *Aerosol Age* **9,** 42 (July 1964).
16. Technical Data Sheets, "Macro-Dry" and "Micro-Dry," Reheis Chemical Company.

III
MISCELLANEOUS

22

FLUOROCARBONS IN THE ATMOSPHERE

Richard B. Ward

THE ATMOSPHERE

The earth's atmosphere, schematically shown in Figure 22-1, exists in the form of concentric, spherical layers. The two layers of interest here are the *troposphere* and the next higher layer—the *stratosphere*. The division between the two, not always distinct, is called the *tropopause*. Superimposed on Figure 22-1 is a temperature profile and an indication of the depth to which various wavelengths of sunlight penetrate.

The troposphere is characterized by a decrease in temperature as altitude increases. This temperature gradient is maintained by conversion of the sun's radiation to heat upon absorption at the earth's surface. With such a thermal structure, the troposphere is subject to *rapid vertical mixing*. This can be seen when a thunderhead builds in the sky. We may conveniently visualize the troposphere as a churning, well-mixed air mass constantly washed (or scrubbed) by moisture condensation and subsequent precipitation. The northern and southern hemisphere troposphere do not mix rapidly across the equator due to the pattern of natural air circulation.

The decrease in temperature with altitude ceases at the tropopause. Above the tropopause, the temperature at first remains constant and then increases. The altitude of the tropopause is quite variable, being highest in the tropics and also higher in summer than in winter at mid and high latitudes. At times, the tropopause may be indistinct or apparently "folded" in a north–south direction.

In the stratosphere, the temperature increases with altitude, creating strata or layers which retard vertical mixing. Rapid air movements occur horizontally. Vertical mixing in the atmosphere may be estimated by "eddy diffusion coefficients," which provide a measure of the rate of vertical transport at a given altitude. Several such descriptions exist, some of which are shown in Figure 22-2. The Hunten eddy diffusion coefficient profile describes relatively slow mixing

415

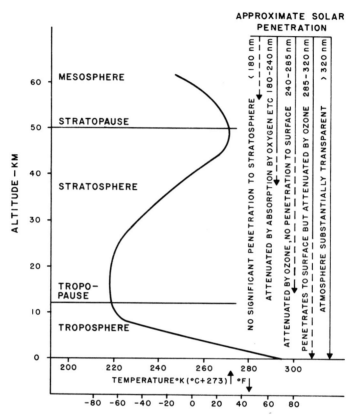

Figure 22-1 Schematic diagram of atmosphere showing temperature profile and solar radiation penetration.

and is supported by observations of carbon 14 concentrations in the lower stratosphere, while the faster eddy diffusion coefficient profile of Crutzen is based upon measurements of other trace species, particularly methane and hydrogen. The profile due to Chang is derived from heat-flux data. All are clearly slower in the lower stratosphere than in the troposphere. The rate of transport governs the speed with which tropospheric gases will mix into the stratosphere and the rate at which their decomposition products diffuse out of the stratosphere.

TROPOSPHERIC EFFECTS

Preliminary measurements[1] showed that fluorocarbons 11 and 12 could be detected in the troposphere. Continued measurements showed that their levels were increasing and that the amount in the troposphere was comparable to, but

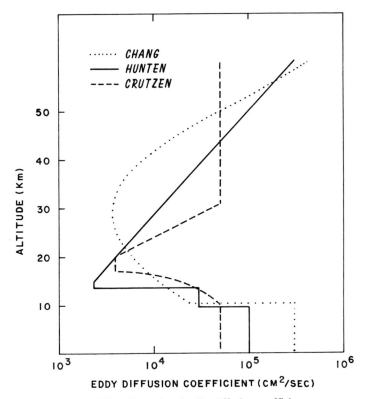

Figure 22-2 Examples of eddy diffusion coefficients.

rather less than, that calculated from accumulated release. In 1972, industry representatives met to consider the following objective:

"Fluorocarbons are intentionally or accidentally vented to the atmosphere worldwide at a rate approaching one billion pounds per year. These compounds may be accumulating in the atmosphere, returning to the surface, land, or sea [unchanged] or as decomposition products. Under any of these alternatives it is prudent that [the fluorocarbon industry] investigate any effects which the compounds may produce on plants or animals now or in the future."

The fluorocarbon industry responded by establishing an International Technical Panel for Fluorocarbon Research under the auspices of the Manufacturing Chemists Association.* Three programs were initially funded:

*The Manufacturing Chemists Association (MCA) is a trade association of chemical manufacturing companies. One function is to permit cooperative industry funding of research needs having broad application to the chemical industry.

1. Study of atmospheric concentrations of fluorocarbons and related species (J. E. Lovelock, University of Reading, England).
2. Ultraviolet photochemistry of fluorocarbons (C. Sandorfy, University of Montreal, Canada).
3. Atmospheric reactions of fluorocarbons (O. C. Taylor, University of California, Riverside).

These studies, published prior to the ozone depletion theory, have been reported by the Manufacturing Chemists Association.[2] Early measurements of fluorocarbon 11 were substantiated by later studies over wider areas and time spans. These studies also demonstrated that other halocarbons, particularly carbon tetrachloride and methyl chloride, were important, and perhaps largely natural, carriers of chlorine to the stratosphere. Fluorocarbons were found to be transparent to the ultraviolet light that penetrates to sea level, but to show absorption at wavelengths below about 210 nm, wavelengths at which photodissociation of the fluorocarbon molecule would occur.[3]

Fluorocarbons neither enhance nor inhibit simulated smog reactions and are themselves unaffected.[4] This and other requirements for long tropospheric life were established and the resultant accumulation explained. Stratospheric photolysis was considered a possible removal process, but at that time the immediate consequence appeared to be decomposition to halogen acids and carbon dioxide (Reaction 22-1). Downward diffusion of these compounds would contribute negligibly to the normal levels of chloride, fluoride, and carbon dioxide found in the troposphere:

$$CCl_3F \xrightarrow{\ h\nu\ } CCl_2F \cdot + Cl \cdot \qquad \text{(Reaction 22-1)}$$

oxidation
and hydrolysis $\quad \longrightarrow CO_2 + HCl + HF$

STRATOSPHERIC EFFECTS

The characteristic thermal profile of the stratosphere is created by various processes absorbing the shorter wavelengths of sunlight. Sunlight having a wavelength shorter than about 180 nm does not penetrate to the stratosphere. Sunlight of 180-260 nm is absorbed by molecular oxygen, which breaks down into oxygen atoms. The oxygen atoms, in the presence of a third body, M (typically nitrogen), to absorb the excess energy, can combine with molecular oxygen to yield ozone (Reactions 22-2 and 22-3):

$$O_2 \xrightarrow{\ h\nu\ } 2O \qquad \text{(Reaction 22-2)}$$

$$O + O_2 \xrightarrow{\ M\ } O_3 \qquad \text{(Reaction 22-3)}$$

Ozone itself absorbs ultraviolet radiation, with a maximum absorption at 250 nm, extending to above 300 nm. These absorbance characteristics are shown in

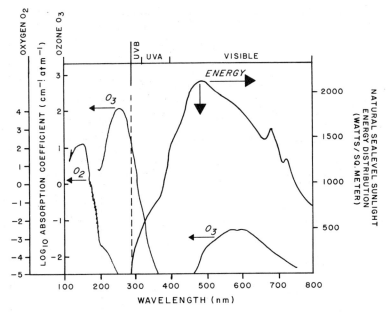

Figure 22-3 Solar sea level energy distribution and the absorption spectra of oxygen and ozone.

Figure 22-3. Thus, ozone limits ultraviolet penetration to the troposphere, as shown in Figures 22-1 and 22-3. Heat from these and other absorption processes in the upper stratosphere creates the increase in temperature with altitude.

The concentration of ozone in the atmosphere is markedly variable, being almost twice as great near the poles as at the equator. The atmospheric content may be measured from the ground using the Dobson Spectrophotometer and from space by satellite techniques utilizing the ultraviolet and infrared spectral characteristics of ozone. Figure 22-4 shows the total global ozone distribution in milliatmosphere-cm ("Dobson units").

The amount of ozone at a location undergoes substantial change. Weather fronts may produce short-term variations of up to 30%. A seasonal process (maximum in spring, minimum in fall) may fluctuate 30%; longer cycles up to 11 years have been noted with changes of 5%.

Statistical methods may be used to evaluate ozone records, both locally and globally. Several such analyses have not detected abnormal trends (either increasing or decreasing) in the ozone content of the atmosphere.[5]

CIAP Research

The Climatic Impact Assessment Program (CIAP)[6] was initiated by Congress in 1971 to study possible deleterious effects from the proposed supersonic transport (SST). The large inflow of Federal research funds, supporting a broad range of

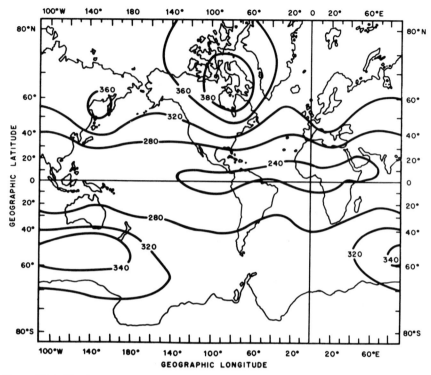

Figure 22-4 Total global ozone distribution (from Reference 17). Ozone column isopleths ("contours") are in milliatmosphere-cm ("Dobson units").

efforts, resulted in a marked increase in our understanding of stratospheric composition and processes. Processes leading to ozone formation had been previously described. Ozone concentrations, as measured, were much less than initially calculated. At first, only odd-oxygen processes had been proposed as ozone removal mechanisms (Reactions 22-4 and 22-5):

$$\text{Odd-oxygen}\begin{cases} O_3 \xrightarrow[\lambda \sim 210-310 \text{ nm}]{h\nu} O + O_2 & \text{(Reaction 22-4)} \\ O_3 + O \longrightarrow 2O_2 & \text{(Reaction 22-5)} \end{cases}$$

Later, other depletion processes were recognized. The odd-nitrogen and odd-hydrogen cycles are central to the SST concern, since nitrogen oxides and water vapor from the exhaust would be injected into the stratosphere:

$$\text{Odd-nitrogen}\begin{cases} NO + O_3 \longrightarrow NO_2 + O_2 & \text{(Reaction 22-6)} \\ NO_2 + O \longrightarrow NO + O_2 & \text{(Reaction 22-7)} \end{cases}$$

$$\text{Odd-hydrogen}\begin{cases} HO + O_3 \longrightarrow HO_2 + O_2 & \text{(Reaction 22-8)} \\ HO_2 + O \longrightarrow HO + O_2 & \text{(Reaction 22-9)} \end{cases}$$

$$\text{Sum process} \quad O_3 + O \longrightarrow 2O_2 \qquad \text{(Reaction 22-10)}$$

Both the odd-nitrogen (Reactions 22-6 and 22-7) and the odd-hydrogen cycle (Reactions 22-8 and 22-9) total to Reaction 22-10, thus converting odd oxygen (ozone and oxygen atoms, the latter being the precursor of ozone in Reaction 22-3) to molecular oxygen and thereby reversing the process of Reactions 22-2 and 22-3. The chlorine oxide cycle, first described in 1974,[7] was of concern because of possible introduction of chlorine into the stratosphere from the exhaust of the space shuttle, the fuel of which will contain chlorine compounds:

$$\text{Chlorine}\begin{cases} Cl + O_3 \longrightarrow ClO + O_2 & \text{(Reaction 22-11)} \\ \text{oxide cycle} \quad ClO + O \longrightarrow Cl + O_2 & \text{(Reaction 22-12)} \end{cases}$$

Summing Reactions 22-11 and 22-12 again yields Reaction 22-10.

An ozone "budget" can now be constructed for the stratosphere[8] (Table 22-1).

These cycles are catalytic, since the initiating compound (for instance, the chlorine in Reaction 22-11) is regenerated, as in Reaction 22-12. The active catalysts can be removed by diffusion and by various reactions which act as temporary or permanent sinks. Hydroxyl and nitrogen dioxide, for example, can form nitric acid in the presence of a third body, M (Reaction 22-13), and the nitric acid either subsequently degrades in the presence of radiation or radicals, regenerating nitrogen oxides (temporary sink), or diffuses to the troposphere, where it is removed as nitrate in rain. While combined as HNO_3, the OH and NO_2 do not react in their respective catalytic cycles:

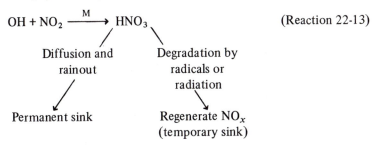

$$OH + NO_2 \xrightarrow{\text{M}} HNO_3 \qquad \text{(Reaction 22-13)}$$

Diffusion and rainout → Permanent sink

Degradation by radicals or radiation → Regenerate NO_x (temporary sink)

The problem addressed by the CIAP was assessment of the effects of incremental addition of ozone-depleting chemicals to the stratosphere as a result of operating the proposed SST's. The conclusions were that the effects would warrant improvements in engines and fuel; that other pollutants may significantly deplete ozone; and that ozone depletion may be expected to increase ultraviolet radiation with potential biological and even climatological effects. Among the

TABLE 22-1   STRATOSPHERIC PRODUCTION AND REMOVAL OF OZONE

Production		Removal (Estimated)	
1. Radiation (essentially all solar)	100%	1. Nitrogen oxide cycle	50–70%
		2. Reaction with oxygen (Chapman reactions)	15–20%
		3. Hydrogen oxide cycle	10–15%
		4. Chlorine oxide cycle	20–0%
		5. Other, including downward transport to troposphere	1–2%

potential biological effects was increased incidence of skin cancer in humans resulting from increased incident levels of ultraviolet radiation. The debate and search for quantitatively accurate assessments of future effects still continues for the nitrogen oxides.

THE OZONE DEPLETION THEORY

In 1974, a theory[9] which brought together the CIAP and industry research programs just described was proposed. Utilizing mathematical models developed in the CIAP research,[10] a quantitative estimate of effects was made.[11-14] The theory included the following:

1. Fluorocarbons 11 and 12 released into the troposphere are not removed by tropospheric processes and thus will accumulate and diffuse to the stratosphere.
2. In the stratosphere, they will be photodissociated by solar ultraviolet radiation, producing chlorine atoms.
3. The chlorine atoms will enter into a catalytic or chain reaction (Reactions 22-11 and 22-12), resulting in ozone depletion.
4. Depleted ozone will result in an increased level of ultraviolet radiation at the earth's surface.
5. An increased level of ultraviolet radiation will result in climatological and biological damage, one facet of which is an increased incidence of skin cancer.

The industrial and Federal research programs were expanded and rather complete agreement was reached on the research needs, which were spelled out first by the Manufacturing Chemists Association[2] and subsequently by the Interdepartmental Committee for Atmospheric Sciences (ICAS)[15] and the National

Academy of Sciences (NAS).[16] A Federal report[17] provided a review of the question and concluded that:

> ". . . fluorocarbon releases to the environment are a legitimate cause for concern. Moreover, unless new scientific evidence is found to remove the cause for concern, it would seem necessary to restrict uses of fluorocarbons 11 and 12 to replacement of fluids in existing refrigeration and air conditioning equipment and to closed recycled systems or other uses not involving release to the atmosphere."

The Federal report also indicated that the National Academy of Sciences would complete an in-depth scientific study and recommended that if the National Academy of Sciences confirmed the ozone depletion calculations, Federal regulatory agencies should ". . . initiate rulemaking procedures for implementing regulations to restrict fluorocarbon uses." A target date for regulations of January 1978 was proposed.

THE NATIONAL ACADEMY OF SCIENCES'* REPORT

Prediction, Uncertainties, and Recommendations

The National Academy of Sciences' report was issued September 13, 1976, in two parts: a technical assessment of relevant stratospheric chemistry made by a Panel on Atmospheric Chemistry (PAC)[18] and an assessment of effects with policy recommendations made by the Committee on Impacts of Stratospheric Change (CISC).[19]

One of the most important contributions made by the CISC report is a clear differentiation between the *ultimate effects* (calculated to be produced if the problem is "ignored" for about a century or more) and the *rate* at which such effects develop. The former dictates the *eventual* level of control which may be needed, while the latter defines the *immediacy* of the problem, the extent to which society can measure its response, and the delay time during which society can seek an optimum solution that minimizes economic dislocation. Much of the disagreement on how to handle major environmental questions is created when demands for immediate and extreme action are based solely on ultimate effects or when obstruction of any action is based on slow deterioration. It is important

---

*The National Academy of Sciences[20] consists of outstanding scientists from a very wide range of physical and biological sciences. Although it is an independent organization, it exists to provide the Government with a source of technical knowledge on scientific matters. Committees are selected to report on technical questions, with members picked to provide as wide a range of disciplines as needed.

to keep this differentiation and its significance clear, as did the National Academy of Sciences.

The National Academy of Sciences limited its study to chlorofluoromethanes (CFMs) 11 and 12 and used as their main scenario a world situation in which use and release, in the absence of regulation, would remain constant at 1973 levels. Under such conditions, ozone depletion would slowly increase, reaching an "ultimate" level in several hundred years, at which time removal of fluorocarbons and other chlorine compounds by stratospheric photolysis would equal surface release and a new steady state would be established.

The first factor considered was the reliability of the theory, and the CISC concluded that:

> "The scientific evidence for the proposed consequences of CFM releases when carefully examined is coherent but leaves us with a substantial range of uncertainty."

This conclusion was distinguished from the theory being quantitatively right or wrong.

The conclusion was reached that the most likely calculated "ultimate depletion" was 7%, but that this was associated with a tenfold range of uncertainty — from 2% to 20% — interpreted as a range from tolerable to intolerable. The uncertainty arises from the following sources:

1. Measurements of seven important reaction rates (fivefold range).
2. Approximations due to one-dimensional modeling (threefold range).
3. Treatment of photochemical processes (twofold range).
4. Measurements of concentrations of natural stratospheric species (twofold range).

However, there are additional uncertainties which cannot be quantified. These include the following:

1. Omission of unrecognized but essential chemical reactions.
2. Unexpected effects of tropospheric sinks.
3. Possible inadequacies in one-dimensional transport models.

It was noted that the calculated rate at which depletion developed is very slow, about 0.07% per year.

On these bases, short- and long-term policies were developed. The CISC concluded that within two years the uncertainties could be reduced, and that if ultimate ozone reductions of more than a few percent then remained a major possibility, *selective* regulation based on usage should be undertaken. A recommendation against immediate regulation was made on the basis of three important factors:

1. Inadequacies in the calculations.
2. Improved calculations will result from existing research programs.

3. Only slight increases in ozone depletion would result from a delay of one or two years.

The bases for possible future selective regulation would be the importance of individual uses to human life and the magnitude of the release associated with each individual use. Home refrigeration was noted as important and associated

TABLE 22-2 ESTIMATED ANNUAL WORLDWIDE
RELEASES OF CFMs FOR 1975[a]

Categories of Use Associated with Release	Release ($\times 10^6$ lb)	%
Aerosols–1115.1 $\times 10^6$ lb (74.5%)		
Personal Care–934.8 $\times 10^6$ lb (62.5%)		
Antiperspirants/deodorants	458.4	30.6
Hair care products	401.5	26.8
Medicinal products	37.3	2.5
Fragrances	2.3	0.2
Shaving lathers	0.9	0.1
Others	34.4	2.3
Household–69.1 $\times 10^6$ lb (4.6%)		
Room deodorants	17.7	1.2
Cleaners	9.6	0.6
Laundry products	23.4	1.6
Waxes and polishes	9.2	0.6
Others	9.2	0.6
Miscellaneous–111.2 $\times 10^6$ lb (7.4%)		
Insecticides	33.3	2.2
Coatings	22.9	1.5
Industrial products	39.0	2.6
Automotive products	8.0	0.5
Veterinary and pet products	2.3	0.2
Others	5.7	0.4
Air Conditioning/Refrigeration–204.7 $\times 10^6$ lb (13.7%)		
Mobile air conditioners	89.8	6.0
Chillers	42.9	2.9
Food store refrigeration	33.1	2.2
Beverage coolers	5.8	0.4
Home refrigerators and freezers	5.8	0.4
Others	27.3	1.8
Plastic Foams–176.5 $\times 10^6$ lb (11.8%)		
Open cell	100.0	6.7
Closed cell	76.5	5.1
	1496.3	100.0

[a]The data exclude Russia and Eastern Europe, whose annual release is estimated at an additional 11% (165 $\times 10^6$ lb).
From Reference 19.

with relatively little release. On the other hand, aerosols were noted as major sources of release but, in most cases, replaceable by other delivery systems, although with some loss of convenience, efficiency, and safety. The CISC estimates of global release rates are shown in Table 22-2.

While stressing that regulation is a political decision and not the responsibility of the CISC, the report made recommendations which were based on the technical aspects of the environmental question. While recommending delay (but not for more than two years) so that uncertainties could be resolved, the CISC urged that other countries be encouraged to follow actions in the United States so that the global issue would be handled on a global basis. In suggesting that some selective regulation of uses appeared almost certain, the CISC recommended (1) elimination of any inadequacies in the legislative authority that might be needed to regulate use and release of CFMs and (2) a system of informative labeling for aerosol and some other CFM-containing packages *intended to release* CFMs.

In their technical assessment of effects, the CISC noted the accumulation of CFMs in the atmosphere and expressed three areas of concern:

1. The potential ozone depletion would lead to an increase in the levels of ultraviolet radiation in the 290–320 nm range with possible biological effects.
2. Since ozone depletion is predicted to be most pronounced at higher altitudes, some modification of the stratospheric temperature profile should be anticipated (with possible climatological effects).[21]
3. Accumulation of CFMs throughout the atmosphere will have the effect of increasing infrared absorption in the atmosphere, thus contributing to the "greenhouse" effect. A similar effect is attributed to the accumulation of carbon dioxide from the burning of fossil fuels. Such processes are predicted to increase global temperatures and hence have climatic consequences.[22]

The warmer earth radiates infrared (IR) radiation into the cooler outer space. The atmosphere is transparent to some wavelengths; for example, it is transparent throughout much of the 700–1,250 μm band (8-14 cm^{-1}). At other wavelengths (for example, at 1,700 μm–6 cm^{-1} –due to water vapor), this radiation is absorbed, thus warming the atmosphere. The atmosphere reradiates IR, but does so at a lower temperature. This slows the loss of radiation from the planet, which effectively loses energy as though it were at a lower temperature. The main natural IR absorbing gases in the atmosphere are carbon dioxide, water vapor, and ozone, and the effect maintains surface temperatures some 30°C (54°F) higher than would otherwise be the case. Rising concentrations of carbon dioxide due to the burning of fossil fuels threaten to increase this effect. While the increasing carbon dioxide levels are well documented, detecting a warming trend against the background climatic variability has not been possible. The *calculations* cited by the CISC[19] suggest that a doubling of carbon dioxide levels in the next 50-100 years (considered reasonably probable) from about 300 parts per million (ppm) to about 600 ppm would result in an increase in the mean global surface temperature of about 3°C (5°F). The steady state, or ultimate concentrations (by about 2075, assuming continued release at 1973 rates) of fluorocarbons 11 and 12 were calculated to be 700 parts per trillion (ppt) and 1,900 ppt, respectively, or about seven times the present levels, and would be associated with an increase in global surface temperature of about 0.5°C (0.9°F). This latter temperature increase is about equal to that due to the present calculated increase in atmospheric carbon dioxide levels believed to have occurred since 1890 (about a 50 ppm increase). While the effects from CFM accumulation are small compared with those from carbon dioxide, they appear to be additive, with CFMs producing a much larger effect per unit of concentration than carbon dioxide. A number of factors are omitted from these calculations, and the actual effects could be greater (positive feedback processes omitted) or smaller (negative feedback processes omitted) than currently calculated.

Careful emphasis was placed on the slowness with which all these processes may be occurring and the fact that uncertainties associated with quantitative estimates of climatological effects were even greater than those associated with estimates of ozone depletion. It was also noted that the processes could be reversed by terminating or limiting CFM release, since if release were terminated, the CFMs would, in due course, be removed by the stratospheric photolytic sink and by any other sinks which might exist. The latter two temperature effects noted by the CISC (see p. 426) do not presently require more immediate action than that recommended for the ozone depletion effect. The CFM "greenhouse" effect is presently much smaller than the similar effect calculated for carbon dioxide.

SCIENTIFIC DEVELOPMENTS

The CISC[19] attempted, *on a scientific basis*, to estimate "tolerable" depletion. (Of course, "tolerable" eventually must be decided on a political basis, the scientific basis just being one important factor.) Based upon the relationship between cancer incidence and increased levels of ultraviolet radiation, it was concluded that an ultimate ozone depletion of 2% or less would be tolerable, although not desirable.

Atmospheric Inventory and Tropospheric Lifetimes

Although current atmospheric measurements are not sufficiently accurate to prove the existence or absence of significant tropospheric removal processes, the ozone depletion theory assumes their absence. Thus, if the amount of released fluorocarbon, the rate of removal by stratospheric photolysis, and the average atmospheric concentration are known, then the rate of removal by other processes (if any) and hence lifetime can be calculated. Estimates of release are considered relatively accurate (±5%) and, up to the end of 1975, were 6.5 billion pounds and 9.7 billion pounds, respectively, for fluorocarbons 11 and 12.[23] Stratospheric photolysis is reasonably quantified by the computer model. The difficulty arises in the accuracy of estimates of average atmospheric concentrations. Arguments for fluorocarbon 11 lifetimes as short as 15 years have been advanced,[24,25] while other arguments for the absence of any significant alternative sink[26] have been made. A 25-year lifetime for fluorocarbons 11 and 12 would reduce ozone depletion calculations by about *a factor of three.*[24]

Tropospheric Sinks

Recent work at the National Bureau of Standards[27] has shown that fluorocarbons 11 and 12, adsorbed on sand particles, can be photolyzed by radiation of 300–400 nm. These wavelengths, available down to ground level, do not photolyze these compounds in the unadsorbed gaseous state. The research is too preliminary to assess whether such a process would be significant in the atmosphere. The degradation on silica yielded dichlorofluoromethane from fluorocarbon 11 in the Bureau of Standards experiment.

Dichlorofluoromethane has been reported[28] in atmospheric analyses in quantities much too large (~10 ppt) to be explained by the small amount of this chemical produced and released, but the analyses are erratic, and careful additional work is needed to distinguish between its existence as an atmospheric degradation product of fluorocarbon 11 or as an analytical artifact. The question of tropospheric sinks remains an intriguing challenge.

Chlorine Nitrate

Chlorine nitrate, formed by Reaction 22-14 in the presence of a third body (M), constitutes a sink analogous to nitric acid (Reaction 22-13), and inclusion of chlorine nitrate in the calculations almost halved the ultimate ozone depletion values obtained (from 13% to 7%):[19]

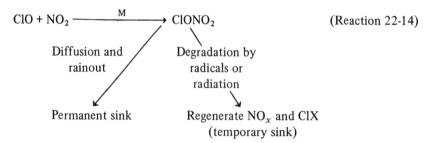

$$ClO + NO_2 \xrightarrow{M} ClONO_2 \qquad \text{(Reaction 22-14)}$$

Diffusion and rainout

Degradation by radicals or radiation

Permanent sink

Regenerate NO_x and ClX (temporary sink)

Predicted concentrations of chlorine nitrate are comparable to the present lowest limit of detection,[30] preventing, for now, confirmation of its existence in the stratosphere.

Chlorine Species

The stratospheric concentration profiles for hydrogen chloride (HCl), chlorine (Cl), and chlorine oxide (ClO) may be readily calculated, and methods for their measurement are now beginning to yield results. All three have been observed in the stratosphere, but measurements cannot be reconciled with predicted values. Hydrogen chloride concentrations decrease above about 30 km, in marked contrast to models, and the concentration at 30 km (2 parts per billion, ppb) is about twice that expected. While the *ratio* of chlorine oxide to chlorine is close to the predicted value, actual concentrations of these two species are again higher than predicted and exceed the total calculated stratospheric chlorine concentration.

Such deviations from theory raise doubts about the reliability of quantitative predictions of ozone depletion. The deviations can be attributed, for instance, to analytical error, to underestimation of the atmosphere's chlorine budget, or to chemical processes not yet recognized as important (so-called "missing chemistry"). The latter two explanations raise the possibility of significant revisions to estimates of the ozone depletion resulting from fluorocarbon release when more complete knowledge becomes available.

Reaction Rates

Seven important reaction rates, defined by the National Academy of Sciences Panel on Atmospheric Chemistry,[18] have been the subject of recent changes. The net effect is to reduce estimates of ultimate ozone depletion by about a factor of two.

Diurnal Modeling

The PAC's[18] modeling utilized constant half-intensity solar irradiation and did not include computation of effects from the day/night (diurnal) changes. Recent attempts have been made to model a more realistic on/off diurnal irradiation cycle. Initial indications suggest that ultimate ozone depletion effects will be increased by including a diurnal effect in the calculations.

Release Rates

Use of fluorocarbons, particularly in aerosols, in the U.S. has decreased. Estimates indicate that global release rates have been roughly constant from 1973 to 1977.

Hydroperoxyl Reactions

Until 1977, a rate for Reaction 22-15 was only estimated:

$$HO_2 + NO \longrightarrow HO + NO_2 \qquad \text{(Reaction 22-15)}$$

Recent measurements of the rate of Reaction 22-15 have yielded a value about 30 times higher than previously assumed. This higher rate reduces the predicted impact of nitrogen oxides from the supersonic transport and simultaneously predicts larger ozone depletions from chlorine in the stratosphere. Howard and Evenson[31] have estimated that an increase amounting to 35% of the calculated ozone depletion could result. Ozone depletion due to chlorine is predicted to increase, since a product of Reaction 22-15—hydroxyl (HO)—returns chlorine to the cycle via Reaction 22-16:

$$HCl + HO \longrightarrow Cl + H_2O \qquad \text{(Reaction 22-16)}$$

Reaction 22-15 is perhaps most significant in that it reveals the continuing uncertainties associated with predicted ozone depletion from nitrogen oxides and the complex interrelations in stratospheric chemistry.

LEGISLATIVE AND REGULATORY ACTIVITY

Federal Legislative Activity

A substantial amount of Federal legislative action developed after the Inadvertent Modification of the Stratosphere (IMOS) report was issued (see p. 423). Federal legislative efforts initially were centered on the Clean Air Act Amendments (CAAA), proposed in 1976 and finally passed in 1977. Authority to be provided by the CAAA was, however, provided in a somewhat different form in the Toxic Substances Control Act, which became effective January 1, 1977.

Federal Regulatory Activity

Regulatory authority over aerosols, particularly those propelled by fluorocarbons, resides in the Food and Drug Administration (FDA) (for approximately 80% of the products, principally cosmetics), the Consumer Products Safety Commission (CPSC) (for about 13%, mainly household products), and the Environmental Protection Agency (EPA) (for about 5%, mainly insecticide and pesticide products).

Additionally, the Toxic Substances Control Act (TSCA) provides the EPA with broad regulatory authority over all fluorocarbon uses and establishes the EPA as a logical lead agency to coordinate Federal activity.

The EPA and FDA proposed a schedule for the phase-out of fully halogenated chlorofluoroalkanes in aerosols, with limited exceptions for a few products. The timetable proposed is as follows:[32]

Stop manufacture of specified fluorocarbons for affected uses	October 15, 1978
Stop distribution and processing of specified fluorocarbons for affected uses	December 15, 1978

Stop interstate shipment of   April 15, 1979
finished aerosol products contain-
ing the specified fluorocarbons as
propellants (no recall)

The CPSC indicated that no further action on the phase-out was necessary on
its part due to the action of the FDA and EPA.[32]

Federal efforts have been made[33] to encourage other countries to follow the
Federal program. However, these countries, in general, have stressed their
desire to obtain and assess additional scientific data before acting.

State Activity

Many states have considered legislation against fluorocarbons, generally against
their use in aerosols. Such legislation has been opposed by industry, since the
issue requires a national policy and such a policy is under development and
implementation.

ALTERNATIVE FLUOROCARBONS—WHY AND HOW

The commonly used chlorofluorocarbons (such as fluorocarbons 11 and 12)
provide valuable and necessary properties, but concern over possible stratospheric impact may lead to restrictions. Therefore, the development of replacement or alternative chemicals has become desirable contingency research.

Such alternatives must show improved environmental acceptability, but
otherwise be comparable with the fluorocarbons they might replace, for instance, in such properties as low toxicity, nonflammability, stability, and vapor
pressure.

The concern over possible ozone depletion, defined by the National Academy
of Sciences,[19] is based on *chlorine*-containing halocarbons with such extreme
tropospheric stability that significant transport of chlorine to the stratosphere
occurs. Thus, removal of chlorine from the molecule or reduction of tropospheric
stability would reduce stratospheric effects.

It has been calculated that fluorine atoms liberated in the stratosphere are
rapidly converted to hydrogen fluoride.[34] In contrast to hydrogen chloride,
which can re-form chlorine following radical attack or photolysis, the high
strength of the H-F bond insures that hydrogen fluoride is not readily attacked
in the stratosphere; thus, its impact on ozone levels is insignificant compared to
that of chlorine.

Fluorocarbons containing no chlorine are well known, but have found limited
use as aerosol propellants. Fluorocarbons 152a (CH_3CHF_2) and C-318 (cyclic-
C_4F_8) are examples (the former is flammable, while the latter is quite costly).
To achieve nonflammability and the vapor pressures required for aerosols, both

a relatively high level of fluorine substitution in the molecule and a relatively high molecular weight are needed, factors which contribute significantly to cost.

Reduced stability in the troposphere can be achieved by introducing unsaturation or hydrogen into the molecule. Olefinic fluorocarbons are commonly more toxic than the saturated compounds,[35] so attention has generally focused on hydrogen-containing chlorofluorocarbons. It has been recognized[36-38] that hydrogen-containing chlorofluorocarbons are subject to attack by the hydroxyl radical in the troposphere, with hydrogen abstraction (leading to the formation of water) the first step (Reaction 22-17):

$$CHClF_2 \xrightarrow{HO} H_2O + CClF_2 . \qquad \text{(Reaction 22-17)}$$

Government studies[15,17,39] have concluded that such compounds may be replacements. A study for the EPA[40] listed a number of hydrogen-containing chlorofluorocarbons as potential alternatives for aerosol propellant uses and included most of those presented in Table 22-3.

An assessment of a compound's environmental advantage over existing fluorocarbons can be made by estimating the tropospheric lifetime (τ) based on reaction with hydroxyl (τ is actually the time necessary to reduce the concentration to e^{-1}, that is, to about 37% of its initial value). This estimate requires knowledge of the reaction rate of a compound with hydroxyl as a function of temperature, the tropospheric hydroxyl concentration, and the mean tropospheric temperature ($\sim +8°C$).

While tropospheric hydroxyl measurements are still preliminary,[41,42] recent estimates of the global mean hydroxyl concentration have ranged from 3×10^5 to 6×10^6 radicals/cm^3. Laboratory measurements of the rate at which hydroxyl attacks alternative fluorocarbons have been made,[42-44] and the data are presented in Table 22-3, which also gives the tropospheric lifetime τ. Calculations of relative ozone depletion depend upon the assumed atmospheric lifetime of fluorocarbon 11 (τ_{FC-11} is taken as 91 years) and on the number of chlorine atoms per molecule, as well as on the hydroxyl reaction rate. Thus, values for relative effect can be calculated (Table 22-3).[25]

At constant use rates (based on 1973), the minimum improvement factor (fluorocarbon 142b versus $\tau_{FC-11} = 91$ years) is 0.09, which would lower an ultimate ozone depletion estimate of 7% to 0.6%—well below the 2% level described earlier as tolerable. Other alternative compounds with smaller factors result in an even smaller depletion.

In addition to ozone depletion projections for alternative fluorocarbons, calculations must be made to estimate the "greenhouse" effect. The CFMs and alternative compounds containing the C–F bond have characteristic infrared absorption in the wavelength range 8–12 μm. This range coincides with a region in which the atmosphere is nearly transparent. Consequently, potential "greenhouse" effects will be comparable, within a factor of about three, *at equivalent atmospheric concentrations*. The shorter tropospheric lifetimes of alternative

TABLE 22-3 CANDIDATE ALTERNATIVE FLUOROCARBONS

Chemical Formula	Name	Fluorocarbon Number	Boiling Point (°F)	Number of Chlorine Atoms in Molecule	Hydroxyl Reaction Rate Constant $K_{265°K}$ ($cm^3 \cdot molecule^{-1}\ sec^{-1}$)	Calculated Tropospheric Lifetime τ (years)	Relative Ozone Depletion versus Fluorocarbon 11 $\tau_{FC\text{-}11} = 91$ years
Perhalogenated Chlorofluoromethanes							
CCl_3F	Trichlorofluoromethane	11	74.8	3	$<5 \times 10^{-16}$	>70	1.0
CCl_2F_2	Dichlorodifluoromethane	12	−21.6	2	$<5 \times 10^{-16}$	>70	⩾1
Hydrogen-Containing Candidate Alternatives							
$CHCl_2F$	Dichlorofluoromethane	21	48.1	2	1.8×10^{-14}	2	0.01
$CHClF_2$	Chlorodifluoromethane	22	−41.4	1	2.4×10^{-15}	15	0.05
CH_2ClF	Chlorofluoromethane	31	15.6	1	2.5×10^{-14}	1.4	0.005
CH_2F_2	Difluoromethane	32	−61.0	0	No chlorine transport		
$CHCl_2CF_3$	1,1-Dichloro-2,2,2-trifluoroethane	123	82.3	2	1.1×10^{-14}	3	0.02
$CHClFCF_3$	1-Chloro-1,2,2,2-tetrafluoroethane	124	12.2	1	5.8×10^{-15}	6	0.02
CHF_2CF_3	Pentafluoroethane	125	−55.3	0	No chlorine transport		
$CH_2ClCClF_2$	1,2-Dichloro-1,1-difluoroethane	132b	115.5	2	Data not available, probably comparable to FC-123		
CH_2ClCF_3	1-Chloro-2,2,2-trifluoroethane	133a	44.5	1	5.8×10^{-15}	6	0.02
CH_2FCF_3	1,1,2-Tetrafluoroethane	134a	−15.7	0	No chlorine transport		
CH_3CClF_2	1-Chloro-1,1-difluoroethane	142b	14.4	1	1.5×10^{-15}	24	0.09
CH_3CF_3	1,1,1-Trifluoroethane	143a	−53.7	0	No chlorine transport		
CH_3CHF_2	1,1-Difluoroethane	152a	13.0	0	No chlorine transport		
Example Chlorocarbons							
CH_2Cl_2	Methylene chloride	—		2	8.7×10^{-14}	0.4	0.003
CH_3Cl	Methyl chloride	—		1	3×10^{-14}	1.2	0.004
CH_3CCl_3	Methyl chloroform	—		3	15×10^{-15}	2	0.02
C_2HCl_3	Trichloroethylene	—		3	Very rapid, several processes as well as hydroxyl reaction		

fluorocarbons result in lower equilibrium atmospheric concentrations for a given release rate. At 1973 release rates, equilibrium concentrations of alternative fluorocarbons would be much less than for fluorocarbons 11 and 12, and this limitation would reduce any "greenhouse" effect by a large factor relative to that of fluorocarbons 11 and 12.

Alternative fluorocarbons as aerosol propellants are further discussed in terms of commercial accessibility, functional utility, and toxicology in Chapters 3 and 25.

ADDENDUM

This addendum discusses recent advances in the rapidly developing area of calculated future effects of chlorofluorocarbons on stratospheric ozone.

The most important regulatory development is the promulgation of the final regulation requiring the phase-out of fully halogenated chlorofluorocarbons in nonessential aerosols, estimated at about 98% of the original fluorocarbon propellant market. The final rule[45] differs but marginally from the proposed rule.[32] A Federal agency work group has initiated public meetings to gather information on which to base future regulatory decisions on all other (nonpropellant) uses of chlorofluorocarbons, which would include refrigerants, blowing agents, solvents, and others.

These rapid and relentless regulatory developments have been paralleled by steady scientific progress, which, while answering some questions, has raised others. Original expectations that the uncertainties would be substantially reduced within two years of the date of the National Academy of Sciences' reports[18, 19] can now be seen to have been optimistic.

Two further technical reviews[46,47] have confirmed that computer calculations of future ozone depletion from fluorocarbons are almost doubled by the new rate constant for Reaction 22-15 (p. 429), while at the same time the predicted effect of Concorde-type SST's has shrunk to insignificance. These effects are largely due to coupling, or reciprocal interference, between the catalytic cycles involved. It is now clearly recognized that all the major chemical catalytic cycles are closely coupled and cannot be studied separately.

Further stratospheric analyses[46-48] tend to confirm the unpredicted decrease in concentration of hydrogen chloride with increasing altitude above about 30 km. Other analyses continue to show chlorine oxide concentrations which are both higher and more variable than calculated.[46,47,49]

One tentative explanation for such discrepancies has been offered—Reaction 22-18:[50, 51]

$$ClO + HO_2 \longrightarrow HOCl + O_2 \qquad \text{(Reaction 22-18)}$$

Such reactions exemplify coupling between the chlorine and hydrogen catalytic cycles and result in a downward revision of subsequent calculations for ozone depletion for a given amount of chlorine. Recent work has indicated that hypochlorous acid (HOCl) may have much greater photolytic stability than originally believed. This raises the possibility that hypochlorous acid, and perhaps higher oxyacids of chlorine, are significant temporary sinks for chlorine (see Reaction 22-14, p. 428). The possibility of such sinks places a high priority on the need for simultaneous measurements of the stratospheric concentrations of photochemically important species.[47]

Further work on the statistical trend analysis of ozone records[5,52] has suggested that an annual trend of ±0.25% persisting for six years (±1.56% total) should be quantitatively detectable. No such trend has been found in recent records. Since the estimates of ultimate ozone depletion have now increased, the calculated ozone depletion to date should also be correspondingly larger. These calculations for present-day depletion are now almost contradictory to the most recent analyses of ozone observations.

Measurements of ozone at high altitudes (~40 km) suggest a 12% increase over the last decade, while model calculations predict a 5% decrease at such altitudes.[47] The concentration of ozone at 40 km is predicted to be much more sensitive to chlorine from chlorofluorocarbon photolysis than is the total ozone column. The significance of this discrepancy is limited, since little is known about the natural variability of ozone concentration at these altitudes.

An important experiment to measure fluorocarbon concentrations at four widely separated monitoring stations has been initiated through the industry research program administered by the Manufacturing Chemists Association. The experiment is expected to provide, over a period of three to four years, the extremely critical values for the tropospheric lifetimes of fluorocarbons 11 and 12. Lifetimes are not reliably determinable from tropospheric measurements to date. Assumptions that the values are large (>100 years) is a basic tenet of the ozone depletion theory.

Release of fluorocarbons 11 and 12 remains the major, but not the only, calculated cause for potential future ozone depletion. Recent research has:

1. Increased computed estimates of future depletion from fluorocarbon releases but without reducing the range of uncertainty.
2. Demonstrated the interdependent nature of catalytic cycles.
3. Revealed inconsistencies between various model calculations and actual observations, thereby raising substantial suspicion that the chemical basis of the models is incomplete and that further significant chemical processes remain to be introduced before stratospheric modeling can provide reliable estimates of environmental effects.

REFERENCES

1. J. E. Lovelock, *Nature* **230**, 379 (1971).
2. Manufacturing Chemists Association, "Effect of Fluorocarbons on the Atmosphere: Summary of Fluorocarbon Research Program," Revision 6, March 31, 1977.
3. A. Hardin and C. Sandorfy, *Atmos. Environ.* **10**, 343 (1976).
4. N. E. Hester, E. R. Stephens, and O. C. Taylor, *J. Air Pollution Control Assoc.* **24**(6), 591 (1974).
5. W. J. Hill and P. N. Sheldon, *Geophys. Res. Lett.* **2**, 541 (1975).
6. Department of Transportation: (a) "The Effects of Stratospheric Pollution by Aircraft," DOT-TST-75-50 (1974); (b) T. M. Hard and A. J. Broderick (Eds.), "Proceedings of the 4th CIAP (Climatic Impact Assessment Program) Conference," DOT-TSC-OST-75-38 (1975).
7. R. S. Stolarski and R. J. Cicerone, *Can. J. Chem.* **52**, 1610 (1974).
8. E. Bauer and F. R. Gilmore, in T. M. Hard and A. J. Broderick (Eds.), "Proceedings of the 4th CIAP (Climatic Impact Assessment Program) Conference," DOT-TSC-OST-75-38 (1975).
9. M. J. Molina and F. S. Rowland, *Nature* **249**, 810 (1974).
10. M. B. McElroy and J. C. McConnell, *J. Atmos. Science* **30**, 1465 (1973).
11. F. S. Rowland and M. J. Molina, *Rev. Geophys. Space Phys.* **13**, 1 (1974).
12. R. J. Cicerone, R. S. Stolarski, and S. Walters, *Science* **185**, 1165 (1974).
13. P. J. Crutzen, *Geophys. Res. Lett.* **1**, 205 (1974).
14. S. C. Wofsy, M. B. McElroy, and N. D. Sze, *Science* **187**, 535 (1975).
15. "The Possible Impact of Fluorocarbons and Halocarbons on Ozone," Interdepartmental Committee for Atmospheric Sciences, National Science Foundation, May 1975 (ICAS 18a-FY75).
16. "Interim Report of the Panel on Atmospheric Chemistry," Climatic Impact Committee, National Academy of Sciences, July 1975.
17. "Fluorocarbons and the Environment," Federal Task Force on Inadvertent Modification of the Stratosphere, Council on Environmental Quality, June 1975 (GPO stock #038-000-00226-1).
18. "Halocarbons: Effects on the Stratospheric Ozone," Panel on Atmospheric Chemistry, National Academy of Sciences, 1976.
19. "Halocarbons: Environmental Effects of Chlorofluoromethane Release," Committee on Impacts of Stratospheric Change, National Academy of Sciences, 1976.
20. W. Spindel, *Chemtech.*, 744 (December 1976).
21. R. A. Reck, *Science* **192**, 557 (1976).
22. V. Ramanathan, *Science* **190**, 50 (1975).
23. R. L. McCarthy, F. A. Bower, and J. P. Jesson, *Atmos. Environ.* (in press).
24. N. D. Sze and M. F. Wu, *Atmos. Environ.* **10**, 1117 (1976).
25. J. P. Jesson and R. L. McCarthy, "8th International Symposium on Fluorine Chemistry," Kyoto, Japan (1976).
26. F. S. Rowland and M. J. Molina, *J. Phys. Chem.* **80**, 2049 (1976).
27. P. Ausloos, R. E. Rebbert, and L. Glasgow, In preparation for *Natl. Bur. Stand. (U.S.) J. Res.*
28. H. B. Singh, *Geophys. Res. Lett.* **4**, 101 (1977).
29. F. S. Rowland, J. E. Spencer, and M. J. Molina, in "Abstracts of Proceedings of the 12th International Symposium on Free Radicals," Laguna Beach, Ca. (1976).
30. D. G. Murcray, unpublished data.
31. C. J. Howard and K. M. Evenson, American Geophysical Union Meeting, Washington, D.C. (May/June 1977).

32. **42** FR 24536, May 13, 1977 (FDA).
    42 FR 24542, May 13, 1977 (EPA).
    42 FR 24550, May 13, 1977 (CPSC).
33. *Environ. Sci. Technol.* **11**, 438 (1977).
34. R. S. Stolarski and R. D. Rundel, *Geophys. Res. Lett.* **2**, 443 (1975).
35. J. W. Clayton, Jr., *Fluorine Chem. Rev.* **1**, 197 (1967).
36. F. S. Rowland, *New Scientist*, 720 (December 1974).
37. M. J. Molina, F. S. Rowland, and C. C. Chou, in "Abstracts of Proceedings of the 12th International Symposium on Free Radicals," Laguna Beach, Ca. (1976).
38. C. Seigneur, H. Caram, and R. W. Carr, *Atmos. Environ.* (in press).
39. "Chlorofluorocarbons and Their Effect on Stratospheric Ozone," Pollution Paper No. 5, U.K. Dept. of the Environment, HMSO, London, 1976.
40. "Technical Alternatives to Selected Chlorofluorocarbon Uses," EPA Office of Toxic Substances, 1976 (EPA-560/1-76-002).
41. C. C. Wang and L. I. Davis, *Phys. Rev. Lett.* **32**, 349 (1974).
42. J. Chang, D. D. Davis, and D. Wuebbles, American Chemical Society National Meeting, San Francisco, Ca. (1976).
43. C. J. Howard and K. M. Evenson, *J. Chem. Phys.* **64**, 197 (1976).
44. R. Atkinson, D. A. Hansen, and J. N. Pitts, *J. Chem. Phys.* **63**, 1703 (1975).
45. **43** FR 11301, March 17, 1978.
46. "Chlorofluoromethanes and the Stratosphere," National Aeronautics and Space Administration Reference Publication 1010, 1977.
47. "Response to the Ozone Protection Sections of the Clean Air Act Amendments of 1977: An Interim Report," Committee on Impacts of Stratospheric Change, National Academy of Sciences, 1977.
48. J. R. Eyre and H. K. Roscoe, *Nature* **266**, 243 (1977).
49. J. G. Anderson, J. J. Margitan, and D. H. Stedman, *Science* **198**, 501 (1977).
50. R. L. Jaffe and S. R. Langhoff, American Chemical Society Meeting, Chicago, Ill. (September 1977).
51. J. P. Jesson, L. C. Glasgow, D. L. Filkin, and C. Miller, American Geophysical Union Meeting, San Francisco, Ca. (December 1977).
52. W. J. Hill, P. N. Sheldon, and J. J. Tiede, *Geophys. Res. Lett.* **4**, 21 (1977).

23

TECHNICAL PROGRAMS IN THE ATMOSPHERIC SCIENCES

Frank A. Bower

INTRODUCTION

Traditionally, research in the atmospheric sciences has been carried out by the United States Government to support weather forecasting and military operations in the atmosphere. Much of the research has been carried out under contract by university laboratories. Most of these are located within the United States, but a significant number in Europe and Asia have been involved in the American program.

Governments of other countries also have well-developed programs in meteorology and atmospheric chemistry. A number of international programs, developed in recent years, have fostered cooperation between different nations in the gathering of scientific data and the support of global measurement programs.

The current technical program on the atmospheric sciences can be viewed as taking place in three arenas:

1. Federal research.
2. Industry research.
3. International research.

These are discussed in the following sections.

FEDERAL PROGRAMS

When the question of the interaction of fluorocarbons with stratospheric ozone was initially raised (1974), the Federal Government responded by holding Congressional hearings and by forming an *ad hoc* task force on Inadvertent Modification of the Stratosphere (IMOS). The IMOS task force consisted of 14 separate agencies (9 of which were already involved in atmospheric research), plus the

Interdepartmental Committee on Atmospheric Sciences. The 14 member agencies were:

Department of Agriculture
Department of Commerce
Department of Defense
Department of Justice
Department of Health, Education and Welfare
Department of State
Department of Transportation
Energy, Research and Development Administration
Environmental Protection Agency
Consumer Products Safety Commission
National Aeronautics and Space Administration
National Science Foundation
Council on Environmental Quality
Office of Management and Budget

The group was co-chaired by representatives from the Federal Council for Science and Technology and the Council on Environmental Quality.

The IMOS task force produced an early assessment[1] of the fluorocarbon/ ozone controversy, as well as of the possible effects of a change in ozone concentration. They attempted to define, in a preliminary manner, some of the characteristics of the fluorocarbon industry. The IMOS task force concluded that there was a legitimate cause for concern and that a more comprehensive technical assessment of the situation should be made by the National Academy of Sciences (NAS).

The NAS, after studying the various aspects of atmospheric chemistry, physics, and transport, reported to the IMOS task force that judging from the data available in September 1976, the most likely effect of chlorofluoromethanes on ozone would be an ultimate steady-state depletion of about 7.5%, with the possibility that the true value could lie between 2% and 20%. The NAS was careful to point out that unknown processes which might be occurring in the stratosphere could have important effects on this depletion estimate and suggested a long-term research program directed toward obtaining a more complete understanding of the stratosphere.

Federal programs are generally classified under the following categories:

1. Aeronomy.
2. Planetary atmospheres.
3. Meteorology.

Total budgets for Federal programs in recent years, as reported by the Inter-departmental Committee on Atmospheric Sciences (ICAS), were as follows:

Year	Millions
1961	106.4
1962	144.0
1963	200.7
1964	225.0
1965	207.6
1966	208.5
1967	207.9
1968	248.5
1969	236.3
1970	302.5
1971	209.6
1972	222.7
1973	242.9
1974	257.3
1975	290.4

Brief descriptions of the programs of the various agencies are listed below. Full details can be obtained from the publication, "National Atmospheric Sciences Program," ICAS 18-FY 75, May 1974, National Science Foundation.

Department of Agriculture

The Department of Agriculture carries out a broad research program in the atmospheric sciences that is designed to cover major agricultural weather problems and atmospheric processes involved in the management and protection of forests and rangelands.

Department of Commerce

To advance comprehension and beneficial use of the physical environment, the National Oceanic and Atmospheric Administration (NOAA) conducts a comprehensive program of meteorology and a program of aeronomy.

Department of Defense

Army

The army research is directed toward defining and interpreting the ionization phenomenology of the ambient and naturally disturbed atmosphere. The objectives of this research are to provide detailed and representative information on the properties of the atmosphere within the battlefield area and to furnish predictions of changes over related short distances and/or time intervals.

Air Force

The main objectives in the Air Force meteorological research program are greater understanding of atmospheric processes and better methods of observing, processing, displaying, forecasting, and, in some cases, modifying meteorological elements to facilitate safe and efficient operations.

The Air Force research programs in aeronomy are concerned with the studies of the upper atmosphere, the ionosphere, and the near-earth space environment in the quiet state and in the perturbed state, whether the latter is induced by natural or man-made phenomena.

Environmental Protection Agency

Observations and descriptions of the state of the atmosphere receive major emphasis. Approximately half of the observational funds are directed toward defining air pollution on a regional scale and providing the basic input data for development of descriptive simulation models.

Department of the Interior

The Department of the Interior's weather modification research program will concentrate research into six major projects:

1. The High Plains Cooperative Program.
2. The Colorado River Basin Pilot Project.
3. A comprehensive summary analysis and combined evaluation of 14 field experiments completed through 1974.
4. The Sierra Cooperative Pilot Project.
5. The North Platte Pilot Project.
6. The Pyramid Lake Pilot Project.

Energy Research and Development Administration

The atmospheric sciences research program of the Atomic Energy Commission is concerned with the development of predictive and descriptive models for the transport and fate of radioactive materials and other effluents released into the atmosphere from varied nuclear activities and electrical power generation. These studies range from local to global scales and will provide essential information for assessing the impact of these activities on human health and welfare and, as advanced electrical power generation technology evolves, for maintaining the environmental quality.

Department of Transportation

Office of the Secretary

A Climatic Impact Assessment Program (CIAP) is under way to assess the environmental and meteorological effects of the projected world high-altitude aircraft fleet, including subsonic and supersonic vehicles. The CIAP addresses the complex interactions between the engine emissions exhausted into the upper atmosphere, the natural composition of the stratosphere, and the dynamic processes of the atmosphere. It examines the effects of projected changes in the upper atmosphere upon climate close to the earth's surface.

Federal Aviation Administration

The Federal Aviation Administration (FAA) Aviation Weather Program is designed to provide the technical and operational developments required for improving the performance and utilization of existing components of weather data acquisition, transfer, processing, and display equipment to relieve immediate problems and to modify specific components for integration into the modernized configuration of the National Airspace System.

National Aeronautics and Space Administration

The National Aeronautics and Space Administration (NASA) conducts a broad program in the atmospheric sciences in support of earth and space exploration that will lead to applications benefiting all mankind. The work includes the analysis and interpretation of data from flight research experiments, the investigation and exploration in the laboratory of new concepts for flight experiments, the laboratory determination of physical characteristics necessary for the interpretation of flight data, and theoretical studies explaining phenomena of planetary atmospheres.

National Science Foundation

Meteorology

The Meteorology Program supports broad-based research into the physical, chemical, and dynamic processes in the lower atmosphere and stratosphere by means of research grants, primarily to university scientists. The objective of the Foundation's meteorology research is to acquire and maintain fundamental knowledge and expertise in the nation applicable to the behavior of the atmosphere of the earth and the atmospheres of other planets.

Aeronomy

The Aeronomy Program funds investigations that aim at a better understanding of the physical, chemical, and dynamic processes in the upper atmosphere of the earth and other planets. These investigations include field experiments, laboratory experiments on atomic and molecular interaction processes relevant to the atmosphere, and theoretical studies that include modeling the complex time-varying phenomenology in the high atmosphere.

Global Atmospheric Research Program

The highlight of the Global Atmospheric Research Program (GARP) during 1975 was the GARP Atlantic Tropical Experiment (GATE), conducted off the western coast of Africa in the tropical North Atlantic during the summer of 1974. This experiment was one of the largest international scientific cooperative ventures undertaken to date. Some 15 nations are participating significantly in the experiments, and at least 4 more nations have indicated their desire to participate.

National Center for Atmospheric Research

The National Center for Atmospheric Research (NCAR) at Boulder, Colorado is operated by the 44-university nonprofit University Corporation for Atmospheric Research (UCAR) with the purpose of extending and supplementing the capabilities of universities in attacking atmospheric research problems of importance to the nation. Its two major missions are as follows:

1. Plan and carry out research programs of the highest quality on scientific problems that are of great national and international importance and scope and that are characterized by their central importance to society and by the requirement for large-scale, coordinated thrusts by teams of scientists from a number of institutions.
2. Develop and make accessible selected major research services and facilities required by the NCAR and universities for effective progress in atmospheric research programs.

All of these Federal programs are coordinated by the Interdepartmental Committee on Atmospheric Sciences (ICAS), chaired by Edward P. Todd of the National Science Foundation. The Federal Government applied the resources of several of the above agencies to the question of the effects of the supersonic transport. This effort was coordinated by the Department of Transportation, and its basic outline was described earlier.

INDUSTRY RESEARCH

Historically, the global aspects of atmospheric science have not attracted extensive participation by the chemical industry. A significant industry effort has focused on the immediate effects of effluent materials from manufacturing plants and mobile sources on nearby flora and fauna.

Since 1970, when the aircraft industry was drawn into the debate over the effect of high-flying aircraft on stratospheric ozone, industry has been more and more involved in the broad aspects of atmospheric chemistry and physics.

In 1972, the worldwide fluorocarbon industry began a cooperative research program into the possible environmental effects of fluorocarbons in the atmosphere. The original emphasis was directed toward local effects on air quality, for example, smog formation in the Los Angeles Basin. The question of a possible interaction of chlorofluoromethanes with the stratosphere had not yet been stated.

Late in 1973, when the large program on the effects of stratospheric aviation (CIAP) was nearly complete, the possible importance of chlorine compounds on stratospheric chemistry was suggested.[2] This hypothesis prompted redirection of the fluorocarbon industry's existing program into an active effort aimed at resolving the question of whether or not chlorofluoromethanes do indeed diminish the average global ozone levels and, if they do, to what extent?

The technical program is administered through the Manufacturing Chemists Association (MCA). Technical direction is provided by the Technical Panel for Fluorocarbon Research, consisting of one representative from each supporting company. The technical program is directed toward determining the fate of fluorocarbons in the lower atmosphere and in the stratosphere, as well as toward assessing the true nature of chlorine chemistry in the stratosphere. The express purpose of the industry research is to determine precisely what effect the release of chlorofluoromethanes may have on stratospheric ozone.

The fluorocarbon/ozone relationship attracted the attention of many scientists in academic and government laboratories, legislative and regulatory bodies, and the press.

To strengthen the overall effort to find the answer, efforts have been coordinated with others working on the same or related problems, for example, the SST and the space shuttle. All of these concern the Federal Government, and interactions with a number of agencies have been especially helpful in the following:

1. Taking advantage of the knowledge and experience gained in the Climatic Impact Assessment Program.
2. Coordinating funding of programs addressing the halogen/ozone problem.
3. Planning joint experiments with government research groups.
4. Helping to set priorities for industry-sponsored research.

About 190 research proposals were reviewed through mid-1977 and projects totaling over $3,600,000 were funded. Calendar 1977 commitments are expected to exceed $2,000,000, with total expenditures through 1977 expected to exceed $5,000,000. About $790,000 has been committed to studies concerning the measurement of reactive chlorine species in the stratosphere, a difficult and important area. A promising method was developed at the University of Michigan with MCA support, and funding for the follow-up program for development of equipment and stratospheric measurements has been provided by the National Aeronautics and Space Administration. This illustrates an important contribution by industry: prompt funding of promising work that, if successful, can be taken over by a government agency for incorporation into its more comprehensive program.

The research programs funded by the fluorocarbon industry and administered by the Manufacturing Chemists Association are listed in Tables 23-1 and 23-2 (completed programs are shown in Table 23-1, and those still in progress are given in Table 23-2).

In addition to supporting a technical program, the U.S. producers of fluorocarbon compounds and the U.S. aerosol industry formed a nontechnical organization to address legislative, regulatory, and public affairs. This group, the Council on Atmospheric Sciences (COAS), provides a procedure for exchanging information with legislative and regulatory bodies.

INTERNATIONAL RESEARCH

Because chlorofluoromethanes which are released into the atmosphere mix rapidly and are rapidly transported to other parts of the globe, the question of fluorocarbons in the environment is an international issue. Many countries have significant technical programs which are pertinent to a resolution of the fluorocarbon/ozone controversy. In France, for example, the program of Centre Nationale d'Etudes Spatiales (CNES) conducts research on the effects of stratospheric aviation. In the United Kingdom, the Science Research Council (SRC) performs a similar function for the national government. In Canada, the Atmospheric Environmental Service (AES) carries out a comprehensive program of stratospheric measurement and laboratory chemistry. In Australia, Commonwealth of Scientific and Industrial Research Organization (CSIRO) carries out an extensive program on meteorology, ozone monitoring, and heterogeneous chemistry. The West German government expanded their atmospheric research program in 1977 to place additional emphasis on atmospheric chemistry.

The World Meteorological Organization (WMO) is a focal point where the activities of the various international governments and research institutes can be coordinated. The World Health Organization (WHO) provides a forum and a vehicle for data collection on the possible biological effects of changes in ozone concentration. Specific programs are actively discussed within the framework of the United Nations Environment Program (UNEP). Regulatory procedures and philosophy are normally discussed within the framework of the Organization for Economic Cooperation and Development (OECD).

TABLE 23-1   RESEARCH FUNDED BY THE FLUOROCARBON INDUSTRY AND ADMINISTERED BY THE MANUFACTURING CHEMISTS ASSOCIATION

Program	Investigator	Organization	Proposal Number	Completion Date
	Work Completed			
Measurement of fluorocarbons in the atmosphere[a]	Lovelock	U. of Reading	73-1	10/27/74
Monitoring of fluorocarbons in the atmosphere and simulation of atmospheric reactions of fluorocarbons	Taylor	U. of Calif.-Riverside	73-3	10/16/74
Investigation of spectroscopy of and photochemical changes in fluorocarbons	Sandorfy	U. of Montreal	73-2	10/11/74
Continuation of 73-1[a]	Lovelock	U. of Reading	74-3	12/31/75
Laboratory determination of sensitivity of laser-induced fluorescence for the detection of ClO under atmospheric conditions	Davis	U. of Maryland	74-10	5/31/75
Laboratory investigation of the feasibility of measuring ClO in the atmosphere by the chemical conversion-resonance fluorescence detection method	Stedman	U. of Michigan	74-7	2/28/75
Continuation of 73-3	Pitts	U. of Calif.-Riverside	74-2	12/31/75
Measurement of fluorocarbons and related chlorocarbons in the stratosphere and upper troposphere[a]	Rasmussen	Washington State U.	75-2	4/15/76
Laboratory and theoretical studies of the ultraviolet and visible electronic spectra of ClO[a]	Nicholls	York U.	75-11	6/14/76
Continuation of 74-2	Pitts	U. of Calif.-Riverside	75-12	4/15/76
Ground-based millimeter wavelength observations of stratospheric ClO	Ekstrom	Battelle Northwest	75-27	5/24/76

Title	Investigator	Institution	Contract	Date
Investigation of the destruction of chlorofluoromethanes by naturally occurring ions	Campbell	Washington State U.	75-53	4/23/76
Development of an instrument to measure O, ClO, O_3, and total Cl in the stratosphere	Young	Xonics, Inc.	75-50	4/7/76
Investigation of ion-molecule reactions involving chlorofluorocarbons	Mohnen	State U. of N.Y.-Albany	75-64	4/1/76
Modeling of the fluorocarbon/ozone system[a]	Sze	ERT, Inc.	75-32	8/18/76
Critique of models used to estimate chlorofluorocarbon effects on ozone[a]	Cunnold, Alyea, Prinn	CAP Associates	75-24	9/10/76
Studies of reactions of HO_2 by laser magnetic resonance[a]	Thrush	U. of Cambridge	75-58	11/8/76
Continuation of 75-50	Young	Xonics, Inc.	75-86	11/15/76
Measurement of stratospheric distribution of fluorocarbons and related species by infrared absorption spectroscopy[a]	Murcray	U. of Denver	75-13	1/24/77
Laboratory studies of the infrared vibration–rotation spectrum of ClO	Nicholls	York U.	75-30b	3/25/77
Absolute calibration of fluorocarbon measurements	Stedman	U. of Michigan	76-132	4/1/77

[a]Work is continuing in a follow-up contract.

TABLE 23-2 RESEARCH FUNDED BY THE FLUOROCARBON INDUSTRY AND ADMINISTERED BY THE MANUFACTURING CHEMISTS ASSOCIATION

Work in Progress

Program	Investigator	Organization	Proposal Number	Contract Date	Contract Period
Measurement of reaction rates relevant to the fluorocarbon/ozone problem	Birks	U. of Illinois	75-1	4/17/75	15 mo.[a]
Laboratory determination of the feasibility of laser magnetic resonance for ClO detection and reaction studies	Howard	NOAA-Boulder	75-47	11/13/75	12 mo.[a]
Collection and analysis of Antarctic ice cores	Rasmussen	Rasmussen Associates	75-84	11/29/75	3 mo.[a]
Measurement of fluorocarbon content of "antique" air samples	Rasmussen	Washington State U.	75-71	12/3/75	5 mo.[a]
Exploration for unidentified factors in the fluorocarbon/ozone problem	Lovelock	Private	75-67	12/4/75	12 mo.[a]
Laboratory measurement of spectroscopic absorption cross sections of ClO	Davis	U. of Maryland	75-73	2/5/76	8 mo.[a]
Continuation of program for ground level monitoring of ultraviolet solar radiation	Berger	Temple U.	75-62	2/5/76	12 mo.[a]
Continuation of 75-2 (see Table 23-1)	Rasmussen	Washington State U.	75-59	2/13/76	9 mo.[a]
Measurement of OH in the stratosphere by laser induced fluorescence	Davis	U. of Maryland	75-87	2/24/76	12 mo.[a]
Laboratory measurement of high resolution infrared spectra of chlorine-containing molecules of stratospheric interest	Murcray	U. of Denver	75-92	3/4/76	12 mo.[a]
Laboratory investigation of the heterogeneous interaction of Cl and ClO with H_2SO_4	Martin	Aerospace Corporation	75-81	3/22/76	12 mo.[a]

Description	Investigator	Institution	Number	Date	Duration
Measurement of HCl, HF, ClO, etc., in the stratosphere by high resolution infrared spectroscopy	Girard	ONERA-France	75-88	3/29/76	12 mo.[a]
Measurement of fluorocarbons and related chlorocarbons in the stratosphere by collection and analysis	Ridley	York U.	76-102	3/31/76	12 mo.[b]
Continuation of 75-13 (see Table 23-1)	Murcray	U. of Denver	76-101	4/1/76	12 mo.[b]
Construction of Fourier-transform spectrometer	Buijs	Bomem, Inc.	75-90	4/1/76	12 mo.[b]
Measurement of HCl and HF in the stratosphere by Fourier-transform spectroscopy	Buijs	Bomem, Inc.	75-98	4/9/76	5 mo.[a]
Continuation of 75-32 (see Table 23-1)	Sze	ERT, Inc.	76-115	6/18/76	11 mo.
Total world ozone level: statistical analysis	Parzen, Pagano	Frontier Science and Technology Research Foundation, Inc.	76-106	7/15/76	12 mo.
The electron capture detector as a reference standard in the analysis of atmospheric halocarbons	Lovelock	Private	76-120	8/16/76	12 mo.
Continuation of 75-11 (see Table 23-1)	Nicholls	York U.	75-11 II	8/18/76	12 mo.
Continuation of 75-1	Birks	U. of Illinois	76-117A	8/26/76	12 mo.
Studies of heterogeneous reactions	Birks	U. of Illinois	76-117B	9/29/76	12 mo.
Studies of compounds of sulfur, oxygen, and chlorine	Kaufman	Emory U.	76-126	10/7/76	12 mo.
Meteorological and multidimensional modeling considerations relating to atmospheric effects of halocarbons	Cunnold, Alyea, Prinn	CAP Associates	76-122	10/8/76	12 mo.
Photochemical and chemical kinetic measurements of stratospheric importance with respect to the fluorocarbon issue	Timmons	Catholic U.	76-129	10/29/76	12 mo.
Total chlorine measurements in the troposphere and stratosphere	Eggleton	AERE Harwell	76-116	11/2/76	4 mo.[a]
Laser magnetic resonance study of HO_2 chemistry	Howard	NOAA-Boulder	76-100	11/6/76	12 mo.

TABLE 23-2 (Continued)

Work in Progress

Program	Investigator	Organization	Proposal Number	Contract Date	Contract Period
Photochemistry of small chlorinated molecules	Wiesenfeld	Cornell U.	76-128	11/8/76	12 mo.
Stratospheric measurement of ClO and OH	Murcray	U. of Denver	76-135	12/9/76	4 mo.
Climatic effects of fluorocarbons	Cunnold, Alyea, Prinn	CAP Associates	76-122S	12/10/76	12 mo.
Lower stratospheric measurement of nonmethane hydrocarbons	Rasmussen	Private	76-140	1/3/77	4 mo.
Reactions of the HO_2 radical studied by laser magnetic resonance	Thrush	U. of Cambridge	75-58 II	1/28/77	12 mo.
Interlaboratory comparisons of fluorocarbon measurements	Rasmussen	Private	76-142	1/26/77	3 mo.
Millimeter wave observations of chlorofluoro-methane by-products in the stratosphere	Solomon	SUNY Stony Brook	76-130	Pending	12 mo.
Submillimeter-infrared balloon experiment	Bonetti, Carli, Harries	U. of Florence, National Physical Laboratory U.K.	76-137	Pending	24 mo.
Ground-based infrared measurements	Zander	U. of Liege	76-141	Pending	12 mo.
Continuation of 75-67	Lovelock	Private	77-144	Pending	12 mo.
Electron spin resonance detection of stratospheric radicals	Ehhalt	Nuclear Research Establishment-Jülich	76-145	Pending	6 mo.
Continuation of 75-92	Murcray	U. of Denver	77-152	Pending	12 mo.

[a] Contract extended.
[b] Estimate.

REGULATORY

The IMOS task force recommended, in addition to a scientific review, that the regulatory process should begin at once, so that necessary preliminary work would be complete when the National Academy of Sciences made its report. This recommendation was acted upon by Environmental Protection Agency (EPA), Consumer Products Safety Commission (CPSC), and Food and Drug Administration (FDA).

Regulatory activities of various agencies are fully discussed in Chapter 22.

REFERENCES

1. "Halocarbons: Effects on Stratospheric Ozone," Panel on Atmospheric Chemistry, National Academy of Sciences, 1976.
2. M. J. Molina and F. S. Rowland, *Nature* **249,** 810 (1974).

24

SAMPLING
AND ANALYSIS
OF AEROSOL
PRODUCTS

Thomas D. Armstrong, Jr.

The components of typical aerosol products are amenable to analysis by numerous proven methods: chemical, physical, and instrumental. However, the aerosol package presents the analyst with a unique problem: a pressurized and sealed sample comprised of a mixture of components with widely varying vapor pressures. Because of vapor pressure differences, the major components are distributed unevenly in different phases of the sample. Obtaining representative samples and preventing loss of volatile components are major problems in the sampling of aerosol containers. The aerosol valve itself is poorly designed for analytical sampling, and it must be modified or other sampling devices must be found in order to obtain a representative sample from the aerosol package. The following discussion is primarily concerned with sampling techniques and analysis of the total aerosol package or of the volatile propellants. Details of the analyses of the components of the aerosol concentrates are left to the ample technical literature available in the various fields encompassed by aerosol products.

SAMPLING TECHNIQUES

The most obvious way to sample is through the aerosol valve. The valve is adequate for qualitative work, but it is difficult to effect a quantitative transfer through the valve into an instrument or suitable medium for analysis. Brook and Joyner[1] described a type of adapter that could be used to sample through some types of aerosol valves into evacuated bottles. It is possible to invert the can, purge the dip tube, and draw a vapor sample into a gas syringe for air analysis. The ratio of propellant to concentrate can also be determined for some products by discharging part of the aerosol through the valve and into a container to recover the concentrate. The aerosol is weighed and, after evaporation of the propellant, the weight of the recovered concentrate is determined. In

Figure 24-1 Piercing valve, clip, and hose clamps.

general, however, it is better to find sampling methods that give better control and recovery than can be obtained with the aerosol valve.

If the contents of the aerosol container are chilled to below the boiling points of the propellants, the container may be opened and the contents transferred to a more adaptable one. Quantitative transfers are difficult, and the method has limited application for analyses involving the propellant. Jenkins and Amburgey,[2] however, reported pipetting samples for propellant analysis from chilled aerosol containers. The method is satisfactory for obtaining concentrate samples, and it is very useful for sampling foam products.

Greater sampling versatility can be achieved by the use of puncturing devices, several types being available. The easiest to use is the piercing valve employed in the refrigerant industry for emptying disposable cans of refrigerant. These valves are available at refrigerant supply stores. The valve is equipped with an adapter for attachment to the cap of the refrigerant can. The flanges on the adapter can be straightened, and it can then be fastened to the side of an aerosol can with hose clamps. The piercing valve is screwed into the can until it punctures the side. The piercing needle is enclosed in an elastomeric gasket that prevents leakage around the puncture. The valve is joined to $\frac{1}{4}$-in. copper tubing. Samples can be obtained from the liquid or vapor phase depending upon the orientation of the aerosol container. Figures 24-1 and 24-2 illustrate this device. In some instances, particularly with small aluminum cans, it is better to puncture the bottom of the aerosol container. Figure 24-3 illustrates a device used for this purpose. The metal plates are securely fastened over the ends of the can by means of the three screw posts. A piercing device is then screwed through an opening in the center of the bottom plate until the can is punctured.

ANALYSIS

Gas Chromatography

Gas chromatography can be used for a number of the more important analyses performed on aerosol products. Root and Maury[3,4] and Sciarra and Gaglione[5]

Figure 24-2 Piercing valve attached to aerosol container.

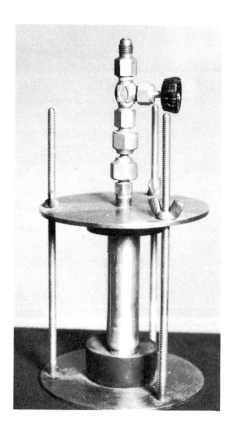

Figure 24-3 Bottom-puncturing device for obtaining aerosol samples.

have discussed the principles of gas chromatography and its application to aerosol analysis. Briefly, the chromatograph is an instrument that separates the components of a gas or vapor mixture. The separated components are passed through a detector that gives a response proportional to concentration. The samples may be vapor or liquid. Liquid samples are introduced into a heated injection port where they are instantly vaporized, and they are kept in the vapor state during their passage through the instrument.

The sample is carried through the instrument by a carrier gas, normally helium. Separation into individual components occurs during passage through the column, which is a length of tubing packed with a fine-mesh solid material called the support. The support is coated with a nonvolatile liquid. Separation occurs because of differences in the absorbancies of the sample components in this liquid coating. The selection of the liquid is critical to obtaining separation. For propellants, di-n-butyl maleate, di-2-ethylhexyl sebacate, silicone oils, Nujol, and similar partitioning liquids can be used. Several of the porous polymer column-packing materials also are suitable for propellant analysis.

The separated components of the sample then pass through a detector, which gives a signal that can be amplified and recorded, usually as a peak on a strip chart. The most common detectors measure the difference between the thermal conductivity of the pure gas and that of the carrier gas–sample component mixture. A number of other types of detectors exist. Flame ionization detectors pass the sample through a hydrogen flame and into an electrical field. The flame ionizes the sample, and in the electrical field the ions migrate to collectors. The current resulting from the neutralization of the collected ions is amplified and recorded. Electron capture detectors pass the sample through an electron beam, and some of the electrons are held by the sample components. As with the ionization detector, the charged sample components are collected, and the resultant current is amplified and recorded. A detector that may be of interest for aerosol analysis detects the difference between the density of the carrier gas and that of the carrier gas–sample mixture. Other detectors, which are more or less specific for compounds containing certain elements such as chlorine, phosphorous, and sulfur, have been developed.

A number of methods can be used to introduce an aerosol sample into a chromatograph. Instruments can be equipped to take vapor, liquid, or liquefied gas samples, and the method chosen depends upon the analysis desired, the components of the sample, the accuracy required, and the equipment available to the analyst.

Most analyses require samples from the liquid phase. Materials which are liquids at STP conditions can be injected into a chromatograph with a syringe or by use of a liquid sampling valve, which is a valve that injects a measured volume of liquid sample into the instrument. Problems arise with mixtures containing liquefied gases. The samples are under pressure, and vaporization tends to occur whenever any of the sample is transferred from the container. Syringes cannot be recommended for this type of sample, but it is possible to use liquid sampling valves. Sampling must be done in a manner insuring that the calibrated volume in the valve is liquid-full. Cannizzaro[6] described a method using a liquid sampling valve for aerosol propellant analysis.

Jenkins and Amburgey[7] applied Henry's Law to sampling aerosol containers for chromatographic analysis. Henry's Law states that the vapor pressure of a solution is proportional to the mole fraction of the solute in solution. In this method, the aerosol container is chilled in dry ice and then punctured. A sample of the chilled aerosol is pipetted into a volumetric flask containing a chilled non-volatile solvent. The weight of the pipetted sample is determined, and the volumetric flask is filled to volume. The partial pressures of the dissolved propellants are low enough to permit the solution to be syringed into a chromatograph without fractionation. Calibration curves plotting weight of propellant versus peak height are prepared, and the weights (in milligrams) of the propellants in the solution are determined from these curves. The concentrations of the propellants in the aerosol are calculated using the following equation:

$$C = \frac{A}{B} \times 100$$

where

C = Concentration of propellant (wt. %) in the aerosol
A = Weight of propellant in solution (determined from the calibration curve)
B = Weight of pipetted sample

Another method based on Henry's Law is to sparge a vaporized sample into a relatively nonvolatile solvent. The aerosol container is punctured with a piercing valve, and a fritted glass sparger tube is connected to the valve. Figure 24-4 shows an aerosol container fitted with a piercing valve and sparger tube. A total of 85 ml of solvent is placed in a 4-oz glass bottle. The sparger tube is placed below the surface of the solvent, the valve is opened slightly, and the sample is sparged into the solvent for 15 sec. Samples that clog the fritted glass can be run through straight tubing, but better contact with the solvent is obtained with the fritted glass, and this method is preferred. The bottle is then sealed with a serum cap. A standard mixture of approximately the same composition as the sample is prepared and sparged in the same manner as the sample. Aliquots of the solutions are then withdrawn through the serum cap into a syringe and injected into the chromatograph. The peak heights of the sample components and of the standard components are measured, and the composition of the sample is calculated by means of the following equations:

$$X, Y, Z, \text{etc.} = \frac{(Ax, y, z, \text{etc.}) \times (Bx, y, z, \text{etc.})}{(Cx, y, z, \text{etc.})}$$

Figure 24-4 Aerosol container with piercing valve and sparger tube attached.

$$Px = \frac{X}{X + Y + Z + \text{etc.}} \times 100$$

$$Py = \frac{Y}{X + Y + Z + \text{etc.}} \times 100$$

where

X, Y, Z, etc. = Relative concentrations of components X, Y, Z, etc., in the sample

Ax, y, z, etc. = The peak heights of X, Y, Z, etc., from the chromatogram of the sample solution

Bx, y, z, etc. = The peak heights of X, Y, Z, etc., from the chromatogram of the standard solution

Cx, y, z, etc. = The concentrations (wt. %) of X, Y, Z, etc., in the standard

Px, y, z, etc. = Percent components X, Y, Z, etc., in the sample

Hexane, heptane, methylene chloride, cyclohexane, acetone, and trifluorotrichloroethane have been used as solvents.

The success of the method depends upon treating the sample and the standard identically with respect to sparge rate, sparge time, and sample injection.

The methods employing dilute sample solutions are particularly useful for sampling aerosol products such as hair sprays and paints that contain nonvolatile material, which would soon foul an instrument. The use of dilute solutions allows a large number of samples to be run before enough nonvolatile material accumulates to cause instrument problems.

It is possible to allow a sample to vaporize in an evacuated volume and then to introduce an aliquot of the vapors into a chromatograph. Brook and Joyner[1] used this method in their work with aerosol propellants. An adapter equipped with a hypodermic needle was fitted on the valve of the aerosol container. The needle was inserted through a serum cap into an evacuated gas bulb. The gas bulb was connected to a manometer and a gas sampling valve, which is a valve that injects a measured volume of gas sample into a chromatograph. The sample was discharged into the volume, the pressure was adjusted to 10 ± 0.1 cm Hg, and the sample was introduced into the chromatograph. Numerous modifications of this procedure are possible.

Figure 24-5 is a line diagram of a gas chromatography system that can be used for sampling completely volatile samples or vapor phase samples. The container is punctured with a piercing valve (valve A in the diagram) and connected to the gas sampling valve as shown in Figure 24-6. The sampling valve (valve E) is positioned so as to connect the sample container with the line from valve C. Valves

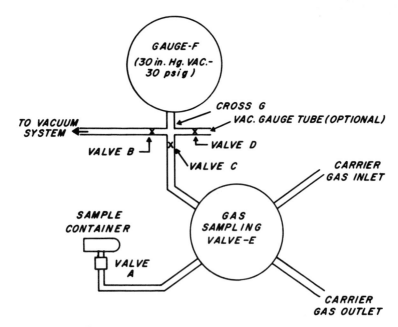

Figure 24-5 Gas chromatography sampling system.

Figure 24-6 Aerosol container with piercing valve connected to a gas sampling valve on a chromatograph.

B, C, and D are opened, valve A remains closed, and the system is evacuated back to valve A. When the system is evacuated, valves B and D are closed, and valve A is opened slightly to admit sample up to the desired pressure, which is read on gauge F. When sampling from the liquid phase, valve A acts as a vaporizing valve, and caution must be exercised to be sure the sample is vaporizing at the valve. When the desired sample pressure is attained, valve A is closed, and the gas sampling valve (valve E) is turned to inject the sample into the chromatograph.

Standards are prepared and run, and the sample concentrations are calculated in the same manner as in the sparging method.

It is occasionally necessary to determine the amount of air in the vapor space of an aerosol product. This determination can be made with the system diagrammed in Figure 24-5. The aerosol is oriented to remove the sample from the vapor space, and a sample is run in the manner just described. An air sample of the same size is run, and the concentration of air in the sample is calculated using the following equation:

$$C = \frac{A}{B} \times 100$$

where

   C = Concentration of air (vol. %) in the sample
   A = Area of air peak in the sample (mm^2)
   B = Area of air peak in pure air sample (mm^2)

The concentration of other noncondensible gases (such as CO_2) can be determined in the same way.

   With most columns, the main components of air—oxygen, and nitrogen—are not separated but appear as one peak on the chromatogram. If it is necessary to know the oxygen concentration in the vapor space of an aerosol, it can be determined using a chromatograph with a column packed with #13X molecular sieves. The sample is run in the same way as in the air analysis just described, and an air sample of the same size is run as a standard. The air will be separated into its components, and two peaks will show on the chromatogram. The first and smaller peak is oxygen. The concentration of oxygen in the sample is calculated by means of the following equation:

$$A = \frac{B \times C}{D}$$

where

   A = Concentration of oxygen (vol. %) in the sample
   B = Area of oxygen peak in the sample chromatogram (mm^2)
   C = 21 vol. % (concentration of oxygen in air)
   D = Area of oxygen peak in the chromatogram of the air standard

It should be noted that the concentration of noncondensible gases in the vapor space will be lowered as samples are withdrawn, allowing more of the liquefied propellants to vaporize into the head space.

   Aerosol products that produce unstable foams can be analyzed by the sparging method, although it may be necessary to use a straight tube rather than a fritted glass sparger. Products producing stable foams are difficult to sample by any of the methods already discussed. Stable foams usually can be collapsed by contacting them with dilute acid. The sample is connected to a volume filled with 6 N hydrochloric acid. The piercing valve is opened slightly, and the foam sample is allowed to slowly displace part of the acid solution. The acid and foam are then agitated until the foam collapses. The vapor space above the acid is then sampled for propellant analysis by gas chromatography.

The retention time of a compound in a chromatograph is constant as long as the conditions of analysis are kept constant. As a result, retention times can be used as a means of identifying the components of a sample. Caution should be exercised, however, since some compounds will have identical or very similar retention times. In case of doubt, the identification should be verified by infrared or mass spectrometer analysis. It is possible to trap the different components of a sample as they exit from a chromatograph. The trapped fractions can then be identified by their infrared or mass spectra.

Propellant-to-Concentrate Ratio

One of the most important determinations made in aerosol analysis is the measurement of the amount of propellant and amount of concentrate in the aerosol package.

Chromatography can be used for products that can be completely vaporized or for products in which the concentration of nonvolatile material in a volatile solvent is accurately known. It is possible to add to the aerosol a measured amount of a volatile material, termed an internal standard, which is foreign to the mixture already present. The ratio of the propellant to the internal standard is determined by chromatography. Knowing the original weight of the aerosol, the amount of internal standard added, and the ratio of the propellant to the internal standard, the concentration of the propellant in the original aerosol can be calculated.

In most instances, however, methods other than chromatography are used.

Brook and Joyner[1] passed a measured aerosol sample through a weighed calcium chloride solution, and the bath solution was then reweighed. Assuming that all of the sample except the propellant was retained in the solution, the amount of concentrate in the aerosol was determined from the weight of sample passed through the bath and the weight of sample retained in the bath.

Clapp[8] reported a procedure in which volatile and nonvolatile fractions of aerosol formulations were determined by comparing the amount of propellant expelled from an aerosol container after each of four steps with that expelled from a standard treated in the same manner.

Sciarra[9] described several gravimetric methods and a densimetric method for determining the volatile content of aerosol products. In the densimetric method, the sample container is chilled to a constant temperature below the boiling point of the propellant, the container is opened, and the contents are transferred to a chilled cylinder. The density is determined with a hydrometer, and the concentration of propellant is determined from a graph of density versus propellant concentration that has been prepared from the density determinations made at the same temperature for carefully prepared standards.

A general method that can be varied to meet the requirements of different types of samples is to transfer the contents of the aerosol container to a distillation flask and distill off the propellant. The aerosol package is weighed, chilled in dry ice, and opened. The entire contents are transferred to a weighed distillation flask, and the empty aerosol container is reweighed so the sample weight can be determined. The distillation flask is fitted with a reflux condenser, and the propellant is allowed to boil away. The distillation is discontinued when temperature measurements show the propellant to be gone. The distillation flask is reweighed, and the weight of sample remaining is determined. The difference between the original sample weight and the weight of material remaining is the weight of propellant. The temperature of the distillation flask can be controlled by immersion in a thermostated bath, and loss of volatile solvent can be controlled by use of chilled water in the reflux condenser.

Moisture and Other Analyses

Moisture in aerosol products is determined best by Karl Fischer titration. Downing and Reed[10] developed a procedure in which a measured amount of standardized Karl Fischer reagent was titrated under pressure with the aerosol sample. The weight of sample necessary to neutralize the known amount of reagent was determined, and the moisture content calculated by means of the following equation:

$$\text{ppm } H_2O = \frac{A \times B \times 10^6}{C - D}$$

where

A = Volume of Karl Fischer reagent (ml)
B = Standardization factor (g H_2O/ml Karl Fischer reagent)
C = Original weight of aerosol sample (g)
D = Final weight of aerosol sample (g)

Reed[11] also developed a method for moisture determinations in refrigerant/oil mixtures that is particularly adaptable to aerosol products. In this procedure, a methanol/chloroform mixture is titrated to dryness with Karl Fischer reagent, and the titration vessel is then chilled with dry ice. The aerosol container is punctured with a piercing valve, weighed, and connected to the titration vessel. A sample of the aerosol is let into the vessel, where it condenses. The solvent/ sample mixture is gradually brought back to room temperature, the propellant escaping through a vent, and the remaining solution is titrated with Karl Fischer reagent.

It is possible to modify automatic Karl Fischer analyzers[12] for use with propellants, and the same apparatus and procedure can be used with aerosol samples. The sample is slowly passed into a titration vessel containing dried methanol or a dried methanol/methyl chloroform solution. The sample is titrated as it enters the titration vessel. When a sufficient titer of Karl Fischer reagent has been used, the weight of the aerosol sample is determined, and the moisture content is calculated as before.

The success of any of these methods depends upon the prevention of atmospheric moisture contamination. Dry air is used to keep a positive pressure in the reaction vessel, and the vents are protected with driers.

Acidity in an aerosol sample can be determined by sparging the sample into previously neutralized isopropanol. The isopropanol is then titrated with base to the desired end point.

Many inorganic determinations can be made on aqueous solutions obtained by sparging the sample through water. If two liquid phases result, they are separated. The organic phase is washed several times with water. The water washes are combined, and the analysis is performed by an appropriate method. Samples prepared in this way are suitable for rapid analysis with the specific ion electrodes now available for chloride, fluoride, sodium, potassium, and a number of other inorganic ions. By knowing the weight of the sample sparged and the amount of water in the washes, these results can be calculated as concentrations in the original aerosol product.

Air Monitoring

It is evident that monitoring the workplace environment has become a permanent requirement for operations where chemicals are used and where vapors, mists, fumes, or dusts may be in the atmosphere.

Monitoring is used for the following purposes:

1. Industrial hygiene, that is, determining and controlling worker exposures.
2. To determine the efficiency of recovery and control measures.
3. To determine the degree of discharge to the environment.
4. To audit losses.
5. To assure compliance with standards set by governmental regulatory agencies.

An excellent review of industrial hygiene principles and practices with many references is found in "The Industrial Environment—Its Evaluation and Control," U.S. Department of Health, Education and Welfare, Public Health Service, National Institute for Occupational Safety and Health (NIOSH), 1973.

Recommended exposure limits for some 500 substances can be found in "Threshold Limit Values for Chemical Substances and Physical Agents in the Workroom Environment with Intended Changes," issued annually by the American Congress of Governmental Industrial Hygienists (ACGIH), P.O. Box 1937, Cincinnati, Ohio.

Standards set under the Occupational Safety and Health Act (OSHA) are contained in the "Code of Federal Regulations," Title 29, Part 1910, Occupational and Environmental Health Standards. They were first published in the *Federal Register*, **26**, No. 105 (May 29, 1971). They are subject to revision, but most are still the ACGIH threshold limit values of 1970. It should be noted that the different state governments are promulgating their own occupational health standards, and some are more restrictive than the OSHA standards.

Sampling and analytical procedures to determine compliance with standards are being developed by NIOSH. The "NIOSH Manual of Analytical Methods," issued in 1974, is available from the Superintendent of Documents, U.S. Printing Office, Washington, D.C. (Stock No. 017-033-00041-7, $3.90). Sets of methods developed since 1974 are available from the U.S. Department of Commerce, National Technical Information Service, Springfield, Va.

The threshold limit value (TLV) of the ACGIH or the time-weighted average (TWA) of the OSHA standards is the maximum average concentration it is believed nearly all workers can experience 8 hours a day for a 40-hour work-week during a lifetime without adverse effect. Ceiling levels and short-term exposure levels (STEL) are concentrations to which a worker may be exposed momentarily or for limited time periods.

The NIOSH analytical procedures emphasize integrated samples obtained by personnel monitors—usually small pumps pulling measured volumes of air through collection media, which are subsequently analyzed. The monitors are worn by the workers. Collection media may be absorbents such as charcoal or alumina, or they may be solutions which react quantitatively with the substance to be measured. The integrated sample obtained with a personnel monitor gives a true time-weighted average exposure.

Because of their volatility and general inertness, most propellants cannot be collected quantitatively by absorption methods. For these materials, integrated samples can be obtained by using the personnel monitoring pumps to collect air samples in impervious bags worn by the workers in backpacks. The air collected in the bags can then be analyzed for propellant vapor concentrations.

To determine if ceiling or STEL concentrations are approached or exceeded, short-term or grab sampling techniques are used. Grab samples may be taken in gas sample bulbs, plastic bags, evacuated cylinders, or other vessels, and they are then analyzed by gas chromatography. Results can be calculated using calibration curves prepared from known standards.

Monitors which continuously record concentrations and give warning when high concentrations are encountered are available. For propellants, monitors with infrared detectors are particularly useful. Both fixed and portable models are available from a number of instrument manufacturers. These instruments can be made to monitor specific compounds by proper selection of wavelength.

Other continuous monitors with thermal conductivity, flame ionization, electron capture, or halogen detectors may be useful. However, for the most part, these instruments are not specific, and in the presence of a mixture of vapors results may be misleading.

REFERENCES

1. R. J. Brook and B. D. Joyner, *J. Soc. Cosmet. Chem.* **17**, 401 (1966).
2. J. W. Jenkins and J. M. Amburgey, *Proc. Sci. Sect. Toilet Goods Assoc.* **31**, (1959); *Aerosol Age* **4**, 35 (June 1959).
3. M. J. Root and M. J. Maury, *Proc. 43rd Ann. Meet. Chem. Spec. Manuf. Assoc.*, 44 (December 1956); *Soap Chem. Spec.*, Part I, **33**, 101 (January 1957); *Amer. Perfum. Aromat.* **69**, 50 (1957).
4. M. J. Root and M. J. Maury, *J. Soc. Cosmet. Chem.* **8**, 92 (1957).
5. J. J. Sciarra and O. G. Gaglione, *Paint Varn. Prod.*, Part I, **54**, 63 (1964); Part II, **54**, 77 (1964).
6. R. D. Cannizzaro, *Aerosol Technicomment* **11**, No. 1 (1968).
7. J. W. Jenkins and J. M. Amburgey, *Proc. Sci. Sect. Toilet Goods Assoc.* **31** (1959); *Aerosol Age* **4**, 35 (June 1959).
8. C. Clapp, *Proc. 37th Mid-Year Meet. Chem. Spec. Manuf. Assoc.*, 21 (May 1951).
9. J. J. Sciarra, *Paint Varn. Prod.* **55**, 68 (1965).
10. R. C. Downing and F. T. Reed, *Proc. 38th Ann. Meet. Chem. Spec. Manuf. Assoc.*, 41 (December 1951).
11. F. T. Reed, *Refrig. Eng.* **62** (1954).
12. Freon® Technical Bulletin B-23, E. I. du Pont de Nemours & Company (1956).

25

TOXICITY OF FLUOROCARBON PROPELLANTS

Ruth R. Montgomery and Charles F. Reinhardt

INTRODUCTION

Toxicity and Hazard

Derived from the Greek word for poison, *toxicity* can be briefly defined as the capacity of a substance to cause adverse effects within living organisms. In modern usage, toxicology has evolved into a multidisciplinary science which seeks to analyze and evaluate chemically induced injury.[1,2]

In contrast, *hazard* is described as a measure of the probability of risk of injury.[3] This probability depends upon a combination of factors: inherent toxicity, degree and route of exposure, type of use, and physical properties such as volatility, viscosity, solubility, etc. The probability of a toxic effect from a given chemical can either be increased or decreased by its physical properties and end-use.

ACUTE TOXICITY

In medicine, the term "acute illness" suggests that the onset of symptoms is sudden and that the illness rapidly progresses to either recovery, death, or chronic disability. In toxicology, the term "acute dose" means a dose of a substance administered within 24 hours or less, usually at a single time. Results of acute toxicity tests are often expressed in terms of the LD_{50} or LC_{50}: the lethal dose or lethal atmospheric concentration for 50% of the exposed animals. A statistically valid LD_{50} or LC_{50} is based[4,5] upon an animal experiment that has yielded a series of mortality ratios at varying dose levels. The ED_{50} or EC_{50} is defined as the effective dose or effective concentration required to produce a given biologic response other than mortality in 50% of the exposed animals.

The "approximate lethal dose" (ALD) or "approximate lethal concentration"

(ALC) is defined as the lowest lethal dose or concentration obtained by a simplified testing procedure.[6] Since this determination provides a rapid, relatively inexpensive estimate of the LD_{50} or LC_{50}, it is particularly useful in screening a series of compounds.

Certain volatile organic chemicals, including some propellants, can cause cardiac sensitization. When the heart is sensitized, it loses its normal rhythm and develops an abnormal one (sometimes specifically identified as ventricular arrhythmia). This abnormal rhythm may be reversed without residual effect, or it may dramatically progress to cardiac arrest and death within minutes. Cardiac arrhythmia occurring in people anesthetized with various halocarbon anesthetics has been known for several decades.[7] "Aerosol sniffing" became notorious in the late 1960's.[8,9] The dangers of this practice, involving intentional misuse of chemicals, have received widespread publicity, but abusive chemical inhalation continues.

PROLONGED AND CHRONIC TOXICITY

In humans, "a chronic illness is characterized by symptoms or disease of long duration or frequent recurrence."[10] In experimental work, "chronic toxicity" has been defined as an exposure period of at least three months but current usage often implies a year or more. A "lifetime" study with rats generally extends for two years or longer; some studies with dogs have been conducted for more than five years. Dosing schedules intermediate between acute and chronic are usually called "prolonged," "subacute," or "subchronic." Objectives of chronic toxicity testing have been comprehensively described in several reviews[1-3] and will only be highlighted here:

1. Assess carcinogenic potential and cumulative toxic effects.
2. Determine no adverse effect and/or minimal effect levels in one or more species.
3. Identify pathways and kinetics of tissue assimilation, retention, metabolism, and detoxification.

HUMAN EXPOSURE ROUTES

Chemicals may enter the body via the following natural routes:

1. Inhalation.
2. Ingestion.
3. Skin and eyes—topical.

Chemicals entering the body may produce local or systemic effects, or both. Local effects develop at the site of direct chemical contact, for example, tooth erosion from sulfuric acid vapor, silicosis from deposits of silica in lung tissue, or an alkali burn of the skin or eyes. Systemic effects develop at one or more sites remote from the point of contact, for example, liver injury from carbon tetrachloride poisoning.

INDUSTRIAL AND CONSUMER EXPOSURE*

In the United States, acceptable limits of exposure in the workplace are recommended by the American Conference of Governmental Industrial Hygienists (ACGIH) or prescribed by the Occupational Safety and Health Administration (OSHA). The ACGIH, founded in 1938, formulated the concept of the "threshold limit value" (TLV®):

> "Threshold limit values refer to airborne concentrations of substances and represent conditions under which it is believed that nearly all workers may be repeatedly exposed day after day without adverse effect."[11]

In 1971, OSHA adopted various consensus standards, including those of the American National Standards Institute (ANSI) and the 1968 TLV values then used under the Walsh-Healey Act. Thus, time-weighted average (TWA) values for an 8-hour day or 40-hour week are frequently identical in the OSHA regulations and these consensus standards. The ACGIH reaffirms its values annually; if new data are developed, values are modified accordingly. The OSHA regulations have remained the same for the past seven years (since 1971), except for new values on some 20 compounds.

Both TLVs and TWAs are expressed in ppm (parts of vapor, gas, or mist per million parts of air) or mg/m^3 (milligrams of substance per cubic meter of air), determined under ambient conditions. Except for carbon dioxide, which is considered a special case, the highest TLV or TWA recommended as a guide in the control of health hazards is 0.1% or 1,000 ppm.

Assuming a static system, the release of 10 g (0.3 oz) of fluorocarbon 12 into an $85\text{-}m^3$ ($3,000 \ ft^3$) space would, after mixing, yield a concentration of 20 ppm. With dynamic equilibrium, the concentration of compound in a given area is dependent upon the size of the area or space in question, the ventilation rate, and the rate of compound release. Based upon accepted principles,[12] the following relationship to the TLV has been formulated (M.W. = molecular weight):

$$\text{Cubic feet air/pound solvent} = \frac{400}{\text{M.W.}} \times \frac{10^6}{\text{TLV}}$$

---

*We thank James F. Morgan and David L. Van Lewen for their help in developing this section.

In the case of the commonly used fluorocarbon propellants, a TLV of 1,000 ppm results in a required ventilation rate that is well within the limits generally desirable for good ventilation. The 1,000-ppm TLV far exceeds the ambient fluorocarbon concentrations found after the release of hair sprays and anti-perspirant products and also in background concentrations from the street. (See Chapter 24 for details of analyses and Reference 13 for an extended discussion.)

SCOPE OF THIS CHAPTER

The 19 fluorinated propellants reviewed here are identified in Table 25-1. Space limitations do not permit detailed analysis of the many reports and discussions of propellants now available. Rather, this chapter presents highlights of the known toxicologic characteristics of these compounds and provides a guide to the background literature. Data on mixtures have been omitted when available literature on the pure material is extensive. Many important early references or references presenting various viewpoints are provided in the Supplemental References.

TOXICITY INVESTIGATIONS

Historically, systematic toxicity investigations of fluorocarbon propellants date from the extensive experiments conducted by Underwriters Laboratories in the 1930's. These experiments led to a classification of comparative hazard,[14] as shown in Table 25-2, which has proved valuable. When aerosols were introduced, candidate propellants were selected from Groups 5 and 6—the classes of least toxicity. More recently, critical comparison of propellants in drug applications has led to the proposed guide of Aviado,[15] also presented in Table 25-2. This guide separates propellants on the basis of vapor pressure as well as toxicity to the circulatory or respiratory system in animal experiments. Definition of the limits varies somewhat,[16,17] but, overall, compounds that influence cardio-vascular and bronchopulmonary function at concentrations of $\leqslant 5\%$ are considered to show a *high* level of toxicity; those that influence these functions at 10-20% are called *intermediate* in toxicity; and *low* toxicity refers to compounds that produce no toxicity when inhaled at concentrations up to 20%.[16]

Like the general public, industry has become increasingly concerned about the possible carcinogenic potential of some chemicals. To assess this potential, tests are conducted to determine whether or not a given chemical produces a significant increase in the number and types of tumors.[18-20] These long-term tests may extend through the life span of the exposed animals. *In vitro* mutagen tests have recently been suggested for short-term screening of carcinogenic

TABLE 25-1 FLUOROCARBONS INCLUDED IN THIS REVIEW[a]

Group	Fluorocarbon Number	Formula	Chemical Name	CAS[b] Registry Number
Perhalogenated fluorocarbons	11	CCl_3F	Trichlorofluoromethane	75-69-4
	12	CCl_2F_2	Dichlorodifluoromethane	75-71-8
	113	CCl_2FCClF_2	1,1,2-Trichloro-1,2,2-trifluoroethane	76-13-1
	114	$CClF_2CClF_2$	1,2-Dichloro-1,1,2,2-tetrafluoroethane	76-14-2
	115	$CClF_2CF_3$	Chloropentafluoroethane	76-15-3
	C-318	C_4F_8	Octafluorocyclobutane	115-25-3
Hydrogenated halomethanes	21	$CHCl_2F$	Dichlorofluoromethane	75-43-4
	22	$CHClF_2$	Chlorodifluoromethane	75-45-6
	31	CH_2ClF	Chlorofluoromethane	593-70-4
	32	CH_2F_2	Difluoromethane	75-10-5
Hydrogenated haloethanes	123	$CHCl_2CF_3$	2,2-Dichloro-1,1,1-trifluoroethane	306-83-2
	124	$CHClFCF_3$	2-Chloro-1,1,1,2-tetrafluoroethane	2837-89-0
	125	CHF_2CF_3	Pentafluoroethane	354-33-6
	132b	$CClF_2CH_2Cl$	1,2-Dichloro-1,1-difluoroethane	1649-08-7
	133a	CH_2ClCF_3	2-Chloro-1,1,1-trifluoroethane	75-88-7
	134a	CF_3CH_2F	1,1,1,2-Tetrafluoroethane	811-97-2
	141b	CCl_2FCH_3	1,1-Dichloro-1-fluoroethane	1717-00-6
	142b	$CClF_2CH_3$	1-Chloro-1,1-difluoroethane	75-68-3
	152a	CHF_2CH_3	1,1-Difluoroethane	75-37-6

[a] Boiling points and other physical properties are given in Chapter 3.
[b] CAS = Chemical Abstracts Service.

TABLE 25-2 COMPARATIVE LIFE HAZARD OF GASES AND VAPORS

Underwriters Laboratories—Group Number and Definition	Chemical	Aviado's Rating of Toxicity
Group 1: Gases or vapors which, in concentrations of 0.5 to 1.0% for duration of exposure of 5 min., are lethal or produce serious injury.	Sulfur dioxide	
Group 2: Gases or vapors which, in concentrations of 0.5 to 1.0% for duration of exposure of 0.5 hr, are lethal or produce serious injury.	Ammonia	
	Methyl bromide	
Group 3: Gases or vapors which, in concentrations of 2.0 to 2.5% for duration of exposure of 1 hr, are lethal or produce serious injury.	Carbon tetrachloride	
	Chloroform	
	Methyl formate	
Group 4: Gases or vapors which, in concentrations of 2.0 to 2.5% for duration of exposure of 2 hr, are lethal or produce serious injury.	Dichloroethylene	
	Methyl chloride	
	Ethyl bromide	
Between Groups 4 and 5: gases or vapors less toxic than Group 4.	Ethyl chloride	
	Methylene chloride	High
	Fluorocarbon 113	High
	Fluorocarbon 21	High
Group 5a: gases or vapors much less toxic than Group 4, but more toxic than Group 6.	Carbon dioxide	High
	Fluorocarbon 11	Intermediate
	Fluorocarbon 22	Intermediate
	Fluorocarbon 142b	
Group 5b: gases or vapors which available data indicate would be classified as either Group 5 or Group 6.	Ethane	Intermediate
	Isobutane	Intermediate
	Propane	
Group 6: gases or vapors which, in concentrations up to at least 20% for duration of exposure of 2 hr, do not appear to produce injury.	Fluorocarbon 12	Intermediate
	Fluorocarbon 114	Intermediate
	Fluorocarbon 115	Low
	Fluorocarbon 152a	Low
	Fluorocarbon C-318	Intermediate

potential.[18,21-23] These tests are considered useful in evaluating priorities for determining the carcinogenic potential of chemicals (solutions, suspensions, and gases) by means of long-term animal tests. At present, procedures for gases are in a state of flux, but an exposure period of 48 hr is favored. Little comparative long-term animal test data will be available before 1980.

A related field, teratology,[18,24-26] achieved wide prominence when it was learned that ingestion of thalidomide in early pregnancy resulted in the birth of children with misshapen or missing arms and legs. The potential for harm to the unborn child under certain industrial conditions is now recognized.

Additionally, environmental concerns have led to increased interest in aquatic toxicity.[27-29]

Commercial propellants must show little hazard under conditions of actual use. Also, the general population includes special risk groups, such as children, pregnant women, and aged or infirm persons. Emphasis is placed on the following:

1. Acute and chronic inhalation.
2. Chronic oral effect where exposure might occur, for example, with food products propelled by a fluorocarbon.
3. Acute topical effect.
4. Carcinogenic and teratogenic potential.

Perhalogenated Fluorocarbons (Fluorocarbons 11, 12, 113, 114, 115, and C-318)

Toxicity data for the perhalogenated fluorocarbons are summarized in Table 25-3. These compounds, particularly fluorocarbons 11 and 12, have been extensively investigated and account for the majority of the literature on fluorocarbon propellant toxicity. The first four have been assigned a TLV of 1,000 ppm, which is also the OSHA standard. It is anticipated that a TLV of 1,000 ppm will be recommended for fluorocarbon 115.

Table 25-3 includes data from several tests for cardiac sensitization on unanesthetized animals. These tests may be considered particularly stressful acute exposures. For example, in the technique developed by Reinhardt and his co-workers,[9] dogs breathing a known concentration of test chemical through a mask (Figure 25-1) also receive an injection of epinephrine estimated to be approximately ten times the adrenal output in a person under stress.

From a practical standpoint, the most likely adverse health effects from these propellants at concentrations above the TLV are cardiac sensitization and central nervous system (CNS) depression.

Fluorocarbon 11 (Trichlorofluoromethane)

Although fluorocarbon 11 has low acute inhalation toxicity, Table 25-3[9,14,30-46] shows that the 4-hr LC_{50} of 2.62% for rats is appreciably less than acute lethal concentrations of homologous compounds. Clinical signs—weakness and hyper-

Figure 25-1 Dog mask with intake and exhaust ports, as used in cardiac sensitization technique of Reinhardt and his co-workers.[9] Reproduced with permission.

activity followed by inactivity, tremors, lethargy, and unconsciousness—suggest weak narcotic action. Lethal concentrations produce progressive depression of the central nervous system and death within hours. Fluorocarbon 11 is definitely the most toxic (specifically, the most cardiotoxic) of the common fluorinated aerosol propellants.[17,47] However, at the TLV of 1,000 ppm, dogs exposed to the triple stress of intravenous epinephrine, exposure to fluorocarbon 11, and a healed myocardial infarction did not show evidence of cardiac sensitization.[37] Repeated exposures, both by inhalation and by the oral route, have not shown evidence of cumulative toxicity.

Deliberate "sniffing" cases involving fluorocarbon propellants led to widespread concern[48,49] and research. One noteworthy series of experiments with anesthetized dogs[50] showed that a 10-min. exposure with fluorocarbon 11 at concentrations below 1.5% was not lethal. Another question arose concerning a possible link between fluorocarbon 11 and increased mortality from asthma. Several tests showed that asthmatics using an aerosol inhaler[51] containing fluorocarbon 11 and normal subjects using household aerosols containing fluorocarbon 11[52] had blood levels of fluorocarbon 11 much lower than those found in dogs overexposed to the point of cardiac sensitization.

TABLE 25-3 TOXICITY OF HALOGENATED FLUOROCARBONS

Test	Animal	Experimental Data	Reference
		PERHALOGENATED FLUOROCARBONS	
		I. FLUOROCARBON 11	
Acute Toxicity			
Inhalation	Rat	4-hr exposure, $LC_{50} = 2.62\%$	30
	Rat, Guinea Pig	10–20% lethal in 1–2 hr,[31] 10% may be fatal to rats as soon as 20 min.[32]	31, 32
	Guinea Pig	Survived 10% for 2 hr, evidence of lung congestion with some hemorrhage.	14
Oral	Dog, Rabbit	Slight biochemical changes from 5% for 20 min.	33
	Rat	Survived 5 ml/kg of 1:1 mixture of fluorocarbon 11 in liquid paraffin.	
Eye	Rabbit	Single doses of spray[35] or 15–40% solution in dimethyl phthalate,[36] very mild conjunctival irritation.	34, 35, 36
Cardiac Sensitization	Dog[a]	With epinephrine injection–evidence of arrhythmia, some deaths at 0.5–1%, not at 0.1%[9] even in dogs with scars from old myocardial infarctions.[37] With exercise–no evidence of arrhythmia at 0.5–1%.[38]	9, 37, 38
	Hamster	Arrhythmias, some deaths at 2–10%.	39
	Rabbit	Minor arrhythmia at 5%, some other effects.	40
Prolonged Toxicity			
Inhalation	Monkey, Dog, Rat, Guinea Pig	A series at 1.025% (10,250 ppm) × 30 6-hr exposures in 6 weeks and also a continuous 90-day exposure at 1,000 ppm: no effects attributed to fluorocarbon 11 in rats, guinea pigs, and monkeys. Dogs showed elevated blood urea nitrogen levels (significance of this elevation unclear in absence of any other biochemical effect or any specific pathologic lesion of the liver or kidneys).	41
	Rat, Rabbit	2% or 5% × 1 hr twice daily for 15 days: slight temporary metabolic changes at higher level only.	

	Species	Effects	Ref.
	Dog, Cat, Rat, Guinea Pig	2.5% (dogs at 1.25%), 3.5 hr daily, 5 days/week × 4 weeks, no organic damage.	31
	Rat, Guinea Pig, Rabbit, Mouse	4,000 ppm × 28 6-hr exposures in 6 weeks: no effects attributed to fluorocarbon 11.	42
Oral	Dog, Rat	Rats fed 450 mg/kg for three months, transient equivocal effect on urinary fluoride excretion; no effects in rats or dogs at lower levels.	43
Skin	Rat	40% in sesame oil, 12 applications in 16 days, no effect.[31] 10-sec spray at 10 cm, 5 days/week × 6 weeks, slight changes at microscopic study of tissue; retarded healing of injured tissue.[44]	31, 44
	Rabbit	5-sec spray at 20 cm, 5 days/week × 1 month, some conjunctival inflammation but no changes in eyeball.	44
Mutagen, Teratogen	Rat, Rabbit	Mutagen: Ames bacterial test negative.	45
		Teratogen: Inhalation, 20% of propellant mixture (10% fluorocarbon 11, 90% fluorocarbon 12) × 2 hr daily, rats 4th–16th days of gestation and rabbits 5th–20th days of gestation, no adverse effects.	46

II. FLUOROCARBON 12

	Species	Effects	Ref.
Acute Toxicity Inhalation	Rat	Survived 6-hr exposures at 80%.	32
	Rat, Guinea Pig	Survived 2 hr at 20%.	31
	Guinea Pig	Survived 2 hr at 20%.	14
	Mouse	3-hr exposure, LC_{50} = 62%.[58] Survived 24-hr exposure at 10%, microscopic examination showed tissue changes in lungs.[59]	58, 59
Eye	Rabbit	70:30 fluorocarbon 11/fluorocarbon 12 mixture, no significant ocular irritation	60
Cardiac Sensitization	Dog[a]	With epinephrine injection—evidence of arrhythmia, single death at 5%, no effect at 2.5%.[9] With exercise —one case of arrhythmia at 10%.[38]	9, 38

TABLE 25-3 (Continued)

Test	Animal	Experimental Data	Reference
Prolonged and Chronic Toxicity			
Inhalation	Rat, Guinea Pig, Rabbit, Dog, Monkey	810 ppm 24 hr daily × 90 days, some liver change ("submassive necrosis") noted in guinea pigs.	61
	Rat, Guinea Pig, Dog, Cat	10%, 3.5 hr daily, 5 days/week × 4 weeks, no organic damage.	31
	Guinea Pig, Dog, Monkey	20%, about 40 hr weekly average × 10–12 weeks, generalized tremors, other nervous system effects, slight blood changes.	62
Oral	Rat	Dose level 160 to 379 mg/kg daily, 5 days/week × 18 weeks, resulted in slight biochemical effect on liver function (slight elevation of plasma alkaline phosphatase).	63
	Rat, Dog	Dose equivalent to dietary level of 3,000 ppm for 2 years resulted in slightly decreased weight gain for rats, no other adverse effects.	64
Skin	Rat	40% in sesame oil, 12 applications in 16 days, no effect.[31] Repeated spraying generally well tolerated, but retarded healing of injured skin.[44] (Data from series with fluorocarbons 11, 12, and 114, also described in Reference 65.)	31, 44, 65
Eye	Rabbit	5-sec spray at 20 cm, 5 days/week × 1 month, some conjunctival inflammation but no changes in eyeball.	44, 65
Carcinogen, Mutagen, Teratogen		Carcinogen: See negative results of 2-year oral tests, *Prolonged and Chronic Toxicity*, above.	64
		Mutagen: Ames bacterial test negative.	66
		Teratogen: Inhalation, 90% of propellant mixture (90% fluorocarbon 12, 10% fluorocarbon 11) × 2 hr daily, rats 4th–16th days of gestation and rabbits 5th–20th	46, 64, 67, 68

days of gestation, no adverse effects.[46] Repeated oral dosing by intubation of pregnant[67] rats at a dose level ≤171 mg/kg/day × 10 and also of offspring in 2-year test[64] without effect. Inhalation in aerosol formulation without adverse effect from aerosol.[68]

III. FLUOROCARBON 113

			References
Acute Toxicity			
Inhalation	Rat	4-hr exposure, LC$_{50}$s vary from 5.2% to 6.8%, depending on strain of rat and grade of fluorocarbon 113.	73–76
	Rabbit	2-hr exposure, LC$_{50}$ = 6.0%.	77
	Rat, Rabbit	11,000–12,000 ppm, 2 hr daily, 12–24 months: no changes attributed to fluorocarbon 113. See also References 15, 31, 78, and 79. Earlier literature indicates deaths in guinea pigs and rats at ca. 5.0 and 8.7%, respectively.	85
Oral	Rat	LD$_{50}$ = 43 g/kg. (Rabbits may be more susceptible—see under Teratogen below.)	80
Skin	Rabbit	Graded practically nonirritating.	81
Eye	Rabbit	Minimal corneal and conjunctival changes.	81, 82
Aquatic	Fish	Slightly toxic to fathead minnows[83] and several estuarine species.[84]	83, 84
Cardiac Sensitization	Dog	With epinephrine injection—evidence of arrhythmia, some deaths at 0.5–1%.	85
Prolonged and Chronic Toxicity			
Inhalation	Rat, Guinea Pig, Dog	5.1%, 6 hr daily × 20 = "no significant changes."	86
	Rat, Mouse, Dog, Monkey	2,000 ppm, "continuous," for 14 days, no adverse effects.	87
Oral		See also References 15, 31, 78, and 79.	
Skin	Rabbit	See under Teratogen below and also References 15 and 80. 20 weeks of applications without visible effect, whereas trichloroethylene used as comparison showed erythema with beginning ulceration at the end of the first week.	88

TABLE 25-3 (Continued)

Test	Animal	Experimental Data	Reference
Mutagen, Teratogen	Mouse	Mutagen: Ames bacterial test negative.[89] Dominant lethal test on mice negative.[90]	89, 90
		Teratogen: Oral, 1 g/kg/day × 4 = 2 of 8 adult rabbits died; 5 g/kg/day × 4 = 4 of 8 adult rabbits died.[91] Inhalation, 2-hr exposure of rabbits at 2% and 0.2%, 8th–16th days of gestation, produced no effects other than partial or complete eye closure at 2% during exposure only.[92]	91, 92
		IV. FLUOROCARBON 114 (See also Addendum)	
Acute Toxicity			
Inhalation	Rat, Guinea Pig	Survived 2-hr exposure at 60%.	31
	Mouse	30-min. exposure lethal to 5 of 10 at 70%, 8 of 10 at 80%.	102
Cardiac Sensitization	Dog[a]	With epinephrine injection—evidence of arrhythmia, some deaths at 5% and occasional arrhythmia at 2.5%.	9
Prolonged Toxicity			
Inhalation	Rat	20%, 2.5 hr daily, 5 days/week × 2 weeks, decreased growth rate, some lung and blood effects.	102
	Mouse	4%, 30 min. daily × 3–6 weeks, changes in lung function.	103
	Rat, Rabbit	1%, 2 hr each working day × 8–9 months, no effect. See also References 31, 105, and 106. Earlier literature indicates tremors, convulsions, and death in dogs receiving 3–4 exposures of 8–16 hr daily at 20%.	104
Oral	Rat	Daily intubations of fluorocarbon 114 in peanut oil, 100–200 mg/kg, 5 days/week × 3 weeks, caused probable metabolism of the fluorocarbon, no evidence of harmful effects.	107

	Species	Description	Reference
	Rat, Dog	Dietary levels of 250–450 mg/kg (rats) and 150–380 mg/kg (dogs) × 3 months, both no effect.	108
Skin	Rat	10-sec spray at 10 cm, 5 days/week × 6 weeks, inflammatory reaction visible microscopically.	44, 65
Eye	Rabbit	5-sec spray at 20 cm, 5 days/week × 1 month, some conjunctival inflammation but no change in eyeball.	44, 65

V. FLUOROCARBON 115
(See also Addendum)

	Species	Description	Reference
Acute Toxicity			
Inhalation	Rat	Survived 4-hr exposure at 80%.	112
	Rat, Guinea Pig	Survived 2-hr exposure at 60%.	113
Cardiac Sensitization	Dog	With epinephrine injection—evidence of arrhythmia at 25% and occasionally at 15%.	9
Prolonged Toxicity			
Inhalation	Rat, Mouse, Rabbit, Dog	10% for 90 6-hr exposures, no effect.	112
	Rat, Guinea Pig, Dog, Cat	20%, 3.5 hr daily, 5 days/week × 4 weeks, no effect.	113
Oral	Rat	139–172 mg/kg fluorocarbon 115 in cottonseed oil, 5 doses/week for 2 weeks, no differences between test rats and oil-dosed control rats. See also Reference 114.	112

VI. FLUOROCARBON C-318

	Species	Description	Reference
Acute Toxicity			
Inhalation	Rat	Survived 4 hr at 80% plus 20% O_2.	118
	Rat, Guinea Pig	Survived 2 hr at 60%.	113
	Mouse	No anesthetic effect during 10-min. exposure.	118
Cardiac Sensitization	Dog	With epinephrine injection—evidence of arrhythmia at 25–50% and occasionally at 10%.	9

TABLE 25-3 *(Continued)*

Test	Animal	Experimental Data	Reference
Prolonged Toxicity			
Inhalation	Rat, Mouse, Rabbit, Dog	10% × 90 6-hr exposures, without effect.	118
	Rat, Guinea Pig, Dog, Cat	20%, 3.5 hr daily, 5 days/week × 4 weeks, no effect.	113
HYDROGENATED HALOMETHANES			
VII. FLUOROCARBON 21			
Acute Toxicity			
Inhalation	Rat	4-hr exposure, LC_{50} = 4.99%.	120
	Rat, Guinea Pig	2-hr exposure lethal at 10%, no clinical signs at 1%.	113
Skin	Guinea Pig	Negligible irritation, no evidence of sensitization after series of intradermal exposures.	121, 122
Eye	Rabbit	40% solution of fluorocarbon 21 in propylene glycol, moderately irritating; corneal, iritic, and conjunctival injuries resolved < 10 days.[36,121] Undiluted fluorocarbon 21 chilled to temperature of dry ice, mild irritation.[123] Aerosol spray, mild effects only on day of treatment.[35]	35, 36, 121, 123
Cardiac Sensitization	Dog	With epinephrine injection[b]—evidence of arrhythmia at 1%, not at 0.5%.	124
Prolonged Toxicity			
Inhalation	Rat	1%, 6 hr daily, 5 days/week × 2 weeks, resulted in biochemical changes, moderate to severe liver damage.	125
	Rat, Dog	0.5%, 6 hr daily, 5 days/week × 13 weeks, caused: in rats, excess mortality, severe liver damage, and other changes; in dogs, only mild liver damage. At 0.1%, same time schedule, rats also showed excess mortality but dogs were without significant change.	126

		Ref.
Puppies	40:60 propellant mixture of fluorocarbon 21 and fluorocarbon 22, 5 min. twice daily, 5 days/week × 2 weeks, caused "sedated and ataxic" appearance after 2–3 min. exposure; normal appearance within a few minutes after removal from inhalation box. See also Reference 113 and *Teratogen* (directly below).	127
Teratogen Rat	Pregnant rats inhaling 1%, 6 hr daily × 10 exposures (days 6–15 of 21-day gestation) had decreased maternal weight; no abnormalities in full-term offspring but interference with early development (15 of 25 presumably pregnant females without implants or viable fetuses at end of test).	128

VIII. FLUOROCARBON 22

		Ref.
Acute Toxicity **Inhalation**		
Rat	40% lethal within $1/2$ hr; 4 of 4 rats survived 2 hr at 30%, 1 of 4 died during 2-hr exposure at 20%.	113
Guinea Pig	40% lethal within 1 hr; 2 of 2 survived 2 hr at 30% and also 20%.	113
Mouse	Minimum fatal level = 1,300 mg/liter (37%).	131
Guinea Pig	Survived 2-hr exposure at $\simeq$ 20%.	132
Dog	Anesthesia at $\geq$ 25%, 1 death from hypoxia at 70%. See also reviews in References 101, 105, and 134.	133
Cardiac Sensitization		
Dog	With epinephrine injection—evidence of arrhythmia at 5%, not at 2.5%.	9
Prolonged Toxicity **Inhalation**		
Rat, Guinea Pig, Dog, Cat	5%, 3.5 hr daily, 5 days/week × 4 weeks, no effect.	113
Rat, Mouse	50 mg/liter (14,000 ppm), 6 hr daily except Sunday × 10 months = chemical, blood, and tissue effects; similar exposure at 7 mg/liter (2,000 ppm), no effects. See *Teratogen* below, also fluorocarbon 21, *Prolonged Toxicity*, inhalation test on puppies.[127]	131

TABLE 25-3 (*Continued*)

Test	Animal	Experimental Data	Reference
Skin	Rat	Mixture of fluorocarbon 22 and fluorocarbon 11, presumably 60:40, 10-sec spray from 10 cm, twice daily × 30 = very slight reaction but more than fluorocarbon 11 alone.	44
Eye	Rabbit	Mixture as directly above, 5-sec spray, 5 days/week × 1 month = transitory effect (hyperemia).	44
Mutagen, Teratogen		Mutagen: Ames bacterial test positive.	135
		Teratogen: Pregnant rats inhaling 100, 300, 500, 1,000 10,000, or 20,000 ppm, 6 hr daily × 10 exposures (days 4 or 6 through 13 or 15 of 21-day gestation) showed a very low incidence of eye abnormalities (9 of 1,996 fetuses with microphthalmia or anophthalmia, including one affected fetus at 100 ppm and none at 300 ppm).[136] The incidence of eye abnormalities was not dose-related and was consistently below the level of statistical significance in the extended series of the three individual tests, but it did achieve statistical significance when compared to historical controls. Data on spontaneous incidence of this abnormality in control rats are limited. Overall, fluorocarbon 22 appears weakly teratogenic in rats in view of these results and comparable results[137] from a recent study conducted by Imperial Chemical Industries, Ltd., in England. The incidence of eye abnormalities in rats was statistically significant only at the highest concentration tested, 5%, which was not associated with this effect in comparably exposed rabbits.	136, 137

IX. FLUOROCARBON 31

Acute Toxicity			
Inhalation	Rat	4-hr exposure, LC_{50} = 4.5%.	138
	Guinea Pig	"cannot be tolerated . . . in concentration greater than a few percent."	139
Cardiac Sensitization	Dog	With epinephrine injection[b]—evidence of arrhythmia at 5–10%, "questionable" arrhythmia at 2.5–3.75%.	140
Prolonged Toxicity			
Inhalation	Rat	1%, 6 hr daily, 5 days/week × 2 weeks, resulted in "moderate damage to kidneys, adrenals, testes, epididymis, and hemopoietic tissues."	125
Mutagen		Ames bacterial test, positive. Based on a 6-hr[141] rather than usual 48-hr exposure; mutagenic activity greater than that observed with most homologous halocarbon gases.[142]	141, 142

X. FLUOROCARBON 32

Acute Toxicity			
Inhalation	Rat	4-hr exposure, LC_{50} > 76% (highest feasible concentration, which necessitated supplementary oxygen to maintain chamber oxygen concentrations of ~20%); clinical signs (such as lethargy, loss of mobility in hind legs, and spasms) rapidly reversible after exposure.	143
Cardiac Sensitization	Dog	With epinephrine injection[b]—evidence of occasional arrhythmia at 25%; the 1 susceptible dog in a group of 12 did not react at 20%.	144
Prolonged Toxicity			
Inhalation	Rat	1%, 6 hr daily, 5 days/week × 2 weeks, resulted in no effect.	125

TABLE 25-3 (Continued)

HYDROGENATED HALOETHANES
XI. FLUOROCARBON 123

Test	Animal	Experimental Data	Reference
Acute Toxicity			
Inhalation	Rat	LC_{50} = 3.2%.[145] Behavioral test failure[146] consistently found at 0.5–1.0% during 60-min. exposure; at 0.5%, failures in vertical bar and pinna reflexes first observed 15 min. into exposure.	145, 146
	Mouse	10 min.[146,147] or 30-min.[148] exposures = median anesthetic concentration of 2.4–2.7%, LC_{50} = 7.4–7.7%. Other data[149] consistent.	146–149
Oral	Rat	Lethal dose = 9,000 mg/kg (>7,500 mg/kg) when tested as 50% solution in corn oil.	150
Skin	Guinea Pig	Not irritating, not sensitizing after intradermal injection series.	151
Eye	Rabbit	No corneal or iritic changes when neat fluorocarbon 123 allowed to evaporate, but temporary corneal damage after dosing followed by 1-min. water wash. Moderate, reversible effects from excess of 50% mixture in propylene glycol.	152
Cardiac Sensitization	Dog	With epinephrine injection[b]–limited number of exposures, but evidence of arrhythmia (usually fatal), in all at 4%, most at 2%, and none at 1%.	153
Prolonged Toxicity			
Inhalation	Rat	1%, 6 hr daily, 5 days/week × 2 weeks, caused less activity and responsiveness to noise than controls, but no adverse biochemical or tissue changes.	125, 154 (same test)
Teratogen	Rat	See *Teratogen* directly below. Pregnant rats inhaling 1%, 6 hr daily × 10 exposures	128

(days 6–15 of 21-day gestation) resulted in mild, transient sedation of the pregnant females (no other effects).

XII. FLUOROCARBON 124

	Species		Ref.
Acute Toxicity			
Inhalation	Guinea Pig	Survived 2-hr exposures at ⩽20%.	155
Cardiac Sensitization	Dog	With epinephrine injection[b] arrhythmia in 2 of 2 at 5%, 4 of 10 (1 death) at 2.5%, and 0 of 10 at 1%.	156
Prolonged Toxicity			
Inhalation	Rat	10%, 6 hr daily, 5 days/week × 2 weeks, caused slightly slower rate of weight gain, "mild anesthetic effects" . . . reversible within 15–30 minutes post-exposure," but no major biochemical or tissue changes (slight change in hepatocyte appearance at end of exposure series, absent 2 weeks later).	125, 157 (same test)

XIII. FLUOROCARBON 125

	Species		Ref.
Acute Toxicity			
Inhalation	Rat	4-hr exposure, lethal concentration > 10%.	158
	Mouse	No anesthesia at ⩽ 93%.	149

XIV. FLUOROCARBON 132b

	Species		Ref.
Acute Toxicity			
Inhalation	Rat	Survived 4-hr exposure at 1%, 2-hr exposure at 2%; 2% lethal to 1 of 4 and 2 of 4 rats when exposed for 4 or 7 hr, respectively.	159
	Rabbit	At ≃3%, "convulsions, cardiac arrhythmias and lung lesions."	148
	Mouse	10-min.[147] or 30-min.[148] exposures = median anesthetic effect at 1.3%, LC_{50} = 4.3–4.9%.	147, 148

TABLE 25-3 *(Continued)*

Test	Animal	Experimental Data	Reference
Oral	Rat	Lethal dose = 25,000 mg/kg when tested as 50–70% solution in corn oil.	160
Skin	Guinea Pig	Not irritating. Not sensitizing after intradermal injection series.	161
Eye	Rabbit	Mild corneal injury reversible within 7 days.	162
Aquatic	Fish	Slightly toxic to fathead minnows.	163
Cardiac Sensitization	Dog	With epinephrine injection[b]—arrhythmia in 2 of 2 at 1%, 3 of 10 at 0.5%, and 0 of 10 at 0.25%.	164
Prolonged Toxicity	Rat	1%, 6 hr daily, 5 days/week × 2 weeks, caused irregular respiration, lethargy, incoordination, and occasional tremors. One of 10 rats died 2nd exposure day (pulmonary congestion, edema, hemorrhage; moderate generalized autolysis). Slightly slower rate of weight gain, "mild anesthetic effects . . . reversible within 15–30 minutes post-exposure." Slight effects in tissue of liver, thymus, and testis reversible within 14 days after exposure, but biochemical effect (blood urea nitrogen elevated) in 3 of 5 rats sacrificed at this time.	125, 157 (same test)
Mutagen		Ames bacterial test negative.	165

XV. FLUOROCARBON 133a

Test	Animal	Experimental Data	Reference
Acute Toxicity Inhalation	Mouse	10-min.[147,166] or 30-min.[148] exposures = median anesthetic concentration of 4–8%, LC_{50} = 15–25%.	147, 148, 166
	Dog	Anesthesia at 8–13%, respiratory arrest at 23–41%.	166
	Rat	Mutagen: Ames bacterial test negative.	167
Mutagen, Teratogen		Teratogen: Inhalation of 0.2%–5%, 6 hr daily × 10	168

exposures (days 6–15 of 21-day gestation) was embryotoxic. At 0.2%, more than half the fetuses were resorbed; at exposure levels $\geq$ 0.5%, the resorption was total or almost total. Comparable exposure at 0.05% produced no effect.

XVI. FLUOROCARBON 134a

Acute Toxicity Inhalation	Rat	Single 30-min. exposure at 75% caused death in 2 of 4 rats; lung pathology all 4 rats.	170

XVII. FLUOROCARBON 141b
(See also Addendum)

Acute Toxicity Inhalation	Rat	6-hr exposure, 5.02% lethal to 1 of 6, all survived 4.28%.	171
	Mouse	10-min. exposure = median anesthetic concentration of 2.5%, LC_{50} = 5%.	147
Prolonged Toxicity	Rat	1%, 6 hr daily, 5 days/week × 2 weeks, caused no adverse clinical signs; slight biochemical changes appeared reversible within 14 days. Pneumonitis in test rats increased over that present in controls, suggesting that exposures to fluorocarbon 141b may have exacerbated this disease.	172

XVIII. FLUOROCARBON 142b

Acute Toxicity Inhalation	Rat	4-hr exposure, range-finding data[177] suggests LC_{50} in vicinity of 12.8% concentration (which was lethal to some but not all exposed). 30-min. exposures[32] lethal at 50%, not at 40%.	32, 177
	Mouse	10-min. exposures = median anesthetic concentration of 25%.	147

TABLE 25-3 (Continued)

Test	Animal	Experimental Data	Reference
Eye	Rabbit	Liquid chilled to dry ice temperature = very slight conjunctival reaction first day after treatment.	178
Cardiac Sensitization	Dog	With epinephrine injection—arrhythmia in 5 of 12 at 5%, 0 of 6 at 2.5%. With noise (to stimulate internal epinephrine)—arrhythmia in 5 of 12 at 80%, compared to noise alone producing arrhythmia in 1 of 12.	9
Prolonged Toxicity Inhalation	Rat	10%, 16 hr daily = death in 7–9 days with lungs showing severe irritation.[32] Similar exposure at 1% = mild, diffuse microscopic change in lungs.[32] 2%,[c] 6 hr daily, 5 days/week × 2 weeks, caused no effect.[125,179]	32, 125, 179 (last two—same test)
	Rat, Dog	0.1% or 1%, 6 hr daily, 5 days/week for 90 days = no effect.	126, 180
Mutagen, Teratogen		Mutagen: Ames bacterial test weakly positive.	181
		Teratogen: Pregnant rats inhaling 0.1% or 1%, 6 hr daily × 10 exposures, days 6–15 of 21-day gestation, no maternal or fetal effect attributed to fluorocarbon 142b.	182

XIX. FLUOROCARBON 152a
(See also Addendum)

Test	Animal	Experimental Data	Reference
Acute Toxicity Inhalation	Rat	4-hr exposure, 38.3% lethal to 1 of 6, all survived 31.9%.	183
	Rat	30-min. exposure lethal at 50%, not at 45%.	32
Cardiac Sensitization	Dog	With epinephrine injection—arrhythmia in 3 of 12 at 15%, 0 of 12 at 5%.	9
Prolonged Toxicity Inhalation	Rat	10%, 16 hr daily × 2 months[32] caused mild, diffuse,	32, 125, 184

		(last two—same test)
	microscopic change in lungs. 10%, 6 hr daily, 5 days/week × 2 weeks caused no adverse biochemical or tissue changes; urinary fluoride ~ 2 times that of controls.	
Mutagen, Teratogen	Mutagen: Ames bacterial test negative.	185
	Teratogen: Pregnant rats inhaling ≤ 5%, 6 hr daily × 10 exposures, no effect on behavior and body weight gains of the mothers, outcome of pregnancy, and development of the fetuses.	168

a These data are supported by quite similar experiments of Clark and Tinston (1972), as cited in the Supplemental References.
b Procedure as given in Reference 9.
c Reference 125 (Abstract) erroneously gives the exposure concentration as 1%, rather than 2% as correctly specified in Reference 179.

An 8-hr continuous exposure at 1,000 ppm results (estimate[53] by pharmacokinetic model) in a blood level about $1/4$ of that associated with the threshold for cardiac sensitization in dogs. Fluorocarbon 11 does not appear to undergo significant biotransformation and is excreted somewhat less rapidly than fluorocarbon 12, 113, or 114.[54-57]

The Ames test[45] was negative. No teratogenic effect[46] was evident.

Fluorocarbon 12 (Dichlorodifluoromethane)

Data cited in Table 25-3[9, 14, 31, 32, 35, 38, 46, 58-68] indicate that fluorocarbon 12 is characterized by unusually low acute inhalation toxicity. Clinical signs such as tremors, disturbance of equilibrium, or narcosis have been observed only at concentrations above 20% and suggest the central nervous system depression characteristic of a weak narcotic. In a 90-day study involving 24-hr exposure at 810 ppm,[61] one of five species—guinea pigs—showed microscopic liver injury that the investigators linked to "the continuous nature of the exposure or to the high order of susceptibility of the guinea pig. . . ." Fluorocarbon 12 is significantly less cardiotoxic than fluorocarbon 11 and has been rated moderate in circulatory–respiratory effect.[17] Cardiac sensitization in humans has been readily elicited under conditions of gross misuse.[9,48] An 8-hr continuous exposure at 1,000 ppm results (estimate[53] by pharmacokinetic model) in a blood level roughly $1/20$ of that associated with the threshold for cardiac sensitization in dogs.[58] Studies on human volunteers showed that inhalation of fluorocarbon 12 at the 1% level for 2.5 hr caused a 7% reduction in standardized psychomotor test scores—the only adverse effect.[69] Like fluorocarbon 11, inhaled fluorocarbon 12 is not metabolized appreciably.[54,55] Also, it appears to be very rapidly eliminated.[56,57]

Feeding tests[63,64] indicate a wide degree of safety. Based on oral studies, several metabolism tests,[70,71] and other data, the World Health Organization[72] has evaluated the dietary level of fluorocarbon 12 causing no toxicologic effect in rats as 0.3% (equivalent to 150 mg/kg). For humans, the acceptable daily intake of fluorocarbon 12 has been estimated as 0–1.5 mg/kg body weight.

Skin and eye tests on rats or rabbits have shown little, if any, reaction. The 2-year test data indicate that fluorocarbon 12 is not carcinogenic. The Ames test[66] and the earlier dominant lethal test[64] were negative. Several tests for teratogenic or other reproductive effects[46,67,68] revealed no changes attributable to fluorocarbon 12.

Fluorocarbon 113 (1,1,2-Trichloro-1,2,2-trifluoroethane)

As indicated in Table 25-3,[15,31,73-92] fluorocarbon 113 is more toxic than fluorocarbon 12, but less toxic than fluorocarbon 11 (according to one reviewer, "about half as toxic as FC-11"[15]). Although less potent than fluorocarbon 11, fluorocarbon 113 is a relatively strong cardiotoxic agent,[17] comparable to

1,1,1-trichloroethane and trichloroethylene.[85] Tests with dogs have shown that arterial blood concentrations of fluorocarbon 113 resulting from exposure at 1,000 ppm are appreciably lower than those resulting from the 5,000-ppm level associated with cardiac sensitization.[93] Fluorocarbon 113 appears to be eliminated less rapidly than fluorocarbon 12, but more rapidly than fluorocarbon 11.[57] Experimental human exposures to fluorocarbon 113[94,95] support the lack of effect at the TLV of 1,000 ppm. At the Kennedy Space Center, 50 workers exposed for an average of 2.77 years to an overall average of 0.07% fluorocarbon 113 (average 669 ppm, range 46 to 4,700 ppm, median 435 ppm) showed "no evidence of adverse effects."[96]

Oral exposure, as well as skin and eye contact, is not expected to cause problems in ordinary use. A human under anesthesia who accidentally received about 1 liter of fluorocarbon 113 in the stomach survived, although suffering immediate but transient cyanosis and then severe rectal irritation.[97] Fluorocarbon 113 is capable of penetrating human skin, but absorption is not insidious —a tingling sensation is noticed almost immediately.[98] An experiment involving the swabbing of abdominal skin on human volunteers[99] supports its use as a defatting agent in selected presurgical procedures. In a well-ventilated room, the concentration of fluorocarbon 113 in room air reached a maximum of 585 ppm. Presurgical cleaning with fluorocarbon 113 should be done in well-ventilated rooms and avoided in patients with a medical history of cardiac arrhythmia.

No carcinogenic potential was observed in long-term tests.[88] Synergistic carcinogenicity has been suggested on the basis of subcutaneous injections in mice of fluorocarbon 113 and piperonyl butoxide,[100] but this claim has been challenged.[101] Mutagen tests[89,90] were negative.

Reproduction studies with rabbits have shown no specific action on the fetus, but pregnant animals appeared to show a narcotic effect. One oral dose of 1 g/kg fluorocarbon 113 made rabbits more docile than controls, and 5 g/kg resulted in negligible food and water intake (and consequent weight loss) during the dosing period.

Fluorocarbon 114 (1,2-Dichloro-1,1,2,2-tetrafluoroethane)

Table 25-3[9,31,51,102-108] shows that the acute toxicity of fluorocarbon 114 is similar to that of fluorocarbon 12. According to Aviado,[15] fluorocarbon 114 "would appear to have only ⅕ the predisposition to cardiac arrhythmia of FC-11." Human studies using various inhalation techniques[57,109-111] indicate that fluorocarbon 114 is excreted rapidly. In a study using radiolabeled fluorocarbon 114,[57] only 12% of the compound remained in the animal 30 min. after dosage. This can be compared to the retention of 23%, 10%, and 20%, respectively, for comparable doses of fluorocarbons 11, 12, and 113.

Animals receiving repeated oral doses, including rats and dogs fed for 90 days, showed no obvious toxic effects.

Skin and eye tests indicate low irritation potential, particularly from occasional accidental contact.

An Ames mutagen test was negative.

Fluorocarbon 115 (Chloropentafluoroethane)

Data cited in Table 25-3[9,112-114] explain why fluorocarbon 115 is generally considered to have very low acute, chronic, and cardiovascular toxicity. (For earlier literature, see citations in the references listed above.) Tests with anesthetized dogs show that fluorocarbons 115 and C-318 "do not elicit any change in heart rate and produce only bronchodilation."[115] In a stringent comparison of circulatory and respiratory effects associated with 15 propellants that have been used commercially, fluorocarbon 115 was rated a propellant of low toxicity.[116]

Correlation of blood levels of fluorocarbon 115 obtained after 10-min. and 6-hr inhalation exposures at a concentration of 10% with blood levels of fluorocarbon 115 obtained after a single 12-g oral dose established the very low toxicity potential of fluorocarbon 115 by the oral route.[117] If fluorocarbon 115 blood levels of 2 to 3 μg/ml from inhalation and the absence of a detectable level ($\geqslant$0.06 μg/ml) by the oral route are considered in terms of aerosol packaging, "a person would have to eat at one meal many times his own body weight of an aerosol food preparation to achieve the fluorocarbon 115 blood level reached in a 10-minute (10%) inhalation exposure."[117]

An Ames mutagen test was negative.

Fluorocarbon C-318 (Octafluorocyclobutane)

The toxicity data for fluorocarbon C-318 are summarized in Table 25-3.[9,113,114,118] (For earlier literature, see citations in these references.) Fluorocarbon C-318 has very low toxicity. In acute testing, rats survived in an atmosphere of 80% fluorocarbon C-318 and 20% oxygen. In cardiac sensitization tests, there were no deaths at exposure levels of 25–50%. In chronic tests, there was no evidence of toxicity when rats, mice, and dogs were exposed to 10% fluorocarbon C-318 for 90 days.

Hydrogenated Halomethanes (Fluorocarbons 21, 22, 31, and 32)

Toxicity test data generated to date show wide differences in toxicity for the respective compounds. As of September 1978,[119] a TLV of 10 ppm was recommended for fluorocarbon 21; the documentation for the previously established TLV of 1,000 ppm for fluorocarbon 22 is scheduled for review.

Fluorocarbon 21 (Dichlorofluoromethane)

As shown in Table 25-3,[36,113,120-128] fluorocarbon 21 is definitely more toxic than the six perhalogenated fluorocarbons discussed above. Acute inhalation toxicity is relatively low, but repeated exposures have clearly demonstrated fluorocarbon 21's potential for hepatotoxicity. The exact mechanism is currently unknown; one suggestion from *in vitro* studies (rabbit liver mitochondria) is that fluorocarbon 21 may possess a singular capacity to inhibit glutamate oxidation.[129] Fluorocarbon 21 is slightly less cardiotoxic than fluorocarbon 11. Belej and Aviado[130] (see also Reference 17) call fluorocarbon 21 the "prototype" of hypotensive low-pressure propellants.

Skin tests do not indicate a dermatitis hazard. Fluorocarbon 21 can act as an eye irritant if contact is facilitated. Inhalation tests with pregnant rats caused no fetal malformations or visible abnormalities, but the number of expected fetuses was reduced at the 1% exposure level.

Fluorocarbon 22 (Chlorodifluoromethane)

Available data cited in Table 25-3[9,65,101,105,113,127,131-137] indicate that fluorocarbon 22 has low acute and chronic inhalation toxicity. Skin and eye tests indicate little hazard. Fluorocarbon 22 is positive in the Ames bacterial mutagen test[135] and shows "an extremely weak, atypical response" in rat teratology studies.[136] The significance of these results for the human species is unknown. A recent evaluation[137] concludes that the TLV and OSHA standard of 1,000 ppm satisfactorily protects the fetuses of pregnant women in the workplace. An additional long-term inhalation test is in progress in Europe, but will not be completed until late 1979 or 1980. At present, therefore, du Pont does not recommend use of fluorocarbon 22 in unconfined applications such as aerosol propellants but continues to recommend its use in applications such as air conditioning.

Fluorocarbon 31 (Chlorofluoromethane)

Data cited in Table 25-3[125,138-142] indicate that fluorocarbon 31 is too toxic to be acceptable as an aerosol propellant. Acute inhalation data suggest the possibility of lingering kidney injury, and 2-week data show evidence of cumulative reaction (kidneys, testes, adrenals, and also hematologic effects). Cardiac sensitization potential is considered moderate, likely to be overshadowed by other reactions (nervous effects such as excitation, salivation at 5%, and partial anesthesia at 10%). Among homologous compounds, fluorocarbon 31 is the strongest known positive in the Ames mutagen test.

Fluorocarbon 32 (Difluoromethane)

Available data for fluorocarbon 32, as cited in Table 25-3,[125,143,144] all indicate very low toxicity. Fluorocarbon 32 has a very weak cardiac sensitization potential.

Hydrogenated Haloethanes (Fluorocarbons 123, 124, 125, 132b, 133a, 134a, 141b, 142b, and 152a)

The first toxicologic investigations of some of these materials, such as fluorocarbons 124 and 152a, date back several decades. A number of toxicity studies have been conducted recently; Figures 25-2, 25-3, and 25-4 show inhalation chambers and equipment currently used by Du Pont.

Fluorocarbon 123 (2,2-Dichloro-1,1,1-trifluoroethane)

Tests on rats, as listed in Table 25-3[125,128,145-154] indicate relatively low acute toxicity. However, the lethal concentration is less than $\frac{1}{25}$ of the saturated vapor concentration of 89%, and lethal atmospheres produce immediate

Figure 25-2 Stainless steel and glass inhalation chambers (1.4-m^3 capacity). (*Courtesy of David P. Kelly, Toxicologist, Haskell Laboratory, Du Pont Company*)

Figure 25-3 Sampling atmosphere of 1.4-m^3 chamber used for repeated exposures (2-week, 90-day, and teratology studies). (*Courtesy of David P. Kelly, who is taking sample*)

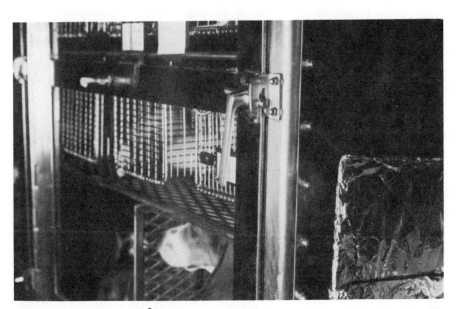

Figure 25-4 Loaded 1.4-m^3 chamber as used during 90-day study with fluorocarbons 21 and 142b. (*Courtesy of David P. Kelly*)

unresponsiveness. Some deaths occurred within minutes after onset of exposure. Cardiac sensitization potential is strong, slightly less than that of fluorocarbon 11. A series[146] of 1-hr exposures (0.1, 0.25, 0.5, or 1% fluorocarbon 123) showed rats failing measurements of unconditioned reflexes, locomotor activity, and coordination at exposure concentrations ⩾0.5% fluorocarbon 123; recovery was rapid, generally within 15 min. after exposure.

Tests with rats indicate low acute oral toxicity. Fluorocarbon 123 is not a skin irritant or sensitizer but is capable of acting as a mild to moderate irritant in rabbit eyes. Tests on pregnant rats revealed no embryotoxic or teratogenic effects.

Fluorocarbon 124 (2-Chloro-1,1,1,2-tetrafluoroethane)

Table 25-3[155-157] shows low inhalation toxicity in acute exposures and a 2-week test. Fluorocarbon 124 is considered a moderate cardiac sensitizer, slightly stronger than fluorocarbon 114.

Fluorocarbon 125 (Pentafluoroethane)

Table 25-3[149,158] lists the limited available data.

Fluorocarbon 132b (1,2-Dichloro-1,1-difluoroethane)

Table 25-3[147,148,157,159-165] reveals fluorocarbon 132b as one of the more toxic of the fluorocarbon propellants reviewed in this series. Acute exposures with rats produced varying degrees of lung, liver, and kidney injury. A 2-week series of repeated exposures indicates that the 1% level is toxic. Fluorocarbon 132b is considered a strong cardiac sensitizer. Skin, eye, and mutagen tests show no particular hazard.

Fluorocarbon 133a (2-Chloro-1,1,1-trifluoroethane)

As indicated in Table 25-3,[147,148,166-168] anesthetic tests with mice suggest low acute toxicity. Fluorocarbon 133a is extremely rapid in anesthetic action; one report[166] describes anesthesia developing at concentrations of 15% in about 45 sec and disappearing about 30 sec after termination of the 10-min. exposure. Experiments with dogs[166] showed anesthetic effects that were reversible in as little as 1 min. when the concentration of fluorocarbon inhalant was sufficiently reduced.

A patent application[169] describes "reduced perception or pain in a human, dog, cat, rabbit, rat or monkey [that] is obtained by circulating through the respiratory system a mixture consisting of 10–20% F_3CCH_2Cl."

Fluorocarbon 133a is embryotoxic to rats at exposure levels ⩾0.2% but not at 0.05%.[168] Exposure of women in the workplace has not been a practical issue and guidelines at present are undefined. An Ames bacterial mutagen test was negative.[167]

Fluorocarbon 134a (1,1,1,2-Tetrafluoroethane)

The very limited available toxicity data are cited in Table 25-3.[170]

Flourocarbon 141b (1,1-Dichloro-1-fluoroethane)

Acute toxicity is considered low on the basis of tests described in Table 25-3.[147,171,172] Repeated exposures at the 1% level may produce biochemical changes. Several Russian reports[173-176] describing acute plus repeated exposure tests on "dichlorofluoroethane" may also be of interest, but chlorine–fluorine positions are not specified. Fluorocarbon 141b is considered a strong cardiac sensitizer and is weakly positive in the Ames mutagen test.

Fluorocarbon 142b (1-Chloro-1,1-difluoroethane)

Tests cited in Table 25-3[9,32,125,126,147,177-181] show considerable variation in acute mortality figures, but all are within the range of low toxicity. (References 173 and 174 may also be of interest; see fluorocarbon 141b.)

Cardiac sensitization tests indicate that fluorocarbon 142b is a moderate sensitizer. Fluorocarbon 142b is similar to fluorocarbon 114 ("intermediate") in overall effect on the respiratory and circulatory systems in various experimental animal models.[15,116]

Repeated exposure to concentrations of 10% for 16 hr daily was lethal, but rats exposed to 1% for 16 hr daily showed only lung changes suggestive of chronic irritation. This early work has been supplemented by 6-hr exposures at 1–2% for 10 and 90 days with complete clinical, hematologic, blood chemical, urine analytic, and histopathologic examination. In both tests, there were no toxic effects attributable to fluorocarbon 142b. It is particularly noteworthy that the 90-day test revealed no significant increase in urinary fluoride in either dogs or rats, which suggests negligible metabolism of this fluorocarbon. With dogs, "FC-142b itself was not detected in any of ten tissues, indicating no long-term retention."[181]

A rabbit eye test[178] and rat teratogen test[182] showed negligible reaction. The significance of the weakly positive Ames test,[181] if any, is unknown at present; it is recommended that exposure of people to fluorocarbon 142b should be minimized pending further investigation.

Fluorocarbon 152a (1,1-Difluoroethane)

Data in Table 25-3[9,32,125,182-185] indicate low toxicity. Fluorocarbon 152a produced "good" anesthesia[133] in a screening test where dogs and humans briefly inhaled 50% fluorocarbon 152a. Fluorocarbon 152a is considered a weak cardiac sensitizer, which is consistent with the weak potential for disturbance of the cardiovascular and respiratory systems found in dogs, monkeys, mice, and rats.[15,17,116]

Two series of repeated exposures with rats at concentrations of 10% showed no marked cumulative effect. Tests conducted for two months (16 hr/day) at 10%[32] revealed lung changes suggestive of mild chronic irritation. A more recent 2-week test (6 hr/day) at 10% showed no clinical or other obvious evidence of toxicity attributable to fluorocarbon 152a, although increased urinary fluoride suggests that "FC-152a may be metabolized to a small extent."[183] Mutagen and teratogen tests revealed no effects.

ADDENDUM

The results of several additional recent tests are listed below:

Fluorocarbon 114:   Negative in the Ames bacterial mutagen test.[186]
Fluorocarbon 115:   Negative in the Ames bacterial mutagen test.[187]
Fluorocarbon 141b: Strong cardiac sensitizer in dogs; with epinephrine injection—evidence of arrhythymia, 1 death at 0.5-1%.[188] Weakly positive in the Ames bacterial mutagen test.[189]

Note. D. F. Krahn, Haskell Laboratory, plans to submit the results of the unpublished mutagen tests cited above for publication in 1979.

REFERENCES

1. T. A. Loomis, "Essentials of Toxicology," 2nd ed., Lea & Febiger, Philadelphia, Pa., 1974.
2. L. J. Casarett and J. Doull, "Toxicology: The Basic Science of Poisons," Macmillan Publishing Company, Inc., New York, 1975.
3. J. W. Clayton, Jr., "Toxicity," in P. A. Sanders, "Principles of Aerosol Technology," Van Nostrand Reinhold Company, New York, 1970.
4. C. I. Bliss, "The Determination of the Dosage-Mortality Curve from Small Numbers," *Quart. J. Pharm. Pha, acol.* **11**, 192 (1938).
5. J. T. Litchfield, Jr., and F. Wilcoxon, "A Simplified Method of Evaluating Dose-Effect Experiments," *J. Pharmacol. Exp. Ther.* **96**, 99 (1949).
6. W. B. Deichmann and T. J. LeBlanc, "Determination of the Approximate Lethal Dose with About Six Animals," *J. Ind. Hyg. Toxicol.* **25**, 415 (1943).
7. H. L. Price, A. A. Lurie, R. E. Jones, M. L. Price, and H. W. Linde, "Cyclopropane Anesthesia. II. Epinephrine and Norepinephrine in Initiation of Ventricular Arrhythmias by Carbon Dioxide Inhalation," *Anesthesiology* **19**, 619 (1958).
8. M. Bass, "Sudden Sniffing Death," *J. Am. Med. Assoc.* **212**, 2075 (1970).
9. C. F. Reinhardt, A. Azar, M. E. Maxfield, P. E. Smith, Jr., and L. S. Mullin, "Cardiac Arrhythmias and Aerosol 'Sniffing'," *Arch. Environ. Health* **22**, 265 (1971).
10. R. R. Montgomery and C. F. Reinhardt, "A Capsule Dose of Toxicology," *Occup. Health Nursing* **24**, 7 (1976).
11. American Conference of Governmental Industrial Hygienists, "TLV®s Threshold

Limit Values for Chemical Substances and Physical Agents in the Workroom Environment with Intended Changes for 1977," Cincinnati, Ohio, 1977 (revised annually).

12. W. C. L. Hemeon, "Plant and Process Ventilation," Industrial Press, New York, 1955.

13. G. Paulet, "Les Fluorocarbones en Question," *Eur. J. Toxicol.* 9, 385 (1976).

14. A. H. Nuckolls, "The Comparative Life, Fire, and Explosion Hazards of Common Refrigerants," Miscellaneous Hazard No. 2375, Underwriters Laboratories, Chicago, Ill., 1933.

15. D. M. Aviado, "Toxicity of Propellants," in "Progress in Drug Research," Vol. 18, Birkhäuser Verlag, Basel, 1974.

16. D. M. Aviado and D. G. Smith, "Toxicity of Aerosol Propellants in the Respiratory and Circulatory Systems. VIII. Respiration and Circulation in Primates," *Toxicology* 3, 241 (1975).

17. D. M. Aviado, "Toxicity of Aerosol Propellants in the Respiratory and Circulatory Systems. X. Proposed Classification," *Toxicology* 3, 321 (1975).

18. H. C. Grice, T. DaSilva, D. R. Stoltz, I. C. Munro, D. J. Clegg, L. R. A. Bradshaw, and J. D. Abbatt, "The Testing of Chemicals for Carcinogenicity, Mutagenicity and Teratogenicity," Authority of M. Lalonde, Minister of Health and Welfare, Ottawa, Ontario, Canada, 1973.

19. J. H. Weisburger, "Bioassays and Tests for Chemical Carcinogens," in C. E. Searle (Ed.), "Chemical Carcinogens," ACS Monograph 173, American Chemical Society, Washington, D.C., 1976.

20. L. Tomatis, C. Agthe, H. Bartsch, J. Huff, R. Montesano, R. Saracci, E. Walker, and J. Wilbourn, "Evaluation of the Carcinogenicity of Chemicals: A Review of the Monograph Program of the International Agency for Research on Cancer (1971 to 1977)," *Cancer Res.* 38, 877 (1978).

21. J. McCann, E. Choi, E. Yamasaki, and B. N. Ames, "Detection of Carcinogens as Mutagens in the *Salmonella*/Microsome Test: Assay of 300 Chemicals," *Proc. Nat. Acad. Sci. U.S.A.* 72, 5135 (1975).

22. J. McCann and B. N. Ames, "Detection of Carcinogens as Mutagens in the *Salmonella*/ Microsome Test: Assay of 300 Chemicals: Discussion," *Proc. Nat. Acad. Sci. U.S.A.* 73, 950 (1976).

23. I. F. H. Purchasc, E. Longstaff, J. Ashby, J. A. Styles, D. Anderson, P. A. Lefevre, and F. R. Westwood, "An Evaluation of Six Short-Term Tests for Detecting Organic Chemical Carcinogens," *Br. J. Cancer* 37, 873 (1978).

24. J. G. Wilson and J. Warkany, "Teratology: Principles and Techniques," University of Chicago Press, Chicago, Ill., 1965.

25. H. Nishimura and J. R. Miller, "Methods for Teratological Studies in Experimental Animals and Man," Igaku Shoin Ltd., Tokyo, Japan, 1969.

26. C. L. Berry and J. P. Germain, "Polygenic Models in Teratological Testing," in C. L. Berry and D. E. Poswillo (Eds.), "Teratology: Trends and Applications," Springer-Verlag, New York, 1975.

27. J. Cairns, Jr., and K. L. Dickson, "Biological Methods for the Assessment of Water Quality," ASTM Special Technical Publication 528, American Society for Testing and Materials, Philadelphia, Pa., 1973.

28. J. Cairns, Jr., K. L. Dickson, and G. F. Westlake, "Biological Monitoring of Water and Effluent Quality," ASTM Special Technical Publication 607, American Society for Testing and Materials, Philadelphia, Pa., 1976.

29. C. E. Stephan, "Methods for Acute Toxicity Tests with Fish, Macroinvertebrates, and Amphibians," National Environmental Research Center Office of Research and Development, U.S. Environmental Protection Agency, Corvallis, Oreg., 1975.

30. C. E. Barras, unpublished data, Haskell Laboratory, E. I. du Pont de Nemours & Company, Wilmington, Del., October 1974.

31. J. Scholz, "Neue toxikologische Untersuchungen einiger als Treibgas verwendeter Frigen-Typen," in "Fortschr. Biol. Aerosol-Forsch. Jahren 1957–1961, Ber. Aerosol-Kongr. 4th.," 1962, p. 420.

32. D. Lester and L. A. Greenberg, "Acute and Chronic Toxicity of Some Halogenated Derivatives of Methane and Ethane," *Arch. Ind. Hyg. Occup. Med.* **2**, 335 (1950).

33. G. Paulet, G. Roncin, E. Vidal, P. Toulouse, and J. Dassonville, "Fluorocarbons and General Metabolism in the Rat, Rabbit, and Dog," *Toxicol. Appl. Pharmacol.* **34**, 197 (1975).

34. T. F. Slater, "A Note on the Relative Toxic Activities of Tetrachloromethane and Trichloro-fluoro-methane on the Rat," *Biochem. Pharmacol.* **14**, 178 (1965).

35. D. B. Hood, unpublished data, Haskell Laboratory, E. I. du Pont de Nemours & Company, Wilmington, Del., September 1964.

36. C. W. Eddy, unpublished data, Haskell Laboratory, E. I. du Pont de Nemours & Company, Wilmington, Del., October 1970.

37. H. J. Trochimowicz, C. F. Reinhardt, L. S. Mullin, A. Azar, and B. W. Karrh, "The Effect of Myocardial Infarction on the Cardiac Sensitization Potential of Certain Halocarbons," *J. Occup. Med.* **18**, 26 (1976).

38. L. S. Mullin, A. Azar, C. F. Reinhardt, P. E. Smith, Jr., and E. F. Fabryka, "Halogenated Hydrocarbon-Induced Cardiac Arrhythmias Associated with Release of Endogenous Epinephrine," *Am. Ind. Hyg. Assoc. J.* **33**, 389 (1972).

39. G. J. Taylor and R. T. Drew, "Cardiomyopathy Predisposes Hamsters to Trichlorofluoromethane Toxicity," *Toxicol. Appl. Pharmacol.* **32**, 177 (1975).

40. G. J. Taylor, "Cardiac Arrhythmias in Hypoxic Rabbits during Aerosol Propellant Inhalation," *Arch. Environ. Health* **30**, 349 (1975).

41. L. J. Jenkins, Jr., R. A. Jones, R. A. Coon, and J. Siegel, "Repeated and Continuous Exposures of Laboratory Animals to Trichlorofluoromethane," *Toxicol. Appl. Pharmacol.* **16**, 133 (1970).

42. S. Nick, unpublished data, Haskell Laboratory, E. I. du Pont de Nemours & Company, Wilmington, Del., June 1964.

43. H. Sherman, unpublished data, Haskell Laboratory, E. I. du Pont de Nemours & Company, Wilmington, Del., February 1972.

44. A. Quevauviller, "Hygiène et sécurité des pulseurs pour aérosols médicamenteux," *Prod. Probl. Pharm.* **20**, 14 (1965).

45. A. Koops, unpublished data (FC-11), Haskell Laboratory, E. I. du Pont de Nemours & Company, Wilmington, Del., August 1977.

46. G. Paulet, S. Desbrousses, and E. Vidal, "Absence d'effet tératogène des fluorocarbones chez le rat et le lapin," *Arch. Mal. Prof. Med. Trav. Secur. Soc.* **35**, 658 (1974).

47. D. M. Aviado, "Toxicity of Aerosol Propellants in the Respiratory and Circulatory Systems. IX. Summary of the Most Toxic: Trichlorofluoromethane (FC 11)," *Toxicology* **3**, 311 (1975).

48. W. M. Nicholas, "Are Freon Propellants Inert?" *N.Y. State J. Med.* **11**, 1939 (1974).

49. F. A. Charlesworth, "Cardiac Responses to Fluorocarbon Propellants," *J. Cosmet. Toxicol.* **12**, 552 (1974).

50. N. C. Flowers, R. C. Hand, and L. G. Horan, "Concentrations of Fluoroalkanes Associated with Cardiac Conduction System Toxicity," *Arch. Environ. Health* **30**, 353 (1975).

51. C. T. Dollery, F. M. Williams, G. H. Draffan, G. Wise, H. Sahyoun, J. W. Paterson, and S. R. Walker, "Arterial Blood Levels of Fluorocarbons in Asthmatic Patients Following Use of Pressurized Aerosols," *Clin. Pharmacol. Ther.* **15**, 59 (1974).

52. G. Marier, H. MacFarland, G. S. Wiberg, H. Buchwald, and P. Dussault, "Blood Fluorocarbon Levels Following Exposure to Household Aerosols," *CMA J.* **111**, 39 (1974).
53. J. Adir, D. A. Blake, and G. M. Mergner, "Pharmacokinetics of Fluorocarbons 11 and 12 in Dogs and Humans," *J. Clin. Pharmacol.* **15**, 760 (1975).
54. G. W. Mergner, D. A. Blake, and M. Helrich, "Biotransformation and Elimination of ^{14}C-Trichlorofluoromethane (FC-11) and ^{14}C-Dichlorodifluoromethane (FC-12) in Man," *Anesthesiology* **42**, 345 (1975).
55. D. A. Blake and G. W. Mergner, "Inhalation Studies on the Biotransformation and Elimination of ^{14}C-Trichlorofluoromethane and ^{14}C-Dichlorodifluoromethane in Beagles," *Toxicol. Appl. Pharmacol.* **30**, 396 (1974).
56. G. Paulet, J. Lanoe, A. Thos, P. Toulouse, and J. Dassonville, "Fate of Fluorocarbons in the Dog and Rabbit after Inhalation," *Toxicol. Appl. Pharmacol.* **34**, 204 (1975).
57. A. Morgan, A. Black, M. Walsh, and D. R. Belcher, "The Absorption and Retention of Inhaled Fluorinated Hydrocarbon Vapours," *Int. J. Appl. Radiat. Isotop.* **23**, 285 (1972).
58. V. A. Shugaev, "Toxicity of Freon-12," *Gigiena Sanit.* **28**, 95 (1963). *Chemical Abstracts* **59**:9230e.
59. A. Quevauviller, M. Chaigneau, and M. Schrenzel, "Étude expérimentale chez la Souris de la tolérance du poumon aux hydrocarbures chlorofluorés," *Ann. Pharm. Fr.* **21**, 727 (1953).
60. D. B. Hood, unpublished data, Haskell Laboratory, E. I. du Pont de Nemours & Company, Wilmington, Del., June 1961.
61. J. A. Prendergast, R. A. Jones, L. J. Jenkins, Jr., and J. Siegel, "Effects on Experimental Animals of Long-Term Inhalation of Trichloroethylene, Carbon Tetrachloride, 1,1,1-Trichloroethane, Dichlorodifluoromethane, and 1,1-Dichloroethylene," *Toxicol. Appl. Pharmacol.* **10**, 270 (1967).
62. R. R. Sayers, W. P. Yant, J. Chornyak, and H. W. Shoaf, "Toxicity of Dichlorodifluoro Methane: A New Refrigerant," Department of Commerce, United States Bureau of Mines, R.I. 3013, May 1930.
63. H. Sherman and J. R. Barnes, unpublished data, Haskell Laboratory, E. I. du Pont de Nemours & Company, Wilmington, Del., January 1966.
64. H. Sherman and J. R. Barnes, unpublished data, Haskell Laboratory, E. I. du Pont de Nemours & Company, Wilmington, Del., January 1974.
65. A. Quevauviller, M. Schrenzel, and V. N. Nuyen, "Tolérance locale (peau, muqueuses, plaies, brûlures) chez l'animal, aux hydrocarbures chlorofluorés," *Therapie* **19**, 247 (1964).
66. A. Koops, unpublished data (FC-12), Haskell Laboratory, E. I. du Pont de Nemours & Company, Wilmington, Del., August 1977.
67. R. Culik and H. Sherman, unpublished data, Haskell Laboratory, E. I. du Pont de Nemours & Company, Wilmington, Del., May 1973.
68. B. S. Brar, B. A. Jackson, C. E. Traitor, D. E. Rodwell, C. R. Boshart, and J. F. Noble, "Triamcinolone Acetonide (1,4-pregnadiene-3,20-dione, 9α-fluoro-11β, 21 dihydroxy 16α,17α-(isopropylidenedioxy)): Aerosol Inhalation and Teratology Studies in Rabbits," Paper presented at the 15th Annual Meeting of the Society of Toxicology; abstract published *Toxicol. Appl. Pharmacol.* **37**, 151 (1976).
69. A. Azar, C. F. Reinhardt, M. E. Maxfield, P. E. Smith, Jr., and L. S. Mullin, "Experimental Human Exposures to Fluorocarbon 12 (Dichlorodifluoromethane)," *Am. Ind. Hyg. Assoc. J.* **33**, 207 (1972).
70. J. R. Barnes and H. Sherman, unpublished data, Haskell Laboratory, E. I. du Pont de Nemours & Company, Wilmington, Del., September 1966.

71. F. D. Griffith, unpublished data, Haskell Laboratory, E. I. du Pont de Nemours & Company, Wilmington, Del., December 1969.
72. World Health Organization, "Toxicological Evaluation of Certain Food Additives; Some Food Colours, Thickening Agents, and Certain Other Substances," World Health Organization Technical Report Series No. 576, Geneva (1975).
73. J. W. Hiddemen, III, and R. S. Waritz, unpublished data, Haskell Laboratory, E. I. du Pont de Nemours & Company, Wilmington, Del., June 1969.
74. O. L. Dashiell, unpublished data, Haskell Laboratory, E. I. du Pont de Nemours & Company, Wilmington, Del., February 1971.
75. J. W. Sarver, unpublished data, Haskell Laboratory, E. I. du Pont de Nemours & Company, Wilmington, Del., April 1971.
76. T. K. Bogdanowicz, unpublished data, Haskell Laboratory, E. I. du Pont de Nemours & Company, Wilmington, Del., March 1973.
77. K. J. Leong, unpublished data, Hazleton Laboratories, Inc., Falls Church, Va., June 1967 (sponsored by Haskell Laboratory).
78. American Conference of Governmental Industrial Hygienists, "1,1,2-Trichloro, 1,2,2-Trifluoroethane (TCTFE, Freon® 113)," in "Documentation of the Threshold Limit Values," 3rd ed., 2nd printing, Cincinnati, Ohio, 1974.
79. American National Standards Institute, "American National Standard: Acceptable Concentrations of 1,1,2-Trichloro-1,2,2-trifluoroethane (Fluorocarbon 113)," Z37.35, New York, 1973.
80. J. B. Michaelson and D. J. Huntsman, "Oral Toxicity Study of 1,2,2-Trichloro-1,1,2-trifluoroethane," *J. Med. Chem.* 7, 378 (1964).
81. P. Duprat, L. Delsaut, and D. Gradiski, "Pouvoir irritant des principaux solvants chlores aliphatiques sur la peau et les muqueuses oculaires du lapin," *Eur. J. Toxicol.* 9, 171 (1976).
82. R. E. Reinke, unpublished data, Haskell Laboratory, E. I. du Pont de Nemours & Company, Wilmington, Del., June 1962.
83. E. J. Smith, unpublished data, Haskell Laboratory, E. I. du Pont de Nemours & Company, Wilmington, Del., September 1977.
84. B. H. Sleight, III, unpublished data, Bionomics, Inc., Wareham, Mass., October 1971 (sponsored by Haskell Laboratory).
85. C. F. Reinhardt, L. S. Mullin, and M. E. Maxfield, "Epinephrine-Induced Cardiac Arrhythmia Potential of Some Common Industrial Solvents," *J. Occup. Med.* 15, 953 (1973).
86. M. Steinberg, R. E. Boldt, R. A. Renne, and M. H. Weeks, "Inhalation Toxicity of 1,1,2-Trichloro-1,2,2-trifluoroethane (TCTFE)," Edgewood Arsenal, Md., January 1969 (AD 854705).
87. V. L. Carter, Jr., P. M. Chikos, J. D. MacEwen, and K. C. Back, "Effects of Inhalation of Freon 113 on Laboratory Animals," Wright-Patterson Air Force Base, Ohio, December 1970 (AD 727524).
88. H. Desoille, L. Truffert, A. Bourguignon, P. Delavierre, M. Philbert, and C. Girard-Wallon, "Étude expérimentale de la toxicité du trichlorotrifluoroéthane (fréon 113)," *Arch. Mal. Prof. Med. Trav. Secur. Soc.* 29, 381 (1968).
89. J. F. Russell, Jr., unpublished data, Haskell Laboratory, E. I. du Pont de Nemours & Company, Wilmington, Del., November 1977.
90. S. S. Epstein, E. Arnold, J. Andrea, W. Bass, and Y. Bishop, "Detection of Chemical Mutagens by the Dominant Lethal Assay in the Mouse," *Toxicol. Appl. Pharmacol.* 23, 288 (1972).
91. W. M. Busey, unpublished data, Hazleton Laboratories, Inc., Falls Church, Va., February 1967 (sponsored by Haskell Laboratory).

92. W. M. Busey, unpublished data, Hazleton Laboratories, Inc., Falls Church, Va., December 1967 (sponsored by Haskell Laboratory).
93. H. J. Trochimowicz, A. Azar, J. B. Terrill, and L. S. Mullin, "Blood Levels of Fluorocarbon Related to Cardiac Sensitization. Part II," *Am. Ind. Hyg. Assoc. J.* **35**, 632 (1974).
94. G. J. Stopps and M. McLaughlin, "Psychophysiological Testing of Human Subjects Exposed to Solvent Vapors," *Am. Ind. Hyg. Assoc. J.* **28**, 43 (1967).
95. C. F. Reinhardt, M. McLaughlin, M. E. Maxfield, L. S. Mullin, and P. E. Smith, Jr., "Human Exposures to Fluorocarbon 113," *Am. Ind. Hyg. Assoc. J.* **32**, 143 (1971).
96. H. R. Imbus and C. Adkins, "Physical Examinations of Workers Exposed to Trichlorotrifluoroethane," *Arch. Environ. Health* **24**, 257 (1972).
97. B. L. Odou, personal communication (1963), as cited in J. W. Clayton, Jr., "The Mammalian Toxicology of Organic Compounds Containing Fluorine," *Handbuch der experimentellen Pharmakologie* **XX/1**,459 (1966).
98. C. F. Reinhardt and L. R. Schultze, unpublished data, Haskell Laboratory, E. I. du Pont de Nemours & Company, Wilmington, Del., April 1968.
99. M. E. Maxfield, A. Azar, and L. S. Mullin, unpublished data, Haskell Laboratory, E. I. du Pont de Nemours & Company, Wilmington, Del., April 1971.
100. S. S. Epstein, S. Joshi, J. Andrea, P. Clapp, H. Falk, and N. Mantel, "Synergistic Toxicity and Carcinogenicity of 'Freons' and Piperonyl Butoxide," *Nature* **214**, 526 (1967).
101. P. H. Howard, P. R. Durkin, and A. Hanchett, "Environmental Hazard Assessment of One and Two Carbon Fluorocarbons," Syracuse University, Syracuse, N.Y., September 1975 (PB 246419).
102. G. Paulet and S. Desbrousses, "Le dichlorotétrafluoroéthane," *Arch. Mal. Prof. Med. Trav. Secur. Soc.* **30**, 477 (1969).
103. T. Watanabe and D. M. Aviado, "Subacute Inhalational Toxicity of Aerosol Propellants," *Pharmacologist* **17**, 192 (1975).
104. H. Desoille, L. Truffert, C. Girard-Wallon, J. Ripault, and M. Philbert, "Recherche expérimentale de la toxicité chronique éventuelle du dichlorotétrafluoroéthane," *Arch. Mal. Prof. Med. Trav. Secur. Soc.* **34**, 117 (1973).
105. J. W. Clayton, Jr., "The Mammalian Toxicology of Organic Compounds Containing Fluorine," *Handbuch der experimentellen Pharmakologie* **XX/1**,459 (1966).
106. J. W. Clayton, Jr., "Fluorocarbon Toxicity and Biological Action," *Fluorine Chem. Rev.* **1**, 197 (1967).
107. F. D. Griffith and H. Sherman, unpublished data, Haskell Laboratory, E. I. du Pont de Nemours & Company, Wilmington, Del., December 1969.
108. H. Sherman, unpublished data, Haskell Laboratory, E. I. du Pont de Nemours & Company, Wilmington, Del., January 1972.
109. G. Paulet, R. Chevrier, J. Paulet, M. Duchêne, and J. Chappet, "De la rétention des fréons par les poumons et les voies aériennes," *Arch. Mal. Prof. Med. Trav. Secur. Soc.* **30**, 101 (1969).
110. G. Paulet, "Action of fluoro chloro hydrocarbons used in aerosols. Problems of their retention by the organism after inhalation," *Trib. CEBEDEAU (Cent. Belge Etude Doc. Eaux)* **23**, 324 (1970); as cited in *Chemical Abstracts* 94:97243v.
111. C. T. Dollery, D. S. Davies, G. H. Draffan, F. M. Williams, and M. E. Conolly, "Blood Concentrations in Man of Fluorinated Hydrocarbons after Inhalation of Pressurized Aerosols," *Lancet* **II**, 1164 (1970).
112. J. W. Clayton, Jr., D. B. Hood, M. S. Nick, and R. S. Waritz, "Inhalation Studies on Chloropentafluoroethane," *Am. Ind. Hyg. Assoc. J.* **27**, 234 (1966).

113. W. Weigand, "Untersuchungen uber die Inhalationstoxizität von Fluorderivaten des Methan, Athan und Cyclobutan," *Zentralbl. Arbeitsmed. Arbeitsschutz* **2**, 149 (1971).

114. J. W. Clayton, Jr., "Fluorcarbon Toxicity: Past, Present, Future," *J. Soc. Cosmet. Chem.* **18**, 333 (1967).

115. D. M. Aviado and J. Drimal, "Five Fluorocarbons for Administration of Aerosol Bronchodilators," *J. Clin. Pharmacol.* **15**, 116 (1975).

116. D. M. Aviado, "Toxicity of Aerosols," *J. Clin. Pharmacol.* **15**, 86 (1975).

117. J. B. Terrill, "Arterial Venous Blood Levels of Chloropentafluoroethane: Inhalation versus Oral Exposures," *Am. Ind. Hyg. Assoc. J.* **35**, 269 (1974).

118. J. W. Clayton, Jr., M. A. Delaplane, and D. B. Hood, "Toxicity Studies with Octafluorocyclobutane," *Am. Ind. Hyg. Assoc. J.* **21**, 382 (1960).

119. T. R. Torkelson, Dow Chemical Company, Personal communication to R. R. Montgomery, September, 1978.

120. C. H. Tappan and R. S. Waritz, unpublished data, Haskell Laboratory, E. I. du Pont de Nemours & Company, Wilmington, Del., November 1964.

121. D. B. Hood, unpublished data, Haskell Laboratory, E. I. du Pont de Nemours & Company, Wilmington, Del., August 1964.

122. N. C. Goodman, unpublished data, Haskell Laboratory, E. I. du Pont de Nemours & Company, Wilmington, Del., January 1976.

123. M. R. Brittelli, unpublished data, Haskell Laboratory, E. I. du Pont de Nemours & Company, Wilmington, Del., January 1976.

124. L. S. Mullin, unpublished data, Haskell Laboratory, E. I. du Pont de Nemours & Company, Wilmington, Del., November 1975.

125. H. J. Trochimowicz, B. L. Moore, and T. Chiu, "Subacute Inhalation Toxicity Studies on Eight Fluorocarbons," presented at the 16th Annual Meeting of the Society of Toxicology; abstract published *Toxicol. Appl. Pharmacol.* **41**, 198 (1977).

126. H. J. Trochimowicz, J. P. Lyon, D. P. Kelly, and T. Chiu, "Ninety-Day Inhalation Toxicity Studies on Two Fluorocarbons," presented at the 16th Annual Meeting of the Society of Toxicology; abstract published *Toxicol. Appl. Pharmacol.* **41**, 200 (1977).

127. J. K. Smith and M. T. Case, "Subacute and Chronic Toxicity Studies of Fluorocarbon Propellants in Mice, Rats and Dogs," *Toxicol. Appl. Pharmacol.* **26**, 438 (1973).

128. R. Culik and D. P. Kelly, unpublished data, Haskell Laboratory, E. I. du Pont de Nemours & Company, Wilmington, Del., April 1976.

129. O. W. Van Auken, A. O. Henderson, R. T. S. Lee, R. H. Wilson, and J. N. Bollinger, "Functional and Structural Changes in Isolated Rabbit Liver Mitochondria Induced by Fluorodichloromethane," *J. Pharmacol. Exp. Ther.* **193**, 729 (1975).

130. M. A. Belej and D. M. Aviado, "Cardiopulmonary Toxicity of Propellants for Aerosols," *J. Clin. Pharmacol.* **15**, 105 (1975).

131. B. C. Karpov, "Material on Toxicology of Chronic Effect of Freon-22," *Trans. Leningr. Sanit.-Gig. Med. Inst.* **75**, 241 (1963); as translated in Joint Publications Research Service 28, 721, TT65-30312, Washington, D.C., 1965, p. 118. *Chemical Abstracts* **61**:3601a (B. D. Karpov).

132. R. E. Dufour and A. J. Perkins, "The Comparative Life, Fire, and Explosion Hazards of Difluoromonochloromethane ('Freon-22')," Miscellaneous Hazard No. 3134, Underwriters Laboratories, Chicago, Ill., 1940.

133. A. Van Poznak and J. F. Artusio, Jr., "Anesthetic Properties of a Series of Fluorinated Compounds. I. Fluorinated Hydrocarbons," *Toxicol. Appl. Pharmacol.* **2**, 363 (1960).

134. American Conference of Governmental Industrial Hygienists, "Chlorodifluoro-

methane," in "Documentation of Threshold Limit Values," 3rd ed., 2nd printing, Cincinnati, Ohio, 1974.

135. A. Koops, unpublished data (FC-22), Haskell Laboratory, E. I. du Pont de Nemours & Company, Wilmington, Del., August 1977.

136. R. Culik and C. D. Crowe, unpublished data, Haskell Laboratory, E. I. du Pont de Nemours & Company, Wilmington, Del., June 1978.

137. C. F. Reinhardt, Letter to S. D. Jellinek, Environmental Protection Agency, August 1978.

138. D. P. Kelly, unpublished data, Haskell Laboratory, E. I. du Pont de Nemours & Company, Wilmington, Del., October 1974.

139. A. L. Henne, "Fluorinated Derivatives of Methane," *J. Am. Chem. Soc.* **59**, 1400 (1937).

140. L. S. Mullin and H. J. Trochimowicz, unpublished data, Haskell Laboratory, E. I. du Pont de Nemours & Company, Wilmington, Del., April 1973.

141. F. C. Barsky, unpublished data, Haskell Laboratory, E. I. du Pont de Nemours & Company, Wilmington, Del., August 1976.

142. D. F. Krahn, Haskell Laboratory, Personal communication to R. R. Montgomery, August 1977.

143. B. L. Moore, unpublished data, Haskell Laboratory, E. I. du Pont de Nemours & Company, Wilmington, Del., November 1975.

144. L. S. Mullin and H. J. Trochimowicz, unpublished data, Haskell Laboratory, E. I. du Pont de Nemours & Company, Wilmington, Del., June 1973.

145. B. L. Moore and G. T. Hall, unpublished data, Haskell Laboratory, E. I. du Pont de Nemours & Company, Wilmington, Del., July 1975.

146. L. S. Mullin, unpublished data, Haskell Laboratory, E. I. du Pont de Nemours & Company, Wilmington, Del., December 1976.

147. B. H. Robbins, "Preliminary Studies of the Anesthetic Activity of Fluorinated Hydrocarbons," *J. Pharmacol. Exp. Ther.* **86**, 197 (1946).

148. J. Raventós and P. G. Lemon, "The Impurities in Fluothane: Their Biological Properties," *Brit. J. Anaesth.* **37**, 716 (1965).

149. T. H. S. Burns, J. M. Hall, A. Bracken, and G. Gouldstone, "Fluorine Compounds in Anaesthesia (5). Examination of Six Heavily Halogenated Aliphatic Compounds," *Anaesthesia* **17**, 337 (1962).

150. J. E. Henry, unpublished data, Haskell Laboratory, E. I. du Pont de Nemours & Company, Wilmington, Del., October 1975.

151. N. C. Goodman, unpublished data, Haskell Laboratory, E. I. du Pont de Nemours & Company, Wilmington, Del., December 1975.

152. M. R. Brittelli, unpublished data, Haskell Laboratory, E. I. du Pont de Nemours & Company, Wilmington, Del., January 1976.

153. H. J. Trochimowicz and L. S. Mullin, unpublished data, Haskell Laboratory, E. I. du Pont de Nemours & Company, Wilmington, Del., March 1973.

154. D. P. Kelly, unpublished data, Haskell Laboratory, E. I. du Pont de Nemours & Company, Wilmington, Del., March 1976.

155. R. E. Dufour and A. J. Perkins, "The Comparative Life, Fire, and Explosion Hazards of Tetrafluoromonochloroethane ('Freon-124')," Miscellaneous Hazard No. 3135, Underwriters Laboratories, Chicago, Ill., 1940.

156. L. S. Mullin, unpublished data, Haskell Laboratory, E. I. du Pont de Nemours & Company, Wilmington, Del., March 1976.

157. G. T. Hall, unpublished data, Haskell Laboratory, E. I. du Pont de Nemours & Company, Wilmington, Del., November 1976.

158. S. Nick, unpublished data, Haskell Laboratory, E. I. du Pont de Nemours & Company, Wilmington, Del., May 1964.

159. T. R. Torkelson, C. D. Kary, M. B. Chenoweth, and E. R. Larsen, "Single Exposure of Rats to the Vapors of Trace Substances in Methoxyflurane," *Toxicol. Appl. Pharmacol.* **19**, 1 (1971).

160. J. E. Henry, unpublished data, Haskell Laboratory, E. I. du Pont de Nemours & Company, Wilmington, Del., November 1975.

161. N. C. Goodman, unpublished data, Haskell Laboratory, E. I. du Pont de Nemours & Company, Wilmington, Del., January 1976.

162. M. R. Brittelli, unpublished data, Haskell Laboratory, E. I. du Pont de Nemours & Company, Wilmington, Del., January 1976.

163. E. J. Smith, unpublished data, Haskell Laboratory, E. I. du Pont de Nemours & Company, Wilmington, Del., September 1977.

164. L. S. Mullin, unpublished data, Haskell Laboratory, E. I. du Pont de Nemours & Company, Wilmington, Del., February 1976.

165. J. F. Russell, Jr., unpublished data, Haskell Laboratory, E. I. du Pont de Nemours & Company, Wilmington, Del., February 1976.

166. M. Shulman and M. S. Sadove, "Safety Evaluation and Anesthetic Properties of 1,1,1-Trifluoroethyl Chloride," *Toxicol. Appl. Pharmacol.* **7**, 473 (1965).

167. A. Koops, unpublished data, Haskell Laboratory, E. I. du Pont de Nemours & Company, Wilmington, Del., September 1977.

168. R. Culik, Haskell Laboratory, Personal communication to R. R. Montgomery, September 1978.

169. M. Shulman, "Anesthesia Method and Compositions Using 1,1,1-Trifluoroethyl Chloride," U.S. Patent 3,325,352 (Cl. 167-52), June 13, 1967, appl. March 19, 1965, as cited in *Chemical Abstracts* **67**:84848b.

170. S. B. Rissolo, unpublished data, Haskell Laboratory, E. I. du Pont de Nemours & Company, Wilmington, Del., November 1967.

171. C. Doleba-Crowe, unpublished data, Haskell Laboratory, E. I. du Pont de Nemours & Company, Wilmington, Del., February 1977.

172. C. Doleba-Crowe, unpublished data, Haskell Laboratory, E. I. du Pont de Nemours & Company, Wilmington, Del., September 1977.

173. B. C. Karpov, "Lethal and Threshold Concentrations of Freons," *Trans. Leningr. Sanit.-Gig. Med. Inst.* **75**, 241 (1963); as translated in Joint Publications Research Service 28, 721, TT65-30312, Washington, D.C., 1965, p. 128. See Reference 131.

174. T. K. Nikitenko and M. S. Tolgskaya. "Toxico-Pathomorphological Alterations Appearing in Animals under the Effect of Freons 141, 142, and 143," *Gigiena Truda Prof. Zabol.* **9** (10), 37 (1965); as cited in *Chemical Abstracts* **64**:7260h (excerpt translated by R. Culik, Haskell Laboratory).

175. T. K. Nikitenko, "Toxicity of Dichlorofluoroethane (F-141) in a Prolonged Exposure of Test Animals," *Toksikol. Novykh. Prom. Khim. Veshchestv*, No. 8, 83 (1966); as cited in *Chemical Abstracts* **67**:93737s.

176. T. K. Nikitenko and T. A. Kochetkova, "Effect of Some Fluorochlorohydrocarbons on the Thyroid (Experimental Study)," *Toksikol. Novykh. Prom. Khim. Veshchestv*, No. 9, 175 (1967); as cited in *Chemical Abstracts* **70**:6416g.

177. C. P. Carpenter, H. F. Smyth, Jr., and U. C. Pozzani, "The Assay of Acute Vapor Toxicity, and the Grading and Interpretation of Results on 96 Chemical Compounds," *J. Ind. Hyg. Toxicol.* **31**, 343 (1949).

178. M. R. Brittelli, unpublished data, Haskell Laboratory, E. I. du Pont de Nemours & Company, Wilmington, Del., January 1976.

179. B. L. Moore, unpublished data, Haskell Laboratory, E. I. du Pont de Nemours & Company, Wilmington, Del., February 1976.
180. D. P. Kelly, unpublished data, Haskell Laboratory, E. I. du Pont de Nemours & Company, Wilmington, Del., June 1976.
181. A. Koops, unpublished data (FC-142b), Haskell Laboratory, E. I. du Pont de Nemours & Company, Wilmington, Del., August 1977.
182. R. Culik and D. P. Kelly, unpublished data, Haskell Laboratory, E. I. du Pont de Nemours & Company, Wilmington, Del., September 1976.
183. B. L. Moore, unpublished data, Haskell Laboratory, E. I. du Pont de Nemours & Company, Wilmington, Del., November 1975.
184. B. L. Moore, unpublished data, Haskell Laboratory, E. I. du Pont de Nemours & Company, Wilmington, Del., March 1976.
185. A. Koops, unpublished data (FC-152a), Haskell Laboratory, E. I. du Pont de Nemours & Company, Wilmington, Del., September 1977.
186. J. F. Russell, Jr., unpublished data, Haskell Laboratory, E. I. du Pont de Nemours & Company, Wilmington, Del., January 1978.
187. J. F. Russell, Jr., unpublished data, Haskell Laboratory, E. I. du Pont de Nemours & Company, Wilmington, Del., March 1978.
188. L. S. Mullin, unpublished data, Haskell Laboratory, E. I. du Pont de Nemours & Company, Wilmington, Del., November 1977.
189. J. F. Russell, Jr., unpublished data, Haskell Laboratory, E. I. du Pont de Nemours & Company, Wilmington, Del., December 1977.

SUPPLEMENTAL REFERENCES

Alarie, Y., Barrow, C., Choby, M. A., and Quealy, J. F., "Pulmonary Atelectasis Following Administration of Halogenated Hydrocarbons," *Toxicol. Appl. Pharmacol.* **31**, 233 (1975).

American National Standards Institute, "American National Standard: Acceptable Concentrations of Trichlorofluoromethane (Fluorocarbon 11)," Z37.36, New York, 1973.

Aviado, D. M., and Belej, M. A., "Toxicity of Aerosol Propellants on the Respiratory and Circulatory Systems. I. Cardiac Arrhythmia in the Mouse," *Toxicology* **2**, 31 (1974).

Aviado, D. M., and Belej, M. A., "Toxicity of Aerosol Propellants in the Respiratory and Circulatory Systems. V. Ventricular Function in the Dog," *Toxicology* **3**, 79 (1975).

Aviado, D. M., and Smith, D. G., "Toxicity of Aerosol Propellants in the Respiratory and Circulatory Systems. VIII. Respiration and Circulation in Primates," *Toxicology* **3**, 241 (1975).

Azar, A., "Cardiovascular Effects of Fluorocarbon Exposure," *Proc. 2nd Ann. Conf. Environ. Toxicol.*, Wright-Patterson Air Force Base, Ohio, AMRL-TR-71-120, 41 (1971).

Azar, A., Zapp, Jr., J. A., Reinhardt, C. F., and Stopps, G. J., "Cardiac Toxicity of Aerosol Propellants," *J. Am. Med. Assoc.* **215**, 1501 (1971).

Banks, A. A., Campbell, A., and Rudge, A. J., "Toxicity and Narcotic Activity of Fluorocarbons," *Nature* **174**, 885 (1954).

Baselt, R. C., and Cravey, R. H., "A Fatal Case Involving Trichloromonofluoromethane and Dichlorodifluoromethane," *J. Foren. Sci.* **13**, 407 (1968).

Belej, M. A., and Aviado, D. M., "Acute Fluorocarbon Toxicity in Rhesus Monkey," *Pharmacology* 3359:814Abs.

Belej, M. A., Smith, D. G., and Aviado, D. M., "Toxicity of Aerosol Propellants in the Respiratory and Circulatory Systems. IV. Cardiotoxicity in the Monkey," *Toxicology* **2**, 381 (1974).

Bower, F. A., "Propellant Safety," *Aerosol Age* **17**, (12), 28 (1972).

Bower, F. A., "Are Fluorocarbons Really Hazardous?" *Aerosol Age* **18** (11), 22 (1973).

Brenner, C., "Note on the Action of Dichloro-Difluoromethane on the Nervous System of the Cat," *J. Pharmacol. Exp. Ther.* **59**, 176 (1937).

Brody, R. S., Watanabe, T., and Aviado, D. M., "Toxicity of Aerosol Propellants on the Respiratory and Circulatory Systems. III. Influence of Bronchopulmonary Lesion on Cardiopulmonary Toxicity in the Mouse," *Toxicology* **2**, 173 (1974).

Brubaker, S., Peyman, G. A., and Vygantas, C., "Toxicity of Octafluorocyclobutane after Intracameral Injection," *Arch. Ophthalmol.* **92**, 324 (1974).

Caujolle, F., *Bull. Inst. Int. Froid*, No. 1, 1 (1964); as cited in Quevauviller, Reference 44.

Chenoweth, M. B., "Ventricular Fibrillation Induced by Hydrocarbons and Epinephrine," *J. Ind. Hyg. Toxicol.* **28**, 151 (1946).

Chiou, W. L., "Aerosol Propellants: Cardiac Toxicity and Long Biological Half-Life," *J. Am. Med. Assoc.* **227**, 658 (1974).

Chiou, W. L., and Hsiao, J.-H., "Fluorocarbon Aerosol Propellants. V: Binding Interaction with Human Albumin," *J. Pharm. Sci.* **64**, 1052 (1975).

Clark, D. G., and Tinston, D. J., "The Influence of Fluorocarbon Propellants on the Arrhythmogenic Activities of Adrenaline and Isoprenaline," *Proc. Eur. Soc. Study Drug Toxicity* **13**, 212 (1972).

Clark, D. G., and Tinston, D. J., "Correlation of the Cardiac Sensitizing Potential of Halogenated Hydrocarbons with Their Physicochemical Properties," *Brit. J. Pharmacol.* **49**, 355 (1973).

Clayton, Jr., J. W., "Fluorocarbon Toxicity and Biological Action," in "Toxicity of Anesthetics," Williams and Wilkins Company, Baltimore, Md., 1968.

Clayton, Jr., J. W., "The Toxicity of Fluorocarbons with Special Reference to Chemical Constitution," *J. Occup. Med.* **4**, 262 (1962).

Conkle, J. P., Camp, B. J., and Welch, B. E., "Trace Composition of Human Respiratory Gas," *Arch. Environ. Health* **30**, 290 (1975).

Constable, I. J., and Swann, D. A., "Vitreous Substitution with Gases," *Arch. Ophthalmol.* **93**, 416 (1975).

Cox, P. J., King, L. J., and Parke, D. V., "The Binding of Trichlorofluoromethane and Other Haloalkanes to Cytochrome P-450 under Aerobic and Anaerobic Conditions," *Xenobiotica* **6**, 363 (1976).

DiPaolo, T., and Sandorfy, C., "Hydrogen Bond Breaking Potency of Fluorocarbon Anesthetics," *J. Med. Chem.* **17**, 809 (1974).

Doherty, R. E., and Aviado, D. M., "Toxicity of Aerosol Propellants in the Respiratory and Circulatory Systems. VI. Influence of Cardiac and Pulmonary Vascular Lesions in the Rat," *Toxicology* **3**, 213 (1975).

Dollery, C. T., "Aerosol Inhalers–How Safe Are They?" *Consultant*, No. 5, 85 (1972).

Downing, R. C., and Madinabeitia, D., "The Toxicity of Fluorinated Hydrocarbon Aerosol Propellants," *Aerosol Age* **5** (9), 25 (1960).

E. I. du Pont de Nemours and Company, "'Freon' Fluorocarbons. Properties and Applications," Freon Product Information G-1, Wilmington, Del., June, 1975.

E. I. du Pont de Nemours and Company, "Toxicity Studies with 1,1,2-Trichloro-1,2,2-trifluoroethane," Freon Technical Bulletin S-24, Wilmington, Del., December, 1967.

Epstein, S. S., Andrea, J., Clapp, P., and Mackintosh, D., "Enhancement by Piperonyl Butoxide of Acute Toxicity Due to Freons, Benzo(α)pyrene, and Griseofulvin in Infant Mice," *Toxicol. Appl. Pharmacol.* **11**, 442 (1967).

Fabel, H., Wettengel, R., and Hartmann, W., "Myokardischämie und Arrhythmien durch den Gebrauch von Dosieraerosolen beim Menschen?" *Dtsch. Med. Wschr.* **97**, 428 (1972).

Flowers, N. C., and Horan, L. G., "Nonanoxic Aerosol Arrhythmias," *J. Am. Med. Assoc.* **219**, 33 (1972).

Foltz, V. C., and Fuerst, R., "Mutation Studies with *Drosophila melanogaster* Exposed to Four Fluorinated Hydrocarbon Gases," *Environ. Res.* **7**, 275 (1974).

Fraser, P., and Doll, R., "Geographical Variations in the Epidemic of Asthma Deaths," *Brit. J. Prev. Soc. Med.* **25**, 34 (1971).

Fraser, P. M., Speizer, F. E., Waters, S. D. M., Doll, R., and Mann, N. M., "The Circumstances Preceding Death from Asthma in Young People in 1968 to 1969," *Brit. J. Dis. Chest*, **65**, 71 (1971).

Friedman, S. A., Cammarato, M., and Aviado, D. M., "Toxicity of Aerosol Propellants on the Respiratory and Circulatory Systems. II. Respiratory and Bronchopulmonary Effects in the Rat," *Toxicology* **1**, 345 (1973).

Fuerst, R., and Landry, M. M., "Mutation Studies of Gas Treated *Escherichia coli* Cells," *Proc. Int. Congr. Genet.* **12**, 82 (1968).

Fuerst, R., and Stephens, S., "Studies of Effects of Gases and Gamma Irradiation on *Neurospora crassa*," *Dev. Ind. Microbiol.* **11**, 301 (1969).

Gandevia, B., "The Changing Pattern of Mortality from Asthma in Australia," *Med. J. Aust.* **1**, 747 (1968).

Gandevia, B., "Pressurized Sympathomimetic Aerosols and Their Lack of Relationship to Asthma Mortality in Australia," *Med. J. Aust.* **1**, 273 (1973).

Gandevia, B., "The Changing Pattern of Mortality from Asthma in Australia. 2. Mortality and Modern Therapy," *Med. J. Aust.* **1**, 884 (1968).

Garrett, S., and Fuerst, R., "Sex-Linked Mutations in *Drosophila* after Exposure to Various Mixtures of Gas Atmospheres," *Environ. Res.* **7**, 286 (1974).

Harris, M. C., "Are Bronchodilator Aerosol Inhalations Responsible for an Increase in Asthma Mortality?" *Ann. Allergy* **29**, 250 (1971).

Healy, J., Lee, R. T. S., Perrotta, D., and Van Auken, O. W., "Effects of Fluorocarbons on Biological Systems," *Tex. J. Sci.* **25**, 1 (1974).

Kamm, R. C., "Fatal Arrhythmia Following Deodorant Inhalation," *Foren. Sci.* **5**, 91 (1975).

Kramer, R. A., and Pierpaoli, P., "Hallucinogenic Effect of Propellant Components of Deodorant Sprays," *Pediatrics* **48**, 322 (1971).

Landry, M. M., and Fuerst, R., "Gas Ecology of Bacteria," *Dev. Ind. Microbiol.* **9**, 370 (1968).

Little, A., "Mortality from Bronchial Asthma," *Can. Med. Assoc. J.* **100**, 485 (1969).

Marier, G., MacFarland, H., and Dussault, P., "Canadians Study Fluorocarbon Levels after Exposure to Household Aerosols," *Aerosol Age* **18** (12), 30 (1973).

McClure, D. A., "Failure of Fluorocarbon Propellants to Alter the Electrocardiogram of Mice and Dogs," *Toxicol. Appl. Pharmacol.* **22**, 221 (1972).

Molk, L., and Falliers, C. J., "Isoproterenol Aerosol in Perspective," *Am. Fam. Physician* **5** (6), 88 (1972).

Morgan, A., Black, A., and Belcher, D. R., "Studies on the Absorption of Halogenated Hydrocarbons and Their Excretion in Breath Using ^{38}Cl Tracer Techniques," *Ann. Occup. Hyg.* **15**, 273 (1972).

Morgan, A., Black, A., Walsh, M., and Belcher, D. R., "The Absorption and Retention of Inhaled Fluorinated Hydrocarbon Vapours," *Mammalian Pathol. Biochem.* **77**, 343 (1972).

Muacevic, G., "New Apparatus and Method for the Toxicological Investigation of Metered Aerosols in Rats," *Arch. Toxicol.* **34**, 1 (1975).

Murphy, J. P. F., Van Stee, E. W., and Back, K. C., "Effects of Fluorocarbons on Hepatic Microsomal Enzymes," *Proc. 4th Ann. Conf. Environ. Toxicol.*, Wright-Patterson Air Force Base, Ohio, AMRL-TR-73-125, 85 (1973).

Niazi, S., and Chiou, W. L., "Partition Coefficients of Fluorocarbon Aerosol Propellants in Water, Normal Saline, Cyclohexane, Chloroform, Human Plasma, and Human Blood," *J. Pharm. Sci.* **63**, 532 (1974).

Niazi, S., and Chiou, W. L., "Fluorocarbon Aerosol Propellants. IV: Pharmacokinetics of Trichloromonofluoromethane Following Single and Multiple Dosing in Dogs," *J. Pharm. Sci.* **64**, 763 (1975).

Oujesky, H., and Bhagat, I., "Response of *Staphylococcus aureus* to Atmospheres of Freons and Genetrons," *Dev. Ind. Microbiol.* **14**, 229 (1973); as cited in *Chemical Abstracts* **82**:52092u.

Oujesky, H., and Fuerst, R., "Effects of Freons and Genetrons on DNA Polyermase from *Escherichia coli*," *Dev. Ind. Microbiol.* **15**, 405 (1974); as cited in *Chemical Abstracts* **82**:134794p.

Paulet, G., "The Fluorohydrocarbons at Issue," *Aerosol Rep.* **16**, 22–30 (1977).

Paulet, G., Desbrousses, S., and Sorais, J., "Toxicité chronique à moyen terme de deux hydrocarbures chlorofluorés: R_{11} et R_{12}," *Arch Mal. Prof. Med. Trav. Secur. Soc.* **28**, 464 (1967).

Paulet, G., and Lessard, Y., "De l'action du difluorodichlorométhane (FC 12) sur le coeur isolé de Rat et de Lapin." *C. R. Seances Soc. Biol. Ses Fil.* **169**, 1048 (1975).

Paulet, G., and Lessard, Y., "Action des fluorocarbones 12 (difluorodichlorométane) et 11 (monofluorotrichlorométane) sur la musculature lisse," *C. R. Seances Soc. Biol. Ses Fil.* **169**, 665 (1975).

Paulet, G., Rochcongard, P., and Desbrousses, S., "Axe hypophysosurrénalien et fluorocarbones," *Arch. Mal. Prof. Med. Trav. Secur. Soc.* **35**, 662 (1974).

Paulet, G., Thos, A., and Dassonville, J., "Distribution dans l'organisme de gaz étrangers inertes après leur inhalation," *J. Physiol.* (Paris) **62**, Suppl. 3, 425 (1970).

Peng, G. W., and Chiou, W. L., "Fluorocarbon Aerosol Propellants. VII: Interaction Studies with Human and Bovine Globulins Using Partition Coefficient Method," *J. Pharm. Sci.* **64**, 1577 (1975).

Peyman, G. A., Vygantas, C. M., Bennett, T. O., Vygantas, A. M., and Brubaker, S., "Octafluorocyclobutane in Vitreous and Aqueous Humor Replacement," *Arch. Ophthalmol.* **93**, 514 (1975).

Press, E., and Done, A. K., "Physiologic Effects and Community Control Measures for Intoxication from the Intentional Inhalation of Organic Solvents. I," *Pediatrics* **39**, 451 (1967).

Press, E., and Done, A. K., "Physiologic Effects and Community Control Measures for Intoxication from the Intentional Inhalation of Organic Solvents. II," *Pediatrics* **39**, 611 (1967).

Quevauviller, A., "La tolérance cutanée des chloro-fluoro-méthanes utilisés comme pulseurs en cosmétologie," *Parfum. Cosmet. Savons* **3**, 228 (1960).

Quevauviller, A., Chaigneau, M., and Schrenzel, M., "Etude expérimentale chez la Souris de la tolérance du poumon aux hydrocarbures chlorofluorés," *Ann. Pharm. Fr.* **21**, 727 (1953).

Rauws, A. G., Olling, M., and Wibowo, A. E., "The Determination of Fluorochlorocarbons in Air and Body Fluids," *J. Pharm. Pharmacol.* **25**, 718 (1973).

Read, J., "The Reported Increase in Mortality from Asthma: A Clinico-Functional Analysis," *Med. J. Aust.* **1**, 879 (1968).

Scott, A., "Report of a Survey of Deaths Due to Asthma in the Republic of Ireland," *J. Ir. Med. Assoc.* **64**, 132 (1971).

Shargel, L., and Koss, R., "Determination of Fluorinated Hydrocarbon Propellants in Blood of Dogs after Aerosol Administration," *J. Pharm. Sci.* **61**, 1445 (1972).

Silverglade, A., "Evaluation of Reports of Deaths from Asthma," *J. Asthma Res.* **8**, 95 (1971).

Silverglade, A., "Changing Mortality from Bronchial Asthma," *Aerosol Age* **17** (1), 24 (1972).

Simaan, J. A., and Aviado, D. M., "Hemodynamic Effects of Aerosol Propellants. I. Cardiac Depression in the Dog," *Toxicology* **5**, 127 (1975).

Slater, T. F., "A Note on the Relative Toxic Activities of Tetrachloromethane and Trichlorofluoro-methane on the Rat," *Biochem. Pharmacol.* **14**, 178 (1965).

Speizer, F. E., Wegman, D. H., and Ramirez, A., "Palpitation Rates Associated with Fluorocarbon Exposure in a Hospital Setting," *New Eng. J. Med.* **292**, 624 (1975).

Standefer, J. C., "Death Associated with Fluorocarbon Inhalation: Report of a Case," *J. Foren. Sci.* **20**, 548 (1975).

Stephens, S., DeSha, C., and Fuerst, R., "Phenotypic and Genetic Effects in *Neurospora crassa* Produced by Selected Gases and Gases Mixed with Oxygen," *Dev. Ind. Microbiol.* **12**, 346 (1971).

Stolley, P. D., "Asthma Mortality," *Am. Rev. Resp. Dis.* **105**, 883 (1972).

Strobach, D. R., "New Developments in Aerosol Fragrance Products," *Cosmet. Perfum.* **89**, 46 (1974).

Struck, H. C., and Plattner, E. B., "A Study of the Pharmacological Properties of Certain Saturated Fluorocarbons," *J. Pharmacol. Exp. Ther.* **68**, 217 (1940).

Taylor, IV, G. J., and Drew, R. T., "Cardiovascular Effects of Acute and Chronic Inhalations of Fluorocarbon 12 in Rabbits," *J. Pharmacol. Exp. Ther.* **192**, 129 (1975).

Taylor, IV, G. J., Harris, W. S., and Bogdonoff, M. D., "Ventricular Arrhythmias Induced in Monkeys by the Inhalation of Aerosol Propellants," *J. Clin. Invest.* **50**, 1546 (1971).

Terrill, J. B., "Determination of Fluorocarbon Propellants in Blood and Animal Tissue," *Am. Ind. Hyg. Assoc. J.* **33**, 736 (1972).

Thompson, E. B., and Harris, W. S., "Time Course of Epinephrine-Induced Arrhythmias in Cats Exposed to Dichlorodifluoromethane," *Toxicol. Appl. Pharmacol.* **29**, 242 (1974).

Trochimowicz, H. J., and Reinhardt, C. F., "Studies Clarify Potential Toxicity of Aerosol Propellants," *Du Pont Innovation* **6** (3), 12 (1975).

Uehleke, H., Greim, H., Kraemer, M., and Werner, T., "Covalent Binding of Haloalkanes to Liver Constituents, but Absence of Mutagenicity on Bacteria in a Metabolizing Test System," *Mutat. Res.* **38**, 114 (1976).

Van Auken, O. W., and Healy, J., "Comparison of the Effects of Three Fluorocarbons on Certain Bacteria," *Can. J. Microbiol.* **21**, 221 (1975).

Vasil'ev, G. A., Tiunov, L. A., and Kustov, V. V., "Resistance of Animals to the Toxic Action of Some Gases after Adaptation to Anoxia," *Probl. Kosm. Biol.* **16**, 178 1971); as cited in *Chemical Abstracts* **76**:10838n.

Vygantas, C. M., Peyman, G. A., Daily, M. J., and Ericson, E. S., "Octafluorocyclobutane and Other Gases for Vitreous Replacement," *Arch. Ophthalmol.* **90**, 235 (1973).

Waritz, R. S., "The Toxicology of Some Commercial Fluorocarbons," *Proc. 2nd Ann. Conf. Environ. Toxicol.*, Wright-Patterson Air Force Base, Ohio, AMRL-TR-71-120, 85 (1971).

Willis, J. H., Bradley, P., Kao, H., Grace, H., Hull, W., Griffin, T. B., Coulston, F., and Harris, E. E., "Sensitization of the Heart to Catecholamine Induced Arrhythmia by Haloalkanes," *Toxicol. Appl. Pharmacol.* **22**, 305 (1972).

Zapp, Jr., J. A., "Fluorocarbons," in "Encyclopedia of Occupational Health and Safety," Vol. 1, International Labour Office, Geneva, 1971, p. 560.

INDEX